TRANSACTIONS OF THE INTERNATIONAL ASTRONOMICAL UNION

VOLUME XXVIIIB

PROCEEDINGS OF THE TWENTY EIGHTH GENERAL ASSEMBLY

BEIJING 2012

COVER ILLUSTRATION

A pictorial reconstruction of the astronomical clock-tower built by Su Sung and his collaborators at Kaifeng in 1090. Original drawing by John Cristiansen in Needham et al (1960) *Heavenly Clockwork. The Great Astronomical Clocks of Medieval China.* Cambridge University Press, Cambridge.

IAU SYMPOSIUM PROCEEDINGS SERIES

INTERNATIONAL ASTRONOMICAL UNION

UNION ASTRONOMIQUE INTERNATIONALE

TRANSACTIONS OF THE INTERNATIONAL ASTRONOMICAL UNION VOLUME XXVIIIB

PROCEEDINGS OF THE TWENTY EIGHTH GENERAL ASSEMBLY BEIJING, CHINA, 2012

Edited by

THIERRY MONTMERLE
Editor-in-Chief

Shaftesbury Road, Cambridge CB2 8EA, United Kingdom

One Liberty Plaza, 20th Floor, New York, NY 10006, USA

477 Williamstown Road, Port Melbourne, VIC 3207, Australia

314–321, 3rd Floor, Plot 3, Splendor Forum, Jasola District Centre, New Delhi – 110025, India

103 Penang Road, #05–06/07, Visioncrest Commercial, Singapore 238467

Cambridge University Press is part of Cambridge University Press & Assessment, a department of the University of Cambridge.

We share the University's mission to contribute to society through the pursuit of education, learning and research at the highest international levels of excellence.

www.cambridge.org
Information on this title: www.cambridge.org/9781107078833

First published 2015

A catalogue record for this publication is available from the British Library

ISBN 978-1-107-07883-3 Hardback

PRESIDENTS OF THE INTERNATIONAL ASTRONOMICAL UNION

Présidents de l'Union Astronomique Internationale

Robert Williams
2009–2012

Norio Kaifu
2012–2015

Table of Contents

Preface xii

CHAPTER I - INAUGURAL CEREMONY

1. Opening Event 1
2. Opening Address by President Robert Williams 1
3. Welcome Address by the Vice-President of the People's Republic of China Xi Jinping 2
4. Presentation of Partners, Sponsors, and Exhibitors 4
5. Ceremony Event 6
6. The Gruber Foundation: Presentation of the Cosmology Prize 2012 and TGF Fellowship 2012 6
7. General Astronomy Talk by Jocelyn Bell-Burnell 8
8. Chinese Astronomy Talk by Past President of the Chinese Astronomical Society Su Dingqiang 8
9. Ceremony Event 11
10. Welcome Address by the President of the Chinese Astronomical Society Cui Xiangun 11

- Astronomy: amazing subject, amazing universe: Jocelyn Bell-Burnell 13

CHAPTER II - TWENTY EIGTH GENERAL ASSEMBLY BUSINESS SESSIONS . . . .

BUSINESS SESSIONS: First Session 15

1. Opening and Welcome 15
2. Representatives of IAU National Members 15
3. Adoption of the Agenda 16
4. Reminder of Voting Rules 16
5. Appointment of Official Tellers 16
6. Admission of New National Members to the Union 16
7. Revisions of Statutes and Bye-Laws 16
8. Report of the Executive Committee 16
9. Report of the Special Nominating Committee 16
10. Presentation of the Proposed Changes to Divisional Structure 17
11. Proposal to host the XXXth General Assembly in 2018 - Closure of Session . 18

CHAPTER II - TWENTY EIGTH GENERAL ASSEMBLY BUSINESS SESSIONS . . . .

BUSINESS SESSIONS: Second Session 20

1. Opening and Welcome 20

2. Individual Membership 20

3. Deceased members 20

4. Tribute to Franco Pacini (IAU President, 2000-2003) 20

5. Appointment of the Official Tellers 21

6. Proposed Changes to the Divisional Structure 21

7. Resolutions 21

8. Proposed Division Presidents and Vice-Presidents 21

9. Proposed Commission Presidents and Vice-Presidents 22

10. Financial Matters 23

11. Election of the Finance Committee and Membership Committee 25

12. Appointment of the Resolutions Committee 25

13. Appointment of the Special Nominating Committee 25

14. Election of the Executive Committee 26

15. Place and Date of the XXXth General Assembly in 2018 - Closure of Session 26

CHAPTER III - CLOSING CEREMONY . . . .

1. Welcome by Norio Kaifu, IAU President 27

2. Invitation to XXIXth General Assembly, Hawaii, August 2015 27

3. Address by the Retiring President 27

4. Address by the Retiring General Secretary 27

5. Address by the incoming President 29

6. Address by the Incoming General Secretary 31

7. Traditional Chinese "Thousand Hands" Dance 33

8. Closing Remarks 33

9. Handing out the IAU flag 34

10. Image Archive 34

CHAPTER IV - RESOLUTIONS . . . .

1. Resolution Committee 2009–2012 35

2. Approved Resolutions 35

Resolution B1 35

Resolution B2 37

Resolution B3 39

Resolution B4 40

3. Résolutions Approuvées 42

Résolution B1 42

Résolution B2 43

Résolution B3 46

Résolution B4 46

CHAPTER V - REPORT OF EXECUTIVE COMMITTEE 2009-2012

1. Introduction 48

2. Composition of the Executive Committee 48

3. Secretariat 50

4. Membership of the Union 50

5. Divisions, Commissions, Working Groups/Program Groups 50

6. International Year of Astronomy 2009 51

7. Strategic Plan and OAD 52

8. Educational Activities 54

9. Public Outreach Coordinator 54

10. Scientific Meetings 54

11. IAU Publications 56

12. Relationships with other Organizations 56

13. Financial Matters 57

14. Gruber Foundation 58

15. Norwegian Academy of Science and Letters and the Kavli Prize in Astrophysics 58

- Annex: Agreement concerning the Hosting of the Office of Astronomy for Development 59

- Addendum to the Agreement 66

- Executive Committee Report: Appendix I - Simplified Accounts 2010-2012 . . . 67

- Executive Committee Report: Appendix II - Proposed Income and Expenditure 2013-2015 71

CHAPTER VI - REPORTS on DIVISION, COMMISSION, and WORKING GROUP MEETINGS ....

Division I Commission 4 77

Division I Commission 7 83

Division I Commission 8 87

Division I Commission 19 95

Division II Commission 10 106

Division II Commission 12 109

Division II Commission 49 112

Division III Commission 15........ 115

Division III Commission 22........ 120

Division IV Commission 36........ 124

Division V Commission 42 126

Division VII Commission 37 128

Division IX Commission 30........ 132

Division XII Commission 6 134

Division XII Commission 14 135

Division XII Commission 46 137

Division XII Commission 55 140

CHAPTER VII - STATUTES, BYE-LAWS, and WORKING RULES

1. IAU Statutes 146

2. Statuts de l'UAI (Version Française) 151

3. IAU Bye-Laws 157

4. IAU Working Rules 162

CHAPTER VIII - NEW AND DECEASED MEMBERS AT THE GENERAL ASSEMBLY ....

1. New Members Admitted at the General Assembly........ 176

2. Deceased members (2009-2012) 187

CHAPTER IX - DIVISIONS, COMMISSIONS, AND WORKING GROUPS

Division I Fundamental Astronomy 189

Division II Sun & Heliosphere........ 190

Division III Planetary Systems Sciences........ 190

Division IV Stars 191

Division V Variable Stars 191

Division VI Interstellar Matter 191

Division VII Galactic System 192

Division VIII Galaxies & the Universe 192

Division IX Optical & Infrared Techniques 193

Division X Radio Astronomy 193

Division XI Space & High Energy Astrophysics 194

Division XII Union-Wide Activities 194

Division Working Groups 195

Inter-Division Working Groups 195

Inter-Commission Working Group 195

CHAPTER X - DIVISIONS MEMBERSHIP ....

Division I Fundamental Astronomy 197

Division II Sun & Heliosphere 202

Division III Planetary Systems Sciences 208

Division IV Stars 215

Division V Variable Stars 223

Division VI Interstellar Matter 229

Division VII Galactic System 236

Division VIII Galaxies & the Universe 241

Division IX Optical & Infrared Techniques 253

Division X Radio Astronomy 256

Division XI Space & High Energy Astrophysics 263

Division XII Union-Wide Activities 270

CHAPTER XI - COMMISSIONS MEMBERSHIP ....

Division I Commission 4 Ephemerides 277

Division XII Commission 5 Documentation & Astronomical Data 279

Division XII Commission 6 Astronomical Telegrams 282

Division I Commission 7 Celestial Mechanics & Dynamical Astronomy 283

Division I Commission 8 Astrometry 286

Division II Commission 10 Solar Activity 289

Division II Commission 12 Solar Radiation & Structure 294

Division XII Commission 14 Atomic & Molecular Data 297

Division III Commission 15 Physical Study of Comets & Minor Planets 299

Division III Commission 16 Physical Study of Planets & Satellites 302

Division I Commission 19 Rotation of the Earth 305

Division III Commission 20 Positions & Motions of Minor Planets, Comets & Satellites 307

Division IX Commission 21 Galactic & Extragalactic Background Radiation . . . 309

Division III Commission 22 Meteors, Meteorites & Interplanetary Dust 311

Division IX Commission 25 Astronomical Photometry & Polarimetry 313

Division IV Commission 26 Double & Multiple Stars 316

Division V Commission 27 Variable Stars 318

Division VIII Commission 28 Galaxies 323

Division IV Commission 29 Stellar Spectra 333

Division IX Commission 30 Radial Velocities 337

Division I Commission 31 Time 339

Division VII Commission 33 Structure & Dynamics of the Galactic System 340

Division VI Commission 34 Interstellar Matter 344

Division IV Commission 35 Stellar Constitution 352

Division IV Commission 36 Theory of Stellar Atmospheres 355

Division VII Commission 37 Star Clusters & Associations 358

Division X Commission 40 Radio Astronomy 361

Division XII Commission 41 History of Astronomy 369

Division V Commission 42 Close Binary Stars 372

Division XI Commission 44 Space & High Energy Astrophysics 376

Division IV Commission 45 Stellar Classification 384

Division XII Commission 46 Astronomy Education & Development 386

Division VIII Commission 47 Cosmology 390

Division II Commission 49 Interplanetary Plasma & Heliosphere 397

Division XII Commission 50 Protection of Existing & Potential Observatory Sites 399

Division III Commission 51 Bio-Astronomy 400

Division I Commission 52 Relativity in Fundamental Astronomy 402

Division III Commission 53 Extrasolar Planets (WGESP) 403

Division IX Commission 54 Optical & Infrared Interferometry 406

Division XII Commission 55 Communicating Astronomy with the Public 408

Preface

Traditionally, the IAU *Transactions B* series report the Proceedings of General Assemblies, for their non-scientific, administrative activities. These activities are essential to the life of the IAU: business sessions, admission of new members (countries and individuals), finances, committees, Division and Commission reports, resolutions, etc.

But each General Assembly (GA) has its own specificity that everyone present will remember. At the 2006 GA in Prague, there were famous Resolution on the definition of planets and dwarf planets (B5), and the status of Pluto (B6), that still remains controversial to a very vocal minority. The 2009 GA in Rio de Janeiro took place in the middle of the IAU-UNESCO International Year of Astronomy (nicknamed "IYA2009"), which drew the attention of hundreds of millions across the globe. It was also the place where the IAU "Strategic Plan", with the creation of the Office of Astronomy for Development (OAD) as its backbone, was approved (Resolutions A1 and B1) and launched, and with the adoption of a Resolution to defend "the right to starlight" (B5).

From that point of view, the business meetings highlights of the 2012 GA in Beijing are different, but nonetheless crucial for the future of the Union: a Resolution was voted almost unanimously to approve a new Divisional structure for the IAU, reducing their number from 12 to 9, and, more importantly, giving a clearer image of the spectrum of activities of the IAU, including the creation of a new Division "Education, Outreach and Heritage", also in contact with the public. This vote opens the door to an even deeper reform for Commissions and Working Groups, to take place in the present triennium (2012-2015).

Another highlight has been the welcome address by Mr. Xi Jinping, Vice-President of the People's Republic of China at the time, and who is now President of his country.

The Executive Committee report is very rich in achievements over the past triennium, in particular the consolidation and aftermath of the IYA2009, and the first steps of the OAD and the selection of its location in Cape Town (South Africa). In parellel, a new position of Public Outreach Coordinator, co-funded by the IAU and several other institutes, has been created in Tokyo: the IAU goes global !

This *Transactions B* volume is the third and last companion volume published by Cambridge University Press related to the Beijing General Assembly. The other volumes are the *Transactions A* volume, also called *Reports of Astronomy*, edited by my predecessor Ian Corbett (2012), which contains the triennal reports (2009-2012) of Divisions and Commissions, and *Highlights of Astronomy*, vol.16, edited by myself (2015), which contain the papers presented at the Special Sessions and Joint Discussions of the GA.

The eight Symposia that took place during the GA (IAU Symposia 288 to 295) are published separately by Cambridge University Press.

To finish, a sincere word of thanks to our Chinese colleagues, who organized an exceptionally rich General Assembly, and who demonstrated that the millenary Chinese tradition of astronomy is still very much alive today, with a vibrant and young community, and exciting telescopes and instrumental developments on the ground and in space.

Thierry Montmerle
General Secretary

Transactions IAU, Volume XXVIIIB
Proc. XXVIII IAU General Assembly, August 2012
Thierry Montmerle, ed.

doi:10.1017/S1743921315005426

CHAPTER I

TWENTY EIGHTH GENERAL ASSEMBLY

INAUGURAL CEREMONY

Tuesday, 31 August 2012, 14:00-16:00

China National Convention Centre, Beijing

1. Opening Event

The XXVIIIth General Assembly was opened by a Chinese Drum Performance.

2. Opening Address

Prof. Robert Williams, IAU President

Welcome to Beijing and IAU XXVIII General Assembly. Although the world economic situation continues to show stress, the science of astronomy is flourishing on many fronts. True, we are not immune to the negative effects of sharply reduced funding for projects and positions, yet our do-main sees increased international collaboration, pioneering facilities and techniques in development, and significant discoveries that are changing the way humanity thinks about the universe and our place in it. Programs that the IAU has undertaken such as the United Nations International Year of Astronomy 2009 and the creation of the Office of Astronomy for Development in Cape Town have been hugely successful.

The IAU is going through a period of transition from an organization that historically has maintained a largely internal focus emphasizing meetings and events for its members to one that is becoming more involved in education and outreach to the general public. We are also making efforts to foster international collaboration on large facility projects. Several changes to the Statutes are being proposed that will allow broader input and oversight from members on issues such as membership and finances. And most importantly we are proposing for the future that scientific resolutions be presented and discussed at the GA and on the IAU website but that voting be open to all members and conducted electronically following the GA. In addition, the Executive Committee is recommending a change to the divisional structure of the IAU that should organize the divisions more in line with current major research themes in astronomy. As the first large astronomical meeting of this nature to be held in China the present General As-sembly is a historic occasion for Chinese astronomy. It represents the tremendous growth in science in this country and increased funding that has enabled new projects and departments, and has attracted large numbers of students to undertake post-graduate studies in astrophysics.

We will see evidence of this activity in the talks and posters, of which many will be authored by young investigators.

The astronomical community in China has joined together to create an excellent venue and program of activities that make the General Assembly an ideal place to exchange ideas. There is the old and the new. The ancient traditions and scientific advances of past dynasties have provided a solid foundation for China?s innovative projects LAMOST, FAST, and AMiBA. You can experience excellent examples of both the old and the new in one place that you must visit while you are at the GA: the Beijing Planetarium. It combines the most modern facilities for astronomy education and outreach with a stunning collection of ancient instruments from the Han and Ming dynasties that long pre-date the instruments of Tycho Brahe on the island of Hven. Like many attractions in the Beijing area, it should not be missed. You should also take the opportunity to make the 10-minute walk over to the Chinese Academy of Sciences compound off Datun Road to visit the National Astronomical Observatories of China.

The "Inquiries of Heaven" daily newspaper will provide useful information for GA attendees. In addition to the daily schedule it will call attention to opportunities that you may find interesting, in addition to posting official information needed for official business meetings.

It is a pleasure to thank our Chinese colleagues who have worked to make the GA a success, and we look forward to a scientific program that will cause you to leave here full of ideas and enthusiasm.

3. Welcome Address by the Vice-President of the People's Republic of China

Mr. Xi Jinping,

Exploring the vast universe hand in hand, Working together toward a better future for humankind

Dear honored chairman, ladies, gentlemen, and friends,

Today, more than 2000 astronomers from all over the world gather together in Beijing to attend the 28th General Assembly of the International Astronomical Union. This is a grand event for astronomy. It is the first time for China to host an IAU General Assembly since China joined the IAU in 1935. On behalf of President HU Jintao, the Chinese Government, and the Chinese people, I am here to express our warm congratulations to this General Assembly, and express our sincere gratitude and cordial welcome to all attendants.

Astronomy, as the science to explore the universe, is one of the most important and the most active scientific frontiers that has pushed forward natural sciences and technology, and led to the advances of modern society. It has tremendously important influences on the progress of other branches of natural science and the development of technology. The vast expanse of space always stirs the curiosity of human beings on the earth, fascinates them, and has attracted generations after generations to devote themselves to the exploration of the universe. As the science to study the position, distribution, motion, morphology, structure, chemical composition, physical properties, origin, and evolution of the celestial bodies and matters in the universe, astronomy occupies an important position in the humans activity of understanding and transforming the world. As we see, every major discovery in astronomy has deepened our understanding of the mysterious universe, every significant achievement in astronomy has enriched our knowledge repository, and every breakthrough in the cross-disciplinary research between astronomy and other sciences has exerted both immediate and far-reaching impacts on fundamental science and even human civilization.

As one of the ancient civilizations in the world, the ancient Chinese used to work after sunrise and rest after sunset, and started to gaze at the starry sky from very early on. At the end of the Warring States period more than 2300 years ago, the great romantic poet Qu Yuan in his "Inquiries of Heaven" queried "Whoever has conveyed to us, Stories of the remotest past? Who can verify the shapeless, Beginning time has overcast?" Our ancestors already built their astronomical observatories as early as 13th century BCE or even earlier, and we have kept the longest and most comprehensive records of astronomical phenomena in the world. Modern astronomy in China started 90 years ago, with the Chinese Astronomical Union being founded in 1922, the Chinese Astronomical Research Institute founded in 1928, and the Purple Mountain Observatory built in 1934. Since the founding of the People's Republic of China, especially since its reforming and opening up, Chinese Academy has established the systematic operating mechanism of modern astronomical observatories, after building the large sky area multi-object fiber spectroscopic survey telescope (LAMOST), now is constructing the five-hundred-meter spherical radio telescope (FAST), and is also making progress in space astronomy and Antarctic astronomy.

The advancement of astronomy is the result of the efforts of all humankind, and manifests the wisdom of humanity. The history of its development has offered us very valuable and profound enlightenment.

First, the development of science and technology is the driving force for humankind's exploration and transformation of the materialistic world. Science and technology are the most active, most revolutionary factor in eco-social development. Every grand advancement of human civilization is closely related to the revolutionary breakthrough in science and technology. The development of science and technology has profoundly changed the way people live and work, and science and technology are becoming the main driving forces for eco-social progress. To achieve sustainable eco-social development and wholesome development of human beings, it is critical to rely on scientific progress and technological innovation.

Secondly, the development of science and technology requires persistent exploration and long term accumulation. The exploration of the mysterious universe, just like the explorations of other science branches, should be endless. Science and technology, as the achievements of humankind in their exploration and transformation of the world, are the creative products of scientists only after their persistent exploration and long term accumulation. Only working in full devotion, exploring with never-ceasing steps, furthering continuously on the shoulders of giants, can one reach the pinnacles of science and drive the progress of humankind.

Thirdly, the development of science and technology requires to continuously emphasize and strengthen basic research. Astronomy as an observational science is a very crucial field of basic research. Such a field requires strategic plans for deployment in advance, with full respect to the internal logic of research activities and their long-term benefits. We will make larger and larger investments in such a field and ensure their execution, provide long-term and stable support to scientists, so that the scientists can discover, invent, create, and advance constantly, and make more and more achievements that will benefit humankind.

Fourthly, the development of science and technology requires broad and sound support from the public. Science and technology are a noble course that both benefit and rely on society, and the full development requires not only public understanding from all sides, but also the active participation of the public. Public outreach should be given equal emphasis as scientific research to educate the public, so as to create a positive atmosphere for the public to respect, love, learn and use science, and to inspire the creativity for science and technological innovation among the public.

Fifthly, the development of science and technology requires extensive international cooperation. Science and technology have no nationality! The vast expanse of space is the common home of all humankind; to explore this vast universe is the common goal of all humankind; astronomy in fast development is the shared fortune of all humankind. Nowadays the challenges for science and technology are more and more globalized, and all humankind are faced with the same problems in energy and resources, ecological environments, climate change, natural disasters, food security, public health, and so on. Both basic research such as astronomy and these common problems require scientific and technological exchanges and cooperation in various forms between different nations and districts, in order to push forward science and technological innovation, human civilization, sustainable development, and to benefit all humankind.

Today's world is an open world, and countries are depending on each other more and more heavily. In the past 30 years, China opened its gate not only for economic development, but also for exchanges and cooperation in science and technology. Especially since the advent of the 21st century, China has hosted a series of important international conferences in natural sciences and engineering disciplines, such as the international congress of mathematicians and the World's Engineers' Convention and so on. This has greatly broadened the international horizon of the Chinese science and technology community, deepened the worlds understanding of China, promoted mutual exchanges and cooperation between the Chinese and international science and technology communities, and created favorable conditions for the Chinese community to make their contributions to the world.

The convening of the 28th IAU General Assembly in China, I believe, will certainly promote the friendship between Chinese astronomers and astronomers from other countries, promote the exchanges and cooperation between the Chinese and international astronomy communities, and promote the development of China's astronomy and other related sciences. This convention, I believe, will inspire curious youngsters from all over the world including China to cast their attention and desire to the vast universe, and motivate them to devote themselves to the observations and studies in astronomy, and to science and technological innovations.

Finally, I wish this General Assembly a great success, and wish astronomers from all countries to explore the vast universe hand in hand, and to work together toward a better future for humankind.

Thank you, all!

4. Presentation of Partners, Sponsors and Exhibitors

The IAU and Organising Committees acknowledge the invaluable support of the following institutions and organisations:

Partners
Air China
Beijing Science Video Network
H3C

Microsoft Research
Star Alliance

Sponsors

Associated Universities, Inc.
Astronomy & Astrophysics
Chinese Association for Science and Technology
Chinese Academy of Sciences
National Astronomical Observatory of Japan (NAOJ)
National Natural Science Foundation of China
Wiley-Blackwell
World Wide Telescope Academy Program

Exhibitors

ALMA, the Atacama Large Millimeter/submillimeter Array
American Astronomical Society
Andor Technology
ASTRON Netherlands Institute for Radio Astronomy
Betop Multimedia
Cambridge University Press
Chinese Astronomical Society
Copernicus Meetings & Publications
CSIRO Astronomy and Space Science
Division for Particle Astrophysics, Institute of High Energy Physics, Chinese Academy of Sciences
European Southern Observatory (ESO)
European VLBI Network
Fulldome.pro
Finger Lakes Instrumentation
GREAT - Gaia Research for European Astronomy Training
Institute for the History of Natural Sciences, Chinese Academy of Sciences
Instituto de Astrofsica de Canarias
International Astronomical Union Office of Astronomy for Development
The International Centre for Radio Astronomy Research
Korea Astronomy and Space Science Institute
Cherenkov Telescope Array, HESS/MAGIC/VERITAS
Nanjing Institute of Astronomical Optics & Technology, NAOC
Nanjing iOptron Scientific, Inc.
National Astronomical Research Institute of Thailand (NARIT)
National Astronomical Observatories, Chinese Academy of Sciences
National Hellenic Research Foundation
National Radio Astronomy Observatory
National Time Service Center, Chinese Academy of Sciences
Oceanside Photo & Telescope
Officina Stellare SRL, Italy
Purple Mountain Observatory, Chinese Academy of Sciences
Santa Barbara Instrument Group
PlaneWave Instruments

School of Astronomy & Space Science, Nanjing University
School of Space Science and Physics, Shandong University at Weihai
Shanghai Astronomical Observatory, Chinese Academy of Sciences
SKA Organisation
Springer
The Kavli Institute for Astronomy and Astrophysics at Peking University
Thirty Meter Telescope (TMT)
Xinjiang Astronomical Observatory, NAOC
Yunnan Astronomical Observatory, NAOC

5. Ceremony Event

Classic Dance of Qin Dynasty
Ancient Traditional Musical Instrument Performance

6. The Gruber Foundation: Presentation of the Cosmology Prize 2012 and TGF Fellowship 2012

The President of the IAU, Robert Williams, opens the Ceremony, in the presence of

- Patricia Murphy Gruber, President Emeritus, The Gruber Foundation
- Charles Bennett, 2012 Cosmology Prize Recipient
- Wendy Freedman, Chair, Cosmology Selection Advisory Board
- Ian Corbett, General Secretary, IAU
- Anna Lisa Varri, recipient of the 2012 Gruber Fellowship

Robert Williams:

In year 2000 the Gruber Foundation created its International Prize Program with its inaugural prize in Cosmology. The intent of the Prizes is to call attention to the importance of accomplishments that benefit mankind in special, important areas of endeavor. The IAU is pleased to have been chosen by the Gruber Foundation to collaborate with them on the Cosmology Prize since its inception. The IAU has an advisory role in the constitution of the selection committee and as part of our collaboration we are very fortunate to receive an annual grant of $50,000 from the Gruber Foundation to be awarded to a postdoctoral fellow.

It goes without saying that the prestige and reputation of any award is determined by the quality of its awardees. By this criterion the Gruber Cosmology Prize has done superbly. During the past decade the Cosmology Prize recipients alone could write as authoritative a history of the universe as any group. The careers of the Gruber Fellows have also benefitted greatly from the resources made available to them by the Gruber Foundation stipend.

In the years of the IAU General Assembly the tradition has been for the Gruber Cosmology Prize to be presented in this Inaugural Ceremony. It is my great pleasure to present to you the President of The Peter & Patricia Gruber Foundation and co-Founder & President Emeritus of The Gruber Foundation, which will continue the Gruber Prize Program as part of Yale University, Patricia Gruber, who will introduce this year?s Cosmology Prize awards ceremony.

Patricia Gruber:

Welcome to the presentation of the 13th annual Cosmology Prize? honoring a leading cosmologist, astronomer, astrophysicist or scientific philosopher for theoretical, analytical, or conceptual discoveries leading to fundamental advances in our understanding of the universe. On behalf of my husband, Peter Gruber, myself, and all of us at the Foundation, we are pleased to be here in Beijing to present this Prize at the 28th General Assembly of the International Astronomical Union. Thank you, Bob Williams, for your warm welcome.

The Cosmology Prize was established in 2000 as the first Gruber international prize, and I would like to acknowledge the vision and leadership of Peter Gruber in establishing this and the other prizes. The Cosmology Prize is presented in conjunction with the International Astronomical Union. It is my pleasure to introduce Dr. Ian Corbett, Secretary General of the IAU, who will say a few words about this fruitful collaboration.

Ian Corbett:

The primary goal of the IAU is the development of astronomy world-wide. To this end, the IAU is pleased to collaborate with the Gruber Foundation on the Cosmology Prize.

The collaboration between the Gruber Foundation and the IAU consists not only of the Cosmology Prize, but also an annual $50,000 Fellowship. The fellowship is administered by the IAU and awarded competitively to a postdoctoral researcher—the stipend is to be used to further his or her research.

Awards are presented to promising young scientists of any nationality to pursue education and research at a center of excellence in their field; the IAU selects recipients from applications received from around the world. The fellowship has been awarded to scientists from Poland, India, Spain, Greece, the Russian Federation, Mexico, the UK, Colombia, and the United States.

The 2012 Gruber Fellow is Anna Lisa Varri, from the Universita' degli Studi di Milano (Italy). Her work focuses on the dynamics of globular clusters, with the aim of providing a more realistic dynamical paradigm for this class of stellar systems. I am happy to introduce her on this occasion and invite her to say a few words.

Anna Lisa Varri: expresses thanks and makes brief remarks.

Patricia Gruber:

Thank you Anna Lisa Varri.

We are here to honor the achievements of Charles Bennett and the Wilkinson Microwave Anisotropy Probe team. But first let me tell you a little about the company they are keeping.

The Foundation's prize program, established in 2000, now presents three annual $500,000 prizes in the fields of: Cosmology; Genetics; and Neuroscience.

Each prize recognizes achievements and discoveries that produce fundamental shifts in human knowledge and culture. Until 2011, the Foundation also awarded prizes in Justice and Women?s Rights. Under our newly enacted succession plan with Yale University, these two prizes are now part of an exciting new program at Yale Law School.

On October 14th, at the annual meeting of the Society for Neuroscience, the Neuroscience Prize will be presented to Lily and Yuh Nung Jan, and the Genetics Prize will be presented at the annual meeting of the American Society of Human Genetics on November 9th, to Douglas Wallace.

Returning to Cosmology, the 2012 Prize recipients were was selected by a distinguished Cosmology Prize advisory board:

- Wendy Freedman (Chair)
- Andrew Fabian
- Gerhard Huisken
- Helge Kragh
- Andrei Linde
- Julio Navarro
- Sadanori Okamura

Owen Gingerich and Martin Rees also serve as special Cosmology Prize advisors to the Foundation. Peter and I deeply appreciate the knowledge, commitment, and enthusiasm that the advisors bring to the judging process. Let me now invite the advisory board Chair, Wendy Freedman, to present the official citation and introduce the scientific accomplishments of our Cosmology Prize recipients.

Wendy Freedman:
The Recipients of the 2015 Prize are Charles L. Bennett and the WMAP Team.

The official citation reads:
The Gruber Foundation proudly presents the 2012 Cosmology Prize to Charles Bennett and the Wilkinson Microwave Anisotropy Probe team for their exquisite measurements of anisotropies in the relic radiation from the Big Bang—the Cosmic Microwave Background. These measurements have helped to secure rigorous constraints on the origin, content, age, and geometry of the Universe, transforming our current paradigm of structure formation from appealing scenario into precise science.

Charles Bennett: Expresses thanks, very brief remarks.

Patricia Gruber:
Please note that Charles Bennett will give a public lecture entitled "From Ancient Light to Modern Cosmology: The WMAP Mission" at 12:45 pm tomorrow in this room. Thank you for attending the 2012 Cosmology Prize ceremony. This concludes our presentation.

7. General Astronomy Talk

Dr. Jocelyn Bell-Burnell presented a talk entitled "Astronomy - Amazing Subject, Amazing Universe". See at the end of this Chapter.

8. Chinese Astronomy Talk

Prof. Ding-qiang Su, Past President of the Chinese Astronomical Society

Understanding Astronomy in China through recent major projects

Dear colleagues,
Good afternoon!

When I was President of the Chinese Astronomical Society, our application to host the IAU 28th General Assembly in Beijing became successful. Today I am glad to be here to attend the opening ceremony of IAU 28th GA. I wish this meeting a success and I wish all participants a happy and memorable time in China.

China learned its modern science and technology from the West. However, any nation with some self-respect is not satisfied with always following advanced countries. We hope that one day we can catch up with them, and even surpass them in some fields. In the 20th century, because of Japanese invasion, the Chinese civil wars, and various political campaigns, until 1976 Mainland China was poor and undeveloped. By now, however, much of this situation has changed. China′s GDP has risen to the second highest in the world. China is now one of the most advanced countries with respect to the supercomputer, steelmaking, hydraulic engineering, the high-speed train, the modern bridge, and so on. In addition, China has carried out manned space missions and lunar exploration, and has set up a space lab.

Please allow me to limit my talk to Mainland China′s projects since 2006.

I. The ground-based optical-infrared projects

1. LAMOST - Large Sky Area Multi-Object Fiber Spectroscopic Telescope

The Large Sky Area Multi-Object Fiber Spectroscopic Telescope (LAMOST) is an innovative reflecting Schmidt telescope. LAMOST has a clear aperture of 4.3 meters and a field of view of 5°. It is the largest wide field-of-view telescope in the world. 4000 optical fibers are put on its focal surface. Such a large-scale slit spectroscopic survey is unprecedented. At present, several other projects around the world are being planned along this direction. LAMOST was completed in October 2008. LAMOST is an 8m-class telescope. Through the development of LAMOST China has proved the basic ability to develop a 30m-class telescope.

2. NEOST - Near-Earth Object Survey Telescope NEOST is a 1m Schmidt telescope.

3. NVST - New Vacuum Solar Telescope NVST, with a 1m aperture, is the largest vacuum solar telescope in the world.

4. ONSET - Optical and Near-Infrared Solar Eruption Tracer ONSET can obtain white light, Hα and 10830 $\dot{A}$ images at the same time. All these projects are completed.

II. The ground-based radio projects

1.21CMA - 21 Centimeter Array

Exploration of the neutral hydrogen reionization may be the last frontier in observational cosmology. 21CMA is a radio array for this goal. It was constructed in the Tianshan Mountains in west China. A total of 10287 antennas have been deployed along two perpendicular baselines of 10 km. 21 CMA was completed in 2007. It is the first facility of this kind in the world to start the search for the epoch of reionization.

2. CSRH - Chinese Spectral Radioheliograph

Images over centimeter and decimeter wavelengths are important for addressing fundamental problems in solar flares and coronal mass ejections. The Chinese Spectral Radioheliograph (CSRH) will be the first new-generation instrument of this kind in the world. It can obtain images with high temporal, spatial, and spectral resolutions simultaneously. It includes two arrays. The CSRH-I has been installed. The CSRH-II will be finished next year.

3. Shanghai 65m radio telescope

This is Shanghai 65m radio telescope. It is also a bridge for China to develop larger steerable radio telescopes. It will be completed this year or next year.

4. FAST - Five-hundred-meter Aperture Spherical Radio Telescope

The Five-hundred-meter Aperture Spherical Radio Telescope (FAST) can be seen as a modified "Arecibo" type telescope, with an illuminated area of 300 meters in diameter, twice as large as the Arecibo. In FAST the main innovation is that the shape of the basic spherical reflector is changed to keep the illuminated area a paraboloid in real time. It will be completed in 2016. FAST will be the largest single-aperture radio telescope in the world.

III. The space projects

1. HXMT - Hard X-ray Modulation Telescope

The Hard X-ray Modulation Telescope (HXMT) works in 1-250 keV. It will obtain sky maps with high spatial resolution and sensitivity. HXMT also has a unique capability to study short time-scale variations with high spectral resolution.

2. DAMPE - Dark Matter Particle Explorer

A Chinese astronomer, through international cooperation by balloon observation, found "an excess of cosmic ray electrons at energies of 300-800 GeV." These electrons may arise from the annihilation of dark matter particles. This find has claimed worldwide attention, and China has decided to develop a satellite named Dark Matter Particle Explorer (DAMPE) to further detect the high energy electrons and gamma ray. In energy resolution (1.5% at 1TeV), energy detection range (5-10,000 GeV), and background level, DAMPE surpasses other similar projects.

3. DSO - Deep Space Solar Observatory

DSO includes a 1m telescope, mainly to study the solar magnetic field.

4. SVOM - Space Variable Object Monitor (China and France)

SVOM is a Chinese and French joint mission to study gamma-ray burst.

5. POLAR - Gamma-ray Burst Polarization Observation Experiment (an international collaboration project led by China)

All these projects will be launched between 2014 and 2016.

Dome A is the summit of the Antarctic icecap, where the Chinese Expedition Team made the first recorded human arrival in January 2005. Antarctica has the best astronomical sites on the ground.

A second-generation Chinese Antarctic Survey Telescope AST3-1 with an aperture of 50 cm was installed there in January this year. At present, it is the largest optical telescope in Antarctica.

The following projects are being planned or conceived:

1. Antarctic Astronomical Observatory

The third-generation Antarctic telescopes include a 2.5m optical-infrared telescope and a 5m THz telescope.

2. 20-30m Optical-Infrared Telescope

China has no large-aperture telescope for fine observation. So it is important to develop such a telescope.

3. 110m Steerable Radio Telescope

4. Large Solar Telescope (aperture $\geqslant$ 4m)

5. LAMOST South
6. 2m Space Optical Survey Telescope
7. XTP - X-ray Timing and Polarization Mission

and others.

Other countries are welcome to participate in all Chinese projects and China will also be pleased to join the projects of other countries.

At present China is not yet one of the leading countries in astronomy, but it is approaching this goal. China has a brilliant ancient civilization, it has the largest population, and has recently made great economic achievements. China should make greater contributions to science and to humankind.

Thank you!

9. Ceremony Events

Chinese Long Ribbon Dance - Flying Apsaras
Tibet Dance
Chinese Silk Acrobatics - Butterfly Love, Winner of Italian Golden Circus Festival and China National Acrobatic Contest
Performance by NAOC staff - (Dream of Heaven)

10. Welcome Address by the President of the Chinese Astronomical Society

Prof. Xiangqun Cui

Respected Dr. Robert Williams,
Ladies and Gentlemen,

I am much honored to welcome you all to the IAU 28th General Assembly in Beijing on behalf of the Chinese Astronomical Society.

The Society is very happy and encouraged to host this IAU General Assembly for the first time in China, especially in this festival year when we celebrate the 90th anniversary of the establishment of the Chinese Astronomical Society.

As it is known, the Chinese Astronomical Society started to make preparations for the IAU 28th General Assembly as early as August 2006 at the 26th General Assembly in Prague when it succeeded in its bid for hosting this meeting. Early in 2007, the National Advisory Committee, the National Organizing Committee and the Local Organizing Committee were established to secure smooth planning and preparations for the meeting. Throughout the planning process, the close collaboration with IAU leadership and staff and the precious experiences from the 26th and 27th General Assembly meetings have been of great help to us.

I would like to take this opportunity to express our sincere gratitude to our counterparts from all around the world for their continuous guidance and support, in particular to Robert Williams, President of IAU, and Ian Corbett, General Secretary of IAU.

In addition, I would also like to express our sincere gratitude on behalf of the Chinese Astronomical Society to the many Chinese colleagues, scholars and students for their hard work and deep involvement in the preparation, to the Chinese public for their enthusiasm, and to the Chinese government for the serious concern and extensive support.

Chinese astronomy has a very long history. The recordings of the earliest astronomical observatory can be traced back to about 3000 years ago, with eminent ancient astronomers such as Zhang Heng and Guo Shoujing who are still being honored by the Chinese until present day. The Chinese Astronomical Society was initiated and established by Chinese astronomers in October 1922 during the very period of the anti-imperialist and anti-feudal "May 4th Movement". After the founding of the People's Republic of China in 1949, and in particular over the last 30 years since China's opening to the world, Chinese astronomy has experienced rapid development. At the same time, the Chinese Astronomical Society has grown extensively. Currently, the Chinese Astronomical Society consists of 17 Science Committees with branches scattered in different provinces around China. The Society has about 2000 individual members, and its member organizations come from relevant research institutions and universities all over the country. Up to now, the society has more than 400 IAU members since it joined the IAU in 1935.

Every year, the Chinese Astronomical Society holds a national Annual Meeting, as well as various kinds of seminars, workshops and activities. As such, the Chinese Astronomical Society has played a vital role in promoting scientific exchange and public awareness of astronomy.

Chinese astronomy is developing and we still have a long way to go to reach global levels. As such, the IAU 28th General Assembly in Beijing will provide an opportune moment for Chinese astronomers to discuss and exchange views extensively with their international counterparts. It is my sincere wish that this meeting will give an impulse to the further development of Chinese astronomy, and provides a platform for international cooperation in astronomy.

I would like to extend once again a very warm welcome to all of you to the IAU 28th General Assembly, and welcome to Beijing. We hope the meeting will be a great success for all of you, and we wish you a pleasant stay in Beijing.

Thank you.

Astronomy: amazing subject, amazing universe

Jocelyn Bell Burnell
University of Oxford, UK

One hundred years ago

One hundred years ago our astronomical understanding was very different from today. We did not know about the Big Bang or the expansion of the Universe, nor about its age and scale. Inflation and dark energy were beyond our imagination, as was the Cosmic Microwave Background radiation. We barely knew of galaxies beyond the Milky Way and were ignorant of clusters of galaxies and dark matter. Active Galactic Nuclei, black holes and jets were unknown to us. We had not seen neutral hydrogen or Giant Molecular Clouds. Indeed the only astronomy done was in the visual band so the idea that one could observe at other wavelengths was alien to us. Cosmic rays were just being discovered; that we could study the universe through them or through other particles was also a foreign idea. We did not yet know what stars were made of, what their energy source was, or how they evolved. Things like exoplanets, space flight and the exploration of the solar system were only found in science fiction; Pluto had not yet been discovered and trans-Neptunian objects were unknown.

We have come a long way in one hundred years! This has been made possible by funding from our Governments, foundations and private individuals, and the support of industry. Also we have had amongst us some very smart people ? astrophysicists, engineers and ICT specialists ? and there have been brilliant technical innovations which have opened up the universe to us.

"Astronomical Treasures"

I was asked a few years ago to talk about the things that I felt were ?treasures of the universe? or astronomical treasures. Each of us will have our own list of astronomical treasures, but some of the remarkable developments that I believe deserve to be treasured are:
a) the COBE satellite data showing that the Cosmic Microwave Background radiation perfectly fits a blackbody spectrum (with temperature 2.74K);
b) the Ghez ? Genzel infrared observations over a number of years of the motions of stars within one parsec of the Galactic centre, showing that the stars move in curved tracks because they are diverted from straight lines by the gravity of a black hole at Sagittarius A*. The black hole has a mass of 4×10^6 solar masses; this looks large but actually is quite small for a black hole at the centre of a galaxy. You can see the movie at `http://www.astro.ucla.edu/ gehzgroup/gc/pictures/orbitsMovie.shtml`;
c) the staggering number of exoplanets now being found. The rate of discovery gets faster and faster! At the time of writing (August 2012) 786 are known, many of them in multiple systems, and when we look at the night sky we need to remind ourselves that there are as many planets up there as there are stars; and finally
d) the exquisite images now becoming available with wonderful clarity and detail.

Astronomy in our cultures

Astronomy is not and never has been the preserve of the professionals only; there is great public interest and considerable non-professional participation. Amateurs have helped find comets and observe variable stars, and now thousands of people participate in Citizen Science programmes helping us with the flood of data we now experience. So the public help search for pulsars, classify galaxies, identify small craters on the moon and scan Spitzer images (see for example, https;//www.zooniverse.org) The concept is being extended beyond astronomy to the identification of people or places in old photographs, and the scrutiny of old ships? log books to determine the weather in the past.

Chinese astronomers in previous centuries noted ?guest stars? (novae and supernovae) and their records have been invaluable in identifying historic supernovae.

Our ancestors were much more aware of the night sky and different cultures gave different names to the main objects in the night sky. Some alternative names for the Milky Way are: The Silver Street (Celtic); The Celestial River (China); River of Fire (ancient Hebrew); The Backbone of the Night (Kalahari); Silver River Water (Korean); The Long Fish (Maori); The Place where the Lightning Rests (Setswana); Silicon River (Siberia); The Winter Road (Sweden); and The Way of the White Elephant (Thai). Astronomy enters our culture in other ways too ? as inspiration for music, painting and poetry, and it is interesting to explore these portrayals of our subject and revealing to us to see what astronomical topics have caught the attention of poets, painters and musicians.

The next hundred years?

What of the future? The hectic pace will continue, not least because of the telescopes we have operating right now! New telescopes such as ALMA (Atacama Large Millimetre Array) in Chile and the X-ray astronomy satellite NuSTAR will come fully on line. We look forward to large optical telescopes such as LSST (Large Synoptic Survey Telescope), and at least one thirty metre telescope like E-ELT (European Extremely Large Telescope), both of which will probably also be built in Chile. There will be the large radio telescopes SKA (Square Kilometre Array) in Southern Africa and Australia, and the Chinese FAST (being built in Guizhou province). It is hard to imagine where we will be in terms of astrophysical understanding, but we will have a greater awareness of transient objects, pulsars will have tested Einstein?s theory of gravity to destruction (or not) and we will likely have found signs of life on an exoplanet. Understanding dark matter will have brought about a revolution in physics and understanding dark energy will probably have forced a change in the way we think about the universe ? a paradigm shift. In a hundred years? time people will look back and be amazed at our naivety!

Transactions IAU, Volume XXVIIIB
Proc. XXVIII IAU General Assembly, August 2012
Thierry Montmerle, ed.

doi:10.1017/S1743921315005438

CHAPTER II

TWENTY EIGHTH GENERAL ASSEMBLY BUSINESS SESSIONS

FIRST SESSION: Tuesday, 21 August 2012, 16:30-18:00

Chaired by Robert Williams, President of the IAU
China National Convention Centre, Beijing

1. Opening and Welcome

The President of the IAU, Prof. Robert Williams, welcomed the delegates and members to this first business session of the General Assembly. The President invited the General Secretary, Dr. Ian Corbett, to start the business session.

2. Representatives of IAU National Members

The General Secretary listed the representatives of the National Members (IAU member countries). They are given in the table at the end of this Chapter.

The columns list separately:
- the National Members
- the dues category
- the number of votes attached to each National Representative:
    (a) for ordinary votes, membership,, etc.
    (b) for financial matters

Note: "Interim" National Members, or National Members in arrears of their dues, do not vote, even if they have appointed a Representative (countries indicated by a astrisk).
- the National Representatives, in their capacity to vote the budget and other issues on behalf of the National Members;
- the National Representative for Finance, in their capacity to discuss and vote on financial matters, as well as to propose nominations and elect the Finance Committee (formerly known as Finance Sub-Committee) for the triennium;
- the National Representatives for Membership Nominations, in they capacity to discuss and vote on individual membership issues, as well as to propose nominations and elect the Special Nominating Committee (SNC), and the Membership Committee.

For each National Member, the first line gives the names of the Representatives present during the first week of the GA, the second line, if present, gives the names of the Representatives present during the second week of the GA.

3. Adoption of Agenda

The delegates approved the agenda of the meeting as published in the Program Book.

4. Reminder of voting rules

The President reminded the delegates and the members about the voting rules, depending on the nature of the vote.

5. Appointment of Official Tellers

The Official Tellers were duly approved.

6. Admission of New National Members to the Union

The following new National Members were unanimously admitted to the Union:

Country: The Democratic People's Republic of Korea
Category: Interim
Adhering Organization: Pyongyang Astronomical Observatory, Academy of Sciences

Country: Ethiopia
Category: I
Adhering Organization: Ethiopian Space Science Society

Country: Kazakhstan
Category: I
Adhering Organization: National Center of Space Research and Technology

7. Revisions of Statutes and Bye-Laws

The revisions of the Statutes and Bye-Laws, in particular those relevant to the changes introduced in the Division Structure, were presented by the General Secretary.
The revised Statutes and Bye-Laws are presented in Chapter V of these *Transactions*.

8. Report of the Executive Committee (2009-2012)

The report of the Executive Committee was presented by the General Secretary, Dr. Ian Corbett, and is presented in Chapter XX of these *Ttansactions.*

9. Report of the Special Nominating Committee

Nominations of the IAU Executive Committee for the triennium 2012-2015:

President-Elect	Silvia Torres Peimbert (Mexico)
Assistant General Secretary	Piero Benvenuti (Italy)
Vice-President	Renée Kraan-Korteweg (South Africa)
Vice-President	Xiao-Wei Liu (China Nanjing)
Vice-President	Dina Prialnik (Israel)

The Special Nominating Committee proposes the following slate for IAU members for the Officers and Members of the IAU Executive Committee for the triennium 2012-2015:

President	Norio Kaifu (Japan)
President-Elect	Silvia Torres Peimbert (Mexico)
General Secretary	Thierry Montmerle (France)
Assistant General Secretary	Piero Benvenuti (Italy)
Vice-President	Matthew Colless (Australia)
Vice-President	Renée Kraan-Korteweg (South Africa)
Vice-President	Xiao-Wei Liu (China Nanjing)
Vice-President	Jan Palouš (Czech Republic)
Vice-President	Dina Prialnik (Israel)
Vice-President	Marta Rovira (Argentina)
Adviser	Robert Williams (USA)
Adviser	Ian Corbett (UK)

10. Presentation of Proposed Changes to Divisional Structure

The Assistant General Secretary, Dr. Thierry Montmerle, presented proposed changes to the IAU Divisional structure, endorsed by the Division Presidents and the Executive Committee, in the form of a Resolution put to the General Assembly, as follows:

RESOLUTION B4

on the restructuring of the IAU Divisions,

proposed by the IAU Executive Committee

The XXVIII General Assembly of the International Astronomical Union, noting

(a) that both the IAU and astronomy as a whole have evolved considerably since the current Divisions were introduced in 1994 and formally adopted in 1997, and that it is therefore appropriate to consider re-optimising the Divisional Structure,

(b) the report and recommendations of the Task Group established by the Executive Committee to examine the case for restructuring the Divisions, and the Executive Committee response to these recommendations,

(c) that the Commissions, Working Groups and other bodies under the Divisions may also require reform,

(d) that the implementation of the Strategic Plan through the Office of Astronomy for Development (OAD) and other associated programmes requires the Executive Committee to establish appropriate oversight and governance provisions for all Astronomy for Development activities, including the Office of Astronomy for Development, ensuring a strong link between these activities, the Divisions, and the Executive Committee.

approves

the proposal of the Executive Committee to restructure the Divisions as follows:

Division A Space and Time Reference Systems
Division B Facilities, Technologies, & Data Science
Division C Education, Outreach, & Heritage
Division D High Energies & Fundamental Physics
Division E Sun & Heliosphere
Division F Planetary Systems & Bioastronomy

Division G Stars & Stellar Physics
Division H Interstellar Matter & Local Universe
Division J Galaxies & Cosmology

11. Proposal to host the XXX General Assembly in 2018

One proposal was received from Austria to hold the XXX General Assembly in 2018, with a venue in Vienna. This proposal was presented by the NCA of Austria. It was approved by the Executive Committee at its 91th meeting, Part II, on Aug.23, and the decision announced at the Second business meeting of the General Assembly.

Closure of Session

There being no other business items to discuss, the President declared the session closed.

Representatives of National Members (see Sect. 3)

National Member	Dues Category	Votes (a, b)	National Representatives	Finance Committee Nominations	Membership & SNC Nominations
Argentina	II	1, 3	Marta Rovira	Marta Rovira	Marta Rovira
Armenia	I	1, 2	Areg Mickaelian	Areg Mickaelian	Areg Mickaelian
Australia	IV	1, 5	John Dickey	John Dickey	John Dickey
Austria	II	1, 3	Gerhard Hensler	Gerhard Hensler	Gerhard Hensler
Belgium	IV	1, 5	Conny Aerts	Conny Aerts	Conny Aerts
Bulgaria	I	1, 2	-	-	-
Brazil	III	1, 4	Daniela Lazzaro	Daniela Lazzaro	Beatriz Barbuy
Canada	V	1, 6	Samar Safi-Harb Greg Fahlman	Samar Safi-Harb	Greg Fahlman
Chile	II	1, 3	Patricio Rojo	Patricio Rojo	Patricio Rojo
China Nanjing	VI	1, 7	Xiangqun CUI	Zhanwen HAN	Ji YANG
China Taipei	II	1, 3	Yi-Jehng KUAN	Yi-Jehng KUAN	Yi-Jehng KUAN
Costa Rica	interim	0, 0			
Croatia	I	1, 2	-	-	-
Czech Rep.	III	1, 4	Petr Hadrava	Cyril Ron	Richard Wunch
Denmark	III	1, 4	Johannes Andersen	Birgitta Nordström	Johannes Andersen
Estonia	I	1, 2	Laurits Leedjärv	Laurits Leedjärv	Laurits Leedjärv
Finland	II	1, 3	Esko Valtaoja	Esko Valtaoja	Esko Valtaoja
France	VII	1, 8	Daniel Rouan	Daniel Rouan	Daniel Hestroffer
Germany	VII	1, 8	Matthias Steinmetz	Matthias Steinmetz	Matthias Steinmetz
Greece	III	1, 4	V. Charmandaris	V. Charmandaris	V. Charmandaris
Honduras	interim	0, 0			
Hungary	II	1, 3	L. van Driel-Gesztelyi Laszlo Viktor Tóth	L. van Driel-Gesztelyi	Laszlo Viktor Tóth
Iceland	I	1, 2	-	-	-
India	V	1, 6	Rajesh Kochlar	Rajesh Kochlar	Rajesh Kochlar
Indonesia	I	0, 0			
Iran	I	0, 0			
Ireland	I	1, 2	Paul Callanan	Paul Callanan	Paul Callanan
Israel	III	1, 4	Shay Zucker	Shay Zucker	Shay Zucker
Italy	VII	1, 8	Ginevra Trinchieri	Ginevra Trinchieri	Ginevra Trinchieri

National Member	Dues Category	Votes (a, b)	National Representatives	Finance Committee Nominations	Membership & SNC Nominations
Japan	VII	1, 8	Sadanori Okamura	Masatoshi Ohishi	Sadanori Okamura
Korea, Ref. of	II	1, 3	Myung Gyoon LEE	Myung Gyoon LEE	Myung Gyoon LEE
Latvia	I	0, 0			
Lebanon	interim	0, 0			
Lithuania*	I	0, 0	G. Tautvaisiene	G. Tautvaisiene	G. Tautvaisiene
Malaysia	I	1, 2	-	-	-
Mexico	III	1, 4	Hector Bravo-Alfaro	Hector Bravo-Alfaro	Hector Bravo-Alfaro
Mongolia	interim	0, 0			
Morocco*	I	0, 0	Benkhaldoun Zouhair	Benkhaldoun Zouhair	Benkhaldoun Zouhair
Netherlands	V	1, 6	R.A.M.J Wijers	Lex Kaper	Lex Kaper
New Zealand	II	1, 3	Grant Christie	Grant Christie	Grant Christie
Nigeria	I	1, 2	- Okere Bonaventure	-	-
Norway	II	1, 3	Oddbjørn Engvold	Oddbjørn Engvold	Oddbjørn Engvold
Panama	interim	0, 0			
Peru	I	0, 0			
Philippines	I	1, 2	Cynthia P. Celebre	Cynthia P. Celebre	Cynthia P. Celebre
Poland	IV	1, 5	Andrzej Udalski	Andrzej Udalski	Andrzej Udalski
Portugal	II	1, 3	-	-	-
Romania	I	1, 2	-	-	-
Russian Fed.	V	1, 6	Dmitrij Bisikalo	Nikolai Samus	Dmitrij Bisikalo
Saudi Arabia	I	0, 0			
Serbia	I	1, 2	Zoran Kneževič	Zoran Kneževič	Zoran Kneževič
Slovakia	I	1, 2	Augustin Skopal	Augustin Skopal	Augustin Skopal
South Africa	III	1, 4	Patricia Whitelock	Patricia Whitelock	Patricia Whitelock
Spain	IV	1, 5	Rafael Bachiller	Rafael Bachiller	Rafael Bachiller
Sweden	III	1, 4	Dainis Dravins	Dainis Dravins	Dainis Dravins
Switzerland	III	1, 4	Georges Meylan	Georges Meylan	Georges Meylan
Tajikistan	I	1, 2	Subhon Ibadov	Subhon Ibadov	Subhon Ibadov
Thailand	I	1, 2	Utane Sawangwit B. Soonthornthum	Utane Sawangwit	Utane Sawangwit
Turkey	I	1, 2	Talat Saygaç	Talat Saygaç	Talat Saygaç
Ukraine	III	1, 4	Peter P. Berczik	Peter P. Berczik	Peter P. Berczik
UK	VII	1, 8	Mike Cruise	Mike Cruise	Mike Cruise
United States	X	1, 11	Kathie Mathae-Bailey Ed Guinan	Kevin Marvel	Sally Heap
Vatican	I	1, 2	Christopher Corbally	Jose Funes	Guy Consolmagno
Venezuela	I	1, 2	Gustavo Bruzual	Gustavo Bruzual	Gustavo Bruzual

Transactions IAU, Volume XXVIIIB
Proc. XXVIII IAU General Assembly, August 2012
Thierry Montmerle, ed.

doi:10.1017/S174392131500544X

CHAPTER II

TWENTY EIGHTH GENERAL ASSEMBLY BUSINESS SESSIONS

SECOND SESSION: Thursday, 30 August 2012, 16:30-18:00

Chaired by Robert Williams, President of the IAU
China National Convention Centre, Beijing

1. Opening and Welcome

The President of the IAU, Prof. Robert Williams, welcomed the delegates and members to this first business session of the General Assembly. The President invited the General Secretary, Dr. Ian Corbett, to start the business session.

2. Individual Membership

The General Secretary reported that the XXVII General Assembly admitted 1008 new Individual members. Their list is given in Chap. VIII of these *Transactions.*

As of 31 August 2012, the number of IAU Individual Members, as recorded in the database, was 9896.

Thanks for a new, database compliant nomination form, the new Individual Members can be immediately added to the database after their admission.

3. Deceased Members

The Executive Committee regretfully reported the decease of 192 Individual Members of the Union, since August 2009. The General Secretary displayed their names on slides, and the General Assembly observed a one-minute silence in their respectful memory.

Their list is also given in Chap. VIII of these *Transactions.*

4. Tribute to Franco Pacini (IAU President 2001-2003)

A brief tribute was paid to Franco Pacini (born and deceased in Florence, 1939-2012), who died a few months before the General Assembly. He was famous for having predicted the existence of pulsars, and was instrumental in taking the first steps towards the IAU International Year of Astronomy in 2009, which ended in January 2010 to celebrate Galileo's 400th anniversary of the discovery of the satellites of Jupiter.

5. Appointment of the Official Tellers

The Official Tellers were duly appointed, and confirmed that the quorum requirements for the National Representatives were duly satisfied..

6. Proposed Changes to the Divisional Structure

The Assistant General Secretary chaired a discussion/questions session with the participants to the General Assembly, prior to the vote of the corresponding Resolution B4 (see item 10 of the first business session).

7. Resolutions

The Resolutions submitted to the General Assembly were examined by the Resolutions Committee, Chaired by Daniela Lazzaro. They were all "B" type, i.e., with scientific content. They were submitted in English and published in the GA Newspaper ("Inquiries of Heaven", issue 9) before the vote. The French version was made available on the IAU web site after their adoption.

These Resolutions were:

- B1 On guidelines for the designations and specifications of optical and infrared astronomical photometric passbands.
- B2 On the re-definition of the astronomical unit of length.
- B3 On the establishment of an International NEO early warning system.
- B4 On the restructuring of the IAU Divisions.

They were unanimously approved.

8. Proposed Division Presidents and Vice-Presidents

Following the adoption of the new Division structure by the General Assembly, the Division Presidents and Vice-Presidents were appointed by the Executive Committee, with the mission to implement the recommendations of Resolution B4, and submitted to the General Assembly for approval. Their names were unanimously approved.

The list is as follows:

Division A: Space and Time Reference Systems
President: Sergei Klioner (Germany)
Vice-President: Jacques Laskar (France)

[Note that the name for Division A was re-discussed by its members after the General Assembly, and reverted to the name of former Division I: Fundamental Astronomy.]

Division B: Facilities, Technologes, and Data Science
President: David Silva (USA)
Vice-President: PietroUbertini (Italy)

Division C: Education, Outreach, and Heritage
President: Mary Kay Hemenway (USA)
Vice-President: Hakim Malasan (Indonesia)

Division D: High Energy Phenomena and Fundamental Physics
President: Diana Worrall (UK)
Vice-President: Felix Aharonian (Ireland)

Division E: Sun and Heliosphere
President: Lidia van Driel-Gesztelyi (UK)
Vice-President: Yihua Yan (China)

Division F: Planetary Systems and Bioastronomy
President: Giovanni Valsecchi (Italy)
Vice-President: Nader Haghighipour (USA)

Division G: Stars and Stellar Physics
President: Ignasi Ribas (Spain)
Vice-President: Corinne Charbonnel (France)

Division H: Interstellar Matter and Local Universe
President: Ewine van Dishoek (Netherlands)
Vice-President: Jonathan Bland-Hawthorn (Australia)

Division J: Galaxies and Cosmology
President: Françoise Combes (France)
Vice-President: Thanu Padmanabhan (India)

9. Proposed Commission Presidents and Vice-Presidents

As a result of the Division restructuring, the existing Commissions were kept unchanged but reassigned to the new Divisions.

The parent Divisions, Commssions, Presidents and Vice-Presidents are listed in the Table below.

Division	*Commission*	*President*	*Vice-President*
A	C4	C. Hohenkzerk (UK)	J.-E. Arlot (France)
	C7	A. Morbidelli (France)	C. Baug (Argentina)
	C8	N. Zacharias (USA)	A.G.A. Brown (Netherlands)
	C19	C.-L Huang (China N.)	R.G. Gross (USA)
	C30	D. Pourbaix (Belgium)	T. Zwitter (Slovenia)
	C31	M. Hosokawa (Japan)	E.F. Arias (France)
	C52	M.H. Stoffel (Germany)	S.M. Kopeikin (USA)

Division	*Commission*	*President*	*Vice-President*
B	C5	R.J. Hanisch (USA)	M.W. Wise (Netherlands)
	C6	H. Yamaoka (Japan)	D.W.E. Green (USA)
	C14	L. Mashonkina (Russia)	F. Salama (USA)
	C25	A. Walker (Chile)	S. Adelman (USA)
	C40	J. Chapman (Australia)	G. Giovannini (Italy)
	C50	R. Green (USA)	C. Walker (USA)
	C54	G.T. van Belle (USA)	D. Mourard (France)
C	C41	R. Kochhar (India)	X.-C. Sun (China N.)
	C46	J.-P. de Greve (Belgium)	B.E. Garcia (Argentina)
	C55	L. Christensen (Germany)	P. Russo (Netherlands)
D	C44	C. Jones (USA)	-
E	C10	K. Schrijver (USA)	L. Fletcher (UK)
	C12	G. Cauzzi (Italy)	N.G. Shchkina (Ukraine)
	C49	I. Mann (Sweden)	P.K. Manoharan (India)
F	C15	D. Bockele-Morvan (F)	R.A. Gil-Hutton (Argentina)
	C16	M. Lemmon (USA)	-
	C20	S.R. Chesley (USA)	D. Lazzaro (Brazil)
	C22	P.M.M. Jenniskens (USA)	J. Borovicka (Czech Republic)
	C51	P. Ehrenfreund (USA)	S. Kwok (China N.)
	C53	A. Lecavelier des Etangs (France)	D. Minniti (Chile)
G	C26	B. Mason (USA)	Y.Y. Balega (Russia)
	C27	K. Pollard (New Zealand)	C.S. Jeffery (UK)
	C29	K. Cunha (Brazil)	-
	C35	M. Limongi (Italy)	J.C. Lattanzio (Australia)
	C36	J. Puls (Germany)	I. Hubeny (USA)
	C42	M.T. Richards (USA)	T. Pribulla (Slovakia)
	C45	R.O. Gray (USA	C. Soubiran (France)
H	C33	B. Nordström (Denmark)	J. Bland-Hawthorn (Australia)
	C34	S. Kwok (China N.)	B.-C. Koo (Korea Rep.)
	C37	G. Carraro (Chile)	R. de Grijs (China N.)
J	C21	J. Murthy (India)	-
	C28	J.S. Gallagher III (USA)	-
	C47	B. Schmidt (Australia)	-

10. Financial Matters

Report of the Finance Sub-Committee

Presented by Birgitta Nordström (Chair)

1. Accounts 2009-2011

The accounts of the IAU are in good shape. They show the usual pattern during a triennium with a loss during the IAU General Assembly year and during the other two years: surpluses of 584 k€ and 132 k€ in 2010 and 2011. The average over the triennium is roughly in balance except that the surplus is larger than budgeted due to recovery of previous losses and late payments from National Members that were outstanding during previous financial years.

The assets of the Union increased in the period, which gives room for increased educational activities. It is prudent to maintain the assets at the level of one year's turnover, in order to provide a reserve for cash flow during the triennial cycle, to cover late payments by some National Members, and to provide some investment income.

Between 2009 and 2010 the IAU has changed its unit of account from Swiss Francs (CHF) to euros (EUR). Membership dues will for the future be invoiced in euros. The FSC welcomes this change, which reflects the fact that most of the Union's expenditure is in euros and many of the Union's National Members have their currency in euros. The change reduces exchange losses overall (to the Union and to many National Members) and spreads the risk of currency fluctuations. It will also prove to be more effective and convenient that the accounts are held in Paris.

While the accounts have consistently passed the test of professional audit, investigations by the administration discovered in 2008 that the Union had been subject to systematic fraud. This was reported to the FC at the IAU GA in 2009. The administration took steps to recover from its bank unauthorized payments and much of those have now been recovered. It has implemented changes to its accounting systems and now use an electronic accounting system, SAGE, that should make recurrence of fraud more difficult to conceal in the future. Bank transactions are now completed by internet, entered by the Executive Assistant and authorised by the General Secretary. All transactions are visible to both General Secretary and Executive Assistant. Salaries and annual account reconciliation with the SAGE system are prepared by an external independent accountant. The final result is a robust and transparent system. The accounts for 2009-2011 have been examined and certified by the new auditor and no specific issues have been raised.

The administration has kept the FSC fully informed and consulted on the measures to be taken during the triennium. The FSC believes that the measures taken to prevent recurrence are appropriate to the scale of the Union's activities, and will bring increased effectiveness to the Union's operation. The FSC commends the GSs Karel van der Hucht and Ian Corbett as well as the Administrative Assistant, Vivien Reuter, for the work done to recover the losses. The FSC notes with satisfaction that its bank, Credit Lyonnais has during the triennium reimbursed IAU much of the previous losses carried by IAU.

Noting the large sum of outstanding late payments from National Members, and noting the efforts of the Executive to recover what it can, the FSC recommends the acceptance of the accounts 2009-2011.

2. Members' dues

Budgets 2013, 2014 and 2015

In the budgets, the proposed increase in the unit of contribution has been kept within inflation at a modest and realistic forecast of 2%, but the recovery of earlier losses has meant that the assets for the Union have increased. In the proposed budgets this money has been shown as expenditure on matters within the proposed educational programme, developed in the Union's Strategic Plan. The apportionment of expenditure to individual programmes will be a matter of discussion as the Strategic Plan is implemented and will depend on the degree of success of the Union in raising external funding.

The FSC recommends the adoption of the proposed unit of contribution as set out in the budgets, i.e. 2750 EUR in 2013, 2800 EUR in 2014 and 2860 EUR in 2015.

Noting the improved financial position of the Union and the ambition of the Strategic Plan to raise additional revenue for additional activities from outside sources, the FSC recommends that the budgetary provisions for 2013-2014-2015 should be accepted.

The report was submitted to the vote of National Representatives, and the budget for 2012-2015 approved with one abstention.

11. Election of the Finance Committee and Membership Committees

1. The General Assembly unanimously appointed the following Finance Committee (2012-2015):

Beatriz Barbuy (Brazil; Chair)
Joao Alves (Austria)
Kate J. Brooks (Australia)
Ahanwen Han (China Nanjing)
Nikolaos D. Kylafis (Greece)
Tushar P. Prabhu (India)
Lee Anne M. Willson (USA)

2. The General Assembly unanimously appointed the following Membership Committee (2012-2015):

Christian Henkel (Germany; Chair)
Sara Heap (USA)
Myungshin Im (Korea Rep.)
Lex Kaper (Netherlands)
Rene Alejandro Mndez Bussard (Chile)
Helmut O. Rucker (Austria)
Nikolay N. Samus (Russian Federation)
Ramotholo R. Sefako (South Africa)

12. Appointment of the Resolutions Committee

The General Assembly unanimously appointed the following Resolutions Committee (2012-2015):

Ian F. Corbett (UK; Chair)
Bruce G. Elmegreen (USA)
Rene Kraan-Korteweg (South Africa)
Yanchun Liang (China Nanjing)
Karel A. van der Hucht (Netherlands)

13. Appointment of the Special Nominating Committee

The General Assembly unanimously appointed the following Special Nominating Committee (2012-2015):

- *IAU ex-officio members:*

Norio Kaifu (President, Japan; Chair)
Robert Williams (Pasr President; USA)

- *IAU Advisers:*

Thierry Montmerle (General Secretary; France)
Piero Benvenuti (Assistant General Secretary; Italy)

- *Nominated members:*

Françoise Combes (France)
Malcolm S. Longair (UK)
Matthias Steinmetz (Germany)
Edward P.J. van den Heuvel (Netherlands)
Gang Zhao (China Nanjing)

14. Election of the Executive Committee

The General Assembly unanimously elected the following Executive Committee (2012-2015):

President	Norio Kaifu (Japan)
President-Elect	Silvia Torres Peimbert (Mexico)
General Secretary	Thierry Montmerle (France)
Assistant General Secretary	Piero Benvenuti (Italy)
Vice-President	Matthew Colless (Australia)
Vice-President	Renée Kraan-Korteweg (South Africa)
Vice-President	Xiao-Wei Liu (China Nanjing)
Vice-President	Jan Palouš (Czech Republic)
Vice-President	Dina Prialnik (Israel)
Vice-President	Marta Rovira (Argentina)
Adviser	Robert Williams (USA)
Adviser	Ian Corbett (UK)

15. Dates and place of the XXXth General Assembly in 2018

The XXXth General Assembly will take place in Vienna (Austria), August 20-31, 2018.

Transactions IAU, Volume XXVIIIB
Proc. XXVIII IAU General Assembly, August 2012
Thierry Montmerle, ed.

doi:10.1017/S1743921315005451

CHAPTER III

TWENTY EIGHTH GENERAL ASSEMBLY CLOSING CEREMONY

Thursday, 30 August 2012, 16:30-18:00

China National Convention Centre, Beijing

1. Welcome by Norio Kaifu, IAU President

The incoming IAU President, Norio Kaifu, welcomed the participants to the Closing Ceremony of the XXVIIIth General Assembly.

2. Invitation to the XXIXth General Assembly, Hawai'i, USA, August 2015

A video movie was shown on the screens, presenting the venue and attractions of the next General Assembly, and welcomed the participants in advance.

3. Address by the Retiring President

Prof. Robert Williams

The retiring IAU President expressed his warmest thanks and deep appreciation for the welcome and organization of our Chinese colleagues, and noted the healthy state and successes of Chinese astronomy. He congratulated the Local Organizing Committee, under the leadership of Prof. Gang Zhao, for the smooth running of the General Assembly. He also extended an enthusiastic welcome to the participants to the next General Assembly in Honolulu, Hawai'i.

4. Address by the Retiring General Secretary

Dr. Ian Corbett

It has been privilege and a pleasure to serve the Union as General Secretary for the past three years, and I am very grateful to everyone who has helped me in this task - too many to mention by name but I will single out the President and the members of the Executive Committee, the Division, Commission and Working Group Presidents, and the Secretariat, and above all to Vivien Reuter, without whom the job would have been almost impossible. I will return to this later.

In my incoming speech at Rio de Janeiro three years ago, I said there were four challenges facing us in the triennium that has just finished.

The first was to implement the Strategic Plan endorsed so enthusiastically at Rio. Here we have made spectacular progress: we have established the Office of Astronomy for Development in Cape Town under an agreement with South Africa's National Research Foundation, we have appointed its Director, Kevin Govender, we have established three Task Forces exactly as envisaged in the Strategic Plan, we have signed agreements for two Regional Offices, with more to follow, and we have substantially increased the budget for education and development in the 2013-2015 triennium. All this has required the work of many people, but I must make specific mention of George Miley, the architect and 'builder in chief' of the whole enterprise, and Kevin Govender, who as OAD Director has demonstrated unlimited energy and determination in setting up the OAD programme. With people like that behind it, how can it be anything other than a resounding success. I wish to thank them publicly here for all that they have done, and I am delighted that George will continue to be heavily involved throughout the next triennium.

The second was to complete the International Year of Astronomy 2009 and build on its legacy. IYA 2009 was an overwhelming and unprecedented success. It has received great publicity so I will not dwell any further on this other than to record my thanks to the late Franco Pacini, whose idea it was, to Catherine Cesarsky, who as President did so much to make it happen, to Pedro Russo, who did such a great job in the IYA Secretariat, and to everyone else I cannot name who worked tirelessly to ensure we exceeded even our most optimistic expectations. Its legacy continues in visible form in the programmes of the OAD and its Task Forces, and in less visible forms in the many and diverse activities sparked off by IYA which continue to flourish all over the world. It was truly a 'transformational' event.

The third was the more open ended need for the IAU to develop and (I quote) "reflect the needs and aspirations of its community". There is no better demonstration of this than the development of the proposal to restructure the Divisions of the Union to reflect better the state of modern astronomy and the interests of our members. This was an enormous task, initiated by the Executive Committee which established a Task Group led by the AGS, Thierry Montmerle. The process involved extensive consultation with Divisions through their Presidents and active discussion between the EC and the DPs before the final proposal could be put to the GA earlier today. I am delighted that the enabling Resolution was passed and we can move on to the next phase, as always in close cooperation with the new Division Presidents and their members. I would like to thank the members of the Task Group, the Division Presidents, and especially Thierry for their sterling work in bringing this about. This was not the only item of progress - we introduced electronic voting, to enable and encourage wider participation in Division and Commission affairs, and this has proved very popular. I am delighted that the first session approved the changes to the Statutes and Bye-Laws which will allow the EC to put scientific matters at the GA to an electronic vote of all individual members, thus empowering the whole membership and widening the franchise. I only hope it does not reduce the attendance at the plenary sessions of the GA!

Finally, the fourth challenge concerned this 28th GA itself. Here I had no fears.

Every General Assembly is remembered for something - in the past few years, Sydney for kick starting the International Year of Astronomy, Prague for Pluto, Rio de Janeiro for the Strategic Plan. I thought that Beijing would be remembered for the restructuring of the Divisions and for the impressive progress we have made in implementing the Strategic Plan. But I was wrong! Above all Beijing will be remembered for the unobtrusive but impeccable organization, the overwhelming generosity of our hosts, the smiling teams of volunteers, and the impressive CNCC.

Organising a GA starts more than six years in advance. First, the proposal has to be prepared, which involves a lot of background work. Then, once approved, the real work starts and it never stops, reaching a peak in the first few days of the actual GA. There is the National Organising Committee, the Local Organising Committee, the Professional Conference Organisers, the Convention Centre, which all have to blend together into a seamless team. And this has been done to perfection in Beijing.

In parallel, the IAU has to solicit proposals for the scientific programme, select the best, and plan the schedule, all the time maintaining very close relationships with the local organisers to make sure that everything is planned and all eventualities are covered. We also have to prepare, select and distribute the IAU grants, to a record number and amount. These grants are a vital part of IAU's programme: they enable people, especially young people or those from developing countries, to benefit from the "GA experience".

All this work has paid off. Nearly 3000 people have passed through the doors. Everything that I have seen or heard says that the GA has been a great success, scientifically and organisationally. Our hosts astounded us all by securing the Vice-President of the Peoples Republic of China to make the opening speech at the Inaugural Ceremony, and we were delighted by the things he said, all highly relevant and encouraging. The performances which followed at the Inaugural Ceremony were a breathtaking display of skill and artistry.

The science programme was outstanding. We were fortunate in having such excellent speakers for our four Invited Discourses, of course, but there were many other outstanding contributions, as the Proceedings which will appear next year will certainly show. We also managed the Young Astronomers Lunch, the Women in Astronomy day and many other special events.

I cannot thank everyone by name, much as I would like to do so, and so I thank everyone who had anything to do with the organisation and running of this wonderful GA. It has been a memorable experience.

So I conclude where I started. It has been a privilege to serve you as General Secretary. I look forward to seeing you in Hawai'i in 2015.

5. Address by the Incoming President

Prof. Norio Kaifu

It is a great honor to be elected President of the IAU, one of the most time-honored and active international scientific organizations. I have enjoyed my many years of service to the IAU— as a Vice-President, as a member of the Working Group of IYA 2009, and as one of four officers since 2009. I have particularly appreciated my association during these years with members of the IAU Executive Committee: Bob, Ian, Catherine, Karel, Thierry and the six vice presidents.

I especially thank outgoing President Robert Williams for guiding the Executive Committee and Officers in a very powerful and democratic manner. He led the establishment of the IAU strategic Plan of Astronomy for Developing World, and he also strengthened ties with many other organizations. I also deeply thank Ian Corbett, outgoing General Secretary, for his dedicated hard work in serving the IAU by dealing with many complicated tasks. He made also the operating procedures of the IAU simpler and clearer. I extend my sincere thanks to Vivien Reuter, the IAU's Head of Administration, who used her outstanding experience and skills to manage the IAU secretariat office in Paris.

The Division restructuring was just adopted by the General Assembly, Thanks to attendants, and to Thierry Montmerle for his leadership as a Chair of the Task Force. The registration of the Division members, and discussion on Commissions and Working Groups under the new Divisions will start immediately. Next year the Executive Committee and the Division Presidents will discuss the basic plan of updated structure of the IAU. The implementation of this change will produce more flexible IAU activities. Therefore I am very much looking forward to working with new EC members and new Division Presidents throughout the next three years.

As I mentioned at the beginning of this address, the IAU is one of the most active international unions of basic sciences. Among the many remarkable characteristics of the IAU, I will highlight three that fuel its activities. Obviously the first of these is its exciting and continuous progress in understanding the Universe. A second distinctive feature of the IAU is its individual members. Most international scientific organizations have only national members, but the IAU has ten thousand and further growing individual members who firmly support the growing and evolving activities of IAU. A third characteristic of the IAU relates to the appealing nature of astronomy itself. The wonders of the Universe attract the interest of people worldwide. This is why the IAU can attain such broad and global contact with teachers, parents and a vast number of amateur astronomers.

Let me expand on this third point a bit. When the IAU held its first General Assembly in 1922, only two of the 15 national members were affiliated with areas beyond the so-called ?Western World?: Mexico and Japan. In contrast to this earlier time, the IAU now has 70 national members from all over the world. Although we have a rather good regional distribution, there are still large regional gaps. Let us recall that the United Nations has 192 member countries. In the IYA 2009, however, 148 counties and territories formally joined ?a number more than double the national members of the IAU and nearly 80 percent of the total membership of the United Nations. This participation clearly taught us that the wonder of the Universe attracted everyone, and also demonstrated that we could bridge research and education; professions and children; and cutting-edge science and the developing world. The prospect of such possibilities opens the door to a bright future for astronomical research worldwide.

The objective of the IAU since its establishment is to promote astronomy in all its aspects. One of our long-term missions now is to advance astronomy by widening foothold around the world. The IAU has already begun to implement such mission. We have entered the second three-year term of the IAU?s Decadal Strategic Plan, "Astronomy for the Developing World", which aims to promote research, education, and popularization of astronomy in the developing world. The Director of the Office for Astronomy Development, Kevin Govender, has made excellent progress in organizing support teams and task groups, so we are stepping into the second phase of ?Astronomy for the Developing World? with a firm foundation on which to build further implementation. Together with exciting frontiers of astronomy, and rapid growth of astronomical researches in the developing world, we are opening a new era of the IAU and the promotion of astronomy worldwide.

Finally, I express my sincere thanks to all attendees and to all organizers and supporters of this very successful General Assembly in Beijing. I look forward to seeing you all at our next General Assembly in Honolulu, Hawai'i in 2015.

Thank you for attention.

6. Address by the Incoming General Secretary

Dr. Thierry Montmerle

"As a man sows, so shall he reap.
A thousand-li journey is started
by taking the first step."
(Chinese Proverb)

Dear Colleagues,

My term as General Secretary will begin at the end of this week, under the best possible auspices.

First, I have the honor to succeed Ian Corbett, to whom the IAU owes a great deal. Not only has he (and Executive Secretary Vivien Reuter) put the Secretariat back on track after many difficulties, but he also contributed a lot to the organization and success of this General Assembly, along with our Chinese colleagues.

Second, thanks to the support of the Executive Committee and the active participation of the Division Presidents, we have conducted a major restructuring of the IAU Divisions, put before the General Assembly in the form of Resolution B4, which has been approved here almost unanimously.

This talk is perhaps the opportunity to look back and explain in more detail the reasons for this restructuring, and its consequence for the future triennium, the reform of the Commissions.

As a brief historical note, the Divisions were created nearly 20 years ago. At that time, they were introduced as administrative ?umbrellas? to put the 40 IAU Commissions into a limited number of more or less similarly topical "baskets". The twelve resulting Divisions had their own "Organizing Committees", chaired by a President and a Vice-President. Their structural role, as defined in the statutes, was to monitor the evolution of their Commissions over limited time scales, starting with six years, then renewed if justified for three more years (Bye-laws, Art. 23). But the fact is that, in spite of the tremendous evolution of astronomy over this period, and except for very minor changes, the 12 Divisions and the 40 Commissions did not really evolve, even though most of them were indeed active, producing reports, recommendations, etc., and, perhaps more visibly, selecting IAU Symposia each year.

This situation has long been thought to be unsatisfactory, and after the huge and extraordinarily successful effort of the International Year of Astronomy (IYA) which culminated in 2009, the year of the last GA in Rio de Janeiro, the newly elected Executive Committee (EC) decided to attack this problem. An "EC Task Group on Division Restructuring" "TG" for short in what follows), chaired by myself, was established at the first meeting of the new EC after the GA, which took place in Baltimore in May 2010, with the goal of presenting to the Division Presidents (DPs), at the following EC meeting in Prague in Spring 2011, a full-scale project for a new Division structure.

The vision underlying the new structure, as proposed by the TG, was entirely different from the initial approach. Instead of being purely administrative structures, the new Divisions had to reflect the scientific policy of the IAU, and to promote its adaptation to a rapidly changing world on two grounds: astronomy (for professional scientists), as well as society (for the public, to follow up on the IYA). The first draft of the restructuring consisted not only of a reduction in the number of Divisions, but also in the creation of Divisions of a different nature, as explained below. This draft was strongly supported by the DPs, who immediately urged the TG to continue its work and reflect on the opportunity to also reexamine Commissions and Working Groups.

To make an almost three-year story short, the TG worked very closely with the DPs and the EC, refining the names and scope of the Divisions, and preparing suggestions for the future evolutions of Commissions and Working Groups. The result has been Resolution B4, discussed at the last EC meeting, and proposed for a vote here in Beijing, since any modification to the Divisions must be approved by the General Assembly.

What were the proposed changes? Apart from "merging" existing Divisions (sometimes in an obvious fashion, like "Stars" and "Variable Stars"), perhaps the most visible change is the introduction of a new Division: Division "C", on Education, Outreach, & Heritage, areas which were before within the purview of Commissions. After the IYA2009, this "promotion" seemed more than justified, but above all made more visible the increasing involvement of the IAU in Educational projects, in particular with the creation of the Office of Astronomy for Development (OAD) in Cape Town, to implement the Strategic Plan voted by the General Assembly in Rio three years ago.

The other most notable change is the introduction of a large (but not the largest) Division related to Instrumentation in the broadest sense, from laboratory to observation and interpretation; Division B: "Facilities, Technologies & Data Science".

Div.B indeed has a very broad scope. Its ambition is to be the common "forum" where large facilities, multiwavelength instrumentation (without explicit distinction between ground and space), large surveys and databases (data mining, astrostatistics, etc.), as well as computer science and mathematical methods, etc. can be discussed and expertise exchanged.

Altogether, in essence the proposed Divisions will be the visible backbone of the IAU. Their scientific role will be much increased compared to that in the existing structure. With the help of the newly established "Division Steering Committees", the Division Presidents will be the natural points-of-contact between the community and the EC: as a result, they will collectively constitute a de facto "Advisory Committee" to the Executive Committee. This is the approach that was already adopted in practice by the TG and DPs in the present reform initiatives, and it has proved to work very successfully.

In view of the positive GA vote, a slate of new DPs and DVPs, with significant "new blood", has been presented by the EC and approved by National Representatives. These colleagues are committed to making the proposed Restructuring successful.

How will the restructuring be implemented ? The approval of the new Divisions structure implies other important structural reforms.

First, the "main membership" of the IAU will be by Divisions (one or more), and not by Commissions as was the case until now. Membership of Commissions will come next, but it will be perfectly acceptable to belong to a Division, and not to a Commission, if those existing Commissions are not relevant to one's research or activity. This is a good incentive to create new ones (likely starting by a Working Group)! The procedure will be by electronic voting, so easily accessible to everyone.

Second, the Divisions will be run by "Division Steering Committees". Their exact composition will be discussed between Division Presidents and with the Executive Committee, on a case-by-case basis. These new Committees will include "at-large" members, i.e., members from the community, not necessarily with a great experience of the IAU -it will be a good training for future responsibilities within the Union. The Division Steering Committees will have a much more scientific role than the current "Organizing Committees" of the Divisions.

Third, as the name implies, once elected the Division Steering Committees will have the primary task of "steering" the evolution of the Divisions, in particular of their Commissions (reexamining their role, goals, issues, etc.), under the supervision of the Executive

Committee. Note that in the new structure a Division will be larger than the sum of its Commissions (contrary to the founding definition of Divisions).

The Divisions will start working immediately, with the Commissions that are currently existing, reassigned as proposed in the approved new structure. However, in parallel with "steering" their Commissions, a prime task of the Division Steering Committees, in co-operation with the Commission Presidents and Organizing Committees, will be to reform the Commissions, that is, to bring them more in line with the latest scientific developments in astronomy. Perhaps new ones will be created, perhaps some will disappear.

This will open a new chapter in the history of the IAU. We hope to see at the next General Assembly a different Commission structure than now. It will be a challenge, but I have no reason to doubt that the community will be up to it, as it has been the case here in Beijing for the Divisions.

I look very much forward to seeing you again in Honolulu in 2015.

7. Traditional Chinese "Thousand Hands" Dance

Performance by the Beijing University of Aeronautics and Astronautics.

8. Closing remarks

Dr. Gang Zhao, co-Chair, NOC and LOC of the IAU XXVIIIth General Assembly

Good afternoon, dear participants, dear distinguished guests, ladies and gentlemen,

The General Assembly of the International Astronomical Union is first time hold in China since the Chinese Astronomical Society join IAU in 1935. During the past days, near 3,000 astronomers from more than 70 countries or regions have gathered in Beijing, exchanging the latest progresses in all fields of astronomy, and discussing the future development of astronomy worldwide.This is a great event in the history of Chinese astronomy, and it will for sure have a profound impact on the development of Chinese astronomy, and promote and expand the international collaborations.

Many people have made enormous efforts in planning, preparing for, and organizing this General Assembly, and at this special moment, please give me the honor to express my sincere appreciations to the following:

Thanks to the IAU, for the first time, allowing us the opportunity to hold such an important astronomical event in Beijing, China.

Thanks to all participants, you sitting here or those not sitting here, for your sincerely supporting us and for your warmly joining us in all the programs.

Thanks to CNCC for your professional service and friendly work during the assembly.

Thanks to all sponsors, exhibitors and partners for your generous supports to this General Assembly.

Thanks to all MCI colleaguesas our PCO, who have assisted this General Assembly with their heart and energy.

Thanks to all the volunteers for your careful working and warm services at the registration and conference rooms. Your early arrival and late departure, your friendly smile and genuine help have made the assembly a warm family for all participants. Let's welcome them to stage!

Thanks to the medical supporting staff, especially Dr. Sheng Haiying, who have also contributed a lot to the General Assembly. In past days, they have kindly treated about 80 participants who were not feeling well, and helped relieve their symptom efficiently. Please invite them to stage!

Thanks to the supporting staffs of the website and network team led by Dr. Cui Chenzhou, which set up and maintain the website of the General Assembly and the WiFi system. The webpages have been visited more than 750,000 times during the last one year. 120,000 requests have been connected to the WiFi system during the last 8 days, with maximum concurrent connection number of 1000 on August 23. Please welcome them!

Thanks to the newspaper team led by Prof. WANG Jinxiu. Altogether they have published 10 issues, with a total of 68 pages. For the seemingly "small" pages, the editorial team worked till late night every day, sometimes even till 2 hour in the morning. Let's invite them to stage!

Last but not least, thanks to all my colleagues, LOC members, for your hard working and close cooperating all these months in the past years. It is only because all your efforts and devotion, the IAU 28th General Assembly can be going so well and so exciting until now. Now, please join me with the warmest applause to invite all LOC members to stage!

Thank you, everyone on the stage, and at the audience, or those staffs and participants who are not here. Your enthusiasm, devotion and supports have made this general assembly a successful, enjoyable, and great event! Thank you very much!

9. Handing out the IAU flag

The General Assembly was adjourned after a Ceremony to hand the IAU flag, created by the LOC, to the hosts of the next General Assembly, after which Norio Kaifu called for a vote of thanks to the LOC, while the LOC team (about 100 volunteers and professionals) was assembling on the stage.

10. Image Archive

Images taken during the General Assembly can be viewed and dowloaded from the IAU web site:
http://www.iau.org/public/images/archive/category/general_assembly_2012/

Transactions IAU, Volume XXVIIIB
Proc. XXVIII IAU General Assembly, August 2012
Thierry Montmerle, ed.

doi:10.1017/S1743921315005463

CHAPTER IV

TWENTY EIGHTH GENERAL ASSEMBLY

RESOLUTIONS OF THE XXVIIIth GENERAL ASSEMBLY

1. Resolutions Committee (2009-2012)

The members of the Resolutions Committee for the 2009-2012 triennium were:

Daniela Lazzaro (Brazil; Chair)
Martha P. Haynes (USA)
Zoran Knezevic (Serbia)
Busaba Hutawarakorn Kramer (Germany)
Irina I. Kumkova (Russian Federation)
Silvia Torres-Peimbert (Mexico)
Na Wang (China Nanjing)

The report of the Resolutions Committee is given in Chapter II of these *Transactions*.

2. Approved Resolutions

RESOLUTION B1

on guidelines for the designations and specifications of optical and infrared astronomical photometric passbands

Proposed by IAU Commission 25

The XXVIII General Assembly of the International Astronomical Union,

noting

that considerable confusion has existed and continues to exist in the defining and naming of photometric passbands of all spectral widths in the visible and infrared regions of the electromagnetic spectrum,

considering

that minimizing such confusion has been a long-time goal of members of Commission 25 [e.g., see remarks by Wesselink and by Greaves in Transactions of the IAU, VII, pp. 267–273 (1950)],

recommends

1. that proposers of new passband systems should check the IAU Commission 25 website and links therein, especially to `http://ulisse.pd.astro.it/Astro/ADPS/` (extended version of the paper by Moro and Munari 2000, A&AS 147, 361) to ascertain what passband names have already been used, before creating designations for new passbands.[1]

2. that names for new passbands should avoid relatively well known designations, such as UBVRIJHKLMNQ, and the designations ZJHKLMNQ should be used henceforth to refer exclusively to the terrestrial atmospheric windows in the near and intermediate infrared (see Young *et al.* A&AS, 105, 259–279; Milone & Young (2005), PASP, 117, 485–502).

3. that any publication presenting the new passbands should contain the following information, to aid in transformations and standardizations:

a) a measure of central wavelength which is not flux-dependent, such as the pivot wavelength, or mean photon wavelength, as defined, for example, in Bessell & Murphy (2012), PASP, 124, 140–157; [2]

b) an indication of bandwidth, such as FWHM;

c) the spectral profile of the passband, unless it is completely symmetrical, as, for example, triangular passbands, when this shape and the domain in which this is the case (wavelength or wave number/frequency) are stipulated;

d) a clear statement on whether the passband profile includes the spectral sensitivity curve of the detector or not, and, if so, the characteristics of the detector;

e) the temperature at which these specifications apply;

f) such other details (for example, roll-off, pinhole and leakage specifications) as may be needed to obtain a closely matching filter from manufacturers.

4. that a copy of this resolution should be sent to all editors of astronomical and other journals which publish papers relating to astronomical photometry.

---

[1] Well known and accepted nomenclature also appears in the Drilling and Landolt chapter in Cox's "Allen's Astrophysical Quantities", 4th edition, 2000, page 386, Table 15.5, and other information on basic systems appears in V. Straizys' "Multicolor Stellar Photometry" volume, 1995 (second printing), (see `http://www.itpa.lt/MulticolorStellarPhotometry`), among other sources.

[2] For example, "Y" and "iz" are designations that have been applied to passbands in the 1 μm (Z) atmospheric window.

RESOLUTION B2

on the re-definition of the astronomical unit of length

Proposed by the IAU Division I Working Group Numerical Standards and supported by Division I

The XXVIII General Assembly of International Astronomical Union,

noting

1. that the International Astronomical Union (IAU) 1976 System of Astronomical Constants specifies the units for the dynamics of the solar system, including the day (D=86400 s), the mass of the Sun, M_S, and the *astronomical unit of length* or simply the *astronomical unit* whose definition[1] is based on the value of the Gaussian gravitational constant,

2. that the intention of the above definition of the astronomical unit was to provide accurate distance ratios in the solar system when distances could not be estimated with high accuracy,

3. that, to calculate the solar mass parameter, GM_S, previously known as the heliocentric gravitation constant, in Système International (SI) units[2],the Gaussian gravitational constant k, is used, along with an astronomical unit determined observationally,

4. that the IAU 2009 System of astronomical constants (IAU 2009 Resolution B2) retains the IAU 1976 definition of the astronomical unit, by specifying k as an "auxiliary defining constant" with the numerical value given in the IAU 1976 System of Astronomical Constants,

5. that the value of the astronomical unit compatible with Barycentric Dynamical Time (TDB) in Table 1 of the IAU 2009 System (149 597 870 700 m $\pm$3 m), is an average (Pitjeva and Standish 2009) of recent estimates for the astronomical unit defined by k,

6. that the TDB-compatible value for GM_S listed in Table 1 of the IAU 2009 System, derived by using the astronomical unit fit to the DE421 ephemerides (Folkner *et al.* 2008), is consistent with the value of the astronomical unit of Table 1 to within the errors of the estimate; and

considering

1. the need for a self-consistent set of units and numerical standards for use in modern dynamical astronomy in the framework of General Relativity,[3]

2. that the accuracy of modern range measurements makes the use of distance ratios unnecessary,

3. that modern planetary ephemerides can provide GM_S directly in SI units and that this quantity may vary with time,

4. the need for a unit of length approximating the Sun-Earth distance, and

5. that various symbols are presently in use for the astronomical unit,

recommends

1. that the astronomical unit be re-defined to be a conventional unit of length equal to 149 597 870 700 m exactly, in agreement with the value adopted in IAU 2009 Resolution B2,

2. that this definition of the astronomical unit be used with all time scales such as TCB, TDB, TCG, TT, etc.,

3. that the Gaussian gravitational constant k be deleted from the system of astronomical constants,

4. that the value of the solar mass parameter, GM_S, be determined observationally in SI units, and

5. that the unique symbol 'au' be used for the astronomical unit.

References

Capitaine, N., Guinot, B., Klioner, S., 2011, Proposal for the re-definition of the astronomical unit of length through a fixed relation to the SI metre, Proceedings of the Journées 2010 Systèmes de référence spatiotemporels, N. Capitaine (ed.), Observatoire de Paris, pp 20–23

Fienga, A., Laskar, J., Morley, T., Manche, H. *et al.*, 2009, INPOP08: a 4D-planetary ephemeris, A&A 507, 3, 1675

Fienga, A., Laskar, J., Kuchynka, P., Manche, H., Desvignes, G., Gastineau, M., Cognard, I., Theureau, G., 2011, INPOP10a and its applications in fundamental physics, Celest. Mech. Dyn. Astr., Volume 111, on line edition (`http://www.springerlink.com/content/0923-2958`).

Folkner, W.M., Williams, J.G., Boggs, D.H., 2008, Memorandum IOM 343R-08-003, Jet Propulsion Laboratory

International Astronomical Union (IAU), Proceedings of the Sixteenth General Assembly," Transactions of the IAU, XVIB, p. 31, pp. 52–66, (1976)

International Astronomical Union (IAU), Proceedings of the Twenty Seventh General Assembly," Transactions of the IAU, VXVIIB, p. 57, pp. 6: 55–70 (2010)

Klioner, S., 2008, Relativistic scaling of astronomical quantities and the system of astronomical units, A&A 478, 951

Klioner, S., Capitaine, N., Folkner, W., Guinot, B., Huang, T.-Y., Kopeikin, S. M., Pitjeva, E., Seidelmann P.K., Soffel, M., 2009, Units of relativistic time scales and associated quantities, in Proceedings of the International Astronomical Union, IAU Symposium, Volume 261, p. 79–84

Luzum, B., Capitaine, N., Fienga, A., Folkner, W., Fukushima, T., Hilton. J., Hohenkerk, C., Krasinsky, G., Petit, G., Pitjeva, E., Soffel, M., Wallace, P., 2011, The IAU 2009 system of astronomical constants: the report of the IAU working group on numerical standards for Fundamental Astronomy, Celest. Mech. Dyn. Astr., doi: 10.1007s10569-011-9352-4

Pitjeva, E.V. and Standish, E.M., 2009, Proposals for the masses of the three largest asteroids, the Moon-Earth mass ratio and the astronomical unit, Celest. Mech. Dyn. Astr., 103, 365, doi: 10.1007/s10569-009-9203-8

Standish, E.M., 2004, The Astronomical Unit now, in Transits of Venus, New views of the Solar System and Galaxy, Proceedings of the IAU Colloquium 196, D. W. Kurtz ed., 163

---

[1] The IAU 1976 definition is: "The astronomical unit of length is that length (A) for which the Gaussian gravitational constant (k) takes the value of 0.017 202 098 95 when the units of measurements are the astronomical unit of length, mass and time. The dimensions of k^2 are those of the constant of gravitation (G), i.e., $L^3M^{-1}T^{-2}$. The term "unit distance" is also for the length A." Although this was the first descriptive definition of the astronomical unit, the practice of using the value of k as a fixed constant which served to define the astronomical unit was in use unofficially since the 19th century and officially since 1938.

[2] Using the equation $A^3k^2/D^2 = GM_S$ where A is the astronomical unit and D the time interval of one day, and k the Gaussian gravitational constant.

[3] Relativistically a solar system ephemeris, for which the astronomical unit is a useful unit, is a coordinate picture of solar system dynamics. SI units are induced into such a coordinate picture by using the relativistic equations for photons and massive bodies and by relating the coordinates of certain events with observables expressed in SI units.

RESOLUTION B3

on the establishment of an International NEO early warning system

Proposed by IAU Division III Working Group Near Earth Objects

The XXVIII General Assembly of the International Astronomical Union,

recognizing

– that there is now ample evidence that the probability of catastrophic impacts of Near-Earth Objects (NEOs) onto the Earth, potentially highly destructive to life, and for humankind in particular, is not negligible and that appropriate actions are being developed to avoid such catastrophes;

– that for the largest NEOs, thanks to the efforts of the astronomical community and of several space agencies, the cataloguing of the potentially hazardous ones, the monitoring of their impact possibilities, and the analysis of technologically feasible mitigations is reaching a satisfactory level;

– that even the impact of small- to moderate-sized objects may represent a great threat to our civilizations and to the international community;

– that our knowledge of the number, size, and orbital behaviour of smaller objects is still very limited, thus not allowing any reasonable anticipation on the likelihood of future impacts;

noting

that NEOs are a threat to all nations on Earth, and therefore that all nations should contribute to avert this threat;

recommends

that the IAU National Members work with the United Nations Committee on the Peaceful Uses of Outer Space (UNCOPUOS) and the International Council for Science (ICSU) to coordinate and collaborate on the establishment of an International NEO early warning system, relying on the scientific and technical advice of the relevant astronomical community, whose main purpose is the reliable identification of potential NEO collisions with the Earth, and the communication of the relevant parameters to suitable decision makers of the nation(s) involved.

RESOLUTION B4

on the restructuring of the IAU Divisions

Proposed by the IAU Executive Committee

The XXVIII General Assembly of the International Astronomical Union,

noting

(a) that both the IAU and astronomy as a whole have evolved considerably since the current Divisions were introduced in 1994 and formally adopted in 1997, and that it is therefore appropriate to consider re-optimising the Divisional Structure,

(b) the report and recommendations of the Task Group established by the Executive Committee to examine the case for restructuring the Divisions, and the Executive Committee response to these recommendations,

(c) that the Commissions, Working Groups and other bodies under the Divisions may also require reform,

(d) that the implementation of the Strategic Plan through the Office of Astronomy for Development (OAD) and other associated programmes requires the Executive Committee to establish appropriate oversight and governance provisions for all Astronomy for Development activities, including the Office of Astronomy for Development, ensuring a strong link between these activities, the Divisions, and the Executive Committee.

approves

the proposal of the Executive Committee to restructure the Divisions as follows:

Division A	Space & Time Reference Systems
Division B	Facilities, Technologies, & Data Science
Division C	Education, Outreach, & Heritage
Division D	High Energies & Fundamental Physics
Division E	Sun & Heliosphere
Division F	Planetary Systems & Bioastronomy
Division G	Stars & Stellar Physics
Division H	Interstellar Matter & Local Universe
Division J	Galaxies & Cosmology

and requests

the new Divisions, guided by the Executive Committee, to work together to produce initial plans for a revised structure for Commissions, Working Groups and other bodies to be approved, in accordance with the Statutes and Bye-Laws of the Union, by the Executive Committee at its meeting in May 2013.

3. Résolutions Approuvées

RESOLUTION B1

Sur les recommendations concernant les désignations et spécifications des bandes passantes astronomiques en photométrie optique et infrarouge

Proposée par la Commission 25 de l'UAI

La XXVIIIe Assemblée Générale de l'Union Astronomique Internationale,

Notant

Qu'une confusion considérable a existé et existe encore en ce qui concerne la définition et la dénomination des bandes passantes photométriques de toutes largeurs spectrales dans les domaines optique et infrarouge du spectre éectromagnétique,

Considérant

Que la clarification de cette situation a toujours constitué un des objectifs des membres de la Commission 25 [voir les remarques formulées par Wesselink et par Greaves dans les "Transactions of the IAU", VII, pp. 267–273 (1950)],

Recommande

1. Que les propositions pour de nouveaux systèmes de bandes passantes soient compatibles avec le contenu du site Internet de la Commission 25 et des liens qu'il contient, notamment `http://ulisse.pd.astro.it/Astro/ADPS/` (extension de l'article Moro & Minari, A&AS, 147, 361) pour prendre connaissance des noms des bandes passantes actuellement en usage, avant de formuler des nouvelles désignations,[1]

2. Que les noms de nouvelles bandes passantes évitent d'être trop semblables à des désignations connues, telles que UBVRIJHKLMNQ, et en particulier que les désignations ZJHKLMNQ soient réservées exclusivement aux fenêtres atmosphériques terrestres dans le proche et moyen infrarouge (voir Young *et al.* A&AS, 105, 259–279; Milone & Young (2005), PASP, 117, 485–502),[2]

3. Que toute publication annonçant de nouvelles bandes passantes fournisse les informations suivantes, aux fins de transformation et de standardisation:

a. Une mesure de la longueur d'onde centrale qui soit indépendante du flux, comme la longueur d'onde-pivot, ou la longueur d'onde moyenne, telles que définies dans Bessell & Murphy (2012), PASP, 124, 140–157;

b. Une indication de la largeur de la bande, comme la largeur totale à mi-hauteur;

c. Le profil spectral de la bande passante, à moins qu'il ne soit rigoureusement symétrique, par exemple triangulaire, auquel cas la forme et le domaine doivent être explicités (en fonction de la longueur d'onde, ou du nombre d'onde, ou de la fréquence);

d. Une explication claire selon laquelle le profil de la bande passante inclut ou non la courbe de sensibilité du détecteur, et si c'est le cas, fournir les caractéristiques dudit détecteur;

e. La température de référence concernant ces spécifications;

f. Tous les détails de nature à permettre de reproduire un filtre aussi identique que possible au filtre de référence (par exemple, "roll-off", fenêtre d'entrée et pertes lumineuses).

4. Que le texte de la présente résolution soit communiqué â tous les éditeurs de revues astronomiques ou publiant des articles concernant la photométrie astronomique.

---

[1] Une nomenclature bien connue et acceptée figure également dans l'ouvrage de Drilling et Landolt, la 4e édition de Cox "Allen's Astrophysical Quantities" (2000, p. 386, Table 15.5), ainsi que la 2e édition de "Multicolor Stellar Photometry" (V. Straizy, 1995, voir `http://www.itpa.lt/MulticolorStellarPhotometry`).
[2] Par exemple, "Y" et "iz" sont des désignations utilisées pour des bandes passantes de la fenêtre atmosphérique 1μm (Z).

RESOLUTION B2

Proposée par le Groupe de travail UAI sur les "Standards numériques en Astronomie fondamentale" et Soutenue par la Division I

Re-définition de l'unité astronomique de longueur

La XXVIIIe Assemblée générale de l'Union astronomique internationale,

Notant

1. que le Système UAI 1976 de constantes astronomiques de l?Union astronomique internationale précise les unités pour la dynamique du système solaire, comprenant le jour (D=86400 s), la masse du Soleil, M_S, et *l?unité astronomique de longueur*, ou *unité astronomique*, dont la définition[1] est fondée sur la valeur de la constante de Gauss,

2. que le but de la définition de l'unité astronomique était de donner des valeurs exactes des distances relatives dans le système solaire à une époque où il n'était pas possible d'estimer des distances avec une grande exactitude,

3. que, pour évaluer le paramètre de masse solaire, GM_S, appelée précédemment constante héliocentrique de la gravitation, dans le Système International d'unités (SI)[2], on utilise la constante de Gauss k, ainsi qu'une valeur de l'unité astronomique déterminée par l?observation,

4. que le Système UAI 2009 de constantes astronomiques (Résolution UAI 2009 B2) a conservé la définition UAI 1976 de l'unité astronomique, en définissant k comme une

"constante auxiliaire de définition" avec comme valeur numérique celle qui est donnée par le Système UAI 1976 de constantes astronomiques,

5. que la valeur, compatible avec le Temps dynamique barycentrique (TDB), de l'unité astronomique donnée dans la Table 1 du Système UAI 2009 (149 597 870 700 m ± 3 m), est une moyenne (Pitjeva and Standish 2009) d'estimations récentes de l?unité astronomique définie par k,

6. que la valeur de GM_S compatible avec TDB, donnée dans la Table 1 du Système UAI 2009, qui a été calculée en utilisant la valeur de l'unité astronomique ajustée sur les éphémérides DE421 (Folkner *et al.* 2008), est cohérente avec la valeur de l'unité astronomique de la Table 1 dans la limite des incertitudes estimées; et

Considérant

1. le besoin de disposer d'un ensemble cohérent d'unités et de valeurs numériques de constantes pour leur utilisation en astronomie dynamique moderne dans le cadre de la relativité générale[3]

2. que l'exactitude des mesures modernes de distances rend inutile l'utilisation de distances relatives,

3. que les éphémérides planétaires modernes peuvent déterminer GM_S directement en unités SI et que cette quantité peut varier avec le temps,

4. le besoin d'une unité de longueur qui soit approximativement égale à la distance Terre-Soleil, et

5. que différents symboles sont actuellement en usage pour désigner l'unité astronomique,

Recommande

1. que l'unité astronomique soit re-définie comme une unité conventionnelle de longueur égale à 149 597 870 700 m exactement, selon la valeur adoptée dans la Résolution UAI 2009 B2,

2. que cette définition de l'unité astronomique soit utilisée avec toutes les échelles de temps telles que TCB, TDB, TCG et TT, etc.

3. que la constante de Gauss k soit supprimée du système de constantes astronomiques,

4. que la valeur du paramètre de masse solaire, GM_S, soit déterminée en unités SI par l'observation, et

5. que le seul symbole "au" soit utilisé pour l'unité astronomique.

Références

Capitaine, N., Guinot, B., Klioner, S., 2011, Proposal for the re-definition of the astronomical unit of length through a fixed relation to the SI metre, Proceedings of the

Journées 2010 Systèmes de référence spatiotemporels, N. Capitaine (ed.), Observatoire de Paris, pp 20–23

Fienga, A., Laskar, J., Morley, T., Manche, H. et al., 2009, INPOP08: a 4D-planetary ephemeris, A&A 507, 3, 1675

Fienga, A., Laskar, J., Kuchynka, P., Manche, H., Desvignes, G., Gastineau, M., Cognard, I., Theureau, G., 2011, INPOP10a and its applications in fundamental physics, Celest. Mech. Dyn. Astr., Volume 111, on line edition (`http://www.springerlink.com/content/0923-2958`).

Folkner, W.M., Williams, J.G., Boggs, D.H., 2008, Memorandum IOM 343R-08-003, Jet Propulsion Laboratory

International Astronomical Union (IAU), Proceedings of the Sixteenth General Assembly," Transactions of the IAU, XVIB, p. 31, pp. 52–66, (1976)

International Astronomical Union (IAU), Proceedings of the Twenty Seventh General Assembly," Transactions of the IAU, VXVIIB, p. 57, pp. 6: 55–70 (2010)

Klioner, S., 2008, Relativistic scaling of astronomical quantities and the system of astronomical units, A&A 478, 951

Klioner, S., Capitaine, N., Folkner, W., Guinot, B., Huang, T.-Y., Kopeikin, S. M., Pitjeva, E., Seidelmann P.K., Soffel, M., 2009, Units of relativistic time scales and associated quantities, in Proceedings of the International Astronomical Union, IAU Symposium, Volume 261, p. 79-84

Luzum, B., Capitaine, N., Fienga, A., Folkner, W., Fukushima, T., Hilton. J., Hohenkerk, C., Krasinsky, G., Petit, G., Pitjeva, E., Soffel, M., Wallace, P., 2011, The IAU 2009 system of astronomical constants: the report of the IAU working group on numerical standards for Fundamental Astronomy, Celest. Mech. Dyn. Astr., doi: 10.1007s10569-011-9352-4

Pitjeva, E.V. and Standish, E.M., 2009, Proposals for the masses of the three largest asteroids, the Moon-Earth mass ratio and the astronomical unit, Celest. Mech. Dyn. Astr., 103, 365, doi: 10.1007/s10569-009-9203-8

Standish, E.M., 2004, The Astronomical Unit now, in Transits of Venus, New views of the Solar System and Galaxy, Proceedings of the IAU Colloquium 196, D. W. Kurtz ed., 163

---

[1]La définition UAI 1976 est: "L'unité astronomique de longueur ou unité de distance (A) est la longueur pour laquelle la constante de Gauss (k) prend la valeur 0.017 202 098 95 quand les unités de mesure sont les unités astronomiques de longueur, de masse et de temps. Les dimensions de k^2 sont celles de la constante de la gravitation (G), c.-à-d. $L^3M^{-1}T^{-2}$." Bien que cette définition soit la première définition officielle explicite de l'unité astronomique, l'utilisation de k pour définir l'unité astronomique a été en usage depuis le XIXè siècle avant de devenir officielle en 1938.

[2] en utilisant l'équation $A^3k^2/D^2 = GM_S$, où A est l'unité astronomique, D l'intervalle de temps de 1 jour, et k la constante de Gauss.

[3] En relativité, une éphéméride du système solaire, pour laquelle l?unité astronomique est une unité utile, est une représentation coordonnée de la dynamique du système solaire. Les unités SI sont introduites dans cette représentation coordonnée en utilisant les équations relativistes pour les photons et pour les corps massifs et en reliant les coordonnées de certains événements avec les quantités observées exprimées en unités SI.

RESOLUTION B3

Sur l'établissement d'un système international d'alerte avancée des corps géocroiseurs

Proposée par le Groupe de Travail "Corps Géocroiseurs" de la Division III de l'UAI

La XXVIIIe Assemblée Générale de l'Union Astronomique Internationale,

Reconnaissant

- qu'il y a désormais des indications incontestables que la probabilité d'impacts catastrophiques sur la Terre de corps géocroiseurs, potentiellement hautement dangereux pour la vie, et en particulier pour l'humanité, est loin d'être négligeable et que des mesures appropriées sont à l'étude pour éviter de telles catastrophes;

- qu'en ce qui concerne les corps de grandes dimensions, et grâce aux efforts de la communauté astronomique et des agences spatiales, le catalogage de ceux qui représentent un danger potentiel, le suivi de leurs possibilités d'impact, et l'analyse de solutions technologiquement réalisables pour s'en protéger, atteint désormais un niveau satisfaisant;

- que, cependant, dans le cas de corps de taille réduite ou intermédiaire, la menace pour la civilisation et la communauté internationale reste considérable;

- que notre connaissance du nombre, de la taille et de l'orbite de corps plus petits reste trés limitée, et ne permet pas d'anticiper de façon fiable la probabilité d'impacts futurs;

Notant

Que les corps géocroiseurs sont une menace pour toutes les nations de la Terre, et que par conséquent toutes les nations doivent contribuer à se protéger de ces menaces;

Recommnande

Que les Membres Nationaux de l'UAI travaillent de concert avec le Comité des Nations Unies pour l'Usage Pacifique de l'Espace (UN-COPUOS) et avec le Conseil International pour la Science (ICSU), afin de se coordonner et de collaborer en vue de l'établissement d'un système international d'alerte avancée des corps géocroiseurs, se fondant sur l'avis scientifique et technique de la communauté astronomique compétente, et dont l'objectif principal soit l'identification fiable des risques de collisions entre corps géocroiseurs et la Terre, ainsi que la communication aux autorités compétentes des nations concernées des paramètres permettant prendre les mesures nécessaires. ?

RESOLUTION B4

Sur la restructuration des Divisions de l'UAI

Proposée par le Comité Exécutif de l'UAI

La XXVIIIe Assemblée Générale de l'Union Astronomique Internationale,

Notant

(a) que l'UAI et l'astronomie au sens large ont toutes deux connu une évolution considérable depuis que les Divisions ont été introduites en 1994, et formellement adoptées en 1997, et qu'il est donc justifié d'envisager une adaptation de la structure actuelle des Divisions,

(b) le rapport et les recommandations du Groupe d'Etude institué par le Comité Exécutif en vue de la restructuration des Divisions, et l'accueil fait par le Comité Exécutif à ces recommandations,

(c) que les Commissions, Groupes de Travail et autres instances de l'UAI sont également susceptibles d'être réformées,

(d) que la mise en œuvre du "Plan Stratégique" de l'IAU au moyen du "Bureau de l'Astronomie pour le Développement" (OAD) et autres programmes associés requiert de la part du Comité Exécutif une supervision appropriée et des mesures de gouvernance pour toutes les activités liées à l'Astronomie pour le Développement, afin d'assurer un lien fort entre ces activités, les Divisions, et le Comité Exécutif,

Approuve

La proposition du Comité Exécutif de restructurer les Divisions comme suit:

Division A	Systèmes de référence spatio-temporels
Division B	Instruments, Technologie et Traitement de Données
Division C	Education, Vulgarisation et Patrimoine
Division D	Hautes énergies et Physique fondamentale
Division E	Soleil et Héliosphère
Division F	Systèmes Planétaires et Bioastronomie
Division G	Etoiles et Physique Stellaire
Division H	Matière Interstellaire et Univers Local
Division J	Galaxies et Cosmologie

Et Requiert

Que les nouvelles Divisions, sous l'égide du Comité Exécutif, travaillent conjointement à la mise en place d'une réforme de la structure des Commissions, Groupes de Travail, et autres instances, destinée à être approuvée, en conformité avec les Statuts et le R'eglement Intérieur de l'Union, par le Comité Exécutif lors de sa réunion de Mai 2013.

Transactions IAU, Volume XXVIIIB
Proc. XXVIII IAU General Assembly, August 2012
Thierry Montmerle, ed.

doi:10.1017/S1743921315005475

CHAPTER V

REPORT OF THE EXECUTIVE COMMITTEE

1. Introduction

This report outlines IAU business conducted by or through the Executive Committee in the triennium between the Rio de Janeiro General Assembly, 3-14 August 2009, and the Beijing General Assembly, 20- 31 August 2012.
The principal activities addressed by the Executive Committee during the triennium were:

- The conclusion of the International Year of Astronomy 2009,
- The implementation of the Strategic Plan approved at the Rio General Assembly, the establishment of the Office of Astronomy for Development (OAD) and the start of the OAD programme,
- Setting up a "Task Group" to recommend changes to the Divisional Structure of the Union,
- Completing the work of modernising the Union's finances.

Much of the relevant material has appeared in IAU Information Bulletins 105 - 110 or on the IAU web site, and will not be repeated in full in this report.

2. Composition of the Executive Committee

The Executive Committee elected XXVII General Assembly was:

Robert	Williams	President
Norio	Kaifu	President-Elect
Ian F.	Corbett	General Secretary
Thierry	Montmerle	Assistant General Secretary
Matthew	Colless	Vice-President
Martha P.	Haynes	Vice-President
George K.	Miley	Vice-President
Jan	Paloûs	Vice-President
Marta	Rovira	Vice-President
Giancarlo	Setti	Vice-President
Catherine	Cesarsky	Adviser
Karel A.	van der Hucht	Adviser

2.1. *Executive Committee meetings*

The Executive Committee met on five occasions:
EC87 on 15 August 2009 at the Rio de Janeiro XXVII General Assembly
EC88 on 11-13 May 2010 at the Space Telescope Science Institute, Baltimore, USA
EC89 on 24-26 May 2011 at the Academy of Sciences of the Czech Republic, Prague

EC90 on 18-20 April 2012 at the IAU offices, Paris.
EC91 on 19 and 23 August at the Beijing XXVIII General Assembly.

Division Presidents were invited to attend EC89. Reports on meetings EC87-EC90 have been published in IB 105 - 110. A report on EC91 will appear in IB 111, due to be published in January 2013.

2.2. *Officers' Meetings*

The Officers (President, President-Elect, General Secretary, Assistant General Secretary) met on 3 occasions in the IAU Secretariat in Paris: 25-27 January 2010, 24-26 January 2011, and 23-25 January 2012. Brief reports on all meetings can be found in IB 106, 108, and 110 respectively. In addition to these meetings, the Officers held several teleconferences to discuss urgent business.

2.3. *EC Working Groups*

- *International Year of Astronomy (IYA) 2009*

This Working Group remained very active throughout 2009 and well into 2010. The outstanding and unprecedented success of the IYA is down to the very effective overall strategic direction from the Working Group and the dedication and hard work of the IYA Secretariat established at ESO. This Working Group was disbanded at the end of 2010, upon completion of the IYA 2009 programme.

- *Women in Astronomy*

A major study of statistical data on Women in Astronomy was published in IB 106 and appeared in the Proceedings of the Women in Astronomy 2009 Conference,
http://wia2009.gsfc.nasa.gov This Working Group is responsible for the organisation of two Special Activities at the XXVIII General Assembly in Beijing.

- *General Assemblies*

This Working Group, comprising the organisers of previous General Assemblies and the next two Assemblies, assists those planning future assemblies.

- *Future Large Scale Facilities*

Ten years ago the IAU Executive Committee created a Working Group on Future Large Scale Facilities, but its activities have fallen somewhat into abeyance in the last few years. EC87 decided to revive the group, with Roger Davies (Oxford) as Chair.
The revised Terms of Reference were agreed at EC89:
1. To review the status of current planned or proposed large scale ground based and space projects in astronomy.
2. In doing so, to encourage contacts and cooperation between projects.
3. To report on progress against 1. above to the Executive Committee meeting EC90 in April 2012.
4. To organise a Special Session at the 2012 IAU General Assembly to hear presentations on selected projects and to develop a strategic overview of planned and required investment in large scale facilities.
5. To consider what further work might be undertaken by the Working Group, and present a proposal to the meeting of the new Executive Committee at the 2012 General Assembly.

3. Secretariat

The Paris based Secretariat comprises two full time positions - Head of Administration (formerly Executive Assistant) and the Data Base Administrator, plus a part-time Archivist. The smooth running of the Union depends critically on these three individuals, and their contribution cannot be overestimated. The Head of Administration, Vivien Reuter, who joined in 2008, will be leaving the IAU at the end of 2012. She has proved veritable pillar of strength, taking over at a very difficult period following the death of Monique Orine and implementing a series of major changes in the way the Secretariat functions while at the same time earning the trust and respect of everyone who ever had any dealings with her or the IAU. Her replacement has been recruited and will start on 8 October 2012 to work alongside Vivien for almost 3 months.

The Secretariat is housed in offices within the Institut d'Astrophysique de Paris (IAP) under an agreement with the CNRS. The IAU is most grateful to the CNRS and the IAP for this arrangement. The IAP also provides infrastructure and IT support.

The IAU Press Office, and web and database maintenance and support, are provided by the European Southern Observatory under an outsourcing agreement. This is an extremely cost-effective arrangement and the IAU is most grateful to the Director of ESO for agreeing to make this possible

3.1. *Electronic Voting*

Following a competitive tender process, the Mi-Voice electronic voting system was implemented and has been used successfully by many Divisions and Commissions. The changes to the Statutes and Bye-Laws now approved by the General Assembly allow e-voting by Individual Members to be extended to wide range of issues. It will not be extended to votes by National Members at a General Assembly.

4. Membership of the Union

• NMs With the addition of four new members, Costa Rica, Honduras, Panama, and Vietnam, all with Interim status, at the XXVII Rio General Assembly, the total number of National Members is now 70, listed on
http://www.iau.org/administration/membership/national/.

• IMs 887 Individual Members were admitted at the XXVII Rio General Assembly: the total number of individual members currently stands at 9898. An up-to-date list of deceased members will be presented to the XXVIII General Assembly.

5. Divisions, Commissions and W/PG etc.

In addition to the normal revues of Commissions and Working Groups, the major activity carried out by the EC though a special Task Group, chaired by the Assistant General Secretary, was a review of the Divisional structure of the Union and the preparation of a proposal for a revised structure, based on nine Divisions, to be presented to the XXVIII General Assembly. Details of the proposed restructuring and the rationale behind it were given on page 37 of IB109: the core proposal arrived at after many consultations and face-to-face meetings, including the Division Presidents and the Executive Committee, is for the following 9 Divisions. This restructuring is proposed by the Executive Committee to the XXVIII General Assembly with an enabling Resolution, and was approved at the second session of the XXVII GA:

Division A	Space & Time Reference Systems
Division B	Facilities, Technologies, & Data Science
Division C	Education, Outreach, & Heritage
Division D	High Energies & Fundamental Physics
Division E	Sun & Heliosphere
Division F	Planetary Systems & Bioastronomy
Division G	Stars & Stellar Physics
Division H	Interstellar Matter & Local Universe
Division J	Galaxies & Cosmology

6. International Year of Astronomy 2009

A report on the IYA was given in IB106 (July 2010), and a very comprehensive report is available for download at
http://www.astronomy2009.org/static/archives/documents/pdf/iya2009_final_report.pdf.
It is relevant to repeat some sections of that report in order to convey just how successful IYA 2009 was. In the report, "about half of the stakeholder organisations reported the number of people reached by the events they organised, as well as the budgets they had available to implement their activities. Funds equivalent to at least 18 million Euros were devoted to IYA2009 activities - and this financial investment was complemented by enormous in-kind contributions from the amateur and professional astronomers, educators and organisers who helped to run the events".

Reports from the IYA2009 network show that at least 815 million people worldwide were reached by IYA2009 activities. Star parties, public talks, exhibits, school programmes, books, citizen-scientist programmes, science-arts events, IYA2009 documentaries and parades honouring astronomy and its achievements made IYA2009 the largest science event so far in this century.

The global IYA2009 projects have also been more successful than anyone initially dared to imagine. Two worldwide star parties were held in 2009: 100 Hours of Astronomy in April, and Galilean Nights in October. In total more than three million people were involved, with many members of the public seeing night-sky objects such as the planets and the Moon through a telescope for the very first time - a life-changing experience for many. In the framework of the IYA2009 Cornerstone project, Developing Astronomy Globally, more than five thousand telescopes have been distributed to over 30 developing countries, to help promote astronomy education and outreach there. As part of the IYA2009 legacy, the IAU has initiated and is now implementing Astronomy for the Developing World, a pioneering ten-year plan to exploit astronomy in the service of education and capacity building in the developing world. The IAU has recently chosen the South African Astronomical Observatory as the location for its Office for Astronomy Development (OAD). The OAD will coordinate a wide range of activities designed to stimulate astronomy throughout the world. The IYA2009 presence in the new media sphere has been tremendous: the number of IYA2009-related blog entries and tweets reached millions. The IYA2009 Cornerstone Project Cosmic Diary, a blog where 60 professional astronomers from around the world blog about their lives, families, friends, hobbies and interests, as well as their work, had more than 250,000 visitors and more than 2100 blog entries. As another example, more than 10 000 people participated in Meteorwatch on Twitter, making this the first event of its kind, and also one of the biggest mass-participation events of IYA2009. On both nights of the Perseid meteor shower it was the #1 top "trending topic, by far the most-discussed thing on the Twitter network anywhere in the world!"

7. Strategic Plan and OAD

As a first step in the implementation of the decadal strategic plan "Astronomy for the Developing World - Building from the IYA2009", approved at the XXVII GA in Rio de Janeiro, the IAU invited bids to host the Office of Astronomy for Development (OAD) through an Announcement of Opportunity published in October 2009. The IAU received 40 Expressions of Interest followed up by 20 Proposals. This response was much greater than anticipated and showed the very strong support for, and interest in, the IAU Strategic Plan for Astronomy in the Developing World. After very careful consideration, the Executive Committee, meeting at EC88 in Baltimore 11-13 May, selected the South African Astronomical Observatory (SAAO) as the host site for the OAD. The detailed Agreement between the IAU and the SAAO was signed on 30 July 2010. The SAAO has contributed the following short article to describe itself and its plans for the OAD.

"The development of astronomy and the contributions of astronomy to global development have been high on the agenda of the IAU, as emphasised in the last IAU General Assembly in Brazil in 2009 where the decadal plan entitled "Astronomy for the Developing World" was launched. A key component of this visionary plan was the establishment of an Office for Astronomy Development (OAD) which would oversee and coordinate its implementation. At the 88th meeting of the IAU Executive, after the assessment of 20 proposals, South Africa was selected as the host country of the OAD, with the South African Astronomical Observatory (SAAO), a facility of the National Research Foundation (NRF), being selected as the host institution. This note is to communicate to the IAU membership the perspective of the SAAO and what this selection means for the field." Following the signing of the IAU-NRF Agreement, the OAD Steering Committee was set up, Chaired by George Miley, IAU Vice-President, and comprising Khotso Mokhele (Vice-chair - nominated by NRF), Kaz Sekiguchi (nominated by IAU), Megan Donahue (nominated by IAU), Claude Carignan (nominated by NRF) and Patricia Whitelock (nominated by NRF). The OAD web site is http://www.astronomyfordevelopment.org/.

7.1. *Director of the IAU Office for Astronomy Development*

Immediately after signing the Agreement the IAU and the National Research Foundation of South Africa issued an advertisement for an outstanding individual to fill the role of Director for the OAD to provide the necessary leadership and spearhead the effective implementation of the IAU strategic plan. The tasks of the Director were expected to include:

o worldwide management, coordination and evaluation of IAU programmes in the area of development and education and establishment of their annual budgets;

o liaison with the chairs of relevant IAU program groups/sector task forces and other relevant stakeholders in planning and implementing the relevant programs;

o building up IAU regional astronomy development nodes and liaison with the IAU regional coordinators and nodes in planning and implementing relevant programs;

o implementation of new activities, as outlined in the IAU decadal strategic plan;

o managing the OAD and its staff, including recruitment, establishment and control of the OAD budget, interfacing with the SAAO as host organisation and appropriate regular reporting;

o proactive coordination and initiation of fund-raising activities for astronomy-driven capacity building activities.

A large number of excellent candidates responded to the advertisement and a short list of candidates was drawn up and interviewed. Kevindran Govender, Manager of the SALT Collateral Benefits Programme at the South African Astronomical Observatory, Cape Town, was appointed Director of the IAU Office for Astronomy Development. Kevin played an important role in several global IYA activities and was a major contributor in the development of the IAU Decadal Strategic Plan "Astronomy for the Developing World," which the OAD is responsible for leading and coordinating. The Office formally began its work on 1 March, 2011. It organised a major meeting of stakeholders in December 2011.

7.2. *Establishment of Task Forces*

The OAD has now established the three Task Forces which will drive global activities using astronomy as a tool to stimulate development. These Task Forces are: (i) Astronomy for Universities and Research; (ii) Astronomy for Children and Schools; and (iii) Astronomy for the Public. This milestone forms part of the IAU Strategic Plan which aims to build on the momentum of the successful International Year of Astronomy 2009. The plan recognises and seeks to realise the impact that astronomy has on the development of a country and region because of its contribution to: technology and skills; science and research; and culture and society. The OAD was set up to realise these global benefits, and the Task Forces are a step towards this. The OAD will also be coordinating the establishment of regional nodes and language expertise centres across the world to ensure local input and wide inclusiveness into strategic planning and implementation.
The Task Forces are made of groups of experts in their field who give their time voluntarily to advise on and coordinate projects in the respective targeted areas of development.

The Task Groups end their Chairs are:
Task Force 1: Universities and Research: Edward Guinan (Chair - USA)
Task Force 2: Children and Schools: Pedro Russo (Chair - Netherlands/Portugal)
Task Force 3: Public: Ian Robson (Chair - UK)

7.3. *Establishment of Regional Centres*

The IAU Strategic Plan envisages that a number of regional nodes will be established gradually throughout the decade, supported mainly by external funds, such as in-kind contributions by the host organisations. The OAD will work towards establishing such regional nodes across the world. A regional node could be an astronomical institute that is presently engaged both in research and in promoting astronomy education. It is also important to be pragmatic with the definition of a "region" and define it as an area for which a regional node would clearly further the goals of the strategic plan and the OAD. Such a region could vary in size from a continent to a province within a country. A call for Expressions of Interest in hosting Regional Centres and Language Expertise Centres has been issued and the responses are being evaluated. The number of Regional Centres is not fixed but will depend on the offers received and the funding available. It is expected that a start will be made with a few Regional Centres and the number progressively increased with time. *MoUs were signed for two Regional Centres at the XXVIII GA*

8. Educational Activities

The educational activities of the IAU, coordinated by Commission 46 but overseen by the Executive Committee, are an important component of the IAU's overall strategy and will be progressively integrated within the programme under the OAD's Task Forces.

- International School for Young Astronomers (ISYA)

The following ISYA have been organised, with support from the Norwegian Academy of Science and Letters (NASL):
ISYA 2012, 6 - 26 February, 2012, South African Astronomical Observatory, Cape Town, South Africa.
ISYA 2011, 31 March - 21 April, 2011, Lijang, China
ISYA 2010, 12 September - 3 October, 2010, Byurakan Observatory, Armenia
ISYA 2009, 7 - 18 December, 2009, Trinidad and Tobago.

- Teaching Astronomy for Development (TAD)

TAD organises training and education programmes in developing countries. In the triennium these took place in Kenya and Gaza/West Bank (2009), Burkina Faso (2010), Nicaragua, Ethiopia, and Vietnam (2011).

- World-Wide Development of Astronomy (WWDA)

The WWDA team assess and assist local activities in astronomy education and research. Visits have been made to Paraguay, Senegal, Tajikistan, Panama and Costa Rica, Brunei, Kazakhstan, Fiji, Bolivia and Columbia, Tunisia and Algeria.

- Network for Astronomy School Education (NASE)

NASE has organised a large number of courses for primary and secondary school teachers in Latin America.

9. Public Outreach Coordinator

The value of a dedicated Public Outreach Coordinator (POC) became obvious during the International Year of Astronomy. Thanks to the generous offer to host the POC from the National Astronomical Observatory of Japan, with additional contributions from other Asian institutes, it has been possible to establish an office and permanent position. Sarah Reed has been appointed as POC and will take up her post shortly after the Beijing GA. She will also take over the editorship of the Communicating Astronomy to the Public (CAP) journal.

10. Scientific Meetings

As usual, the IAU supported 9 symposia and one Regional Meeting each year. Full details are available on http://www.iau.org/science/meetings/past/ and the post-meeting reports are available on
`http://www.iau.org/science/meetings/past/meeting_report/`.

In summary, the Symposia held since the XXVII GA in Rio de Janeiro were:

2012

IAUS 279 Death of Massive Stars: Supernovae and Gamma-Ray Bursts, March 12 - 16, Nikko Japan

IAUS 287 Cosmic masers - from OH to Ho, January 29 - Feb 3, Stellenbosch South Africa

2011

IAUS 286 Comparative magnetic minima: characterizing quiet times in the Sun and stars October 3 - 7, Mendoza, Argentina

IAUS 285 New Horizons in Time Domain Astronomy September 19 - 23, Oxford, UK

IAUS 284 The spectral energy distribution of galaxies (SED2011) September 5 - 9, Preston UK

IAUS 283 Planetary Nebulae: an Eye to the Future July 25 - 29, Puerto de la Cruz, Tenerife, Spain

IAUS 282 From Interacting Binaries to Exoplanets: Essential Modeling Tools July 18 - 22, Tatranska Lomnica, Slovakia

IAUS 281 Binary Paths to the Explosions of type Ia Supernovae July 4 - 8 Padova, Italy

IAUS 280 The Molecular Universe May 29 - June 3, Toledo, Spain

IAUS 278 Archaeoastronomy and Ethnoastronomy: Building Bridges between Cultures January 5 - 14, Lima, Peru

2010

IAUS 277 Tracing the ancestry of galaxies (on the land of our ancestors) December 13 - 17, Ouagadougou, Burkina Faso

IAUS 276 The Astrophysics of planetary systems - formation, structure, and dynamical evolution October 11 - 15, Torino, Italy

IAUS 275 Jets at all scales September 13 - 17, Buenos Aires, Argentina

IAUS 274 Advances in plasma astrophysics September 6 - 10, Catania, Italy IAUS 273 Physics of Sun and star spots August 23 - 26, Los Angeles, United States

IAUS 272 Active OB stars - structure, evolution, mass loss, and critical limits July 19 - 23, Paris, France

IAUS 271 Astrophysical dynamics - from stars to galaxies June 21 - 25, Nice, France

IAUS 270 Computational star formation May 31 - June 4, Barcelona, Spain

IAUS 269 Galileo's Medicean Moons - their impact on 400 years of Discovery January 6 - 9, Padova, Italy

2009

IAUS 268 Light Elements in the Universe November 9 - 13, Geneva, Switzerland.

In addition, there were the following Regional Meetings:

2011

APRIM 2011 11th Asia-Pacific IAU Regional Meeting July 26 - 29, Chiang Mai. Thailand

MEARIM 2011 2nd Middle East-Africa IAU Regional Meeting April 10 - 15, Cape Town, South Africa

2010

LARIM 2010 XIII Latin American Regional IAU Meeting (LARIM-2010) November 8 -12 Morelia, Mexico

11. IAU Publications

A comprehensive list of IAU Publications can be found at http://www.iau.org/science/publications/iau/. From September 2009 they comprise IAUS 260 - 287, *Highlights of Astronomy 15, Proceedings of the Twenty Seventh General Assembly Transactions XXVIIB,* and *Reports on Astronomy: Transactions XXVIIIA.*

The Proceedings of Symposia, plus Highlights of Astronomy and Transactions A and B are published by Cambridge University Press under an agreement with the IAU which currently terminates on 31 December 2013. The terms of this Agreement will be reviewed in preparation for either renewal or a competitive tender process. The collaboration with CUP has been extremely effective: the quality of the produced volumes has been excellent and the IAU has received significant income from the 50:50 sharing of net revenues.
The IAU Information Bulletins are published twice per year. During the triennium, IB 105 - 110 were published and are available on the IAU web site at:
http://www.iau.org/science/publications/iau/information_bulletin/.
Electronic newsletters and Press Releases are produced as required. A total of 8 electronic newsletters and 285 Press Releases were produced in the triennium. They are available at:
http://www.iau.org/science/publications/iau/newsletters/ and
http://www.iau.org/public_press/news/ respectively.

12. Relationships with other Organisations

The IAU maintains close relationships with many other scientific bodies - a full list of the IAU representatives can be found at http://www.iau.org/administration/executive_bodies/other_organizations/. This list will be updated after the Beijing GA.

- ICSU

The IAU is a member of the International Council of Science (ICSU). Through ICSU, the IAU participates in highly relevant data activities under CODATA (The Committee on Data for Science and Technology) and WDS (The World Data System). Other activities include COSPAR (Space Research), SCOSTEP (Solar-Terrestrial Physics), IUCAF (Radio Astronomy and Space Science) and SCAR (Antarctic Research). Through these bodies the IAU is able to influence worldwide developments and strategy in fields of interest, but is increasingly concerned that the strategic priorities of ICSU are of little direct relevance to the IAU, and indeed to many of the 'basic science' unions.

- UNESCO

The IAU had a successful collaboration with UNESCO in getting UN endorsement of the International Year of Astronomy, and thought to follow this through with activities of joint interest based on the IAU Strategic Plan, but with only limited success. There is, however, a very effective collaboration on Astronomical Heritage through the IAU Working group on Astronomy and World Heritage. The Memorandum of Understanding signed in October 2008 has been extended to the end of 2012 and can be expected to be extended for a further two years beyond that. One of the major outcomes of this collaboration is the Thematic Initiative "Astronomy and World Heritage", a report published at the end of 2011 and available on the UNESCO web site as part of the UNESCO "Report on the World Heritage Thematic Programmes".

• COPUOS (The United Nations Committee on the Peaceful Uses of Outer Space)
The IAU is an active contributor in this body, and particularly on the issue of forecasting, and potentially mitigating, future impacts of Near Earth Objects (NEOs; comets: NECs; asteroids: NEAs) on Earth.

• BIPM (The International Bureau of Weights and Measures)
The BIPM is responsible for maintaining the SI system of units and the IAU advises and assists on maintaining the integrity of the SI system of units. The BIPM has an important role in maintaining accurate worldwide time of day. It combines, analyses, and averages the official atomic time standards of member nations around the world to create a single, official Coordinated Universal Time (UTC). The IAU has participated actively in a (currently unresolved) debate on whether or not the leap second used to synchronize solar time and UTC should be abandoned.

13. Financial Matters

Following the death of the previous Executive Assistant (M. Orine) and the recommendations of the Finance Sub-Committee, a major revision of the IAU financial processes and systems was started in 2008 by the then General Secretary, Karel A van der Hucht, and the new Executive Assistant, Vivien Reuter, and completed in 2010.
The first step was the implementation of the SAGE electronic accounting system and the appointment of new auditors with a well defined and wider reaching remit. At the same time, the decision was made to manage all IAU finances in euros, rather than Swiss Francs, and to move the accounts from UBS in Lausanne to the Paris branch of an international bank.

The budget for 2007-2009 was prepared in Swiss Francs, so the accounts were presented in both Swiss Francs and euros, as was the budget for 2010-2012, with the euro budget being definitive. Membership dues from 2010 onwards were calculated and paid in euros, and from 2010 the accounts have been kept entirely in euros.

After a careful review of options, and bearing in mind the experience with the French bank LCL, it was decided to move all the IAU accounts to HSBC in Paris, apart from a small balance, also in euros, remaining with UBS in Lausanne. HSBC is a very large international bank with a strong presence in most parts of the world. It offered a full range of services including secure internet banking and a dedicated advisor and contact.

The transfer to HSBC was completed in 2010: the account with LCL was closed as soon as settlement had been reached over the cheque encashment issues of the previous Executive Assistant; and the accounts and investments with UBS were transferred to HSBC once the International Year of Astronomy finances were closed.

Essentially all bank transactions are now completed by internet, entered by the Executive Assistant and authorised by the General Secretary. All cheques are signed by the General Secretary against verified invoices, all transactions are visible to both General Secretary and Executive Assistant, and are checked and entered into the SAGE system by the Executive Assistant. Monthly salaries and annual account reconciliation with the SAGE system are prepared by an external independent accountant. The final result is a robust and transparent system which fully meets the needs of the IAU, with the result that the accounts for 2008-2011 have been examined and certified by the new auditor and no specific issues have been raised.

Given the extremely volatile state of the financial markets, the decision was made, in consultation with the Finance Sub-Committee, to keep the IAU's reserve funds, amounting to about one year's operating expenses or 1 million euros, in cash bonds with HSBC. These offer a comparatively low rate of interest (as is the case for all such investments in the Eurozone), but the capital is protected from market fluctuations. The funds are invested in several bonds at any one time, with terms varying from 3 to 12 months, thus giving the IAU flexible access to meet in year requirements.

The arrival of the new Executive Assistant in 2008 and the adoption of all-euro budgeting gave the opportunity to learn from experience in 2009-2012, so that the budget proposed for 2013-2015 is effectively a zero-base budget which draws on previous experience and takes into account policy implementation, such as the Strategic Plan, in the future.

14. Gruber Foundation

The IAU enjoys a close relationship with the Gruber Foundation, nominating members for the Gruber Cosmology Prize and recommending the award of the annual Gruber Fellowship.
The Cosmology Prize Awardees are listed at
http://www.iau.org/grants_prizes/gruber_foundation/, and the Fellows at http://www.iau.org/grants_prizes/gruber_foundation/fellowships/recipients/fellow2012/.

15. Norwegian Academy of Science and Letters and the Kavli Prize in Astrophysics.

The Kavli Prize is awarded every other year for outstanding achievement in advancing knowledge and understanding of the origin, evolution, and properties of the universe. In 2008 the Norwegian Academy of Science and Letters (NASL) and the International Astronomical Union agreed to cooperate on future Kavli Prizes in Astrophysics, and in 2010 the IAU and the NASL agreed to collaborate when choosing the members of the Kavli Prize selection committee. The Academy remains responsible for the announcement of the prize and the nomination of the winners, with advice from the Kavli Foundation and the IAU. The prize winners are listed at www.iau.org/grants_prizes/kavli_prize/.

The NASL supports one IAU International School for Astronomy (ISYA) every year and contributes to the cost of events for Young Astronomers at the General Assemblies.

Annex to Executive Committee Report

AGREEMENT
Between
The International Astronomical Union,
and
The National Research Foundation of South Africa,
concerning the hosting of the

OFFICE FOR ASTRONOMY DEVELOPMENT

of the IAU Strategic Plan "Astronomy for the Developing World"

Considering that

the International Astronomical Union (hereinafter IAU) produced IAU Strategic Plan 2010-2020 "Astronomy for the Developing World";

the IAU XXVII General Assembly, held in Rio de Janeiro, Brazil 3 - 14 August 2009 resolved to proceed with the implementation of this Strategic Plan;

the IAU issued an Announcement of Opportunity (AO) to host the Office for Astronomy Development (hereinafter OAD) in October 2009;

the South African Astronomical Observatory (hereinafter SAAO), a facility of the National Research Foundation (hereinafter NRF), submitted a proposal with the endorsement of the South African Department of Science and Technology (hereinafter DST) on 28 February 2010 in response to this AO, and subsequently responded to supplementary questions from the IAU;

the IAU Executive Committee, at its meeting in Baltimore, USA, on May 11-13 2010, considered the proposals received and selected SAAO as the host for the OAD;

the SAAO accepted the invitation of the IAU to host the OAD according to its proposal;

the IAU for the one part and NRF (acting herein through the South African Astronomical Observatory) for the second part therefore agree as follows:

1. Tasks and objectives of the OAD

The OAD is the lead organization in the worldwide implementation of the IAU "Astronomy for the Developing World" Strategic Plan 2010-2020 with tasks as outlined in Section 4.1 of that Plan. It will provide the global coordination needed for an integrated strategic approach and in addition will be expected to initiate and develop new programs consistent with the goals of the Strategic Plan.

Its tasks will include:

1. Management, coordination and evaluation of the IAU programs worldwide in the area of development and education, including recruiting and mobilizing participating volunteers.

2. Organization of oversight of the IAU development programs and the formulation of their annual budgets.
3. Liaison with the chairs of the various Commission 46 Program Groups and sector task forces in planning and implementing the relevant programs.
4. Liaison with the IAU regional coordinators and IAU Regional Centres in planning and implementing the relevant programs.
5. Provision of administrative support for IAU programs in development, education, and outreach.
6. Coordination of contacts between the IAU and national authorities throughout the world.
7. Establishment of the new IAU endowed lectureship program.
8. Liaison with other international unions and agencies promoting astronomy in the developing world, such as UNOOSA, COSPAR and URSI.
9. Stimulation of communication on IAU development matters between members and associated members through the maintenance of an IAU website for development and education and appropriate forums.
10. Provision of information for astronomers in all developing countries about IAU programs
11. Proactive coordination and initiation of fundraising activities for astronomy development.

2. Legal Status of the OAD

1. The OAD is a project of the IAU operated, on behalf of NRF, by SAAO under this agreement. The OAD is thus bound by the laws of South Africa and the rules and regulations that apply to and within SAAO. Any necessary deviation from these regulations, due to the international nature of the OAD, should first be approved by the OAD Steering Committee after consulting the NRF.

3. Location of the OAD

1. The OAD shall be located at the SAAO headquarters in Cape Town, South Africa.

4. Responsibilities of NRF

NRF will, through the SAAO, be the host and legal persona of the OAD, and the employer of the Director of Development and Education (DDE) and OAD staff. As such the NRF will, through the SAAO:
1. Ensure that the OAD conducts its affairs in accordance with the laws of South Africa and the requirements of the IAU.
2. Provide the facilities and services set out in this Agreement.
3. Support, assist and advise the DDE in the discharge of the tasks of the OAD.
4. Ensure that the DDE and OAD staff can carry out their tasks, as overseen by the OAD Steering Committee, free of other influence or interference.
5. Assist the IAU in seeking additional funding for the activities of the OAD.
The person within SAAO responsible for hosting the OAD on behalf of the NRF is the SAAO Director, acting under the authority of the NRF Corporate Executive.

5. Responsibilities of the IAU

The IAU is responsible for the implementation of the IAU Strategic Plan 2010-2020 "Astronomy for the Developing World" and as such shall:
1. Advertise and promulgate the OAD activities worldwide.
2. Actively seek out additional funding for OAD activities through donations, in kind support, etc.
3. Establish a network of Regional Centres linked to the OAD.
4. Assist the OAD in the coordination of its activities under Divisions XII and Commission 46.
5. Support, assist and advise the NRF and SAAO in the hosting of the OAD, and the DDE in the discharge of the tasks of the OAD.
The person within the IAU responsible for the OAD and the Strategic Development Plan is the General Secretary, acting under the authority of the Executive Committee and the General Assembly.

6. Financial Contributions

(a) By NRF

1. Contributions from NRF may be made in respect of specific programs carried out by the OAD in fulfilment of NRF objectives.
2. Any contributions from DST to NRF in respect of the OAD and its programs shall be the subject of separate arrangements between DST and NRF.

(b) By IAU

1. While this Agreement is in force the IAU will pay NRF annually, upon presentation of an invoice, the sum of 50,000 EUR, or such greater sum as may have been agreed between IAU and NRF. This amount may be paid directly in the SAAO bank account if deemed more practical by all concerned parties.
2. Further contributions from IAU may be made in respect of specific programs carried out or funded by NRF with the approval of the IAU.

7. Services and Facilities provided by NRF

The NRF, through its facility, the SAAO, will provide the following for the OAD:

1. Office space at SAAO headquarters in Cape Town
2. Communications and IT support
3. Administrative support during initial setting up of OAD
4. Financial management support
5. HR management support
6. Auditorium and accommodation facilities for conferences and workshops

The details of the above will be negotiated between the Director of Development and Education, once appointed, and the SAAO.

8. Additional Resources

1. The OAD will be give reasonable access to NRF projects and expertise relevant to the OAD, especially those based at the SAAO.
2. The IAU will provide additional high level administrative assistance where necessary.

The details of the above will be negotiated between the Director of Development and Education, once appointed, and the SAAO and IAU.

9. Oversight

1. The IAU and NRF shall jointly establish a body, to be known as the OAD Steering Committee, which shall exercise general oversight of the activities of the OAD and DDE.
2. The OAD Steering Committee shall be formed upon this Agreement entering into force, and shall comprise 3 members nominated by the IAU and 3 members nominated by the NRF. The Chair of the OAD Steering Committee shall hold office for 2 years and shall alternate between a nominee of the IAU and a nominee of NRF.
3. Members of the OAD Steering Committee shall normally serve a maximum of 4 years unless appointed Chair, in which case the maximum term shall be 6 years.
4. The OAD Steering Committee shall define its own Rules of Procedure.

10. Appointment of Director of Development and Education

1. The IAU and SAAO (as mandated by NRF as the host of the OAD and employer of its staff) shall jointly advertise the position of Director of Development and Education (DDE).
2. The selection of the DDE shall be done by the OAD Steering Committee and shall be subject to the ratification of the IAU Executive Committee and NRF.
3. The DDE shall be employed by the NRF at SAAO, which shall fix the remuneration, benefits, social charges, and terms and conditions of employment, subject to the approval of the OAD Steering Committee.

11. Recruitment of Staff

1. The DDE shall be responsible for recruiting the staff of the OAD in line with the procedures of the NRF, and with due consideration of the international nature of the OAD.
2. All staff shall be employed by the NRF at SAAO, which shall fix the remuneration, benefits, social charges, and terms and conditions of employment, subject to the approval of the OAD Steering Committee.

12. Terms and Conditions of Employment of DDE and OAD Staff

1. Unless special circumstances require different arrangements, to be agreed by the OAD Steering Committee, all staff shall be employed under the NRF standard terms and conditions of employment.

13. Evaluation of Staff Performance and Remuneration

1. The OAD Steering Committee shall assess the performance of the DDE annually, setting objectives for the following year, and shall report in writing to the IAU Executive Committee and NRF. Within the limits of the budget, the OAD Steering Committee may set the remuneration of the DDE.
2. The DDE shall assess the performance of the OAD staff annually, setting objectives for the following year, and shall report in writing to the OAD Steering Committee. Within the limits of the budget, the DDE may set the remuneration of the staff.
3. Termination of contracts of employment shall follow the procedures and rules which apply to equivalent employees of the NRF, with due consideration of the international nature of the OAD and thus high impact of underperformance.

14. Budget and Budget Control

1. The Financial Year of the OAD shall be the same as the financial year of the NRF for audit and reconciliation purposes.
2. The budget of the OAD shall be proposed to the OAD Steering Committee annually by the DDE in such detail as may be decided from time to time by the OAD Steering Committee.
3. Within each Financial Year the DDE shall have total control of the OAD finances, provided that expenditure does not exceed the overall budgeted income and that any unusual or unexpected expenditure is reported promptly to the Steering Committee.
4. Any increase in expenditure in year shall have the source of funding agreed and shall be approved in advance by the OAD Steering Committee.

15. Programme and Financial Reporting

1. The DDE shall report annually in writing to the OAD Steering Committee on the programme completed, future plans, and financial status of the OAD.
2. The DDE shall present annually to the OAD Steering Committee for approval an income and expenditure budget for the following year.
3. The Steering Committee shall present a written report annually to the IAU and NRF on the programme completed, future plans, and financial status of the OAD, including the approved budget.
4. Fundraising activities shall be reported to the IAU separately.
5. A publicly accessible annual report, approved by the OAD Steering Committee, shall be produced and distributed to funders and stakeholders.
6. Regular updates on activities of the OAD shall be made available on the OAD website and distributed via other means of electronic communication as appropriate.

16. Reviews

1. At intervals of not less than 2 years and not more than 4 years the IAU and NRF shall arrange for independent reviews of the performance of the OAD. The terms of reference of such reviews shall be set by the IAU Executive Committee in consultation with NRF and the review body shall report directly to the IAU Executive Committee and NRF.

2. Unless agreed otherwise by the OAD Steering Committee, there shall be a review before each IAU General Assembly timed such that the IAU Executive Committee can report to the General Assembly on the outcome of the review.
3. Unless agreed otherwise by the IAU Executive Committee, the DDE shall present a report on the activities and achievements of the OAD to each IAU General Assembly and which shall be included in the published Transactions of the IAU.

17. Duration of Agreement

1. This agreement shall come into force on the date of signature by the IAU and NRF, and shall remain in force until 31 December 2015. It shall thereafter be renewed by the IAU and NRF at 3-yearly intervals.

18. Termination of Agreement

1. This agreement may be renegotiated or terminated at any time with the agreement of all signatories.
2. The NRF or the IAU may each terminate this agreement unilaterally by giving notice in writing not less than one year in advance of the proposed date of termination. Such notice shall give the reasons for the termination and shall set out how existing obligations will be discharged. During this period the IAU and NRF shall endeavour to honour all obligations and commitments to each other and to any implicated third parties.
3. The NRF and the IAU may mutually agree to terminate the present Agreement in the case of natural hazard, civil unrest (whether the acts causing the unrest are lawful or unlawful), uprising, acts of terrorism, national or international emergency or conflict, labour unrest, the emergence of a risk to public health or safety, or similar events, which make the carrying out of the functions of the OAD impossible and unreasonable for the IAU or NRF. The IAU General Secretary and NRF President shall determine the existence of any of the said or similar events jointly in so far as they present an obstacle to hosting the OAD.
4. Where termination occurs as a result of such a decision, the IAU and NRF agree that they shall not claim indemnities for any of the costs or other responsibilities that arise directly or indirectly from such termination.
5. Should the IAU Executive Committee decide that NRF is not meeting or is unlikely to meet its obligations under this Agreement, or that the OAD is unable to fulfil its objectives while hosted within the SAAO, the IAU General Secretary shall immediately inform NRF in writing of the concerns of the Executive Committee. NRF will then have a reasonable period of time to respond to and address these concerns and present its resolution in writing to the Executive Committee.
6. Following review of the written response from NRF and assessing the impact of the proposed resolutions to the matters of concern, the IAU Executive Committee may unilaterally terminate the present Agreement should it decide that NRF is reasonably likely to fail to meet its obligations under this Agreement for any reason other than those identified in paragraph 3 above, and may, in such circumstances, seek compensation from NRF, taking into account any extenuating circumstances.

19. Disputes

1. Any dispute over the interpretation of this Agreement shall in the first instance be resolved by negotiation between the Director of the SAAO (acting as host on behalf of the NRF) and the General Secretary of the IAU. Should this fail negotiation between the President of the IAU and President of NRF may proceed to a decision, which shall be binding on both the NRF and the IAU.

For the International Astronomical Union

Ian Corbett, General Secretary

For the National Research Foundation

Albert van Jaarsveld, President

ADDENDUM TO THE AGREEMENT

entered into between
The International Astronomical Union (IAU)
and
The National Research Foundation (NRF)

WHEREAS the IAU and the NRF concluded an agreement ("the Agreement") on 30 July 2010 concerning the hosting of the Office for Astronomy Development ("the OAD") at the South African Astronomical Observatory;

AND WHEREAS the OAD Steering Committee and the IAU Executive Committee have decided that the name of the office should be changed to Global Office of Astronomy for Development (OAD);

NOW THEREFORE the parties agree as follows:

1. All references to the name "Office for Astronomy Development" are hereby deleted and replaced with the name Global Office of Astronomy for Development.

2. The acronym "OAD" in the Agreement shall be retained as an abbreviated form of the name Global Office of Astronomy for Development.

3. The title "Director of Development and Education" is hereby deleted and replaced by "Director OAD".

4. Save for the above, the remaining provisions of the Agreement shall remain in full force and effect.

For the International Astronomical Union
Ian Corbett, General Secretary

For the National Research Foundation
Albert van Jaarsveld, President & CEO

EXECUTIVE COMMITTEE REPORT : APPENDIX I

I. IAU – SIMPLIFIED ACCOUNTS FOR 2010 - 2012 (EURO, rounded)

[August 2012]

I	INCOME	2010 (EUR)	2011 (EUR)	2012 (EUR)	2010 - 2012 (EUR)
A	**ADHERING ORGANIZATIONS**				
A1	National Member dues	*689 891*	*892 711*	*824 412*	
B	**GRANTS**				
B1	GA related	57 250	– –	1 138	
B4	NASL/KAVLI grant for ISYA (US$30,000)	20 932	21 277	22 722	
B5	Other grants & Grant refunds	670	6 553	27 587	
C	**PUBLICATIONS & ROYALTIES**				
C2	CUP	51 692	86 594	52 441	
C3	Springer	595	224	362	
C4	other				
	sub-total ROYALTIES	*52 287*	*86 818*	*52 803*	
D	**FUNDS IN TRANSIT**				
D1	IYA2009	261 734	61 697		
D3	PPGF Fellowship (USD 50 000/yr)	35 426	40 358	35 258	
	sub-total FUNDS IN TRANSIT	*37 000*	*102 155*	*35 258*	
E	**ADMIN. & BANK REFUNDS**	7 503	150 693		
F	**BANK REVENUE**				
F2	Exchange rate gain	32 772	3 844		
F3	Savings accounts	7 052	82 667	13 216	
	sub-total BANK REVENUE	*39 824*	*86 511*	*13 216*	
	total INCOME	**1 165 517**	**1 346 718**	**977 136**	**3 489 371**
G	**total INCOME w/o 'In Transit'**	**871 011**	**1 284 921**	**941 875**	**3 097 807**

II. IAU – EXPENDITURE 2010 - 2012 (EURO, rounded)

II	EXPENDITURE	2010 (EUR)	2011 (EUR)	2012 (EUR)	2010 - 2012 (EUR)
M	**GENERAL ASSEMBLIES (GAs)**	272 565	1 076	10 673	
N	**SCIENTIFIC ACTIVITIES**				
N1	Sponsored Meetings	*73 209*	*190 000*	*199 000*	*513 000*
N2	WORKING GROUPS				
N2.1	EC Working Groups			5 961	
N2.1.3	IYA2009	38 756	6 676	35 000	37 800
N2.2	Div. & Comm. Working Groups	21 878	*3 027*	*35 414*	*114 000*
	total SCIENTIFIC ACTIVITIES	*133 843*	*199 703*	*275 375*	*627 000*

	EXPENDITURE (cont'd)	**2010** (EUR)	**2011** (EUR)	**2012** (EUR)	**2010 - '12** (EUR)
O	**EDUCATIONAL ACTIVITIES**				
O1	IAU-NASL International Schools for Young Astronomers	19 650	32 500	18 628	93 000
O2-O7	Other Educational Activities TAD etc.	53 026	59 066	114 503	
	total EDUCATIONAL ACTIVITIES	*72 676*	*91 566*	*133 131*	*433 200*
P	**FUNDS IN TRANSIT**				
P1	IYA2009	216 477	141 701	64 133	
P3	PPGF FELLOWSHIP (US$ 50 000)	37 040	38 479	37 359	111 000
	total FUNDS IN TRANSIT	*253 517*	*180 180*	*101 492*	*111 000*
Q	**COOP. with OTHER UNIONS**				
Q1	DELEGATES, TRAVEL	662	2 266	9 855	
Q2	DUES TO OTHER UNIONS				
Q2.1	ICSU	12 431	12 804	13 188	37 800
Q2.3	IUCAF		8 000	3 100	9 454
	total COOP. with OTHER UNIONS	*13 093*	*23 500*	*25 143*	*70 500*

	Expenditure (cont'd)	**2010** (EUR)	**2011** (EUR)	**2012** (EUR)	**2010 - 2012** (EUR)
R	**EXECUTIVE COMMITTEE**				
R1	Executive Committee Meetings	52 399	19 172	23 078	143 000
R1	Executive Committee Expenses		1 485		143 000
R2	Officers' Meetings	12 115	12 189	10 292	25 500
R3	Officers' expenditure	46 392	36 149	33 678	
R4	Press-Office		12 996	2 949	18 700
	total EXECUTIVE COMMITTEE	*110 906*	*81 991*	*69 997*	*325 650*
S	**PUBLICATIONS**				
S1	IAU Information Bulletin	29 316	10 658	14 284	45 000
T	**SECRETARIAT / ADMIN.**				
T1	Salaries & Charges	142 754	143 281	136 725	357 000
T2	Travel & Training Courses	1 208	762	296	9 900
T3	Outsourced Tasks	21 537	*50 630*	*53 995*	*102 450*
T4	General Office expenditure	28 900	*27 384*	*23 911*	*90 150*
	total SECRETARIAT / ADMIN.	*216 976*	*222 057*	*214 925*	*694 500*
U1	Loss on Internal Transfers	2 117			
	total EXPENDITURE	**1 105 029**	**809 225**	**845 020**	**2 759 274**
V	**total EXPENDITURE w/o** 'In Transit'	**888 552**	**701 427**	**809 762**	**2 399 741**
	INCOME OVER EXPENDITURE	**- 17 541**	**+583 494**	**+ 132 113**	**+698 066**
	ASSETS IN BANK @ YEAR END	**960,699**	**1 528 480**	**1 659 578**	

EXECUTIVE COMMITTEE REPORT : APPENDIX II

I. IAU – PROPOSED INCOME for 2013 - 2015 (EURO, rounded)

[XXVIII GA August 2012]

I	INCOME	2013 (EUR)	2014 (EUR)	2015 (EUR)	2013 - 2015 (EUR)
	unit of contribution	2 750	2 800	2 860	
	adjustment for inflation	2.0%	2.0%	2.0%	
	number of units	303	303	303	
A	**ADHERING ORGANIZATIONS**				
A1	National Member dues	*833 250*	*848 400*	*866 580*	*2 548 230*
B	**GRANTS**				
B4	NASL grant for ISYA (US$30,000)	23 469	23 469	23 469	70 407
C	**ROYALTIES**				
C2	CUP	40 000	40 000	40 000	120 000
D	**FUNDS IN TRANSIT**				
D3	PPGF Fellowship (USD 50 000/yr)	39 115	39 115	39 115	117 345
F	**BANK REVENUE**	20 000	20 000	20 000	60 000
G	**total INCOME**	**955 834**	**970 984**	**989 164**	**2 915 982**

II. IAU – PROPOSED BUDGET OF EXPENDITURE 2013 - 2015 (EURO, rounded)

II	BUDGET OF EXPENDITURE	2013 (EUR)	2014 (EUR)	2015 (EUR)	2013 - 2015 (EUR)
M	**GENERAL ASSEMBLIES**				
M1	PREPARATION COSTS	10 000	15 000	32 000	57 000
M2	GRANTS GA INCL. 6 SYMPOSIA	--	--	360 000	360 000
	total **GENERAL ASSEMBLIES**	*10 000*	*15 000*	[illegible]	*417 000*
N	**SCIENTIFIC ACTIVITIES**				
N1	SPONSORED MEETINGS				
N1.1	Grants IAU Symposia outside GA	180 000	180 000	60 000	420 000
N1.2	Grants Regional IAU Meetings (RIMs)	20 000	20 000	20 000	60 000
N1.3	Co-sponsored Meetings	15 000	15 000	15 000	45 000
	sub-total Sponsored Meetings	*215 000*	*215 000*	*95 000*	*525 000*
N2	WORKING GROUPS				
N2.1	EC Working Groups	3 000	3 000	3 000	9 000
N2.2	Commission Working Groups	6 500	6 5000	6 500	19 500
N2.2.1	CB for Astronomical Telegrams (C6)	5 000	5 500	6 000	16 500
N2.2.2	Minor Planet Center (DIII)	7 000	7 000	7 000	21 000
N2.2.3	Meteor Data Center (C22)	900	900	900	2 700
	total SCIENTIFIC ACTIVITIES	*237 400*	*237 900*	*118 400*	*593 700*

	Expenditure (cont'd)	**2013** (EUR)	**2014** (EUR)	**2015** (EUR)	**2013 - 2015** (EUR)
O	**EDUCATIONAL ACTIVITIES**				
O1	Guaranteed contribution to OAD	50 000	50 000	50 000	150 000
O5	IAU-NASL International Schools for Young Astronomers	31 000	31 000	31 000	93 000
O11	co-sponsoring COSPAR Capacity Buiding Workshops	3 100	3 100	3 100	9 300
O12	Education& development inc. OAD programs, incl. TAD, WWDA, etc	95 000	105 000	115 000	315 000
O13	Public Outreach contribution to	25 000	25 000	25 000	75 000
	total EDUCATIONAL ACTIVITIES	*204 500*	*214 500*	*224 500*	*643 500*
P	**FUNDS IN TRANSIT**				
P3	Greuber fllowship (US$ 50 000)	39 115	39 115	39 115	117 345
Q	**COOP. with OTHER UNIONS**				
Q1	delegates, travel	7 000	7 800	8 100	23 400
Q2	dues to other unions				
Q2.1	icsu	12 600	19 982	20 581	59 963
Q2.3	iucaf	3 100	3 200	3 300	9 600
	total COOP. with OTHER UNIONS	*29 500*	*30 182*	*30 881*	*90 563*

	Budget of Expenditure (cont'd)	**2013** (EUR)	**2014** (EUR)	**2015** (EUR)	**2013 - 2015** (EUR)
R	**EXECUTIVE COMMITTEE**				
R1	Executive Committee Meetings	18 000	32 000	36 000	86 000
R1.1	EC expenses other than meetings	3 000	3 000	3 000	9 000
R2	Officers' Meetings	8 000	8 000	10 000	26 000
R3	Officers' expenditure (other)				
R3.1	General Secretary expenditure				
R3.1.1	GS Paris duty	1 000	1 000	1 000	3 000
R3.1.2	GS Other expenses	6 000	6 000	6 000	18 000
R3.5	Assistant General Secretary	3 500	3 500	3 500	10 500
R3.2	President	6 000	6 000	6 000	18 000
R3.3	President-elect	3 500	3 500	3 500	10 500
R4	Press-Office	3 000	3 000	9 000	15 000
	total EXECUTIVE COMMITTEE	*52 000*	*66 000*	*78 000*	*196 000*
S	**PUBLICATIONS**				
S1	IAU Information Bulletin	10 000	10 000	12 000	32 000

	BUDGET OF EXPENDITURE (cont'd)	**2013** (EUR)	**2014** (EUR)	**2015** (EUR)	**2013 - 2015** (EUR)
T	**SECRETARIAT / ADMIN.**				
T1	SALARIES & CHARGES	151 000	151 000	151 000	453 000
T2	TRAINING COURSES	2 000	2 000	2 000	6 000
T3	OUTSOURCED TASKS				
T3.1	Web/DB Development at ESO	10 000	10 000	10 000	30 000
T3.2	Data Base management at ESO	20 000	20 000	20 000	60 000
T3.3	IT Assistance in Paris	1 500	1 500	1 500	4 5000
T3.4	Personnel Administration	2 000	2 000	2 000	6 000
T3.5	Accounting & SAGE sub.	4 000	4 000	4 000	12 000
T3.6	Auditing	6 000	6 000	6 000	18 000
T3.7	Legal Fees	4 000	2 000	2 000	8 000
T3.8	Mi-Voice e-voting	2 500	2 500	2 500	7 500
	sub-total office support costs	*52 000*	*50 000*	*50 000*	*152 000*
T4	GENERAL OFFICE EXPENDITURE				
T4.1	Post	1 800	1 800	1 800	5 400
T4.2	Telephone and Internet	3 000	3 000	3 000	9 000
T4.3	Rent (INSU/IAP)	4 300	4 450	4 600	13 350
T4.4	IT Software & hardware	1 500	1 500	1 500	4 500
T4.5	Copier/Printer, rental, consumables and maintenance	10 000	10 000	10 000	30 000
T4.6	Office Consumables	1 500	1 500	1 500	4 500
T4.7	Miscellaneous items books, posters, etc	1 500	1 500	1 500	4 500
T5	Bank Charges	6 000	6 300	6 500	18 800
	sub-total General Office expenditure	*29 600*	*30 050*	*30 400*	*90 050*
	total SECRETARIAT / ADMIN.	*232 600*	*231 050*	*231 400*	*694 500*

Budget of Expenditure (cont'd)	**2013** (EUR)	**2014** (EUR)	**2015** (EUR)	**2013 - 2015** (EUR)
total EXPENDITURE	**815 115**	**843 747**	**1 126 296**	**2 785 158**
total INCOME	**955 834**	**970 984**	**989 164 640**	**2 915 982**
INCOME OVER EXPENDITURE	**+ 140 719**	**+ 127 237**	**− 137 132**	**+130 824**

Transactions IAU, Volume XXVIIIB
Proc. XXVIII IAU General Assembly, August 2012
Thierry Montmerle, ed.

doi:10.1017/S1743921315005487

CHAPTER VI

REPORTS on DIVISION, COMMISSION, and WORKING GROUP MEETINGS

DIVISION I
COMMISSION 4 — EPHEMERIDES
(EPHEMERIDES)

PRESIDENT	**George Kaplan**
VICE-PRESIDENT	**Catherine Hohenkerk**
PAST PRESIDENT	**Toshio Fukushima**
ORGANIZING COMMITTEE	**Jean-Eudes Arlot, John A. Bangert, Steven A. Bell, William Folkner, Martin Lara, Elena V. Pitjeva, Sean E. Urban, Jan Vondrak**

PROCEEDINGS BUSINESS SESSIONS, Session 1 of 27 August 2012

The triennial meeting of Commission 4 was attended by 16 people. All of the presentations from the meeting are provided on the commission website at
`http://www.iaucom4.org/c4docs.html`, so this report provides only summaries.

A minute of silence was observed in memory of the four members of the commission who were known to have passed away since the 2009 meeting: Shinko Aoki (1927–2011), Alan Fiala (1948–2010), George Krasinsky (1939–2011), and Nguyen Mau Tung (1926–2011).

George Kaplan, the president of the Commission, gave an overview of the activities of the commission over the 2009–2012 triennium. These included:

- Revising the terms of reference for the commission.
- Reorganizing the commission website (`http://www.iaucom4.org`), now located at the UK Hydrographic Office. Steve Bell is the webmaster.
- Submitting two reports for IAU Transactions.
- Organizing the Working Group on Standardizing Access to Ephemerides, to consider how best to ensure that software developers are able to access planetary/lunar ephemerides from multiple institutions in a simple and unified way. James Hilton is the chair and the working group's report is summarized below.
- Distributing ten "Commission 4 News" e-mails, containing news of scientific and organizational interest.
- Supplying letters of support for (1) sharing spacecraft navigation data important for planetary ephemerides and (2) establishing standard values of some solar system parameters for use by binary star and exo-planet researchers.

It was also mentioned that commission members had been active in organizing and participating in two *Journées* meetings in Europe, three AAS Division on Dynamical Astronomy meetings in the US, and two meetings on the future of the time scale UTC, one in the US and one in the UK. The ephemeris-producing institutions represented within the commission continue to cooperate on various projects and publications.

Dr. Kaplan noted that a proposal to provide, on the Commission 4 website, a tabular comparison of the models, data, and numerical procedures used in three fundamental solar system ephemerides did not come to fruition, and that such a table should be a high priority during the next triennium. The ephemerides to be compared are produced by the Jet Propulsion Laboratory (JPL) in the US, the Institut de Mécanique Céleste et de Calcul des Éphémérides (IMCCE) in France, and the Institute for Applied Astronomy (IAA) in Russia.

The commission gained three new members during the triennium, and 17 prospective members of the IAU have requested membership. These additions together increase the size of the commission by 20%.

The new officers of the commission for the 2012–2015 triennium are: President: Catherine Hohenkerk (UK); Vice-President: Jean-Eudes Arlot (France); Organizing Committee (OC): John Bangert (US), Steven A. Bell (UK), Jose M. Ferrandiz (Spain), Agnès Fienga (France), William Folkner (US), Marina Lukashova (Russia), Elena Pitjeva (Russia), Mitsuru Soma (Japan), William Thuillot (France), and Sean Urban (US).

Summaries of the other reports presented at the meeting follow. **Working Group on**

Standardizing Access to Ephemerides

The Commission 4 Working Group on Standardizing Access to Ephemerides consists of James Hilton, USNO, US (chair); Jean-Eudes Arlot, IMCCE, France; Steve Bell, HMNAO, UK; Nicole Capitaine, Paris Observatory, France; Agnès Fienga, Besançon Observatory, France; William Folkner, JPL, US; Mickaël Gastineau, IMCCE, France; Elena Pitjeva, IAA, Russia; Vladimir Skripnichenko, IAA, Russia, and Patrick Wallace, UK. The purpose of this working group is to provide guidance on a consistent format for distributing ephemerides of solar system bodies to the astronomical community. The working group recommends the use of the *binary* Planetary Constants Kernel (PCK) and Spacecraft and Planet Kernel (SPK) file formats used in SPICELIB, which is written and maintained by the Navigation and Ancillary Information Facility (NAIF) of the Jet Propulsion Laboratory. See
http://naif.jpl.nasa.gov/pub/naif/toolkit_docs/FORTRAN/. It is recommended that ancillary data such as body masses and initial conditions be stored in *text* PCK files.

Both SPK and binary PCK file structures are very similar; SPK is used for the positions of solar system bodies and binary PCK is used for body orientation. NAIF has agreed to some modifications of their standards to accommodate the Working Group's requirements:

- Add a new type to accommodate IAA-style (velocity-based) ephemerides;
- Add new types to accommodate TCB-based ephemerides;
- Add a new ID number (10 000 000) for TT–TDB or TCG–TCB time ephemerides; and
- Set aside type numbers 901–910 for experimental ephemeris types.

The WG thanks the Navigation and Ancillary Information Facility of JPL for their help and cooperation in adapting their PCK and SPK formats to the needs of the international community.

— James Hilton

Institute of Applied Astronomy, Russia

The Institute of Applied Astronomy (IAA) of the Russian Academy of Science deals with the construction of numerical ephemerides of the Sun, planets, and natural satellites. The

IAA uses these ephemerides for calculating data and issuing the Russian Astronomical Yearbook (AY), the Naval Astronomical Yearbook (NAY), and the Nautical Astronomical Almanac (NAA).

Ephemerides of planets and the Moon (EPM) have been constructed by the simultaneous numerical integration of equations of motion for all the major planets, the Sun, the Moon, asteroids, trans-neptunian objects, and the lunar libration over 400 years (1800–2200). The parameters of EPM2011 (65 for the lunar part and about 270 for the planetary part) have been fitted to 17 378 lunar laser ranging measurements 1970–2011, as well as about 680 000 planet and spacecraft observations 1913–2011 of different types. The numerical ephemerides of the main satellites of planets have been constructed taking into account the mutual perturbations of the satellites, the Sun, major planets, figures of the planets, and fitted to modern observations. (See presentation at Joint Discussion 7 at this General Assembly).

Since 2006, ephemerides for the AY, NAY, and NAA prepared at IAA take into account the IAU recommendations of 2000–2006, including new models of precession and nutation, a new algorithm for sidereal time, and catalogs referred to ICRS. The ephemerides are based on the equinox system, but there are also correction tables for transformation to the CIO-based system. IAA releases the ephemerides in electronic form also, using the DE405 planetary ephemerides in addition to its own ephemerides. PersAY, the Personal Astronomical Yearbook (ftp://quasar.ipa.nw.ru/pub/PERSAY/persay.zip), is an electronic version of AY. In addition to the tables of AY, it enables calculation of topocentric astronomical data for any time scale, any position of the observer, and at any time interval. The electronic version of NAA contains examples of 21 astro-navigation tasks. Currently, the next electronic version of NAA (Navigator) is being developed for release on a CD. Its structure consists of four basic parts: observation planning with planetarium; determination of ship and compass correction by astronomical observations; a description of some sections of the NA; and a detailed manual.

— Elena V. Pitjeva

Institut de Mécanique Céleste et de Calcul des Éphémérides, France

IMCCE is a laboratory of Paris Observatory with 60 permanent staff members. The missions are:

- performing research activities on the dynamics of the solar system objects;
- teaching in universities;
- making the French official ephemerides on behalf of Bureau des longitudes; and
- providing special calculations on request for professionals, space agencies, and the public.

An important project is the development of the INPOP series of fundamental solar system ephemerides (see presentation at Joint Discussion 7 at this General Assembly). Distribution of various kinds of ephemeris data takes several forms. The yearly printed publications are:

- *Annuaire du Bureau des longitudes*, similar to *The Astronomical Almanac*
- *Connaissance des temps*, for high precision ephemerides (electronic version available); and
- *Ephemerides nautiques*, the French nautical almanac for the Navy.

Electronic ephemerides are provided through the Internet at http://www.imcce.fr. An ephemeris generator is provided, giving positional and physical ephemerides; also available are predictions of celestial phenomena, calendars, and an astronomical database. Specific web services are provided in the Virtual Observatory framework at http://vo.

imcce.fr/, using VO standard protocols, including a Sky Body Tracker facility, which identifies any solar system object in any field of view, an asteroid search application, and the Miriade ephemeris services.

Several developments are foreseen over the next triennium for the ephemerides service, including upgrade of the software accompanying *Connaissance des temps*, improving access to the Miriade system (http://vo.imcce.fr/webservices/miriade) for distributing VO compliant data, further development of physical and satellite ephemerides, and new online tools for observers. A major revision of the *Annuaire du Bureau des longitudes* content was undertaken in 2011 and will be available for the 2013 edition.

— Daniel Hestroffer

Jet Propulsion Laboratory, California Institute of Technology, US

The Solar System Dynamics group at the NASA Jet Propulsion Laboratory is led by Donald Yeomans and continues to provide updated ephemerides for the planets, planetary satellites, asteroids and comets (see: http://ssd.jpl.nasa.gov).

Planetary ephemerides are developed by William Folkner, Petr Kuchynka and Ryan Park, with primary deliveries to support planetary missions such as MESSENGER to Mercury, the Cassini mission at Saturn, and the New Horizons mission to Pluto. The latest planetary ephemeris, DE425, was delivered to the Mars Science Laboratory project for precision landing on Mars in August 2012. A dedicated campaign of Very Long Baseline Interferometer observations of the Mars Odyssey and Mars Reconnaissance Orbiter spacecraft while in orbit about Mars was performed by James Border and the Deep Space Tracking systems group to tie the planetary ephemeris to the International Celestial Reference Frame to better than 0.1 milliarcsecond accuracy to meet the requirements for the Mars landing. Planetary ephemeris work is done in conjunction with lunar laser ranging analysis by James Williams and Dale Boggs in the Geodynamics and Space Geodesy Group.

Planetary satellite ephemerides are developed by Robert Jacobson and Marina Brozović. They revised the ephemerides of Phobos and Deimos using a dynamical model that included tidal effects and the Phobos figure acceleration due to its forced libration. Their support of the Cassini Project continued with improvements in the ephemerides of the regular as well as irregular Saturnian satellites using Earth-based astrometry and Cassini imaging data. They produced new ephemerides for the irregular Jovian, Uranian, and Neptunian satellites. As part of the work on the irregular satellites, they carried out a campaign of astrometric observations of those bodies at the Palomar Observatory and, in the course of that campaign, discovered two new Jovian irregulars. At the request of the New Horizons project, they developed ephemerides for Pluto's satellites including the two discovered in 2011 and 2012.

Small-body orbits and ephemerides are developed by Steven Chesley, Paul Chodas, Jon Giorgini, and Alan Chamberlin. All available small-body observations are processed with orbits and ephemerides updated and distributed on the Horizons system. Typically, Horizons (http://ssd.jpl.nasa.gov/?horizons) provides over 100 000 ephemeris products per day to the international user community. Near-Earth Objects are a special topic, with emphasis on potential Earth-approaching or Earth impacting asteroids (e.g., 2008 TC3). Comprehensive information is provided on near-Earth orbit and physical characteristics, interactive orbit diagrams, discovery statistics, Earth close approaches, Earth impact probability computations, and information on those objects that are particularly accessible for future, round-trip human exploration spaceflight missions. Information on the Near-Earth Object Program is maintained on the web site http://neo.jpl.nasa.gov/.

— William M. Folkner

Her Majesty's Nautical Almanac Office, UK

Over the last three years, HMNAO has received significant support from the National Hydrographer's group within the United Kingdom Hydrographic Office, particularly in the area of new staff, thereby ensuring the continuity of the office. In particular, Dr. Julia Weratschnig joined the office in November 2009 and Mr. James Whittaker joined in February 2012. Two more staff will join the office at the end of October 2012, bringing the complement up to seven. The collaboration with the US Naval Observatory (USNO) Astronomical Applications Department has grown stronger, celebrating its centenary during 2011, and an exchange program has been established allowing staff to gain experience and share knowledge and expertise.

The joint publications with USNO and those published by HMNAO have continued, the only change being for *The UK Air Almanac*, which is now available for download only. *The Astronomical Almanac* 2009 became fully compliant with the IAU resolutions passed in 2006 and further changes relating to the definition of dwarf planets are in progress. HMNAO staff members have given a number of talks to raise HMNAO's profile within the UKHO and beyond. HMNAO also hosted a meeting in London entitled "Astro-Navigation Solutions for the Future" which brought together many of the parties interested in celestial navigation in the United Kingdom. In view of the possible denial of GNSS, this meeting confirmed the need for celestial navigation as an independent backup to GPS, the continuing requirement for *The Nautical Almanac* and the promotion of *NavPac*, HMNAO's celestial navigation software. It also established the need for better training tools for celestial navigation and star identification including the use of new technology on mobile devices.

— Steven A. Bell

United States Naval Observatory, US

Production of the annual astronomical and navigational almanacs continued as a collaborative effort between the U.S. Nautical Almanac Office and H.M. Nautical Almanac Office (UK). A number of significant changes were made to *The Astronomical Almanac*, including a new table of exoplanets and host stars, a table of the ICRF2 defining sources, and a reorganization of material made necessary by the 2006 IAU definition of a "planet" in the solar system. The third edition of the *Explanatory Supplement to the Astronomical Almanac*—a major update of the previous (1992) edition—was published in late 2012. The Naval Observatory Vector Astrometry Software (NOVAS) in Fortran, C, and a new Python edition was updated to version 3.1 (2011), and the Multiyear Interactive Computer Almanac (MICA) for Windows and Mac OS X computers was updated to version 2.2.2 (2012). Research activities included work on the theory of bodily tides, the long-term orbital evolution of Phobos and Deimos, and a comparison of the three major, generally available solar system ephemerides. The fourth and final release of the USNO CCD Astrograph Catalog (UCAC) was completed in 2012, and the successor to the UCAC project, the USNO Robotic Astrometric Telescope (URAT), commenced observations. The Flagstaff Astrometric Scanning Transit Telescope (FASTT) program was expanded to include the first 5600 numbered asteroids, and completed more than 440 000 observations by mid-2012.

— John. A. Bangert (report presented by James Hilton.)

Comments by the Incoming President

Incoming president Catherine Hohenkerk thanked the outgoing president and members of the OC for their work over the past triennium. She looks forward to working with the new, somewhat enlarged, OC and welcomed all the new members of the Commission.

Ms. Hohenkerk noted that the IAU reorganization may affect how the commissions work and interact with their divisions. She hopes that the recommendations of the WG on Standardizing Access to Ephemerides might be put into action during the new triennium. She is committed to keeping the Commission 4 website up-to-date and relevant. She can be reached at **iaucom4@ukho.gov.uk**.

Report submitted 29 November 2012
George Kaplan, Past President

Transactions IAU, Volume XXVIIIB
Proc. XXVIII IAU General Assembly, August 2012
Thierry Montmerle, ed.

doi:10.1017/S1743921315005499

DIVISION I COMMISSION 7 CELESTIAL MECHANICS & DYNAMICAL ASTRONOMY

(MECANIQUE CELESTE & ASTRONOMIE DYNAMIQUE)

PRESIDENT Zoran Kneževic
VICE-PRESIDENT Alessandro Morbidelli
PAST PRESIDENT Joseph Burns
ORGANIZING COMMITTEE Evangelia Athanssoula, Jacques Laskar, Renu Malhotra, Seppo Mikkola, Stan Peale, Fernando Roig

PROCEEDINGS BUSINESS SESSIONS, 28 August 2012

President of Commission 7 Zoran Knežević opened the business meeting at 18:40 and proposed the agenda which was accepted without change.

Then, he gave a detailed report of the activities in the past period, emphasizing the following issues:

- Successful preparation of the Commission's Terms of Reference, subsequently approved by the Division I Organizing Committee;
- Remake and transfer of the Commission's web site;
- Initiative for co-affiliation of Commission 7 with Division III;
- Proposal for Asteroid Dynamic Site (AstDyS) to become a permanent IAU service;
- Participation of Commission 7 officers in the activities of the Division I Organizing Committee;
- Proposal for an IAU Symposium in Finland in 2012, which, unfortunately, was not accepted by the IAU;
- Election of the new Commission officers;
- Preparation of the Triennial Report and of this Business Meeting.

According to the current list of members maintained by the IAU Secretariat, Commission 7 has 293 members. There are as much as 55 new IAU members who expressed their wish to become members of Commission 7, and the recommendation by the IAU to accept all of them was approved by the Commission members present at the Business meeting.

A minute of silence was called by the President to pay respect to the 8 members of Commission who had passed away in the past three years.

The new officers of the commission for the 2012–2015 triennium are: President: Alessandro Morbidelli (France); Vice-President: Chrisitan Beaugé (Argentina); Organizing Committee (OC): Alessandro Celletti (Italy), Nader Haghighipour (USA), Piet Hut (USA), Jacques Laskar (France), Seppo Mikkola (Finland), Fernando Roig (Brasil). Roig will continue to act as the Commission's secretary. Knežević, on behalf of the members of

Commission 7, expressed gratitude to the outgoing members of the Organizing Committee: Joseph A. Burns, Evangelia Athanassoula, Renu Malhotra and Stanton J. Peale.

Summaries of the other reports presented at the meeting follow.

JOURNAL "CELESTIAL MECHANICS AND DYNAMICAL ASTRONOMY"

The report on the journal "Celestial Mechanics and Dynamical Astronomy" has been given by R. Dvorak, on behalf of the absent Editor-in-Chief S. Ferraz-Mello.

• The mean impact factor was 1.663 in the triennium;

• Two special issues, on "Extrasolar Planetary Systems" and "Resonances in N-body Systems", were published in 2011 and 2012 with C.Beaugé and R.Dvorak as guest editors;

• 50 percent of the submitted papers are accepted for publication. The turnaround time (average time between submission and first decision) is about 9 weeks. The typical time (median) between submission and final decision, for accepted papers, is about 5 months;

• CMDA is currently reaching, through electronic and hard copy subscriptions, more than 7,000 institutions.

— Rudolph Dvorak

FUTURE ACTIVITIES - LETTER OF THE INCOMING PRESIDENT

The incoming president Alessandro Morbidelli was not attending the General Assembly for personal reasons, thus he sent a letter with his address, wjich was read by Knežević. The letter is reproduced here in full:

Dear members of Commission 7

I have decided not to come to this General Assembly in China for the reasons explained on my web-page, that I will not repeat here.

With this letter, I would like to briefly describe my main four priority actions for the upcoming three years, when I will be the president of the Commission, with the help of Christian Beauge', Vice-President, Fernando Roig, secretary and the members of the OC.

But first, I would like to thank the out-going President, Zoran Knežević, for the work done over the past three years. Working with him has been very instructive for me and I hope to use this experience to act equally well and effectively during my term.

The first priority action of my term will be to co-affiliate Commission 7 to Division III. During the past term, the OC of Commission 7 has voted with a large majority in favor of this step. Contacts have been taken with the head of Division III but the dialog has been slow. During my presidency I will renew and intensify this dialog with the new officers of Division III to bring this process to a successful outcome. I am confident that the procedure of co-affiliation we be completed in a short timescale.

The second action of my presidency will be to propose and promote the AstDys service (managed by Andrea Milani at Pisa University) as new IAU service. This action was already the object of some discussions during the past term but was not pursued further because it is convenient to wait until Commission 7 is officially co-affiliated with Division III. In fact, the Minor Planet Center -an already existing IAU service- is promoted by Commission 20 and Division III and therefore we need to discuss and come to an agreement with these bodies on how the AstDys service could become complementary to the MPC rather than in competition with it.

The third action will be to promote an IAU symposium in Dynamical Astronomy in the next term. During the past term, Commission 7 proposed one symposium to be held in Finland this year, but it was not approved. So, there have been no Commission 7 symposia over the past three years. This situation needs to be corrected. It is already too late to propose symposia in 2013 and 2014, but we are on time for 2015. I will propose a dynamical astronomy symposium for the IAU General Assembly in 2015 in Hawaii. I think that it is important that there is a Commission 7 symposium joined to the General Assembly -unlike this year- in order to mark a strong presence of our Commission within the IAU and encourage Commission 7 members to come to the Assembly.

The fourth action will be to promote Commission 7 in the community of dynamical astronomers. There are many very good and high- reputation dynamicists who are not affiliated to our Commission, particularly in the US. We need to approach these people, explain to them the role of our Commission and invite them to join, lifting their reticence, so that our Commission can really become representative of the whole community. I invite all members of Commission 7 to join in this effort, approaching their colleagues who have not yet been affiliated.

Of course, these actions will come on top of other more regular business activities like: updating the terms of reference of our Commission, in particular in view of the co-affiliation to Division III; preparation of the triennial report; contacts, consulting and discussions with General Secretary and Assistant General Secretary of the IAU and with presidents and OCs of Divisions I and III; etc. I hope in an active participation of the members of our Commission and I encourage everybody to make propositions and suggestions of actions by contacting me via email.

Having said this, I wish you a good General Assembly and successful research over the next three years!

— Alessandro Morbidelli

7. CLOSING REMARKS

Two more items were discussed, which were raised by the colleagues present. The first pertained to the meeting in Celestial Mechanics, under the auspices of the IAU, to be organized in the next period. The dilemma was whether it is more promising to ask for an IAU Symposium in 2014, as this may be easier to get, or in 2015, as part of the next General Assembly in Honolulu, as proposed by the incoming President of Commission 7, Alessandro Morbidelli, in order to foster more substantial participation of the C7 members at the next GA . Several possible meetings were mentioned, to be either proposed as an IAU symposium (like the one in 2014 in Namur), co-proposed (in 2015, together with Commissions 20 and 53, for example), or only supported by the C7 (ACM meeting in Helsinki in 2014). The general agreement was that we should put maximum effort in an attempt to have an IAU symposium. It should cover broad enough scientific program, attractive enough for a wide range of astronomers and acceptable for the IAU executives. In this enterprise we shall have a full support of incoming President, Sergei Klioner, and Vice-President, Jacques Laskar, of the IAU Division A.

The second topic had to do with the general dissatisfaction of Commission 7 members with the current name for Division A (former Division I). After a long discussion the present members unanimously adopted a common statement in this regard: “The participants of the business meeting of Commission 7 unanimously disagree with the new name proposed for division A as they are not represented by this name.” Incoming Division officers will work on this in the next period.

After this president Knežević closed the meeting at 8:25 pm.

Report submitted 29 November 2012
Zoran Knežević, Past President

Transactions IAU, Volume XXVIIIB
Proc. XXVIII IAU General Assembly, August 2012
Thierry Montmerle, ed.

doi:10.1017/S1743921315005505

DIVISION I
COMMISSION 8 — ASTROMETRY
(ASTROMETRIE)

PRESIDENT — **Dafydd Wyn Evans**
VICE-PRESIDENT — **Norbert Zacharias**
PAST PRESIDENT — **Irina Kumkova**
ORGANIZING COMMITTEE — **Alexandre Andrei, Anthony Brown, Naoteru Gouda, Petre Popescu, Jean Souchay, Stephen Unwin, Zi Zhu.**

PROCEEDINGS BUSINESS SESSION, 27 August 2012

1. Business session (Chair D. W. Evans)

The business meeting was opened by the President, Dafydd Evans, who presented the agenda, which was approved. It was agreed that Norbert Zacharias should be the secretary of the meeting and take the minutes. This session was attended by about 20 participants.

1.1. *Commission Members*

Evans reported that the Commission has 251 members from 36 countries as of August 14, 2012. A total of 3 new members joined the Commission over the last triennium from existing IAU members and 50 new IAU members chose to join Commission 8 for the next triennium. With a minute of silence, Commission 8 paid tribute to 4 members who had died since the last General Assembly:

Lyssimachos Mavridis
Ernst Raimond
Klaus Guenter Steinert
Hans Walter

1.2. *New Commission Officers*

The result of the selection procedure for the next triennium was:

N. Zacharias (USA)	President
A. Brown (Netherlands)	Vice-president
D. Evans (UK)	Ex-officio
N. Gouda (Japan)	2009–2015
J. Souchay (France)	2009–2015
S. Unwin (USA)	2009–2015
L. Chen (China)	2012–2018
V. Makarov (USA)	2012–2018
O. Shulga (Ukraine)	2012–2018
R. Teixeira (Brazil)	2012–2018

No objections were raised.

Thanks were expressed to the retiring Commission Officers:

Alexandre Andrei (Brazil)
Irina Kumkova (Russia) – past president
Petre Popescu (Romania)
Zi Zhu (China PR)

1.3. *Website, Newsletters and Meetings*

The main activity of the Commission was through the website, newsletters and meetings. It was noted that for the next triennium the Commission website will remain at `http://www.ast.cam.ac.uk/ioa/iau_comm8/`. A total of 6 newsletters were published over the past triennium. The general feeling was that this was about the right number.

Although no specific Commission 8 meetings were held during the last triennium, this was not regarded as an issue, since we hold science sessions at General Assemblies. The one for this meeting was held on 29 August

1.4. *Commission Activities 2009–2012*

The Triennial Report was written in 2011 and will be published in Transactions IAU XXVIIIA. The Commission 8 contribution covered 12 pages. Evans thanked the National Representatives who collated the submissions.

Evans summarized the highlights of the past triennium activity that went into this report:

- Gaia continues development, with a scheduled launch in 2013. The expected parallax accuracies are 10 to 300 μas for 6 to 20 magnitude.
- ICRF2 was adopted by IAU in 2009.
- UCAC4 was completed, with over 100 million stars, Zacharias distributed DVDs of the catalogue during the meeting.
- The JASMINE project continues with the completion of the NanoJASMINE payload.
- The PPMXL and XPM catalogues were released.

Unfortunately, on a negative note, the SIM project was cancelled by NASA.

1.5. *General discussion*

Evans said that the current practice in Commission 8 is that the OC elects the Vice President and OC of the Commission. This Vice President becomes President in the following triennium. This ensures a balanced geographic representation within the OC of the Commission. The IAU would like a wider voting procedure. A recent member-wide vote for the Division I VP had only a 25% turnout, i.e. members in general do not seem to be interested in voting. Other commissions are in a similar position to us. What do we want to do? The following specific opinions were given:

William van Altena: Geographic representation is very important and will likely not occur in the case of general voting.

Brian Mason: Commission 26 (double stars) has all members voting, and up to now gets a good geographic representation. This is because members understand the need for a good geographic representation and vote accordingly.

Nicole Capitaine: it is important to have the same rules for all Commissions, and each Commission needs to have their terms of reference.

Dafydd Evans: considering the overall feeling of the meeting, shall we keep the status quo in Commission 8 for now? No objections were raised. How to proceed will depend on the IAU's instructions and what the situation will be like in 2.5 years, when the next set of voting will occur.

1.6. *Any Other Business*

Evans asked for ideas for new Working Groups. There were no suggestions.

Regarding new meetings, it was felt that astrometry is usually well covered between the IAU GA Commission 8 science sessions and Journées meetings in the years between GAs. However, an IAU Symposium for the next GA will be considered, given the fact that Gaia will be close to publishing first results at that time.

Evans will send notes regarding Commission timeline, newsletters etc. to Zacharias and Brown to assist in running the Commission.

van Altena announced the upcoming book "Astrometry for Astrophysics", Cambridge University Press, Autumn 2012.

2. Science Session (29 August 2012) (chair A. Brown)

The following presentations were given during the meeting. Summaries are given here from the abstracts. All of the presentations can be found on the meeting website:
`http://www.ast.cam.ac.uk/iau_comm8/iau28/`

2.1. *Astrometry with Gaia: what can be expected?*
J.H.J. de Bruijne (ESA/RSSD/SRE-SA)

Gaia is the next astrometry mission of the European Space Agency (ESA), following up on the success of the Hipparcos mission. Gaia's primary science goal is to unravel the kinematical, dynamical, and chemical structure and evolution of the Milky Way. In addition, Gaia's data will touch a wide variety of science topics, e.g., stellar physics, solar-system bodies, fundamental physics, and exo-planets. With a launch in the second half of 2013, the final catalogue is expected in 2021 – the first intermediate data release is envisaged to take place some two years after launch. Gaia will survey the entire sky and repeatedly observe the brightest 1,000 million objects, down to 20th magnitude, during its 5-year lifetime. Parallaxes will be measured with standard errors less than 10 micro-arcsecond (μas) for stars brighter than 12th magnitude, 25 μas for stars at 15th magnitude, and 300 μas at magnitude 20. The properties of the final astrometric catalogue depend, among others, on the adopted scanning law and on the payload-operation and on-ground calibration concepts, in particular the calibration of radiation-induced systematic effects in the data. The importance of these elements is highlighted. In addition, this presentation focuses on expected correlations and systematic errors in the data and on the expected astrometric performance of Gaia in high-density regions on the sky.

2.2. *Resolved Astrometric Binary Stars*
Brian Mason (USNO)

The resolution of binaries first detected astrometrically has a long history. In the early 19th Century Friedrich Wilhelm Bessel found periodic oscillations in the motions of Sirius and Procyon and reported them in a letter to Humboldt in 1834. The large flux ratio and much smaller mass ratio made these the easiest pairs to detect astrometrically. However, the large magnitude difference made resolution difficult and it was not until Alvan Clark and sons built two of their large refractors that this was accomplished. Sirius B was seen by Alvan G. Clark at the end of January 1862 testing the Dearborn

18.5" instrument and Procyon B was first seen by John Schaeberle in 1896 with the Lick 36" telescope. While pairs of this extreme flux ratio will continue to be a problem for resolution, the situation has improved markedly with smaller flux ratios being detected astrometrically with improvements to accuracy and precision of wide-angle astrometry. Also, new techniques and enhanced resolution capability for narrow-angle astrometry has allowed these pairs to be more easily resolved. The complimentary nature of these disparate techniques is exemplified with the new relative solutions of the astrometric binaries kappa For and HIP 42916 recently presented (Hartkopf *et al.* AJ 143, 42; 2012). A single resolution of a binary with an astrometric orbit allows for the determination of the relative orbit by scaling the a_{phot} to a'' appropriately. If the Δm and parallax is known, individual masses will also be forthcoming. Solutions of binaries of these type were presented.

2.3. *Present status of JASMINE projects*
Naoteru Gouda (National Astronomical Observatory of Japan)

The present status of the JASMINE projects were given:

JASMINE is an abbreviation of Japan Astrometry Satellite Mission for Infrared Exploration. Three satellites are planned as a series of JASMINE projects, as a step-by-step approach, to overcome technical issues and promote scientific results. These are Nano-JASMINE, Small-JASMINE and (medium-sized) JASMINE.

Nano-JASMINE uses a very small nano-satellite and is scheduled to be launched in November 2013 at the Alcantara space centre in Brazil by a Cyclone-4 rocket developed in Ukraine. Nano-JASMINE will operate in the zw-band (0.6–1.1 micron) to perform an all sky survey with an accuracy of 3 mas for position, parallaxes and proper motions. Moreover, high-accuracy proper motions (0.1 mas/year) can be obtained by combining the Nano-JASMINE catalogue with the Hipparcos catalogue.

Small-JASMINE will observe towards a region around the Galactic centre and other small regions, which include interesting scientific targets, with accuracies of 10 to 50 μas in an infrared Hw-band (1.1–1.7 micron). The target launch date is around 2017.

(Medium-sized) JASMINE is an extended mission of Small-JASMINE, which will observe towards almost the whole region of the Galactic bulge with accuracies of 10 μas in the Kw-band (1.5–2.5 micron). The target launch date is the first half of the 2020s.

2.4. *Parallaxes of five L dwarfs from a robotic telescope*
Youfen Wang (Shanghai Astronomical Observatory)

A report was given on the parallax and proper motions of five L dwarfs obtained with observations from the robotic Liverpool Telescope. These parallaxes represent new values and they are used to discuss the physical properties of L dwarfs. The derived proper motions are consistent with the published values and have considerably smaller errors. The objects appear to be normal L dwarfs, with space velocities that locate them in the disk and with normal metal abundances according to spectroscopic and model comparisons. For all five objects, effective temperature, luminosity, radius, gravity and mass from evolutional model were derived. The effective temperature were derived combining observational optical and NIR spectra with model synthetic spectra for three of the L dwarfs. The degeneracy of temperature, gravity and metallicity was found in affecting the absorption line strength through comparison among model spectra and among observational spectra. This robotic telescope was convenient in doing the parallax program which need a lot of repeated observations. Such robotic telescopes are able to enhance the efficiency of parallax programs, thus they are continuously needed in future.

2.5. *The NPARSEC Program Data Reduction Procedures.*

Catia Cardoso, NPARSEC Collaboration (Osservatorio Astrofisico di Torino)

The NPARSEC (NTT PARallaxes of Southern Extremely Cool objects) program determines parallaxes of about 80 objects covering the T dwarf spectral range. The areas of research directly impacted by this sample will be widespread. On an individual object basis, distances are key for assignments of binarity, metallicity and gravity and more generally the sample will provide key input for the substellar luminosity and mass functions, the connection to exo-planetary models as well as complex atmospheric processes such as non-equilibrium chemistry and turbulent mixing. Eventually these objects will provide new insights into the history of our galaxy, the kinematics of the solar neighbourhood and our understanding of differing formation scenarios from stars to brown dwarfs to giant planets. Particular attention was paid to the observational and data reduction procedures adopted with an emphasis on the centroiding, which is fundamental to the final astrometric precision.

2.6. *SPM4 - Yale/San Juan Southern Proper Motion Catalog*

W. F. van Altena, T. M. Girard, D. I. Casetti-Dinescu and K. Vieira (Yale University & CIDA)

The fourth instalment of the Yale/San Juan Southern Proper Motion Catalog, SPM4, contains absolute proper motions, celestial coordinates, and B, V photometry for over 103 million stars and galaxies between the south celestial pole and -20° declination. The catalog is roughly complete to V = 17.5 and is based on photographic and CCD observations taken with the Yale Southern Observatory's double astrograph at Cesco Observatory in El Leoncito, Argentina. The proper-motion precision is 2-3 mas/yr for well-measured stars; systematic uncertainties are on the order of 1 mas/yr.

In parallel with the SPM4 construction, and using the same SPM observations, a more accurate catalog of proper motions was made over a 450 sq-deg contiguous area that encloses both Magellanic Clouds. That catalog of 1.4 million objects was used to derive the mean absolute proper motions of the LMC and the SMC and, importantly, to make the most precise determination to date of the proper motion of the SMC relative to the LMC. The absolute proper motions are consistent with the Clouds' orbits being marginally bound to the Milky Way, albeit on an elongated orbit. Combining UV, optical and IR photometry from existing large-area surveys with SPM4 proper motions, we have identified young, OB-type candidates in an extensive 8,000 sq-deg region that includes the LMC/SMC, the Bridge, part of the Magellanic Stream and the Leading Arm. Additionally, a proper-motion analysis has been made of a radial-velocity selected sample of red giants and supergiants in the LMC, shown by Olsen *et al.* (2011) to be a kinematically and chemically distinct subgroup, most likely captured from the SMC. These results help constrain the Cloud-Cloud interaction, suggesting a near collision that took place 100 to 200 Myr ago.

Finally, SPM4 absolute proper motions have been cross-identified with radial velocities from the second release of the Radial Velocity Experiment (RAVE) and the resulting three-dimensional space motions of about 4400 red clump stars used to derive the kinematical properties of the thick disk, including the rotational velocity gradient, dispersions, and velocity-ellipsoid tilt angle.

2.7. *Advert for new book on astrometry*

W. F. van Altena (Yale University)

Astrometry for Astrophysics is intended to fill a serious gap in texts available to introduce advanced undergraduates, beginning graduate students and researchers in related fields to the science of Astrometry. This text provides an introduction to the field with examples of current applications to a variety of astronomical topics of current interest.

Astrometry for Astrophysics is intended for a one-semester introductory course that will hopefully lead to further study by students or serve as a primer on the field for researchers in related astronomical fields. To accomplish the above goals, the book is divided into five parts. Part one provides the impetus to study Astrometry by reviewing the opportunities and challenges of micro-arcsecond positions, parallaxes and proper motions that will be obtained by the new space astrometry missions as well as ground-based telescopes that are now yielding milli-arcsecond data for enormous numbers of objects. Part two includes introductions to the use of vectors, the relativistic foundations of astrometry and the celestial mechanics of n-body systems, as well as celestial coordinate systems and positions. Part three introduces the deleterious effects of observing through the atmosphere and methods developed to compensate or take advantage of those effects by using techniques such as adaptive optics and interferometric methods in the optical and radio parts of the spectrum. Part four provides introductions to selected topics in optics and detectors and then develops methods for analyzing the images formed by our telescopes and the relations necessary to project complex focal plane geometries onto the celestial sphere. Finally, Part five highlights applications of astrometry to Galactic structure, binary stars, star clusters, Solar System astrometry, extrasolar planets and cosmology. I hope that those chapters will stimulate students and researchers to further explore our exciting field.

Astrometry for Astrophysics consists of 28 chapters written by 28 specialists in the field from 15 different countries. The book is edited by van Altena and will be published by Cambridge University Press in November 2012.

2.8. *U.S. Naval Observatory Astrometric Catalogs*
Ralph Gaume (USNO)

Current USNO Astrometry catalogs and products were discussed, including NOMAD and UCAC4. Prospects for future USNO astrometric catalogs were reviewed, including the status of on-going programs such as URAT and UNAC, catalogs derived from large A-Omega programs, and prospects for a future bright-star catalog from the JMAPS space astrometry mission. The fundamental astrometric reference frame is based on the radio interferometric positions of quasars. Prospects for improvement of the fundamental astrometric reference frame were discussed.

2.9. *UCAC4*
N.Zacharias, C.Finch (USNO)

Reduction details, properties and notes for users were presented about the final USNO CCD Astrograph Catalog (UCAC) release number 4 which became public in August 2012. Accurate positions (20 to 100 mas) of 113 million stars to $R = 16$ are given based on over 200,000 CCD images taken by the 20cm astrograph at CTIO and NOFS between 1998 and 2004. Proper motions of most stars are based on SPM and NPM data with average errors of about 4 to 7 mas/yr and smaller errors for stars brighter than 13 utilizing many more catalogs. UCAC4 includes 5-band photometry for about 50 million stars from APASS and near IR photometry for over 100 million stars from 2MASS. FK6, Hipparcos and Tycho2 data are used to supplement bright stars in order to arrive at a complete all-sky catalog.

2.10. *The URAT project*

N.Zacharias, G.Bredthauer, M.DiVittorio, C.Finch, R.Gaume, F.Harris, T.Rafferty, A.Rhodes, M.Schultheis, J.Subasavage, T.Tilleman, G.Wieder (USNO)

The USNO Robotic Astrometric Telescope (URAT) achieved first light in 2011 at USNO in Washington DC and is now deployed at the Naval Observatory Flagstaff Station (NOFS). The red-lens of the UCAC program is again utilized for URAT, however, with a completely new tube assembly, upgraded mount, new electronics and a new 4-shooter camera containing 4 large CCDs (STA1600) each with 10,560 by 10,560 pixels of 9 micrometer size. A single exposure of URAT covers 28 square degrees of sky with a resolution of 0.9 arcsec/pixel. The URAT all-sky survey will reach about magnitude 17.5 in a bandpass between R and I with first data release expected by end of 2013. Several built-in features allow URAT to observe stars as bright as 1st magnitude. Multiple sky-overlaps taken over more than 2 years per hemisphere allow determination of accurate positions (10 mas level), proper motions, and parallaxes.

2.11. *A Preliminary Analysis of the Astrometric Asteroid Observations in the UCAC*

James L. Hilton (USNO)

Included in the UCAC observations made at Cerro Tololo Inter-American Observatory (CTIO) are 5,864 positions of asteroids. The number of observations of individual asteroids varies from 49 observations of (2) Pallas made over three oppositions to 556 asteroids with a single observation each. Analysis of 47 observations of (692) Hippodamia and 10 observations of (755) Sulamitis each made over two oppositions suggest that the accuracy of the these positions is approximately 50 mas in right ascension and 80 mas in declination. The accuracy of the UCAC may be somewhat better than this as the mean apparent diameters at opposition of these two bodies are approximately 60 and 30 mas, respectively, and no adjustments have been made for phase or possible albedo markings on the surface. A preliminary analysis of 41 of the observations of Pallas (mean apparent diameter 410 mas) are in good agreement with those of Hippodamia and Sulamitis. However, the remaining eight observations show a systematic offset in both right ascension and declination. These discrepant observations may indicate an albedo marking on the surface rotating into view.

2.12. *Hipparcos Successors in the 1990s*

Erik Høg (Niels Bohr Institute, Copenhagen University)

The approval in 1980 of the Hipparcos global astrometry mission and the subsequent development gave rise to ideas and work towards a Hipparcos follow-up mission which culminated with the approval of the ESA cornerstone mission Gaia in the year 2000. Ideas for a successor for global astrometry were studied in Russia (then USSR), and ideas for space astrometry by interferometry were studied in the USA, both beginning in the 1980s. The ESA community was, however, fully occupied with Hipparcos and nobody there thought of a follow-up mission. That changed in 1990 when Høg visited Russia, became interested in the Russian ideas and began discussions with Russian colleagues which led to the development in the 1990s, the main subject of the presentation.

2.13. *Astrometry 1960-80: from Hamburg to Hipparcos*

Erik Høg (Niels Bohr Institute, Copenhagen University)

A modest astrometric experiment in Copenhagen in 1925 led to the Hipparcos and Gaia space astrometry missions. Astrophysicists need accurate positions, distances and motions of stars in order to understand the evolution of stars and the universe. Astrometry provides such information, but this old branch of astronomy was facing extinction during much of the 20th century in the competition with astrophysics. The direction forward was shown by observations at the Copenhagen Observatory in 1925 with a new technique: photoelectric astrometry. Digital techniques were introduced in photoelectric astrometry at the Hamburg Observatory in the 1960s by the present author. This development paved the way for space technology as pioneered in France and implemented in the European satellite Hipparcos approved in 1980.

Dafydd Wyn Evans
President of the Commission

Transactions IAU, Volume XXVIIIB
Proc. XXVIII IAU General Assembly, August 2012
Thierry Montmerle, ed.

doi:10.1017/S1743921315005517

DIVISION I
COMMISSION 19

ROTATION OF THE EARTH

ROTATION DE LA TERRE

PRESIDENT	**Harald Schuh**
VICE-PRESIDENT	**Chengli Huang**
SECRETARY	**Florian Seitz**
PAST PRESIDENT	**Aleksander Brzezinski**
ORGANIZING COMMITTEE	**Christian Bizouard, Ben Chao, Richard Gross, Wieslaw Kosek, David Salstein** **IVS representative: Oleg Titov** **IERS representative: Bernd Richter** **IAG representative: Zinovy Malkin**

PROCEEDINGS BUSINESS AND SCIENCE SESSIONS, 29–30 August 2012

1. Introduction

During the XXVIII IAU General Assembly in Beijing IAU Commission 19 - Rotation of the Earth - held a business meeting and a scientific meeting. The business meeting was held on Wednesday, 29 August 2012 during session 1 (08:30-10:00). It was attended by about 35 participants, and six reports were given. First the activities of IAU Commission 19 during the past triennium (2009–2012) were highlighted by the Commission president. Afterwards, the Commission secretary presented the results of the elections for the next triennium (2012–2015) and a list of new members of the Commission. The designated Commission president provided an outlook into the next triennium, before the representatives of the international bodies and services IAG (International Association of Geodesy), IVS (International VLBI Service for Geodesy and Astrometry), and IERS (International Earth Rotation and Reference Systems Service) gave reports about recent activities. A summary of the business meeting is given below in Section 2. The scientific meeting was held on Thursday, 20 August 2012 during sessions 1 and 2 (08:30-12:30). Eleven presentations were given, and about 40 participants attended the sessions. Summaries of the presentations are provided below in Section 3.

2. Business Meeting

2.1. *Activities of IAU Commission 19 - Rotation of the Earth (2009–2012)*

Harald Schuh, Chengli Huang, Florian Seitz

Organizational issues:
- Establishing a Commission's Secretary (Florian Seitz, TU Munich)
- Setting up and maintaining the C19 website: www.iau-comm19.org, including a section on the history of C19.

• Collaboration with other IAU Commissions and with IAG (IAU Resolution 2009 on the ICRF and were also approved as IAG/IUGG Resolutions in 2011)
• Contributing to Division I issues
• Contacts with IAU Bureau
• Update of the Commission 19 member list (see Section 2.2)
• Organizing the C19 OC elections for the term 2012 - 2015 (see Section 2.2).

Conferences and publications
• Co-organisation of the Joint GGOS/IAU Science Workshop 'Observing and Understanding Earth Rotation', Shanghai, October 25–28, 2010 with about 90 international participants
• Co-organisation of a Special Issue on Earth Rotation in the Journal of Geodynamics, 20 papers submitted, 13 papers accepted
• Co-organisation of the Journées Systémes de Référence Spatio-Temporels 2010 in Paris, and 2011 September 19–21 in Vienna, about 100 participants
• Supporting sessions on Earth rotation at EGU 2010, 2011, 2011 in Vienna, at AGU in San Francisco, and at IUGG (2011, Melbourne)
• Preparation of the IAU General Assembly in Beijing, including two own Scientific Sessions on Earth Rotation and co-organisation of the Joint Discussion 7 on Space-Time Reference Systems for Future Research
• Triennial (scientific) report of C19 to IAU (Transactions IAU: Reports on Astronomy 2009–2012)

2.2. *Results of the elections for the new triennium (2012–2015) and new members of the Commission*

Florian Seitz

In February 2012 the Commission Secretary submitted a list of candidates, based on nominations from the members, for the offices of President (1 candidate), Vice-President (2 candidates) and OC (8 candidates). An electronic vote was conducted among all Commission members. Results were submitted to the parent Division for approval.

Offices for the term 2012–2015:
• President: Cheng-Li Huang (PR China)
• Vice-President: Richard Gross (USA)
• Secretary: Florian Seitz (Germany)
• Past President: Harald Schuh (Austria)
• OC member: Christian Bizouard (France), 2nd term
• OC member: Ben Chao (Taiwan), 2nd term
• OC member: Wieslaw Kosek (Poland), 2nd term
• OC member: David Salstein (USA), 2nd term
• OC member: Vladimir Zharov (Russia), 1st term
• IVS representative: Oleg Titov (Australia)
• IERS representative: NN.
• IAG representative: Zinovy Malkin (Russia)

During the last triennium, one active member, Anne-Marie Gontier, sadly passed away. Three members retired from the Commission (Francois E. Barlier, Leslie V. Morrison, Fabian Roosbeek), and 15 young scientists at post-doc level who are very active in the field of Earth rotation were invited to become new members of Commission 19.

2.3. *Outlook into the upcoming triennium (2012–2015)*

Chengli Huang, Richard Gross, Harald Schuh and Florian Seitz

Current achievements of geodesy allow to investigate a lot of smaller effects and geophysical factors and to stimulate further studies of EOP excitation:

- In observation: the precision reaches at 1 mm (or 30 micro-as) and 0.1 mm/year for position (or angle) and velocity respectively, and time resolution reaches hourly even near real time and continuous;
- In services: Very precise and dense EOP and ITRF/ICRF are provided by IERS and other services such as IVS, IGS, ILRS, and IDS.;
- In new techniques and instruments: Ring-laser and optical clocks become more mature and almost ready for services;
- In theoretical models: New precession/nutation models, global geophysical fluids (atmosphere, oceans, hydrology, etc.) models have been considerably improved, but there are still open questions.

Planed activities of C19 (as a bridge, a platform and a service) for the next term:

- Follow new technological achievements with relevance for EOP research (e.g. optical clocks) and make proposals on applications;
- Cooperate with related services (IERS, GGOS, IAG, IVS, ILRS, .) to provide reliable products (EOP, ITRF/ICRF, conventions, ...);
- Promote collaboration with neighbouring disciplines (oceanography, meteorology, hydrology, etc.) to improve the understanding of processes and interactions in the Earth system especially in view of global change;
- Jointly organize workshops/symposia/sessions dedicated to earth rotation;
- Establish of a new IAU/IAG working group on the theory of Earth rotation;
- Update C19's Terms of Reference

2.4. *Report of the IAG to IAU Commission 19 (2009–2012)*

Zinovy Malkin

Three IAG bodies are most closely cooperate with the IAU Commission 19. They are Commission 1 'Reference Frames', Commission 3 'Earth Rotation and Geodynamics' and Global Geodetic Observing System (GGOS).

Commission 1 activities and objectives deal with the theoretical and practical aspects of definition of reference systems and reference frames. Three sub-commissions participated in the IAU C19 related activities.

SC 1.1 'Coordination of Space Techniques' coordinates efforts that are common to more than one space geodetic technique, such as models, standards and formats, as well as combination methods and approaches, common modelling and parameterization standards, and combination strategies.

SC 1.2 'Global Reference Frames' is engaged in scientific research and practical aspects of the global reference frames, and investigation the requirements for the definition and realization of the terrestrial reference systems (TRS) and frames (TRF).

SC 1.4 'Interaction of Celestial and Terrestrial Reference Frames' investigates the impact of astronomical and geophysical modelling on the analysis of space geodetic observations and consistency between the TRF, CRF, and EOP, as well as collocation on Earth and in Space for CRF, and ICRF realization.

Commission 3 studies the entire range of physical processes associated with the motion and the deformation of the solid Earth. Its purpose is to promote, disseminate, and to help coordinate corresponding researches.

The main IAU C19 related activity is performed in the sub-commission 3.3 'Earth Rotation and Geophysical Fluids'. The objective of the SC 3.3 is to serve the scientific community by supporting research and data analysis in areas related to variations in Earth rotation, gravitational field and geocenter, caused by mass re-distribution within and mass exchange among the Earth's fluid sub-systems along with associated geophysical processes.

2.5. *IVS activity highlights 2009–2012*

Dirk Behrend and Harald Schuh

The International VLBI Service for Geodesy and Astrometry (IVS) continued to fulfil its role as a service within the IAU by providing necessary products for the densification and maintenance of the celestial reference frame as well as for the monitoring of Earth orientation parameters (EOP). Here we report on highlights of the service work during the report period focusing on special campaigns, products, and outreach.

• Meetings. During the report period the IVS held two General Meetings, one in Hobart, Tasmania, Australia in February 2010 and the other in Madrid, Spain in March 2012. Further, two Technical Operations Workshops were held at MIT Haystack Observatory in Westford, MA, USA in April 2009 and May 2011, respectively. Another important meeting was the VLBI2010 Workshop on Technical Specifications in Bad Kötzting, Germany in March 2012.

• ICRF2. The Second Realization of the International Celestial Reference Frame (ICRF2) was adopted at the XXVII IAU General Assembly in Rio de Janeiro, Brazil as Resolution B3. The ICRF2 replaced the previously used first realization (ICRF) effective 1 January 2010.

• IYA09. As an activity for the International Year of Astronomy 2009, the IVS organized a very large astrometry session. On 18/19 November 2009, thirty-four VLBI antennas observed the largest astrometry session ever scheduled.

• CONT11. In September 2011, a 15-day continuous VLBI observation campaign called CONT11 was observed. The network consisted of thirteen IVS stations, nine in the northern hemisphere and four in the southern hemisphere, giving the best geographical distribution and coverage in the series of CONT campaigns.

• IVS Live. IVS Live is a generalized version of the IYA09 dynamic Web site, developed to provide an easy access to the entire IVS observing plan. It has grown into a new tool that can be used to follow the observing sessions organized by the IVS, navigate through past or coming sessions, or search and display specific information related to sessions, sources (especially the most recent VLBI images) and stations. The IVS Live user interface and all its functionalities are accessible at the URL: http://ivslive.obs.u-bordeaux1.fr/.

• VLBI2010. The IVS has been developing the next generation VLBI system, commonly known as the VLBI2010 system. A VLBI2010 Project Executive Group (V2PEG) has been created to provide strategic leadership. A number of VLBI2010 projects are underway; several antennas have been erected and construction of about ten antennas is at various stages of completion. Further projects are in the proposal or planning stage. The next generation IVS network is growing, with an operational core of stations becoming available within the next few years, plus further growth continuing into the foreseeable future.

2.6. *Report of the IERS to IAU Commission 19 (2009–2012)*

Wolfgang Dick and Chopo Ma

The International Earth Rotation and Reference Systems Service continued to provide Earth orientation data, terrestrial and celestial references frames, as well as geophysical fluids data to the scientific and other communities. Work on new realizations of the International Terrestrial Reference System (ITRF2008) and the International Celestial Reference System (ICRF2) was finished. Investigations for the next International Terrestrial Reference Frame (ITRF2013) were started. Also discussion started about the next International Celestial Reference Frame (ICRF3), and an IAU Division 1 Working Group on ICRF3 was proposed. In 2009, Bulletin B was revised following a survey which was made among the community. In order to be consistent with ITRF2008, the IERS EOP C04 was revised again in 2011. The new solution 08 C04 is the reference solution which started on 1 February 2011. The system of the Bulletin A was changed to match the system of the new 08 C 04 series. In 2012, 4x/day EOP Combination and Prediction solutions became operational. The IERS Conventions (i.e. standards etc.) have been updated regularly and new revised edition was published at the end of 2010. Work on technical updates to the Conventions (2010) was started.

The Global Geophysical Fluids Centre (GGFC) restructured to allow for the establishment of operational products. It consists now of four Special Bureaus for Oceans, Hydrology, Atmosphere, and Combination. Three new working groups were established in 2009, 2011, and 2012: 1) Working Groups on Combination at the Observation Level; 2) Working Group on SINEX Format; 3) Working Group on Site Coordinate Time Series. The Working Group on Site Survey and Co-location was re-organized in 2012, and 'survey operational entity' was established within the ITRS Center.

The following IERS publications and newsletters appeared in print between 2009 and 2012: A.L. Fey, D. Gordon, and C.S. Jacobs (eds.): The Second Realization of the International Celestial Reference Frame by Very Long Baseline Interferometry, 2009 (IERS Technical Note No. 35); G. Petit and B. Luzum (eds.): IERS Conventions (2010), 2010 (IERS Technical Note No. 36); IERS Annual Reports 2007 and 2008–2009; IERS Bulletins A, B, C, and D (weekly to half-yearly); about 60 issues of IERS Messages. A technical note on ITRF2008 and annual reports for 2010 and 2011 are in preparation. The central IERS web site www.iers.org and about 15 individual web sites of IERS components have been updated, improved and enlarged continually.

3. Scientific Meeting

3.1. *The International Association of Geodesy (IAG) and its Global Geodetic Observing System (GGOS)*

Harald Schuh

In the last 10 years GGOS has been developed as a key component of the IAG. A proposal for GGOS has been developed by the GGOS planning group (2001–2003), and in July 2003, GGOS has been accepted as IAG Project by IAG EC and IAG Council (endorsed by IUGG, Resolution No. 3). Since 2007 GGOS is an integral component of IAG along with Services and Commissions, and during the implementation phase (2009–2011) the organizational structure has been revised, and the Terms of Reference have been set up. GGOS works with the IAG components to provide the geodetic infrastructure necessary for monitoring the Earth system and global change research (IAG Bylaws). In 2009, the fundamental publication 'GGOS: Meeting the Requirements of a Global Society on a Changing Planet in 2020' has been released (Eds. H.-P. Plag, M. Pearlman, Springer) which provides the main arguments for GGOS and its goals:

- GGOS vision: 'Advancing our understanding of the dynamic Earth system by quantifying our planet's changes in space and time.'
- GGOS mission:

1. To provide the observations needed to monitor, map and understand changes in the Earth's shape, rotation and mass distribution.
2. To provide the global frame of reference that is the fundamental backbone for measuring and consistently interpreting key global change processes and for many other scientific and societal applications.
3. To benefit science and society by providing the foundation upon which advances in Earth and planetary system science and applications are built.

- GGOS goals:

1. To be the primary source for all global geodetic information and expertise serving society and Earth system science.
2. To actively promote the sustainment, improvement and evolution of the global geodetic infrastructure needed to meet Earth science and societal requirements.
3. To coordinate with the international geodetic services that are the main source of key parameters needed to realize a stable global frame of reference and to observe and study changes in the dynamic Earth system.
4. To communicate and advocate the benefits of GGOS to user communities, policy makers, funding organizations, and society.

3.2. *Impact of IERS Conventions (2010) on VLBI-derived EOP*

Robert Heinkelmann and Harald Schuh

VLBI is the only technique for directly connecting the celestial and terrestrial reference frames and thus it is a major contributor for the determination of Earth Orientation Parameters (EOP). The VLBI data analysis involves a number of models and other analysis options. The models applied to VLBI and further space-geodetic techniques are specified through the IERS Conventions, while the VLBI-specific models are defined by the International VLBI Service for Geodesy and Astrometry (IVS). The IERS Conventions and its current version (IERS Conventions 2010) are believed to represent state of the art models, which have to be used for the determination of inter-technique products, such as the IERS EOP C04 08. For single technique, e.g. VLBI-based, products, the application of other models might be more appropriate. Quantifying systematic and other effects on the EOP is a major issue for the assessment of the quality and consistency of the EOP and the reference frames. The paper presents the impact of each model update on the EOP given by empirically comparing two solutions; one obtained using the old and one using the new conventions. The impact of the complete convention update can be obtained applying all changes at once. Implications following the new IERS Conventions 2010 are interpreted and commented from the VLBI-analysis point of view.

3.3. *Optical identification of ICRF reference radio sources*

Oleg Titov

We started an optical spectroscopic program to identify ICRF2 reference radio sources in optics and measure their redshifts. Five large optical facilities were involved in this program: The Big Telescope Azimuthal (BTA), Special Astrophysical Observatory, Russia; the New Technology Telescope (NTT), ESO, Chile; Gemini North and South Telescopes, Gemini Consortium, Hawaii and Chile, correspondingly, the Nordic Optical Telescope (NOT), Canary, Spain). More than 250 optical targets were observed, and around 200 red shifts were measured to date. Most of the objects are in the southern hemisphere.

Proper identification of the fiducial objects in different wavelengths is important to establish a reliable link between radio and optical ICRS realizations, once the Gaia mission is launched in the near future. We have identified several radio sources in close proximity of the galactic stars (separation of less than 4 arcseconds), and one radio source (IVS B1946-582) was reported to be perfectly aligned with the foreground optical object (separation is less than 0.3 arcseconds) at the galactic latitude -30 degrees. We believe that the number of similar cases will grow dramatically, when Gaia tracks the areas near the Galactic plane.

This group includes scientists from several countries, who provide theirs contribution in different ways.

3.4. *A new theory of precession and nutation for the Earth*

Enrico Gerlach, Sergei Klioner and Michael Soffel

The rotational motion of the Earth in space is described by the theory of precession and nutation, which gives the motion of the Celestial Pole in the Geocentric Celestial Reference System (GCRS). To include adequately the various processes involved in Earth rotation the currently used models of precession and nutation, most of them being analytical theories, are not sufficient to keep up with the accuracy provided by the observational data.

Therefore our group developed a completely numerical theory to describe the rotational motion of the Earth. This theory, which is based on the model of rigidly rotating multipoles is fully consistent with the post-Newtonian approximation of general relativity and is currently the best available model of precession and nutation for a rigid Earth. It is formulated using ordinary differential equations for the angles describing the orientation of the Earth (or its particular layers) in the GCRS.

In the last 2 years our model was extended towards a more realistic Earth. In detail, we included 3 different layers (crust, fluid outer core and solid inner core) and all important coupling torques, such as gravitational, electro-magnetic and topographic coupling, between them. Further, also the effects of non-rigidity, such as elastic deformation, frequency-dependent tidal deformation are modeled. Relative angular momenta due to atmosphere and ocean are included, using data from the respective state-of-the-art models. In our presentation we discussed all components of our theory in great detail and compared it with the currently used IAU 2000A precession/nutation model.

3.5. *Sub-daily Earth rotation parameters from GNSS and combined GNSS-SLR solutions*

Daniela Thaller, Michel Meindl, Gerhard Beutler, Rolf Dach, Adrian Jäggi and Krzysztof Sośnica

Space-geodetic techniques are capable of determining Earth rotation parameters (ERP) with a high temporal resolution, e.g., with one hour. Especially GNSS with its huge number of observations can determine time series of polar motion (PM) and length of day (LOD) rather well: We achieve a noise level in the time series of PM of about 150-170 micro-arcseconds only. The drawback of estimating sub-daily PM based on satellite techniques is the correlation with the parameters of the satellite orbits. Substantially different revolution periods of the satellites included in the solution affect the sub-daily PM series in different ways. Artificial periods showing up in the PM series are functions of the revolution periods of the satellites and the sidereal day (23.93 h). In the case of GPS two revolution periods exactly equal a sidereal day, whereas two revolution periods of GLONASS are considerably shorter than one sidereal day. This situation gives rise to different artificial signals in the PM series of GLONASS-only and GPS-only PM series. The artificial periods present in the SLR solution based on LAGEOS are fully different from GNSS solutions due to a revolution period of only 3.75 h.

Another problem of estimating PM with sub-daily resolution is the one-to-one correlation of a retrograde-diurnal signal in PM on one hand, and the nutation angles and the orbital parameters on the other hand. In order to handle the correlation between retrograde-diurnal PM and the satellite orbits we tested two methods for GNSS and SLR solutions:

1. Applying a constraint prohibiting a retrograde-diurnal signal in the hourly PM series (i.e., constraining the corresponding amplitude to zero);
2. Introducing (and not estimating) the satellite orbits from a solution with a temporal resolution of 24 h for the ERPs.

Both methods work fine in the sense that no retrograde-diurnal signal is present in the resulting 1-hourly PM time series. There are, however, disadvantages for both methods. In the first method, the constraint affects neighbouring frequencies as well, i.e., their amplitudes are not freely estimated anymore. The range of frequencies around the retrograde-diurnal period which are affected depends on the orbital arc length. SLR solutions have the advantage that they are based on longer orbital arcs, usually on 7 d instead of 1 d or 3 d arcs in GNSS solutions. This advantage is counterbalanced by the fact that the revolution period of the LAGEOS is shorter than diurnal which evokes several artificial signals in the PM series (as mentioned above). These artificial signals cannot be avoided by the retrograde-diurnal constraint.

The opposite is true if the second method is used: the artificial periods can be avoided in GNSS and LAGEOS solutions. The disadvantage is, however, that probably other signals are blocked as well.

In combined GNSS-SLR solutions the situation is complicated by the fact that the orbital arc lengths used for LAGEOS and GNSS satellites do not coincide (7 d and 3 d, respectively). We showed that a retrograde-diurnal constraint (method 1) does not remove all correlations between orbits and sub-daily PM if satellites with different orbital arc lengths are involved. For method 2 we showed that fixing the LAGEOS orbits to the solution with 24-h ERPs removes the correlations in such a way that the GNSS orbits do not need to be fixed in addition and therefore can be estimated together with the combined solution.

3.6. *Short period ocean tidal effects on Earth rotation: Monitoring and modeling*

Sigrid Böhm and Harald Schuh

Short period variations of Earth rotation, quantified as Earth rotation parameters (ERP), are induced predominantly by diurnal and subdiurnal ocean tides. Secondary causes of such variations are thermal tides of the atmosphere, driven by the diurnal solar heating cycle, and the effect of the lunisolar torque on the triaxial figure of the Earth, called libration. Detailed descriptions of diurnal and subdiurnal ERP variations based on a profound geophysical background are essential for various parameter estimation problems in space geodesy, like processing of special VLBI (Very Long Baseline Interferometry) sessions or GNSS (Global Navigation Satellite Systems) orbit determination. Moreover, the accurate removal of ocean tidal signals from high-frequency Earth rotation enables the examination of minor non-harmonic fluctuations of atmospheric or oceanic origin.

At present, the comparison of observational evidence and geophysical modeling still leaves several unexplained gaps. In this study we presented the results of different approaches for capturing short period ERP variations by means of VLBI. We obtained three sets of tidal coefficients (largely congruent with the terms of the model for diurnal and semidiurnal ocean tidal ERP variations of the IERS Conventions 2010) using VLBI observation data from 1984 to 2010. These empirical tidal ERP models were calculated from highly resolved ERP time series, from demodulated ERP time series, and directly within a global solution. Significant deviations of all major VLBI-based tidal terms to the corresponding conventional model terms were observed. Deficiencies of the IERS Conventions 2010 model are suspected to cause aliased signal in GNSS analysis and are problematic for the analysis of special VLBI sessions (Intensives).

A new research project at the Vienna University of Technology, named SPOT (Short period ocean tidal variations in Earth rotation), aims at developing an up-to-date model for the effects of diurnal and subdiurnal ocean tides on Earth rotation, which satisfies the steadily increasing accuracy requirements of science and navigation applications. The strategies to meet this goal are the employment of a recent empirical ocean tide model from multi-mission satellite altimetry, the thorough revision of the transfer functions, and the direct estimation of the contribution of minor tides from satellite altimetry instead of merely using admittance assumptions.

3.7. *New attempts at prediction and verification of AAM and relationship to Earth orientation parameters*

Christian Bizouard, David Salstein and Daniel Gambis

We reported on new efforts to estimate the value of predictions of the atmospheric excitations for polar motion and Earth rotation. Several atmospheric data centers produce predictions out to a time horizon of some 8 days or longer, from which the atmospheric angular momentum (AAM) is calculated; these results are in use in a number of operational settings. The value of such predictions first can be measured against the state of the atmosphere at the verification hour. Such tendencies are measured also against the tendency of UT1, as well as polar motion, measurements of which are modulated by the Chandler wobble period. In particular, weather center data from the US, Japan, the UK, and the European Centre for Medium Range Weather Forecasts are available for atmospheric predictions based on winds and surface pressures. The general skill of weather forecasts have improved over the last many years, and we estimated if the terms required for AAM calculations are significantly more useful than they were in the past.

3.8. *Application of Earth rotation parameters in Earth system science*

Florian Seitz and Stephanie Kirschner

Variations of Earth rotation are associated with the redistribution and motion of mass elements in the Earth system. On seasonal to inter-annual time scales, the largest effects are due to mass redistributions within atmosphere and hydrosphere. In order to study the Earth's reaction on geophysical excitations, the dynamic Earth system model DyMEG has been developed. It is based on the balance of angular momentum in the Earth system which is physically described by the Liouville equation. This coupled system of three first-order differential equations is solved numerically in DyMEG.

Simulations of polar motion and length-of-day variations have been performed with DyMEG for time spans of up to 200 years using angular momentum variations from five ensemble runs of a consistently coupled atmosphere-hydrosphere model as model forcing. Besides, deformations induced by tides, loading and variations of Earth rotation were considered. In particular the contribution focused on the simulation results of the Earth's free polar motion (Chandler oscillation). It was shown that the simulations over 200 years (1860-2059) are capable of exciting realistic variations of the Chandler oscillation. The application of an adaptive Kalman filter on DyMEG allows for the simultaneous simulation of Earth rotation and the estimation of critical model parameters, such as physical Earth parameters (e.g. Love numbers).

3.9. *Mass transport and dynamics in the Earth system: Selected scientific questions and observational requirements*

Richard Gross

The solid Earth is subject to a wide variety of forces including external forces due to the gravitational attraction of the Sun, Moon, and planets, surficial forces due to the action of the atmosphere, oceans, and water stored on land, and internal forces due to earthquakes and tectonic motions, mantle convection, and coupling between the mantle and both the fluid outer core and the solid inner core. The solid Earth responds to these forces by displacing its mass, deforming its shape, and changing its rotation. Geodetic observing systems can measure the change in the Earth's gravity caused by mass displacement, the change in the Earth's shape, and the change in the Earth's rotation. Consequently, geodetic observing systems can be used to study both the mechanisms causing the Earth's shape, rotation, and gravity to change, as well as the response of the solid Earth to these forcing mechanisms. As a result, geodetic observing systems can be used to gain greater understanding of the Earth's interior structure and of the nature of the forcing mechanisms including their temporal evolution. A few selected unsolved scientific questions will be examined here as a way of illustrating the role played by geodetic observing systems in general, and Earth rotation measurements in particular, in understanding the Earth and its interacting systems. For example, extending the theory of the Earth's rotation to a triaxial, deformable body with fluid core and oceans will allow observations of the Earth's rotation to be better modeled; improving the accuracy of the observations by developing next generation observing systems will allow smaller signals such as those caused by earthquakes to be studied; and improving determinations of the period and Q of the Chandler wobble will allow the frequency dependence of mantle anelasticity to be better constrained.

3.10. *Second order effects on the Earth precession and nutation: a brief summary*

Josẽ M. Ferrãndiz Juan Getino, Alberto Escapa, Juan Navarro and Pedro Martĩnez-Ortiz

Using current precession-nutation models IAU 2006/2000 with a convenient FCN model allows fittings to different VLBI observational series with a typical magnitude of residuals about 150 μas in CIP offsets. That is quite satisfactory for present needs, but further progress is necessary to meet future accuracy requirements at the millimetre level (30 μas), e.g. as sought by GGOS initiative.

In this presentation we report on some effects that we call of second order, which provide non-negligible contributions to earth rotation including nutation terms reaching some tens of μas. Among them, there are second order terms in the sense of perturbation theory, non-linear in the dynamical ellipticity H. In other group we gather terms of various physical origins related to time-varying potentials. We consider here direct effects of the actual rotation of the inner core, effects of the observed J2 variations or unaccounted effects of tidal models on precession and nutation.

The Hamiltonian method has been followed to derive all the solutions. For the second order nutations, the differences between two layers earth amplitudes and rigid ones reach the ten μas level for a few terms in the 18 years and semi-annual bands. The indirect effects, due to the induced change in the value of the precession parameter H are larger than 20 μas in obliquity and 40 μas in longitude for the 18 years nutation. The rotation of the inner core produces a deviation of the Hamiltonian with respect to a steady reference considered when deriving the solution to three-layer earth models. In this case an amplitude larger than 40 μas in longitude is got for the semi-annual periods, while annual and fortnightly amplitudes are around 10 μas.

The observed variations of the earth oblateness parameter J2 also affects nutations with periods resulting of the combination of those of rigid nutations and of J2 harmonic contents as well. Some mixed periodic terms are to be accounted, together with some very long period ones which may interact with the value of the dynamical ellipticity H if neglected. Finally, the modelled tidal variations of the solid earth have been proved to effect precession and nutations at a similar level. Those theoretical approaches show that different effects produce both direct and indirect effects on precession and nutation and provide nutation terms with amplitudes of the order of some tens of micro arc seconds, which are not included in current IAU models as far as we know.

In view of their magnitude, it should be analyzed their actual influence on the accuracy of the precession/nutation models and the convenience of adding some of them as new corrections, keeping the consistency of the whole models.

4. Closing remarks

Over the last triennium strong progress has been made in the field of Earth rotation research. Several conferences, workshops and scientific sessions were dedicated to this topic. Many publications, among them a special issue of the Journal of Geodynamics on Earth rotation, have been released. New observatories contributed to a further improvement of the quality of geodetic parameters in terms of accuracy and temporal resolution, and highly precise time series of parameters related to Earth rotation were computed and provided through international services. Many research projects worldwide analysed underlying physical causes of temporal variations of Earth rotation and thus forged a link into the geoscientific community. Promising attempts have been made to incorporate Earth rotation parameters as boundary conditions into models of geophysical fluids. This way, observations of Earth rotation also provide a direct input to Earth system research. Around 15 young scientists have been attracted as new members to Commission 19 which is very promising and warrants fruitful future activities in the field.

Florian Seitz
Secretary of Commission 19

Harald Schuh
President of Commission 19

Transactions IAU, Volume XXVIIIB
Proc. XXVIII IAU General Assembly, August 2012
Thierry Montmerle, ed.

doi:10.1017/S1743921315005529

DIVISION II COMMISSION 10 — SOLAR ACTIVITY

ACTIVITE SOLAIRE

PRESIDENT	**Lidia van Driel-Gesztelyi**
VICE-PRESIDENT	**Karel J. Scrijver**
PAST PRESIDENT	**James A. Klimchuk**
ORGANIZING COMMITTEE	**Paul Charbonneau, Lyndsay Fletcher, S. Sirajul Hasan, Hugh S. Hudson, Kanya Kusano, Cristina H. Mandrini, Hardi Peter, Bojan Vršnak, Yihua Yan**

PROCEEDINGS BUSINESS SESSION, 23 August 2012

The Business Meeting of Commission 10 was held as part of the Business Meeting of Division II (Sun and Heliosphere), chaired by Valentin Martínez-Pillet, the President of the Division. The President of Commission 10 (C10; Solar activity), Lidia van Driel-Gesztelyi, took the chair for the business meeting of C10. She summarised the activities of C10 over the triennium and the election of the incoming OC.

1. IAU meetings

The OC has solicited, reviewed and endorsed IAU Symposium, Special Session, and Joint Discussion proposals, offered advice to the proposers. Members of the OC actively participated in the organisation of IAU meetings and one OC member has served as Editor of the Proceedings of IAUS 286 (Mandrini & Webb, 2012).

2. Triennial Report

The usual team effort by the OC produced a summary of science progress in solar activity over the previous three years. The Triennial Report was published in Transactions IAU, T28A (van Driel-Gesztelyi *et al.*, 2012), and is available on the website of C10. Usually, triennial reports are quite long, e.g. the C10 report was 14 pages long in 2005, and 26 pages long in 2008. This time the OC of C10 made an attempt to observe the nominal 6-page limit on triennial commission reports, but in order to be able to produce a meaningful summary of science progress, the report eventually doubled the nominal length.

3. Website

The C10 website was maintained (see link from $http : //www.lmsal.com/iau_c10/$). It includes recent reports of the Commission and the Division as well as a list of OC members and officers dating back to 1970, and Presidents dating back to 1925. This historical list was revised and updated.

4. Solar Target Naming Convention

In 2009, Carolus J. Schrijver, V-P of C10, proposed a standardised convention for identifying solar events (e.g., flares and CMEs) to be included in publications so that search engines can easily identify other relevant publications in the on-line literature. Three years ago, C10 formally endorsed the proposal, so did the Solar Physics journal. However, the naming convention needs to be widely adopted to become a useful literature-mining tool. It is recommended that the major journals make solar naming convention a standard requirement in solar physics articles. The proposal can be found at $http : //www.iau.org/science/scientific_bodies/divisions/E/objectives/$.

5. Membership and new applicants

C10 has 647 members and received 78 new applications for membership from applicants who became IAU members at the IAU GA – an impressive increase.

6. Remembrance

The names of C10 members deceased during the last triennium, being known to the IAU, were shown: A. Dollfus, S. Enome, G. Gelfreigh, E.A. Tandberg-Hanssen, M. Kundu, J. Parkinson, M. Semel, A. Tlamicha. I. Tuominen, and H. Zirin.

7. Outgoing OC

The president thanked the members of the outgoing OC for their service: Karel (C.J.) Schrijver (USA; Vice-President), Paul Charbonneau (Canada), Lyndsay Fletcher (UK), S. Sirajul Hasan (India), Hugh S. Hudson (USA), Kanya Kusano (Japan), Cristina H. Mandrini (Argentina), Hardi Peter (Germany), Bojan Vršnak (Croatia), Yihua Yan (China), and James A. Klimchuk (USA; Past President).

8. Election of the officers and the OC for 2012-2015

Because electronic voting seemed to be quite complicated, the election was conducted within the OC. First, the new President was elected. Following tradition, the Vice President became President by unanimous approval. Karel (C.J.) Schrijver has long experience in C10: he was Secretary of C10 in the period 2006-2009 and Vice-President of C10 in the period 2009-2012. Second, the new Vice-President was elected from those members of the OC, who wished to be considered. Lyndsay Fletcher, who has been OC member since 2006, became V-P of C10. Third, the new OC was formed in three steps. (i) OC members, who served two terms and/or felt too busy to continue, stepped down, including Hugh Hudson, Kanya Kusano, Cristina Mandrini, Hardi Peter, and Bojan Vršnak. (ii) Based on proposals by the OC, a slate of 18 candidates was assembled for the vacated OC membership positions, and (iii) with secret voting the new OC members were elected. The new OC of C10 is as follows:

President Karel (C.J.) Schrijver (USA)
Vice-President Lyndsay Fletcher (UK)
Past President Lidia van Driel-Gesztelyi (France, Hungary, UK)
Board Ayumi Asai (Japan), Paul Cally (Australia), Paul Charbonneau (Canada), Sarah Gibson (USA), Daniel O. Gomez (Argentina), S. Sirajul Hasan (India), Yihua Yan (China), Astrid Veronig (Austria).

It was remarked that the geographical distribution of the OC remained even, and its gender ratio shows an increasing participation of women (up from 25% to 45%) in the OC compared to the former triennial period.

At the end of the C10 business meeting, the President invited the audience to make suggestions on how the Commission could better serve its members during the next triennium and beyond, but pressing time constraints prevented any substantial discussion.

Lidia van Driel-Gesztelyi
President of the Commission

References

van Driel-Gesztelyi, L., Schrijver, C. J., Klimchuk, J. A., Charbonneau, P., Fletcher, L., Hasan, S. S., Hudson, H. S., Kusano, K., Mandrini, C. H., Peter, H., Vršnak, B., & Yan, Y. 2012, Commission 10: Solar Activity, *Transactions of the International Astronomical Union, Series A*, **28**, 69-80.

Mandrini, C. H. & Webb, D. (eds.) 2012, *Comparative minima characterising quiet times in the Sun and stars.* IAU Symposium **286**, Cambridge, UK: Cambridge University Press.

Transactions IAU, Volume XXVIIIB
Proc. XXVIII IAU General Assembly, August 2012
Thierry Montmerle, ed.

doi:10.1017/S1743921315005530

DIVISION II COMMISSION 12 — SOLAR RADIATION AND STRUCTURE

RAYONNEMENT ET STRUCTURE SOLAIRES

PRESIDENT Alexander Kosovichev
VICE-PRESIDENT Gianna Cauzzi
PAST PRESIDENT Valentin Martinez Pillet
ORGANIZING COMMITTEE Martin Asplund, Axel Brandenburg, Dean-Yi Chou, Jorgen Christensen-Dalsgaard, Weiqun Gan, Vladimir D. Kuznetsov, Marta G. Rovira, Nataliya Shchukina, P. Venkatakrishnan

PROCEEDINGS BUSINESS SESSION, 22 August 2012

Commission 12 held a separate business meeting, on Wednesday August 22, from 14:00-16:00.

The President of C12, Alexander Kosovichev, presented the status of the Commission and its working Group(s). Primary activities included organization of international meetings (IAU Symposia, Special Sessions and Joint Discussion); review and support of proposals for IAU sponsored meetings; organization of working groups on the Commission topics to promote the international cooperation; preparation of triennial report on the organizational and science activities of Commission members. Commission 12 broadly encompasses topics of solar research which include studies of the Sun's internal structure, composition, dynamics and magnetism (through helioseismology and other techniques), studies of the quiet photosphere, chromosphere and corona, and also research of the mechanisms of solar radiation, and its variability on various time scales. Some overlap with topics covered by Commission 10 Solar Activity is unavoidable, and many activities are sponsored jointly by these two commissions. The Commission website can be found at http://sun.stanford.edu/IAU-Com12/, with information about related IAU Symposiums and activities, and links to appropriate web sites.

Commission 12 currently includes 351 members. However, from a cursory inspection it appears that a large number of them is already retired and/or working in fields only marginally related to Solar Physics; the President comments that a revision of the list is in order to redress this situation. Attendees comment that at the time of membership, one is asked to join only one particular Commission, and it is not obvious how to extend membership to other ones (although possible by IAU rules). Both of these issues had been solved after IAU implements the structural changes voted at the GA. Each member was asked to renew their affiliation to one or more Divisions, as well as to the Commissions.

During the past triennium (Sept 2009-Aug 2012), the Commission either organized directly or supported, a total of 5 Symposia and 1 Special Session:

- IAUS 264, Solar and stellar variability - impact on Earth and planets, Rio de Janeiro, Brazil
- IAUS 271, Astrophysical dynamics from stars to galaxies, Nice, France
- IAUS 273, Physics of Sun and star spots, Los Angeles, United States
- IAUS 274, Advances in plasma astrophysics, Catania, Italy
- IAUS 286, Comparative magnetic minima: characterizing quiet times in the Sun and stars, Mendoza, Argentina
- IAUS 294, Solar and astrophysical dynamos and magnetic activity, Beijing, China
- JD11, New advances in helio- and asteroseismology, Rio de Janeiro, Brazil
- SpS6, Science with large solar telescopes, Beijing, China

The Commission members edited 5 volumes of IAUS Proceedings. The triennial report on research activities in the fields of C12 was compiled by the Organized Committee (Kosovichev *et al.* (2012)).

After the discussions at the past GA in Rio de Janeiro, a new Working Group of C12 was established: "Coordination of Synoptic Observations of the Sun", chaired by Alexei Pevtsov (NSO, USA). The WG has the mission to facilitate international collaboration in synoptic long-term solar observations as well as providing a forum for discussion of all issues relevant to synoptic long-term observations of the Sun. The web page of the WG is at http://www4.nso.edu/staff/apevtsov/IAU-Com12/main/wg_synoptic_observations.html

The President mentioned the impressive array of new facilities related to the study of the Sun that have come online in the past triennium. These include the space-based SDO, Koronas, Picard, as well as the ground-based 1.5 m telescopes NST and GREGOR and the balloon-born SUNRISE. Many more are scheduled for construction in the next years, such as IRIS, Solar Orbiter, Solar InterHelioProbe, ATST. The Commission must actively promote the circulation of information in the community about these facilities. In this regard, it is noted that C12 organized a Special Session during the Beijing GA on "Science with Large Solar Telescopes", which was well attended with over 100 participants. Several articles were also published in the GA Newsletter about both the meeting and the individual projects, giving them ample visibility within the astronomical community at large.

The President noted the result of the election for the incoming Organizing Committee:
President: Gianna Cauzzi (Italy)
Vice-President: Nataliya Shchukina (Ukraine)
Past President: Alexander Kosovichev (USA)
Board: Axel Brandenburg (Sweden), Michele Bianda (Switzerland), Dean-Yi Chou (China Taiwan), Sergio Dasso (Argentina), Mingde Ding (China PR), Stuart Jefferies (USA), Natasha Krivova (Germany), Vladimir D. Kuznetsov (Russian Federation), Fernando Moreno Insertis (Spain).

Despite promise of a user-friendly polling system provided by IAU, Commission-wide elections for the renewal of the Organizing Committee were still not possible, due to mishaps in the system. It is widely expected that these problems will be solved over the course of the next triennium, given the restructuring of IAU Divisions and Commissions' role and activities. However, officers of C12 will keep in touch with the relevant IAU bodies to make sure that such a feature is finally implemented.

The President concludes reminding what will be the focus of C12's activity in the next three years: *i)* Continue effort in supporting proposals and organization of IAU symposia, SpS and JD on C12 topics; *ii)* Promote international cooperation and support of solar astronomy in different countries; *iii)* Promote interdisciplinary links and cooperation within the IAU Division II (now Div. E) and with the other IAU Divisions and Commissions; *iv)* further develop cooperation and coordination in the form of Working Groups; *v)* Support young solar astronomers, summer schools, EPO activities.

In particular, C12 already actively sought to support young students and scientists for participation in the GA activities. To this end, two grants were secured from the Tom Metcalf Travel Fund of the Solar Physics Division/AAS, and from the NSF-sponsored program managed by the University of New Mexico. These grants covered expenses for four young scientists, who gave invited talks in S294; JD3; SpS6.

Alexander Kosovichev
President of the Commission

References

Kosovichev, A., Cauzzi, G., Martinez Pillet, V., Asplund, M., Brandenburg, A., Chou, D.Y., Christensen-Dalsgaard, J., Gan, W., Kuznetsov, V., Rovira, M. G., Shchukina, N., Venkatakrishnan, P. 2012 *Commission 12: Solar Radiation and Structure*, In I. Corbett (ed.) *Transactions IAU, Vol. XXVIIIA, Reports on Astronomy*, Cambridge University Press, pp. 81–94

Transactions IAU, Volume XXVIIIB
Proc. XXVIII IAU General Assembly, August 2012
Thierry Montmerle, ed.

doi:10.1017/S1743921315005542

DIVISION II
COMMISSION 49

INTERPLANETARY PLASMA AND THE HELIOSPHERE

PLASMA INTERPLANETAIRE ET HELIOSPHERE

PRESIDENT Natchimuthuk Gopalswamy
VICE-PRESIDENT Ingrid Mann
PAST PRESIDENT Jean-Louis Bougeret
ORGANIZING COMMITTEE Carine Briand, Rosine Lallement, David Lario, P. K. Manoharan, Kazunari Shibata, David F. Webb

PROCEEDINGS BUSINESS SESSION, 23 August 2012

The President of IAU Commission 49 (C49; Interplanetary Plasma and the Heliosphere), Nat Gopalswamy, chaired the business meeting of C10, which took place on August 23, 2012 in the venue of the IAU General Assembly in Beijing (2:00 - 3:30 PM, Room 405).

1. Business Meeting Agenda

1. President's remarks, 2. New organizing committee, 3. Working groups related to commission 49, 4. Interaction with other organizations, 5. Outreach activities, and 6. New organizational structure.

The meeting started with a moment of silence to those commission members deceased during the past triennium. The following names were known to the IAU: Donald Blackwell (1921–2011), John Dyson (1941–2010), Evry Schatzman(1919–2010), Charles P. "Chuck" Sonett (1924–2011).

The president summarized the activities of C49 over the triennium and the election of the incoming OC.

2. IAU meeting proposals

The OC has solicited, reviewed and endorsed IAU Symposium, Special Session, and Joint Discussion proposals, offered advice to the proposers.

3. Triennial Report

The Triennial Report of C49 was published in Transactions IAU, T28A (Gopalswamy *et al.*, 2012), and is available on the Astrophysical Data System: *http : //adsabs.harvard.edu/abs/*2012*IAUTA*..28...95*G*. The president expressed special thanks to the OC for producing the 30-page report. In addition, he thanked John

D. Richardson (MIT, USA) for contributing to the outer heliosphere section. The report was 30-pages long, which reflects the increased activity on the topics the commission is interested in. The report also demonstrates that Commission 49 is a vital component of Division II and will continue to add new knowledge.

4. Membership

C49 has 196 members from 23 countries and territories. The current list can be found in *http : //www.iau.org/science/scientific_bodies/commissions/49/members/*.

5. Outgoing OC

President Nat Gopalswamy (USA)
Vice-President: Ingrid Mann (Sweden)
Past President Jean-Louise Bougeret (France)
Board: Carine Briand (France), Rosine Lallement (France), David Lario (USA), P. K. Manoharan (India), Kazunari Shibata (Japan), David F. Webb (USA).

The president thanked the members of the outgoing OC for their service.

6. Incoming OC for 2012-2015

Following tradition, the outgoing President drafted the new OC list and circulated among the current OC. After discussion and modifications, the list was circulated among the C49 members. There were no additional comments received from the C49 membership. The list of new OC members was presented to the Business meeting. As per tradition, the Vice President became the President. The new OC of C49 is as follows:

President: Ingrid Mann (Sweden)
Vice-President: P. K. Manoharan (India)
Past President: Nat Gopalswamy (USA)
Board: Carine Briand (France), Igor Chashei (Russia), Sarah Gibson (USA), Yoichiro Hanaoka (Japan), David Lario (USA), Olga Malandraki (Greece)

The geographical distribution of the OC remained even, and the female-to-male ration increased from 30% to 44% in the new OC. The president also informed that the commission has been traditionally recommending the appointment of the Bureau member from IAU to the Scientific Committee on Solar Terrestrial Physics (SCOSTEP). After the current Bureau member Nat Gopalswamy was elected as President of SCOSTEP, Mei Zhang has been recommended as the new Bureau member via consensus.

The new President was invited to address the business meeting. She suggested that the following are the goals for the coming 3 years: (i) reflect on the role of the commission and make close connection to other commissions within the Division, (ii)the unique point of the commission 49 is the connection of space science and astronomy and this should be promoted, (iii) consider studying microphysical phenomena in the interplanetary medium, and (iv) keep and strengthen the liaison to the space weather community that was built up by the outgoing President.

7. Outreach Activities

The president summarized the outreach activities performed via SCOSTEP and ISWI in the form of international schools and comic books relevant to C49. O. Malandraki and

C. Briand provided information on the general planning of public outreach activities., i.e. what one should have in mind while preparing and performing a presentation/talk on public outreach. The key elements are:
(1) Explain your work within the frame of the challenge of solving a problem (which is the basis of your research).
(2) Know your audience, i.e. use the appropriate language and knowledge level.
(3) Be prepared to answer all kinds of questions (especially if this talk/presentation is given to the general public) so that the interest of the audience can be sustained.

The outreach strategy needs to be divided and expressed according to the audience:
For public audience (school students, amateur scientists, participants of open science days), presentations, internet updates, and other communications.
For focused audience (audience of university lectures, scientists/colleagues and conferences/meetings) IAU meetings and proceedings are the best options.

Natchimuthuk Gopalswamy
President of the Commission

References

Gopalswamy, N., Mann, I., Bougeret, J.-L.,Briand,C., Lallement, R., Lario, D., Manoharan, P. K., Shibata, K., Webb, D. F. 2012 Commission 49: Interplanetary Plasma and Heliosphere, Transactions of the International Astronomical Union, Series A, 28, 95-124.

Transactions IAU, Volume XXVIIIB
Proc. XXVIII IAU General Assembly, August 2012
Thierry Montmerle, ed.

doi:10.1017/S1743921315005554

DIVISION III COMMISSION 15 PHYSICAL STUDIES OF COMETS AND MINOR PLANETS

(ETUDE PHYSIQUE DES COMETES & DES PETITES PLANETES)

PRESIDENT	**Dominique Bockelée-Morvan**
VICE-PRESIDENT	**Ricardo Gil-Hutton**
PAST PRESIDENT	**Alberto Cellino**
SECRETARY	**Daniel Hestroffer**
ORGANIZING COMMITTEE	**Irina N. Belskaya, Björn J.R. Davidsson, Elisabetta Dotto, Alan Fitzsimmons, Daniel Hestroffer, Hideyo Kawakita, Thais Mothe-Diniz, Javier Licandro, Diane H. Wooden, Hajime Yano.**

PROCEEDINGS BUSINESS SESSION, 29 August 2012

1. Business Session

The business meeting of IAU Commission 15 (C15) took place in Beijing on 29 August 2012, from 14:00 to 18:00, in room 405 of the China National Convention Center. This report of the business meeting of Commission 15 at the 2012 IAU GA is based on the report provided by Alberto Cellino, past president, and on the minutes taken by Daniel Hestroffer, secretary of Commission 15 in the triennium 2009 to 2012, and current secretary.

The President, A. Cellino, opened the meeting, welcoming members and others in the audience interested in the commission activities. The incoming President, Dominique Bockelée-Morvan, the past Division III President, Karen Meech, were in attendance, as were approximately 15 commission members.

The agenda, as presented by Albeto Cellino, was adopted. The first item of the business meeting was a discussion on the name of Commission 15 (Sect. 2). The next items were the presentation of the new officers and Organizing Committee (Sect. 3), the justification for continuation of the Commission Working Groups and Task Groups, and propositions for new Task Groups (Sect. 5).

The outgoing President noted that, in past C15 business meetings, most time was devoted to a report on the new Commission Officers and to reports from the Chairmen of the WGs and Task Groups. This time, the idea has been instead to minimize the time devoted to the above matters, and to devote a longer time to a few scientific presentations devoted to a handful of currently hot topics in our science. The goal was to stimulate discussions and thinking, and to give the participants some time to appreciate how much our discipline is evolving and how it is effective in facing new challenges posed by new discoveries, both from the remote sensing and in situ, and in developing new reassessments of classical topics. The scientific part of the meeting started then, including the following talks:

- Henry Hsieh: Main Belt Comets.
- Dan Boice (chair of WG Physical studies of Comets): What is a Comet? What is an Asteroid?
- Karri Muinonen (Task Group Asteroid magnitudes): On a new photometric system to replace the IAU (H,G).
- Hajime Yano: What we learnt from Hayabusa?
- Peter Jenniskens: Analysis of a sample of the meteorite that fell recently in California in an event which released 4 kTon of energy.

The outgoing President made some comments about the intense scientific activity carried out in recent years in the fields covered by C15. In particular, he noted that the research activity in the field of physical studies of small bodies has been intense during the last triennium, including theory and remote-sensing observations, as well as in situ (Hayabusa, DAWN, Rosetta, Stardust Next, EPOXI) and remote (Herschel, WISE) space observations. This has led to publication of a number of the order of 1,000 scientific papers, a very impressive result. The physical properties of small bodies are recognized to be of outstanding importance to understand the formation and evolution of our Solar System, and more in general of planetary systems. The small bodies are also very important from the point of view of the evolution of the terrestrial biosphere (delivery of water) and for the role played by catastrophic collisions in mass extinction of biological species. The President noted also with satisfaction that during the last triennium new public web facilities have been developed to help researchers in several fields (Asteroid taxonomy, computation of absolute magnitudes, etc.), a much appreciated improvement.

At the end of the meeting, the new President, Dr. Bockelée-Morvan, thanked Dr. Cellino for the good job done during the past triennium. The meeting was officially over at 18:05.

2. Discussion on Commission 15 name

Reminding that the name of the Commission is Physical Studies of Comets and Minor Planets (L'Étude Physique des Comètes et Petites Planètes), Alberto Cellino noted that there are some reasons to think that this name may look somewhat old-fashioned and might be possibly changed, taking into account that the term "minor planets" tends to be immediately associated with the term "asteroid", and much less with the population of Trans-Neptunian objects. Moreover, recent discoveries do suggest that the conventional distinction between asteroids and comets can be much less sharp than previously imagined (existence of so-called Main Belt Comets, and of extinct comets, possible migration of Trans-Neptunian objects). As a consequence, an updated name for this Commission might be something like "Physical Studies of Small Bodies". In the following discussion, somebody noted that the term "comet" should be kept, whereas someone stated that the term "Minor Planet" is perfectly acceptable, and moreover suggested merging C15 and C20. Another participant mentioned that the use of the terminology 'minor planet' is not recommended by the IAU itself; someone else agreed on the possibility to rename the Commission as Physical Studies of Minor Bodies. The President noted, however, that it is always very difficult to reach a general agreement on this kind of issues, and in any case, whatever may be the suggestion issued by this meeting, the final decision is under the responsibility of Division III. The suggestion of C15 is therefore that of inviting Div.III to open an internal discussion about the possibility to change the name of C15 taking into account all pros and cons.

3. Officers and Organizing Committee

The President showed the list of the C15 Officers for the next triennium, as they have been decided by the Organizing Committee (OC):

• *New President :* Dominique Bockelée-Morvan

• *New Vice-President :* Ricardo Gil-Hutton

• *Secretary :* Dr. Bockelée-Morvan informs that she will confirm D. Hestroffer in this task.

• *Members of the Organizing Committee :* Irina N. Belskaya (new), Björn J.R. Davidsson (confirmed), Elisabetta Dotto (confirmed), Alan Fitzsimmons (confirmed), Daniel Hestroffer (new), Hideyo Kawakita (new), Thais Mothé-Diniz (confirmed), Javier Licandro (new), Diane H. Wooden (confirmed).

The outgoing President informed that an additional member of the OC who had been elected for the next triennium, Dr. Shinsuke Abe, has regretfully declined the appointment, because he is already member of the OCs of two other IAU Commissions. Since Dr. Hajime Yano has expressed his willingness to possibly replace Dr. Abe in the C15 OC, Dr. Cellino said that he is personally in favour of this, but since new OC members are elected by the OC, he invited the new President to discuss this issue with the new OC in the near future. Dr. Yano was unanimously supported by the new OC to replace Dr. Abe in C15 OC, via an e-mail consultation carried out in September 2012, and the IAU secretary was informed.

The President thanked the members of the OC who have now ended their duty: D. Bockelée-Morvan, P. M. M. Jenniskens, D. F. Lupishko, and G. Tancredi.

4. Commission Members

The outgoing President informed that IAU C15 includes at present 383 members (according to the IAU web page). The IAU Secretariat has informed that 54 new C15 Members are being appointed for the next Triennium. A few previous members resigned due to health problems. In addition, three other well known colleagues have asked to be included as members of C15: Daniela Lazzaro, Patrick Michel and Hajime Yano. The President approved after having checked that all of them are already IAU members. Their C15 membership will be communicated to the IAU Secretariat.

The President reminded also that during the 2009-2012 triennium, eight eminent scientists, members of C15, passed away: W. Ian AXFORD (1933 - 2010), Zdenek CEPLECHA (1931 - 2009), Tom GEHRELS (1925 - 2011), Charles KOWAL (1940 - 2011), Brian MARSDEN (1937 - 2010), Douglas O. REVELLE (1945 - 2010), Vladimir SHKODROV (1930 - 2010), Andrzej WOSZCZYK (1935 - 2011). The meeting observes a minute of silence.

5. Working Groups and Task Groups

5.1. *Status and activity of current groups*

Commission 15 has two Working Groups (WG):

• Physical Studies of Asteroids;

• Physical Studies of Comets;

As well as four Task Groups (TG):

• Asteroid magnitudes;

- Cometary Magnitudes;
- Asteroid polarimetric Albedo Calibration;
- Geophysical and Geological Properties of near-Earth Objects.

The President summarized the state of the existing C15 WGs and TGs. In particular, he suggested that the two existing WGs (on asteroids and comets) will continue in the next triennium. The President informed that the Working Group Chairmen for the next triennium will be: David Tholen (Physical Studies of Asteroids), who will replace Dr. Gil-Hutton, and Daniel C. Boice, confirmed for the WG on Physical Studies of Comets.

The **Asteroid magnitudes Task Group** has produced a proposal to improve the currently adopted (H,G) system with a new three-parameter system. A 2010 paper published on Icarus has suggested that the new proposed photometric system is significantly better than the (H,G) system. Moreover, a student of Dr. Karri Muinonen has already applied the new photometric system to a large sample of more than 400,000 asteroids. A presentation by Dr. Muinonen clarified the situation in the second part of the business meeting. As a conclusion, it was decided that this TG should continue to exist. The final decision whether to adopt the newly proposed (H, G1, G2) system will be taken by Div. III level. C15 is in favor of adopting the new system, since this proposal is issued by the activity of one of its Task Groups

Task Group Cometary magnitudes: No significant activity during the last triennium. This TG is removed.

Task Group Calibration of polarization-albedo: some results have been obtained, according to the President. Work is still in progress and this TG should continue to exist.

Geophysical and Geological Properties of NEOs: the President noted that the activity of this TG has not been very strong during the last two triennia, in spite of the successful execution of important space missions. The original idea of this TG was to create a large database of geophysical and geological information on NEOs, but this has hardly been done. Moreover, the President noted that there is already a very active Div.III WG on NEOs (although, not totally overlapping). The Chairman of this C15 TG agreed with the President that this TG can be suppressed.

All the above suggestions by the President were approved. The TG on cometary magnitudes and Geophysical and Geological Properties of NEOs are now officially removed.

5.2. *New Task Groups*

A new WG concerning cometary chemistry has been proposed by Dr. Irakli Simonia, who had not a chance to be present at the IAU GA. Dr. Bockelée-Morvan took care of taking contacts with Dr. Simonia in order to understand precisely the details of this proposal. She made a presentation of the proposal, which is aimed at establishing a large database of cometary spectroscopic data in the visible wavelength range, since these data are currently scattered and often difficult to retrieve. Dr. Bockelée-Morvan concluded that this is a very ambitious program requiring a lot of manpower. As a consequence, she thinks that, as a pre-requisite to even begin such a task, it would be advisable to set up first a more limited Task Group with the purpose of judging whether the cometary community is interested in such a program, whether authors having their own data would be willing to contribute, and to assess in general how the database should be designed and what should be requested to include. This suggestion was approved by the C15 members present at the business meeting. The membership of this new Task Group will be decided in the near future.

The outgoing President proposed to create a new C15 TG on "Asteroid Families" to set up some commonly accepted criteria and procedures to be applied to the identification of groupings of objects sharing a common collisional origin. The membership of this TG is still to be defined, but it is clear that it should include specialists from C20 (who also supports this proposal at Div.III level). Due to the interdisciplinary nature of this subject, the C15 suggests that Div.III might discuss the possibility to set up this as a new Div.III WG, including both C15 and C20 members. In the case that this WG is not accepted at Div.III level, C15 will set up a dedicated Task Group on this subject.

6. WEB page of Commission 15

Alberto Cellino noted that during the last triennium there has been an effort to completely re-design the web page of Commission 15. This has been done by Mrs. S. Rasetti at the Observatory of Torino, in collaboration with the Commission President and Secretary. The redesigned web page can be found at URL http://iaucomm15.oato.inaf.it/. He noted that, in the case that IAU Commission web pages are not moved to some fixed IAU web site (a fixed an editable site is preferable than a site moving every 3 years), the Torino Observatory, in particular Mrs. Rasetti, is willing to continue to maintain and possibly improve the C15 web page and Forum for the next triennium, at least. This was approved by the new President. This new web page includes also a link to a brand new Forum, which has been developed with the aim of possibly becoming a useful tool for C15 members. The idea is that, by registering in the Forum, each member will have the possibility to successfully interact with other members on matters of common interest, writing texts containing a variety of useful and always up-to-date information, and will be automatically informed whenever a new message is posted. In this way, it should be possible to create a kind of dedicated network to make it easier the interactions between C15 members, to encourage exchange of ideas, sharing of telescope time, organization of meetings, etc. So far, however, the Forum has not been used very much. Some details were given by a short presentation by the C15 Secretary, D. Hestroffer.

Dominique Bockelée-Morvan
President of the Commission

Transactions IAU, Volume XXVIIIB
Proc. XXVIII IAU General Assembly, August 2012
Thierry Montmerle, ed.

doi:10.1017/S1743921315005566

DIVISION III COMMISSION 22

METEORS, METEORITES AND INTERPLANETARY DUST

MÉTEORES, MÉTÉORITES ET POUSSIÈRE INTERPLANÉTAIRE

PRESIDENT **Peter Jenniskens**
VICE-PRESIDENT **Jiří Borovička**
PAST PRESIDENT **Junichi Watanabe**
ORGANIZING COMMITTEE

Guy Consolmagno, Tadeusz Jopek(Secretary), Jeremie Vaubaillon, Shinsuke Abe, Diego Janches, Galina Ryabova, Masateru Ishiguro, Jin Zhu

PROCEEDINGS BUSINESS MEETING on 24 August 2012

1. Introduction

The business meeting of commission 22 was held at the room 403 on 24 th August 2012 (14:00-15:30) in the China National Convention Center in Beijing.

Nineteen people attended at this meeting: G.Consolmagno, D.Green, P. Jenniskens, J. Watanabe, J. Zhu, T. Jopek, K. Meech, D. Kinoshita, S. Abe, D. Boice, H. Yano, I. Sato, C. Smith, K. Churymov, J.C. Zhang, R. Dayong, G. Valsecchi, A. Ivantsov, C. Hohenkerk.

This meeting was managed by Junichi Watanabe, the current C22 President. The summary of the meeting is described.

2. Membership

The membership of Commission 22 increased from 132 members at the end of the XXVII General Assembly in Rio de Janeiro (2009) to 143 members at the end of the current XXVIII General Assembly. The 25 new members (24 as new IAU members and one member already registered) are:

Benkhoff Johannes Josef (Germany), Busarev Vladimir V. (Russian Fed.), Čapek David (Czech Rep.), Fromang Sebastien (France), Granvik Mikael Matias S. (Finland), HASEGAWA Sunao (Japan), Husárik Marek (Slovakia), Janches Diego (USA), Kaňuchová Zuzana (Slovakia), Kikwaya Jean-Baptiste (Vatican City Sta), KIMURA Hiroshi (Japan), Kornoš Leonard (Slovakia), Koschny Detlef V. (Germany), Kostama Veli-Petri (Finland), Mawet Dimitri Paul (Belgium), Moór Attila (Hungary), Santos-Sanz Pablo (Spain),

SHANG Hsien (China), Shrbený Lukáš (Czech Rep.), Stewart-Mukhopadhyay Sarah T. (USA), Yair Yoav Y. (Israel), Zadnik Marjan G. (Australia), Zender Joe J. (Germany), Masateru Ishiguro (Korea).

Four members deceased since the last General Assembly: Zdenek CEPLECHA (1929–2009), Bertil Anders LINDBLAD (1921–2010), Bernard LOVELL (1913–2012), Douglas O. REVELLE (1945–2010).

All four had an exceptionally strong influence on our field and will be missed. The meeting stood in silence in remembrance of the deceased C22 members.

3. New executive

The new executive of the Commission 22 for the next triennium has been approved as follows:

President:Peter Jenniskens (USA), Vice-President: Jiří Borovička(Czech Republic), Secretary: Tadeusz Jopek (Poland), and Past President: Junichi Watanabe(Japan).

The members of the organizing committee are: Guy Consolmagno(Vatican), Jeremie Vaubaillon(France), Shinsuke Abe(Taiwan), Diego Janches(USA), Galina Ryabova(Russia), Masateru Ishiguro(Korea), Jin Zhu(China).

4. Activities since the last General Assembly

At the C22 business meeting, several activities were introduced and discussed:

4.1. *Organized and supported scientific meetings*

We supported the following meetings during the past triennium:

• 2008 TC3 Workshop in University of Khartoum, Sudan, on December 5–6, 2009.

• METEOROIDS 2010 conference in Breckenridge, Colorado, USA, on May 24–28, 2010.

• Asteroids, Comets, Meteors conference (ACM 2012) in Niigata City, Japan, on May 16–20, 2012.

We organized a Joint Discussion at the XXVIII GA. It was noteworthy that this was the first organized by C22 in the past 20 years of IAU GA history:

• Joint Discussion 5 gFrom meteors and meteorites to their parent bodies: Current status and future developmentsh Beijing, China on August 22–24 during XXVIII GA

4.2. *E-mail newsletter of C22*

In order to share the information of the IAU or related topics of our commission, President Junichi Watanabe started an e-mail newsletter. Twenty newsletters have been issued during the past triennium. These newsletters were found to be an effective way to communicate with commission members.

It was suggested to continue this newsletter in the new triennium. The list of all C22 members is available at the web page of the Commission, but not all email addresses are known by the commission.

4.3. *Web page of the Commission*

During the past triennium, the web page of C22 was maintained by the secretary of C22, J. Zhu, at the address http://www.iau-c22.org/. For practical reasons it would be preferable to move the C22 web page to the IAU Meteor Data Center website maintained by T. Jopek. For that reason, Jopek was assigned to be the Secretary of C22 in this triennium. In future years, however, we may decide to keep the website at the IAU Meteor Data Center website.

5. Commission report

A brief report of Commission 22 was submitted to the Division III president, K. Meech by the president of C22 J. Watanabe. The report was prepared in a short form, concentrating on Commission 22 activities only, instead of giving a full summary of activities in our field.

6. Established meteor showers

P. Jenniskens, on behalf of the Working Group on Meteor Shower Nomenclature, presented the list of 31 showers that were selected by the Working Group to receive the predicate gestablishedh at this GA. Some 85 candidate showers were submitted to the Meteor Data Center and shown in green highlight at the Meteor Data Center website prior to July 1 of this year. These proposed showers were evaluated in the weeks leading up to the meeting. A final evaluation of the candidate showers was made just before the business meeting by the Working Group members present at the GA. They reached a consensus that 31 candidates out of the proposed list do exist. Most others were showers only detected by one radar source. The Working Group consensus was to look for confirmation from other radar or optical sources before adding those showers to the list of established showers. There was no objection to any of the showers in the proposed list of 31. As a result, 31 new established meteor showers were added to the list of established showers (http://www.astro.amu.edu.pl/%7Ejopek/MDC2007/index.php). The total number of established meteor showers is 95. One name correction to an established shower was also approved. The December Comae Berenicids (COM) will from now on be known as the Comae Berenicids (COM). This change was proposed because it was discovered that the January Comae Berenicids are in fact the same shower.

7. Working Groups

The Working Group on Professional-Amateur Cooperation in Meteors reported no activity in the past triennium, possibly because there are generally good collaborations between professional and professional astronomers in the field. To help maintain this situation, Detlef Koschny (Germany) will become the new chairman. The membership of the new group was left undecided. David Asher has agreed to continue to be a member of this working group. Because none of the other current members were present at the business meeting, Koschny was tasked to establish the new membership, working with the new president of C22.

On the Working Group on Meteor Shower Nomenclature for the next triennium, present chair, P. Jenniskens, recommended the new membership as follows, and it was approved.

Tadeusz Jopek (Poland, chair), Peter Brown (Canada, vice-chair), Peter Jenniskens (USA, C22 president), Zuzana Kanuchova (Slovakia), Pavel Koten (Czech), Javor Kac (Slovenia, IMO), Junichi Watanabe (Japan), Jack Baggaley (New Zealand), Gulchekhra Kokhirova (Tajikistan), Joseph M. Trigo-Rodriguez (Spain).

This working group is very active. It maintains the archive of meteor shower nomenclature, considers proposed shower names from new surveys, and is tasked to identify what next set of showers to name officially as being established in the next IAU General Assembly.

8. Future conferences

J. Watanabe informed about the future conferences of interest to C22 members:

- METEOROIDS 2013 conference will be held in Poznań, Poland, on August 26–30, 2013. http://www.astro.amu.edu.pl/Meteoroids2013/index.php
- Asteroids, Comets, Meteors conference (ACM 2014) will be held in Helsinki, Finland, in 2014. http://www.helsinki.fi/acm2014/
- Annual Meeting of the Meteoritical Society in Edmonton, Alberta, July 29–August 2, 2013.
- Annual Meeting of the Meteoritical Society in in Casablanca, Moroccoin, September 7–14, 2014

Junichi Watanabe
Outgoing President of the Commission

Transactions IAU, Volume XXVIIIB
Proc. XXVIII IAU General Assembly, August 2012
Thierry Montmerle, ed.

doi:10.1017/S1743921315005578

DIVISION IV COMMISSION 36 THEORY OF STELLAR ATMOSPHERES

(THÉORIE DES ATMOSPHÈRES STELLAIRES)

PRESIDENT Joachim Puls
VICE-PRESIDENT Ivan Hubeny
PAST PRESIDENT Martin Asplund
ORGANIZING COMMITTEE France Allard, Carlos Allende Prieto, Thomas R. Ayres, Mats Carlsson, Bengt Gustafsson, Rolf-Peter Kudritzki Tatiana A. Ryabchikova

PROCEEDINGS BUSINESS SESSIONS, 27 August 2012

The business meeting of IAU Commission 36 took place during the GA in Beijing on August 27th, and its major topic was the re-structuring of the IAU Divisions and consequences for our Commission. The meeting was conducted by the new president, Joachim Puls, since the past president (still in charge during the GA), Martin Asplund, could not participate.

1. The new Organizing Committee

At first, Jo Puls expressed his thanks to those OC members who were leaving the OC, Svetlana V. Berdyugina, Lyudmila I. Mashonkina, Sofia Randich, and Hans G. Ludwig, for all their work, input and encouragement, and in particular, to Martin Asplund for his last-term presidency.

Then, the new vice-president, Ivan Hubeny, and four new OC members were introduced, namely France Allard, Mats Carlsson, Rolf-Peter Kudritzki, and Tanya Ryabchikova, where these colleagues had been elected by the members of Commission 36 during a very successful poll in spring 2012. (For remaining OC members, see above).

2. Discussion on the new IAU structure

The meeting continued with a presentation given by Jo Puls, concentrating on the changes which had been suggested (and approved at the end of the GA) regarding a new IAU Division structure, and potential consequences for our Commission 36 (see also IAU Information Bulletin 109). Moreover, he summarized the outcome of the business meeting of Div. IV/V, where it had been suggested that the Commissions within Div. IV and V (to be unified within the new Division G, 'Stars and Stellar Physics', with approx. 2100 members) may overthink their present positions and goals, and might consider a merging with other Commissions to strengthen their impact.

Because of the upcoming re-evalution of all present Commissions by the new 'Division Steering Committee' (DSP), Jo Puls pointed out that there is a lot to overthink and discuss within the OC and the Commission, accounting for the fact that within the new concept Commissions can be created and terminated on very short terms, if they turn out to be no longer efficient or useful.

Insofar, Jo Puls expressed his concerns that our Commission, particularly if we would decide to remain independent, needs to focus its activities, and provided related suggestions. Particularly, the present Commission's web-site, `http://www.usm.uni-muenchen.de/IAU-Comm-36/` might need a significant update, which would allow to spread info about our activities also to the outside, and could be particularly helpful for younger people in our community. During the following discussion, it was agreed upon to envisage a long-lasting site, independent from the actual OC members, which of course is a difficult task. Indeed, Dainis Dravins explained that an informative and useful site already existed in the past, but has 'vanished' with a changing OC.

Chris Corbally mentioned a certain chance that such web-sites could be hosted directly by the IAU, which would be ideal. Most important, however, is to have an administrator who is independent from the OC, and Jo Puls suggested to search for a 'volunteer' who is a member of the Commission (not necessarily of the OC) with a permanent position, which would allow for some longer-lasting plans and activities.

There was also agreement on another potential activity of our Commission: the OC might become involved in suggestions for conferences, symposia etc., if there is the feeling that some initiative were required.

The business meeting ended after approx. 1.5 hours with the understanding, that all this needs to be discussed in detail within the OC and the members. Corresponding actions have been already prepared.

Joachim Puls
President of the Commission

Transactions IAU, Volume XXVIIIB
Proc. XXVIII IAU General Assembly, August 2012
Thierry Montmerle, ed.

doi:10.1017/S174392131500558X

DIVISION V
COMMISSION 42 — CLOSE BINARIES
(ETOILES BINAIRES)

PRESIDENT — Ignasi Ribas
VICE-PRESIDENT — Mercedes T. Richards
PAST PRESIDENT — Slavek Rucinski
ORGANIZING COMMITTEE — David H. Bradstreet, Petr Harmanec, Janusz Kaluzny, Joanna Mikolajewska, Ulisse Munari, Panagiotis Niarchos, Katalin Olah, Theodor Pribulla, Colin D. Scarfe, Guillermo Torres

PROCEEDINGS BUSINESS SESSION, 24 August 2012

1. Introduction

Commission 42 (C42) co-organized, together with Commission 27 (C27) and Division V (Div V) as a whole, a full day of science and business sessions that were held on 24 August 2012. The program included time slots for discussion of business matters related to Div V, C27 and C42, and two sessions of 2 hours each devoted to science talks of interest to both C42 and C27. In addition, we had a joint session between Div IV and Div V motivated by the proposal to reformulate the division structure of the IAU and the possible merger of the two divisions into a new Div G. The current report gives an account of the matters discussed during the business session of C42.

2. The Organizing Committee

The new Organizing Committee of C42 was elected through a nomination and e-mail vote procedure carried out by the Organizing Committee during the months prior to the IAU General Assembly in Beijing. The new OC of C42 will consist of Mercedes Richards (President), Theodor Pribulla (Vice-President), Ignasi Ribas (Past President) and members David Bradstreet, Horst Dreschel, Carla Maceroni, Joanna Mikolajewska, Ulisse Munari, Andrej Prša, Colin Scarfe, John Southworth, Shay Zucker and Tomaz Zwitter. The outgoing OC members are Petr Harmanec, Janusz Kaluzny, Panos Niarchos, Katalin Olah, Slavek Rucinski, and Guillermo Torres. All of them are warmly thanked for their dedication and outstanding contributions to the smooth running of C42.

3. Agenda of the business session

Four main points were covered in the business session, namely a brief account of the main activities during the triennium 2009-2012, a recognition of the members who had recently passed away, a review of the Bibliography of Close Binaries publication, the presentation of a proposal for an IAU resolution on solar units, and the possible creation of a new inter-Division Working Group.

The activities in which C42 has participated during 2009-2012 are reported in Transactions XXVIIIA. An important event that is worth mentioning here is IAU Symposium 282 "From Interacting Binaries to Exoplanets: Essential Modeling Tools " held in 2011 in Slovakia, which very successfully explored the synergies between the fields of close binaries and exoplanets.

During the past year, C42 has been reviewing the impact of our sponsored publication Bibliography of Close Binaries (BCB). Continuous advances in all areas of information technologies make it necessary to evaluate whether the time and effort expended into bringing new BCB issues to light every 6 months are rewarded with a significant audience. A critical analysis was performed by Andras Holl, from Konkoly Observatory, who manages the C42 website. The report indicates that BCB is downloaded on average 70 times per month and each BCB issue is downloaded a total of about 500 times. It is clear that BCB continues to play an important role in present time in spite of the numerous astronomical databases and new means of communication. Given the flood of information and the inability to cope with all the new publications, a good classification of research papers going beyond a mere object index is of great value. Thus, C42 will continue to sponsor BCB and encourage the chief editor Colin D. Scarfe and the rest of his team to continue their excellent work.

The advent of photometric data of exquisite precision has created a need to be extra careful with the use of astronomical constants. This was pointed out by Harmanec & Prša (2011), who called the community to adopt a new set of astronomical parameters to improve the determination of stellar physical properties. The authors also approached C42 to explore the possibility of presenting an IAU Resolution for approval at the IAU GA. This possibility was met with strong interest by the OC of C42 but it was suggested to establish contact with several Divisions and Commissions with ties to the numerical quantities that are proposed to be redefined. Most notably, some of the changes affected values for key parameters of the Sun. A round of consultations made it clear that there would not be sufficient time to put together a resolution text that could be agreed on by other Divisions and Commissions. It was then decided that the most appropriate course of action would be the creation of an inter-Division Working Group with the mandate to come up with a new set of parameter values that would achieve broader acceptance. Such a Working Group could prepare a resolution text to be approved at the next IAU General Assembly in 2015. In the event of rapid progress, the relevant Commissions and Divisions could release a document with the new parameter definitions to be used in the interim. All these plans were presented during the business session and a brief discussion ensued. There was a clear consensus that moving forward with this initiative was very important, and there was overall support for the creation of a new inter-Division Working Group.

References

Harmanec, P. & Prša, A., 2011, *PASP*, 123, 976

Ignasi Ribas
President of the Commission

Transactions IAU, Volume XXVIIIB
Proc. XXVIII IAU General Assembly, August 2012
Thierry Montmerle, ed.

doi:10.1017/S1743921315005591

DIVISION VII COMMISSION 37 STAR CLUSTERS AND ASSOCIATIONS *(AMAS STELLAIRES ET ASSOCIATIONS)*

PRESIDENT Giovanni Carraro
VICE-PRESIDENT Richard de Grijs
PAST PRESIDENT Bruce Elmegreen
ORGANIZING COMMITTEE Dante Minniti, Simon Goodwin, Peter Stetson Douglas Geisler, Barbara Anthony-Twarog

PROCEEDINGS OF THE BUSINESS SESSION, 24 August 2012

1. Introduction

The business session for Commission 37 was held on 24 August 2012 at the IAU General Assembly in Beijing. The meeting was attended by about a dozen members of our Comission, including President Carraro, VP de Grijs and several committee members. We introduced ourselves and then went through a powerpoint presentation first prepared by outgoing President Elmegreen and revised by incoming President Carraro. The contents of the powerpoint presentation are given in this summary.

In what follows, we list past, present and future meetings, publications statistics and important surveys, reviews, and databases about clusters, and then we discuss the procedure for the election of new commission officers.

2. Past Meetings

- Encounters and interactions in dense stellar systems: modeling, computing, and observations, Beijing (China), Aug 30–Sep 2, 2010
- IAU Symposium 270: Computational Star Formation, Barcelona (Spain), May 31–Jun 4, 2010
- A Universe of Dwarf Galaxies, Lyon (France), June 14–18, 2010
- Spiral Structure in the Milky Way: Confronting Observations and Theory, Bahia Inglesa (Chile), Nov 7–10, 2010
- RR Lyrae Stars, Metal-Poor Stars and the Galaxy, Pasadena (USA), Jan 23–25, 2011
- Dynamics of Low-Mass Stellar Systems : From Star Clusters to Dwarf Galaxies, Santiago (Chile), Mar 4–8, 2011
- Assembling the Puzzle of the Milky Way, Le Grand-Bornand (France), Apr 17–22, 2011
- Stellar Clusters and Associations, Granada (Spain), May 23–27, 2011

- From Star Clusters to Galaxy Formation - The Virtual Universe, Heidelberg (Germany), Sep 20–23, 2011
- 42nd Saas-Fee Course: Dynamics of Young Star Clusters and Associations, Villars-sur-Ollon (Switzerland), Mar 25–31, 2012
- Galaxies: Origin, Dynamics, Structure, Sochi (Russia), May 14–18, 2012
- EWASS 2012 - Symposium 6: Stellar Populations 55 years after the Vatican conference, Rome (Italy), Jul 2–4, 2012
- The Formation and Early Evolution of Stellar Clusters, Sexten (Italy), Jul 23–27, 2012

3. Meetings at the IAU General Assembly

- The Formation and Early Evolution of Stellar Clusters, IAU GA 2012 SpS1, Beijing (China), Aug 20–24, 2012
- IAU Symposium 289: Advancing the Physics of Cosmic Distances, IAU GA 2012, Beijing (China), Aug 27–31, 2012

4. Upcoming Meetings

- Workshop on the Magellanic Clouds, Perth (Australia), Sep 10–13, 2012
- 30 Doradus: The Starburst Next Door, Baltimore (USA), Sep 16–19, 2012
- Ecology of Blue Straggler Stars, Santiago (Chile), Nov 5–9, 2012
- Small Stellar Systems in Tuscany, Prato (Italy), Jun 10–14, 2013
- Massive Stars: From alpha to Omega, Rhodes (Greece), Jun 10–14, 2012

5. Publications

The topic of star clusters and associations continues to be one of the most widely followed in all of astronomy. It spans the range of interest from stellar properties, to stellar clusters, to star formation and evolution, with considerable overlap in other commissions.

Publications in Refereed Journals in the period from January 2009 to August 2012 tally as follows:

- Globular Clusters: 800 papers (more than 6400 citations)
- Open Clusters: 450 papers (more than 3000 citations)
- Stellar Associations: 65 papers (more than 540 citations)

Some of the **issues** addressed in these publications are:

- the formation and dynamical evolution of star clusters
- stellar evolution and ages
- star clusters as tracers of stellar populations
- studies of specific types of objects within clusters
- nuclear clusters
- extragalactic cluster systems

The authors utilize observations covering an increasing portion of the electromagnetic spectrum, ranging from X-rays to the far-infrared, as well as advanced N-body simulations.

Reviews related to star clusters that appeared in the bibliography from 2009 to August 2012 are:

• A review on Young Massive Star clusters by Simon F. Portegies Zwart, Stephen L.W.McMilland and Mark Gieles, Annual Reviews of Astronomy and Astrophysics, Vol. 48, Issue 1, pp.431-493 (2010)

• Additional reviews regarding a variety of aspects of star cluster research have also appeared in the proceedings of the conferences and meetings mentioned earlier

6. DataBases

Several new cluster **catalogues** have been published:

• M33: Ma, Jun, 2012, AJ, 144, 41

• PHAT Stellar Cluster Survey, Jonhson *et al.* 2012, ApJ 752, 95

• M31 Galex catalog, Kang *et al.* 2012, ApJS 199, 37

• NGC 5128 star clusters, Harris *et al.* 2012, 143, 84

Recent **databases** on clusters are:

• Data on Open Clusters in the Milky Way and the Magellanic Clouds can be found in the WEBDA site (http://www.univie.ac.at/webda/), which was originally developed by Jean-Claude Mermilliod from the Laboratory of Astrophysics of the EPFL (Switzerland) and is now maintained and updated by Ernst Paunzen from the Institute of Astronomy of the University of Vienna (Austria).

• Data on Galactic Globular Clusters can be found in the "Catalog of parameters for Milky Way globular clusters" by W.E. Harris (http://www.physics.mcmaster.ca/Globular), as well as in "The Galactic Globular Clusters Database" at Astronomical Observatory of Rome (INAF-OAR: http://venus.mporzio.astro.it/ marco/gc/).

• A Catalogue of Variable Stars in Globular Clusters developed and maintained by Christine Clement can be found in http://www.astro.utoronto.ca/ cclement/.

7. Ballot Procedure for Election of New Officers

An election of new officers was held in March and April 2012 by the outgoing president, vice president, and organizing committee.

Balloting was carried out for the first time via a web-based tool. A call for nominations was sent by email. A list of Commission 37 members who were nominated and who accepted the nomination was then put to the vote. We had 5 VP candidates and 18 Commission organizing committee candidates The voting took place via the a website set up by IAU. We had 118 votes, which means that 37% of the commission members voted, a record so far!

This demonstrated that the web-based ballot, previously strongly requested, worked very well.

We attempted to maintain a balance in the Organising Committee with respect to theory versus observations, gender and geographical location. This was difficult to achieve, especially gender, in spite of having several female candidates. In the future a better coordination has to be discussed before.

We would have found it useful to have a list of previous members of the Organizing Committee, including previous presidents and vice-presidents. This list could be on the IAU website. Such a list is important because IAU rules do not allow people to serve for more than one term for president and vice-president, or a maximum of two terms for OC members.

We upgraded our Commission web site. We would like to include an updated list of members of the Commission, and a list of former Organizing Committee members, Presidents and Vice-Presidents. We also want updated links to websites of conferences endorsed by the Commission. We should also create a site that is accessible to the Organizing Committee only, to facilitate OC business management (e.g. evaluation of conference proposals).

8. Closing Remarks

Our discussion at the meeting centered on improving communication between members. This includes getting a more complete list of email addresses, and regularly sending out notices of meetings, reviews, databases, and other material of interest to Commission 37 members. We note that the email newsletter, SCYON (edited by H. Baumgardt, E. Paunzen and P. Kroupa), is very successful at spreading important information and research news.

Another discussion we had was how to make the Commission more relevant among astronomers, since so far it is quite disconnetcted from the community. Only a few professianl consult the webpage and address P and VP with questions or requests.

Giovanni Carraro
President of the Commission

Transactions IAU, Volume XXVIIIB
Proc. XXVIII IAU General Assembly, August 2012
Thierry Montmerle, ed.

doi:10.1017/S1743921315005608

DIVISION IX

COMMISSION 30 — RADIAL VELOCITIES

VITESSES RADIALES

PRESIDENT — Guillermo Torres
VICE-PRESIDENT — Dimitri Pourbaix
PAST PRESIDENT — Stephane Udry
ORGANIZING COMMITTEE — Geoffrey W. Marcy, Robert D. Mathieu, Tsevi Mazeh, Dante Minniti, Claire Moutou, Francesco Pepe, Catherine Turon, Tomaz Zwitter

PROCEEDINGS BUSINESS SESSION, 28 August 2012

1. Business

The meeting was attended by the President and Vice-President of the Commission, along with approximately 15 other members. The President reported on the election of new officers that took place at the end of March 2012, for four new members of the Organizing Committee as well as a new Vice-President, and thanked the outgoing members. Tomaz Zwitter (Slovenia) was elected as the new VP (2012–2015), and the new OC members for the period 2012–2018 are Alceste Bonanos (Greece), Alain Jorissen (Belgium), David Katz (France), and Matthias Steinmetz (Germany). The current VP, Dimitri Pourbaix, became the President through 2015.

Ten individuals became new members of the Commission, and another 24 astronomers applied for IAU membership and chose to be listed as Commission 30 members, pending approval of their IAU membership. This brought the total membership of C30 to 167.

The President also reported on the activities of C30 during the previous triennium, highlighting large-scale radial-velocity surveys, studies of the role of RV measurements for investigations of stellar angular momentum evolution and stellar age, radial velocity surveys in open clusters, efforts toward achieving higher radial-velocity precision, applications of high-precision RV studies to binary stars, and research on the Doppler beaming effect.

A discussion was also held about the future of C30 in the context of the restructuring of Divisions within the IAU. The President pointed out that the scientific scope of C30 is quite broad, including RVs of stars, galaxies, and interstellar gas, although in practice the emphasis has been almost exclusively on stars. Past efforts to bring the stellar and galactic communities together have not had much success (e.g., "RVs" versus "redshifts"). The RV technique is now well established: the past 20 years have seen the precision go from $\sim$300 m s^{-1} to $\sim$1 m s^{-1}, driven by exoplanet searches. The small audience expressed their opinions on the various options for the continuation of C30, and in the end there was general consensus that it would be best for the Commission to be attached to a technical Division rather than to one associated with a specific type of object (e.g., stars).

2. Working groups

The President and Vice-President reported on the activities of the three WGs of the Commission. The WG on Radial-Velocity Standards has been mostly inactive in the last three years. Stephane Udry stepped down as the WG chair, and the position was assumed by Gérard Jasniewicz. As it turns out, however, other groups or individuals in the community not affiliated with the WG have effectively been compiling very useful lists of RV standards. One example is the work of Crifo *et al.* (2010), and another is the work of Chubak et al. (2012). The President gave a brief presentation about the latter effort. Despite its small size, the WG on Stellar Radial Velocity Bibliography, headed by Hugo Levato, has continued its tedious work of compiling all bibliographic references with RV data. The Vice-President expressed his support and recognition for the work accomplished by Levato and his team. He then reported that the activities of the WG on the Catalogue of Orbital Elements of Spectroscopic Binary Systems (SB9) that he leads continues, but that the level of completeness of the SB9 is still difficult to assess although the number of daily queries indicates that, even if incomplete, the resource is certainly being used very often by the community.

3. Other matters

Jos de Bruijne gave a brief presentation about his quest, in the framework of the Gaia astrometric mission, to secure at least one accurate radial velocity measurement for about 100 bright stars that do not appear to have ever been measured, based on a careful literature search. The motivation is that their RVs are needed in order to derive highly accurate proper motions for some of the Hipparcos stars that are single, using first-epoch Gaia positions. The level of RV accuracy required varies from star to star. de Bruijne used his presentation as a way of contacting the RV community for help on this project, and it was agreed to post a link to his paper on this subject on the C30 website.

Dainis Dravins reported on his work on large-scale simulations of stellar convection, in order to estimate its effects on the measured radial velocities. He pointed out that regardless of whether the net convective shift turns out to be blue or red, its amplitude is significantly larger than the radial-velocity precision quoted for current instrumentation, making the velocity *accuracy* difficult to quantify.

Finally, Matthias Steinmetz presented the status of the Radial Velocity Experiment (RAVE), a project to measure the velocities of about half a million stars in the southern hemisphere.

Guillermo Torres
President of the Commission

Transactions IAU, Volume XXVIIIB
Proc. XXVIII IAU General Assembly, August 2012
Thierry Montmerle, ed.

doi:10.1017/S174392131500561X

DIVISION XII
COMMISSION 6 ASTRONOMICAL TELEGRAMS
(TELEGRAMMES ASTRONOMIQUES)

PRESIDENT **N. N. Samus**
VICE-PRESIDENT **H. Yamaoka**
PAST PRESIDENT **A. C. Gilmore**
ORGANIZING COMMITTEE **K. Aksnes, D. W. E. Green, B. G. Marsden, S. Nakano, Martin Lara, Elena V. Pitjeva, T. Sphar, J. Ticha, G. Williams**

PROCEEDINGS BUSINESS SESSIONS, 24 August 2012

1. Introduction

IAU Commission 6 "Astronomical Telegrams" had a single business meeting during the Beijing General Assembly of the IAU. It took place on Friday, August 24, 2012. The meeting was attended by five C6 members (N. N. Samus; D. W. E. Green; S. Nakano; J. Ticha; and H. Yamaoka). Also present was Prof. F. Genova as a representative of the IAU Division B. She told the audience about the current restructuring of IAU Commissions and Divisions and consequences for the future of C6.

2. Program

The participants stood to the memory of the C6 members deceased since the Rio de Janeiro General Assembly of the IAU, Jan Hers and Brian Marsden. A presentation on the scientific biography of the outstanding scientist Brian Marsden, many-year Director of the Central Bureau of Astronomical Telegrams (CBAT), was made by Dr. D. Green.

The leaving C6 President Nikolay Samus and the CBAT Director Dan Green reported on the C6 activity in 2009-2012.

3. Discussion

Specially discussed were problems related to the recent change of the Cambridge office of the CBAT and delays of official announcements of GCVS names for Novae.

The Commission elected Hitoshi Yamaoka the new C6 President for 2012-2015. Dan Green was elected the C6 Vice-President. The C6 business meeting also elected Jana Ticha as the candidate for 2015-2018 presidentship, for the case if Dan Green will not agree for this nomination, being the CBAT Director.

New organizing Committee was consisted by N. N. Samus (Pase President); K. Aksnes; A. C. Gilmore; S. Nakano; T. Sphar; J. Ticha; and G. Williams.

Nikolay N. Samus
President of the Commission

Transactions IAU, Volume XXVIIIB
Proc. XXVIII IAU General Assembly, August 2012
Thierry Montmerle, ed.

doi:10.1017/S1743921315005621

DIVISION XII
COMMISSION 14 ATOMIC AND MOLECULAR DATA
(DONNEES ATOMIQUES ET MOLECULAIRES)

PRESIDENT	**Lyudmila I. Mashonkina**
VICE-PRESIDENT	**Farid Salama**
PAST PRESIDENT	**Glenn M. Wahlgren**
ORGANIZING COMMITTEE	**France Allard, Paul Barklem, Peter Beiersdorfer, Helen Fraser, Gillian Nave, Hampus Nilsson**

PROCEEDINGS BUSINESS SESSIONS, 24 August 2012

Present: P. Caselli, N. Christlieb, E. van Dishoeck, F. Genova, S. Kwok, L. Mashonkina, K. Menten, O. Pintado, J. Shi, G.M. Wahlgren (Chair), F. Wang, S. Yu, P. Zhang, G. Zhao, and about two dozen students and researchers from the National Astronomical Observatories of the Chinese Academy of Sciences

The meeting was called to order by the Chair, who followed the agenda that had been sent to the membership prior to the meeting. The membership of the Commission stands at approximately 220 members, excluding the new members who will join the commission at the end of this General Assembly.

A main point of discussion at the Commission Business Meeting was the anticipated realignment of IAU Divisions and Commissions. Commission 14 is expected to be placed in a Division comprised of other Commissions that serve the entire Union, as opposed to being associated with a particular science driven Commission.

Officers: In our commission the Vice President (VP) becomes the President, and a new VP is chosen from among members of the Organizing Committee. However, our Vice President, Ewine van Dishoeck, was selected to serve as President of Division on Interstellar Medium under the new IAU Division structure that was approved at this General Assembly. As a result, the incoming Vice President, Lyudmila Mashonkina, has been elevated to the position of commission President, and an outgoing member of the Organizing Committee, Farid Salama, was selected to become the new Vice President.

Organizing Committee: Our commission's usual practice is for a member of the Organizing Committee (OC) to serve for two consecutive three year terms, with the past President serving on the OC for three years past their term as President. Officers may serve longer than six years if necessary to undertake service as officers. For the new triennium four new members of the Organizing Committee were selected by vote of the commission membership. The new members are France Allard, Paul Barklem, Helen Fraser and Gillian Nave.

Working Groups: The commission's Working Group (WG) structure will be retained for the next triennium and is composed of the WGs Atomic Data, Molecular Data, Collision Processes, and Solids and Their Surfaces. An expanded set of chairpersons for these WGs is being finalized.

Meetings of Interest: The Commission acts to bring together providers and users of atomic and molecular data and to disseminate data. To these goals, a number of meetings serve as forums for these discussions. Meetings of interest to members of the Commission will be posted on its website.

General Assembly Commission 14 Science Meeting: The commission sponsored a science session immediately following the brief business session. This session was comprised of talks in several areas of interest to the Commission. The speakers and their topics were as follows:
Lyudmila Mashonkina: Astrophysical tests of atomic data important for stellar Mg abundance determinations
Gang Zhao: Laboratory astrophysics and stellar spectroscopy in the Chinese Academy of Sciences
Ewine van Dishoeck: Water and organic molecules with Herschel and ALMA: examples of recent laboratory needs
Karl Menten: Molecular data and software needs for the bright new world of (sub)millimeter astronomy
Shengrui Yu: VUV photochemistry of simple molecules in the gas phase
Peiyu Zhang: Quantum dynamics of astrophysically relevant chemical reactions
Paola Caselli: Nitrogen chemistry and isotopic fractionation - laboratory needs to unveil our origins
Sun Kwok: Unexplained spectral phenomena requiring laboratory data

Lyudmila I. Mashonkina and Farid Salama
President and Vice-President of the Commission, 2012-2015, respectively

Transactions IAU, Volume XXVIIIB
Proc. XXVIII IAU General Assembly, August 2012
Thierry Montmerle, ed.

doi:10.1017/S1743921315005633

DIVISION XII COMMISSION 46 — EDUCATION & DEVELOPMENT OF ASTRONOMY

(EDUCATION & DEVELOPPEMENT DE L'ASTRONOMIE)

PRESIDENT — Rosa M. Ros
VICE-PRESIDENT — John Hearnshaw
PAST PRESIDENT — Magda Stavinschi
ORGANIZING COMMITTEE — Beatriz Garcia, Michele Gerbaldi, Jean-Pierre de Greve, Edward Guinan, Hans Haubold, Barrie Jones, Laurence A. Marshall, Jay Pasachoff.

PROCEEDINGS BUSINESS SESSIONS, 23 and 28 August 2012

1. Busines Meeting Part I: August 23rd, 2012: Astronomy Education Development, Report 2010 2012 by President: Rosa M. Ros

C46 is a Commission of the Executive Committee of the IAU under Division XII Union-Wide Activities. Aiming at improvement of astronomy education and research at all levels worldwide (through the various projects it initiates),maintains, develops, as well as through the dissemination of information. C46 has 332 members and it was managed by the Organizing Committee, formed by the Commission President (Rosa M. Ros, from Spain), the Vice-Presiden (John Hearnshaw, from New Zealand), the Retiring President (Magda Stavinschi, from Romania), the Vice-President of the IAU (George Miley, from Netherland) and the PG chairs:

- Worldwide Development of Astronomy WWDA: John Hearnshaw
- Teaching Astronomy for Development TAD: Edward Guinan and Laurence A. Marshall
- International Schools for Young Astronomers ISYA; chair: Jean-Pierre de Greve
- Network for Astronomy School Education NASE: Rosa M. Ros and Beatriz Garcia
- Public Understanding at the times of Solar Eclipses and transit Phenomena PUTSE: Jay Pasachoff
- National Liaison and Newsletter: Barrie Jones
- Collaborative Programs: Hans Haubold

Brief reports of the various PGs as well as of other relevant activities were presented.The first part ended with the nomination of the C46 President and Vice-president. An election on-line was held among the members of C46. The IAU secretariat was informed of the procedure. The result of the vote was: President: Jean Pierre de Greve (Belgium)and Vice-president: Beatriz García (Argentina.).

2. Business Meeting Part II: August 28th, 2012

The meeting was organized according the subjects present in the following subsections

2.1. *Introduction and Newsletter Editor election*

Here was a discussion from what an Editorial Committee was proposed. Larry Marschall was accepted as Editor in chief (ad referendum from the Comm 46 President, who is the person who finally decides on this). John Baruch (present at the meeting) offered his help with the Newsletter.

2.2. *Proposal for the next three year of Com 46: scientific research in methodological transfer of astrophysical knowledge. J-P de Greve/B.Garcia*

The proposal was presented. Overall, the participants expressed their agreement. The main concerns were:

(*a*) No overlap between Comm46 task and OAD, work hand in hand and always under the Strategic Plan premises.

(*b*) Define the content and role of the Newsletter to as avoid create a periodic publication as AER or CAP.

(*c*) Improve relations with other commissions (Comm 55 and Comm. 50) and working groups.

2.3. *Relationship between Com 46 and ODA, Task Forces, Ed Guinean, Rosa M. Ros, Pedro Russo and Kevin Govender*

A round table was organized; each participant talked about the task forces and their mission and about the possible relationship between them and Comm 46.

2.4. *Invited Presentations*

In order to stablish contacts and analyse the posible relationship between different Groups, programs and activities in Astronomy, a series of invited talks were performed. The complete list of subjects and presenters is:

1 Comm 46 and link with Comm 50, Richard Green.
2 Comm 46 and Young Astronomers and Women in Astronomy movements, Kate Brooks.
3 New proposals on Education of Astronomy: open Forum.
4 GTTP, Rosa Doran.
5 UNAWE, Pedro Ruso.
6 Projects on Astronomy Education Research and C46 , Paulo Bretones.
7 News about Galileoscope, Rick Fienberg.

3. Conclusions from the Round Table and open discussion

The general view of organizing the second session by inviting the referents of Commissions and describe the task of the working groups was very positive. The conclusions are:

- There is much to be done, and each group has its own dynamics.

- Invite the Comm. 46 starts thinking how to teach and how to transfer the astronomical knowledge, as one of the main tasks for the future.
- Propose new ideas about the role of Commission 46.
- Work under the agreement of Division C, discuss with Mary Key Hemenway the lines of action.

Report submitted 29 November 2012
Rosa M. Ros, Past President

Transactions IAU, Volume XXVIIIB
Proc. XXVIII IAU General Assembly, August 2012
Thierry Montmerle, ed.

doi:10.1017/S1743921315005645

DIVISION XII
COMMISSION 55 — **COMMUNICATING ASTRONOMY WITH THE PUBLIC**
(COMMUNIQUER L'ASTRONOMIE (AVEC LE PUBLIC

PRESIDENT	**Lars Lindberg Christensen**
VICE-PRESIDENT	**Pedro Russo**
PAST PRESIDENT	**Ian Robson**
ORGANIZING COMMITTEE	**Kimberly Kowal Arcand, Richard Tresch Fienberg, Carolina Ödman-Govender, Kazuhiro Sekiguchi, Pete Wheeler, Jin Zhu.**

PROCEEDINGS BUSINESS SESSIONS, 27 and 30 August 2012

1. Introduction

A good fraction of the Commission 55 (C55) Organizing Committee met in Beijing in August at the XXVIII IAU General Assembly, where C55 organized Special Session 14 (SpS14) entitled "Communicating Astronomy with the Public for Scientists." During our C55 business meeting, and again during an impromptu gathering a few days later, we discussed changes in the IAU's organizational and programmatic structure and how these changes might affect C55. This report summarizes key points and offers some ideas about what we're calling "C55 v2.0." For background and reference, see the C55 website at http://www.communicatingastronomy.org.

2. Ancient History

IAU Commission 55 had its origins in a conference entitled "Communicating Astronomy to the Public," held in Washington, DC, in October 2003. This "CAP" meeting was a successor to a more general one, "Communicating Astronomy," held in Tenerife, Canary Islands, in February 2002. Both meetings brought together an international group of "producers" of astronomical information (research scientists), "public information officers" (communications coordinators and/or spokespersons affiliated with research institutions, funding agencies, space missions, etc.), and "mediators" (science journalists and popular writers; staffers from museums, planetariums, and national parks; operators of commercial websites focused on astronomy; and science educators).

Following the 2003 conference in Washington an IAU Working Group was set up to coordinate further work on three outcomes from the meeting: the Washington Charter for Communicating Astronomy with the Public (note the intentional change of preposition, from "to" to "with"), an online repository of astronomy-communication resources (now the Virtual Astronomy Multimedia Project, or VAMP), and a series of every-other-year CAP conferences. At the XXVI IAU General Assembly in Prague in August 2006, the working group became Commission 55, "Communicating Astronomy with the Public," under Division XII, "Union-Wide Activities." Members of C55 and attendees at the CAP 2005 and CAP 2007 conferences took many leadership roles in planning, coordinating, and executing the International Year of Astronomy 2009 (IYA 2009).

3. Recent History

The enormous impact of the International Year of Astronomy 2009 (IYA 2009) led the IAU to recognize the importance not only of scientific research, but also of science outreach, to the health of the profession. To build on the success of IYA 2009, the IAU adopted a strategic plan that resulted in the establishment of two new institutions: the Office of Astronomy for Development (OAD), based at the South African Astronomical Observatory in Cape Town and led by Kevin Govender, and the Office for Astronomy Outreach (OAO), based at the National Astronomical Observatory of Japan in Tokyo and led originally by Sarah Reed, now by Sze-leung Cheung. The OAD has set up three task forces to "drive global activities using astronomy as a tool to stimulate development." Task Force 3 (TF3), "Astronomy for the Public," will "drive activities related to communicating astronomy with the public" and is led by chair Ian Robson (UK) and vice-chair Carolina Ödman-Govender (South Africa/EU). Several people active in C55 are members of OAD TF3.

In Beijing, to further align the structure of the IAU with its strategic plan and to better match the organization of the Union to the activities of its national and individual members, attendees at the General Assembly approved a sweeping reorganization that replaced the earlier 12 divisions with 9 new ones. C55 now exists within Division C, "Education, Outreach, and Heritage." The president of Division C is Mary Kay Hemenway (USA), and the vice-president is Hakim Malasan (Indonesia).

4. IAU Commission 55, v1.0

C55 was originally organized with this rationale: *It is the responsibility of every practicing astronomer to play some role in explaining the interest and value of science to our real employers, the taxpayers of the world.*

Here is C55's original mission statement:

• To encourage and enable a much larger fraction of the astronomical community to take an active role in explaining what we do (and why) to our fellow citizens.

• To act as an international, impartial coordinating entity that furthers the recognition of outreach and public communication on all levels in astronomy.

• To encourage international collaborations on outreach and public communication.

• To endorse standards, best practices, and requirements for public communication.

Going into the XXVIII IAU General Assembly, C55 had the following working groups:

- Washington Charter (Chair: Dennis Crabtree)
- VAMP – Virtual Astronomy Multimedia Project (Chair: Adrienne Gauthier)
- Best Practices (Chair: Lars Lindberg Christensen)
- CAP Journal (Chair: Pedro Russo)
- New Ways of Communicating Astronomy with the Public (Chair: Michael West)
- CAP Conferences (Chair: Ian Robson)

5. IAU Commission 55, v2.0

The outcome of our discussions at the C55 business meeting in Beijing and the informal discussions is that, despite the creation of OAD and OAO, we feel that Commission 55 still has an important role to play in the IAU. Our rationale and mission statement remain unchanged, but in light of the new division structure, the OAD (especially TF3), and the OAO, we'll need to revise our approach to conducting activities in support of that mission.

The following ideas were formed:

• C55 should be a "think tank" that unites the global astronomy communication community and seeds initiatives to explore new ways to communicate astronomy with the public.

• C55 should focus more on community initiatives and new trends and less on implementing projects, as implementation naturally falls to the OAO and OAD Task Force 3. (Task Force 3 is a panel of experts to advise the OAD and will not be implementing projects.)

• C55 should further the development and improvement of astronomy communication at all levels throughout the world, through stimulating, gathering, and exchanging ideas and practices.

6. C55 Working Groups

Our recommendations for C55 working groups account for the following inputs:

• The IAU has asked that the Editor-in-Chief position for the CAP Journal be moved to the OAO.

• C55 would like to retain responsibility for organizing the CAP conferences.

• Some of C55's earlier working groups have been almost inactive.

• Some of C55's earlier working groups have migrated to other commissions or outside the IAU.

• It was decided to move the Division II working group Communicating Heliophysics to C55.

• The IAU will accept associate member recommendations from commissions (see below).

6.1. *The Working Groups for the 2012-2015 term are listed below:*

6.2. *WG Communicating Astronomy with the Public Journal (CAPj)*

Chair: Sarah Reed (Editor in Chief, CAPj)

Mission: To publish a free (to readers) peer-reviewed journal for astronomy communicators.

Deliverables: 3 issues of CAPjournal per year

6.3. *WG Communicating Astronomy with the Public Conferences*

Chair: Ian Robson

Mission: To regularly convene producers of astronomical information, public information officers, and mediators worldwide for the interchange of ideas and practices.

Deliverables: An international conference at least every 2 years (including sessions during IAU General Assemblies, where applicable).

6.4. *WG Washington Charter for Communicating Astronomy with the Public*

Chair: Dennis Crabtree

Mission: To promote the importance of astronomy outreach and communication by disseminating information about the Washington Charter and to seek endorsements from funding agencies, observatory directors, department heads and deans, and other employers of astronomers.

Deliverables: At least 10 new institutional and/or individual endorsements of the Washington Charter, and some follow-up with previous endorsers to identify/address any problems that may have arisen.

6.5. *WG Outreach Professionalization & Accreditation*

Chair: Rick Fienberg

Mission: To bring a sense of professionalism and professional respect to the field of astronomy communication, to advocate for our needs as professional communicators, and to serve as a means for information sharing and networking.

Deliverables: A procedure to handle requests for IAU associate membership (presumably including several levels of accreditation) and methods/standards/requirements for achieving accreditation.

6.6. *WG Public Outreach Information Management*

Chair: Lars Lindberg Christensen

Mission: To act as a facilitator enabling the small outreach information management community to gather around a common technical framework optimizing the synergy in the community

Deliverables:

- Describe one possible standard for EPO content management systems.
- Decide on standards for astronomy EPO products metadata.
- Write a CAPj article. (Submissions will follow the standard procedure for accepting articles. Proposed articles should be discussed and agreed with the Editor in Chief)
- Bringing the major players in EPO information management together in an informal network.
- Arrange the first workshop for this network.
- Possibly incorporate Mohammad Heydari-Malayeri's online dictionary into a C55 resource.

6.7. *WG New Media*

Chair: Pamela Gay

Mission: To nurture a professional astronomy culture that utilizes social media to disseminate science effectively.

Deliverables:

- Review one social media channel and its effectiveness per issue of CAPj. (Submissions will follow the standard procedure for accepting articles. Proposed articles should be discussed and agreed with the Editor in Chief)
- Create and publish a map of which social media can be accessed where in the world (depending on the target audiences).
- Publish success stories of efficient social media usage.
- Publish discussions on ways to use social media to create professional development experiences for people at underserved universities (e.g., seminars in Second Life, Google hangouts).
- Publish information about best practices in research dissemination on wiki pages.
- Creating an IAU record of social media accounts of professional astronomers (as part of the already existing members' profiles).

- Astro Comm Journal Club? (Twitter, to be adapted from scicommjc.wordpress.com.)

6.8. *WG New Ways of Communicating Astronomy with the Public*

Chair: Michael West

Mission: To facilitate sharing of diverse and effective new ways to communicate astronomy with the public, with a focus on creative alternatives to press releases, public lectures, print and broadcast media, and other traditional ways of bringing astronomy to a wide audience. The WG will serve as a clearinghouse and network for the worldwide community of astronomy communicators to engage the public in distinctly new ways by thinking outside the box.

Deliverables:

- Creation of a New Ways of Communicating Astronomy with the Public Facebook page (or similar social networking engine) where science communicators can join freely to exchange ideas and ask questions about new ways of communicating astronomy with diverse audiences.
- Engage in proactive efforts to inform astronomy communicators around the world of the existence and activities of the New Ways of Communicating Astronomy with the Public WG, with the goal of creating a large and vibrant network of members. Ways to promote this WG and its activities will include creating a new blog, promoting the activities of the WG at national and international conferences, and organizing New Ways sessions as part of the future CAP meetings and/or as separate conferences outside of the CAP meetings.
- Articles will be written regularly for the CAP Journal, Astronomy Education Review, and other similar journals, highlighting noteworthy new methods and assessing their successes and failures. (Submissions will follow the standard procedure for accepting articles. Proposed articles should be discussed and agreed with the Editor in Chief)

6.9. *WG Communicating Heliophysics*

Chair: Carine Briand

Mission: Promote the outreach activities of the heliophysics community and encourage our people to participate in the activities of C55.

Deliverables: One article/year on heliophysics outreach in CAPj (submissions will follow the standard procedure for accepting articles, and proposed articles should be discussed and agreed with the Editor in Chief); attendance by at least one WG member at every CAP meeting; possible/desirable: maintenance of a robust WG website with resources for heliophysics outreach

7. How to Join C55 and/or Become an IAU Associate Member

Once the new divisions and commissions are in place and current individual IAU members have chosen which ones they'll belong to, how do additional people join C55? They contact the president of C55. The leadership of C55 will need to come up with policies and procedures to guide the commission's response. The final decision rests with the C55 president, who, in the case of approval, then passes the new member's information along to the IAU Secretariat.

New members of C55 do *not* have to be regular individual IAU members, that is, they do not have to be PhD astronomers who apply to their national committee for nomination and get elected at the next triennial General Assembly. They can instead become associate members of the IAU via C55, or, more accurately, associates of IAU C55. As Vivien Reuter of the IAU Secretariat explains, the category of "consultant" was recently changed to "associate member" and was established to enable nonmembers of the IAU to participate in the activities of commissions and working groups. Most associates are astronomers who for some reason have simply neglected to become IAU members; the rest are people working in related fields and mainly associated with outreach and educational activities. Qualified associate members are encouraged to apply for full membership at the next triennial opportunity; note that associate members do not have the right to vote at IAU General Assemblies, whereas full members do. IAU General Secretary Thierry Montmerle adds that associate status should not be considered permanent, but linked to a well-defined activity within the IAU, as long as it lasts, and that responsibility for maintaining or dropping this status rests with the relevant commission. Put another way, associate status is not an honorific — associate members are expected to be active in the commission that admits them into the IAU.

8. General Guidelines for C55 Administration

Substantial references to the commissions, including their roles, responsibilities, and functions, can be found on the IAU website (http://www.iau.org), under the ADMINISTRATION menu, Statutes & Rules, under the links to Statutes, Bye-Laws, and Working Rules. Questions should be addressed to the IAU General Secretary and/or IAU Head of Administration.

Transactions IAU, Volume XXVIIIB
Proc. XXVIII IAU General Assembly, August 2012
Thierry Montmerle, ed.

doi:10.1017/S1743921315005657

CHAPTER VII

TWENTY EIGHTH GENERAL ASSEMBLY

IAU STATUTES, BYE-LAWS, AND WORKING RULES

Beijing, 21 August 2012

1. IAU Statutes

I. OBJECTIVE

1. The International Astronomical Union (hereinafter referred to as the Union) is an international nongovernmental organization. Its objective is to promote the science of astronomy in all its aspects.

II. DOMICILE AND INTERNATIONAL RELATIONS

2. The legal domicile of the Union is Paris, France.

3. The Union adheres to, and co-operates with the body of international scientific organizations through the International Council for Science (ICSU). It supports and applies the policies on the Freedom, Responsibility, and Ethics in the Conduct of Science defined by ICSU.

III. COMPOSITION OF THE UNION

4. The Union is composed of:

4.a. National Members (adhering organizations)

4.b. Individual Members (adhering persons)

IV. NATIONAL MEMBERS

5. An organization representing a national professional astronomical community, desiring to promote its participation in international astronomy and supporting the objective of the Union, may adhere to the Union as a National Member.

6. An organization desiring to join the Union as a National Member while developing professional astronomy in the community it represents may do so:

6a. on an interim basis, on the same conditions as above, for a period of up to nine years. After that time, it must apply to become a National Member on a permanent basis or its membership in the Union will terminate;

6b. on a prospective basis for a period of up to six years if its community has less than six Individual Members. After that time it must apply to become a National Member on either an interim or permanent basis or its membership in the Union will terminate.

7. A National Member is admitted to the Union on a permanent, interim, or prospective basis by the General Assembly. It may resign from the Union by so informing the General Secretary in writing.

8. A National Member may be either:

8.a. the organization by which scientists of the corresponding nation or territory adhere to ICSU or:

8.b. an appropriate National Society or Committee for Astronomy, or:

8.c. an appropriate institution of higher learning.

9. The adherence of a National Member is automatically suspended if its annual contributions, as defined in Articles 23c and 23e below have not been paid for five years; it resumes, upon the approval of the Executive Committee, when the arrears in contributions have been paid in full. After five years of suspension of a National Member, the Executive Committee may recommend to the General Assembly to terminate the Membership.

10. A National Member is admitted to the Union in one of the categories specified in the Bye-Laws.

V. INDIVIDUAL MEMBERS

11. A professional scientist who is active in some branch of astronomy may be admitted to the Union by the Executive Committee as an Individual Member. An Individual Member may resign from the Union by so informing the General Secretary in writing.

VI. GOVERNANCE

12. The governing bodies of the Union are:

12.a. The General Assembly;

12.b. The Executive Committee; and

12.c. The Officers.

VII. GENERAL ASSEMBLY

13. The General Assembly consists of the National Members and of Individual Members. The General Assembly determines the overall policy of the Union.

13.a. The General Assembly approves the Statutes of the Union, including any changes therein.

13.b.The General Assembly approves Bye-Laws specifying the Rules of Procedure to be used in applying the Statutes.

13.c. The General Assembly elects an Executive Committee to implement its decisions and to direct the affairs of the Union between successive ordinary meetings of the General Assembly. The Executive Committee reports to the General Assembly.

13.d. The General Assembly appoints a standing Finance Committee to advise the Executive Committee on its behalf on budgetary matters between General Assemblies, and to advise the General Assembly on the approval of the budget and accounts of the Union. The Finance Committee consists of not more than 8 members of different national affiliations, including a Chairperson, proposed by the National Members, and remains in office until the end of the next General Assembly.

13.e. The General Assembly appoints a Special Nominating Committee to prepare a suitable slate of candidates for election to the incoming Executive Committee.

13.f. The General Assembly appoints a standing Membership Committee to advise the Executive Committee on its behalf on matters related to the admission of Individual Members. The Membership Committee consists of not more than 8 members of different national affiliations, including a Chairperson, proposed by the National Members, and remains in office until the end of the next General Assembly.

14. Voting at the General Assembly on issues of a primarily scientific nature, as determined by the Executive Committee, is by Individual Members. Voting on all other matters is by National Member. Each National Member authorises a representative to vote on its behalf.

14.a. On questions involving the budget of the Union, the number of votes for each National Member is one greater than the number of its category, referred to in article 10. National Members with interim status, or which have not paid their dues for years preceding that of the General Assembly, may not participate in the voting.

14.b. On questions concerning the administration of the Union, but not involving its budget, each National Member has one vote, under the same condition of payment of dues as in §14.a.

14.c.National Members may vote by correspondence on questions concerning the agenda for the General Assembly.

14.d. A vote is valid only if at least two thirds of the National Members having the right to vote by virtue of article §14.a. participate in it by either casting a vote or signalling an abstention. An abstention. An abstention is not considered a vote cast.

15. The decisions of the General Assembly are taken by an absolute majority of the votes cast. However, a decision to change the Statutes requires the approval of at least two thirds of all National Members having the right to vote by virtue of article §14.a. Where there is an equal division of votes, the President determines the issue.

15.a. To enable the widest possible participation of Individual Members the Executive Committee may decide that voting on certain issues of a primarily scientific nature, as determined by the Executive Committee, shall be open for electronic voting for not more than 31 days counting from the close of the General Assembly at which the issue was raised.

15.b. The Executive Committee shall give Members not less than 3 months notice before the opening of the General Assembly of the intention to open certain issues to electronic voting after the General Assembly.

16. Changes in the Statutes or Bye-Laws can only be considered by the General Assembly if a specific proposal has been duly submitted to the National Members and placed on the Agenda of the General Assembly by the procedure and deadlines specified in the Bye-Laws.

VIII. EXECUTIVE COMMITTEE

17. The Executive Committee consists of the President of the Union, the President-Elect, six Vice-Presidents, the General Secretary, and the Assistant General Secretary, elected by the General Assembly on the proposal of the Special Nominating Committee.

IX. OFFICERS

18. The Officers of the Union are the President, the General Secretary, the President-Elect, and the Assistant General Secretary. The Officers decide short-term policy issues within the general policies of the Union as decided by the General Assembly and interpreted by the Executive Committee.

X. SCIENTIFIC DIVISIONS

19. As an effective means to promote progress in the main areas of astronomy, the scientific work of the Union is structured through its Scientific Divisions. Each Division covers a broad, well-defined area of astronomical science, or deals with international

matters of an interdisciplinary nature. As far as practicable, Divisions should include comparable fractions of the Individual Members of the Union.

20. Divisions are created or terminated by the General Assembly on the recommendation of the Executive Committee. The activities of a Division are organized by an Organizing Committee chaired by a Division President. The Division President and a Vice-President are elected by the General Assembly on the proposal of the Executive Committee, and are ex officio members of the Organizing Committee.

XI. SCIENTIFIC COMMISSIONS

21. Within Divisions, the scientific activities in well-defined disciplines within the subject matter of the Division may be organized through scientific Commissions. In special cases, a Commission may cover a subject common to two or more Divisions and then becomes a Commission of all these Divisions.

22. Commissions are created or terminated by the Executive Committee upon the recommendation of the Organizing Committee(s) of the Division(s) desiring to create or terminate them. The activities of a Commission are organized by an Organizing Committee chaired by a Commission President. The Commission President and a Vice-President are appointed by the Organizing Committee(s) of the corresponding Division(s) upon the proposal of the Organizing Committee of the Commission.

XII. BUDGET AND DUES

23. For each ordinary General Assembly the Executive Committee prepares a budget proposal covering the period to the next ordinary General Assembly, together with the accounts of the Union for the preceding period. It submits these to the Finance Committee for advice before presenting them to the vote of the General Assembly.

23.a.The Finance Committee examines the accounts of the Union from the point of view of responsible expenditure within the intent of the previous General Assembly, as interpreted by the Executive Committee. It also considers whether the proposed budget is adequate to implement the policy of the General Assembly. It submits reports on these matters to the General Assembly before its decisions concerning the approval of the accounts and of the budget.

23.b. The amount of the unit of contribution is decided by the General Assembly as part of the budget approval process.

23.c. Each National Member pays annually a number of units of contribution corresponding to its category. The number of units of contribution for each category shall be specified in the Bye-Laws.

23.d. A vote on matters under article 23 is valid only if at least two thirds of the National Members having the right to vote by virtue of article §14.a. cast a vote. In all cases an abstention is not a vote, but a declaration that the Member declines to vote.

23.e. National Members having interim status pay annually one half unit of contribution.

23.f. National Members having prospective status pay no contribution.

23.g. The payment of contributions is the responsibility of the National Members. The liability of each National Members in respect of the Union is limited to the amount of contributions due through the current year.

XIII. EMERGENCY POWERS

24. If, through events outside the control of the Union, circumstances arise in which it is impracticable to comply fully with the provisions of the Statutes and Bye-Laws of the Union, the Executive Committee and Officers, in the order specified below, shall take such actions as they deem necessary for the continued operation of the Union. Such action shall be reported to all National Members as soon as this becomes practicable, until an ordinary or extraordinary General Assembly can be convened. The following is the order of authority: The Executive Committee in meeting or by correspondence; the President of the Union; the General Secretary; or failing the practicability or availability of any of the above, one of the Vice-Presidents.

XIV. DISSOLUTION OF THE UNION

25. A decision to dissolve the Union is only valid if taken by the General Assembly with the approval of three quarters of the National Members having the right to vote by virtue of article §14.a. Such a decision shall specify a procedure for settling any debts and disposing of any assets of the Union.

XV. FINAL CLAUSES

26. These Statutes enter into force on 21 August 2012.

27. The present Statutes are published in French and English versions. For legal purposes, the French version is authoritative.

2. Statuts de l'UAI, Pékin, 21 août 2012

I. OBJECTIF

1. L'Union Internationale Astronomique (dénommée ci-aprés "l'Union") est une organisation non gouvernementale internationale. Son objectif est de promouvoir la science de l'astronomie sous tous ses aspects.

II. DOMICILIATION ET RELATIONS INTERNATIONALES

2. Le domicile légal de l'Union est situé á Paris, en France.

3. L'Union adhère à, et coopère avec, l'ensemble des organisations scientifiques internationales à travers le Conseil International pour la Science (ICSU). Elle soutient et applique les directives sur la Liberté, la Responsabilité, et l'Ethique pour la bonne conduite des sciences telles que définies par l'ICSU.

III. COMPOSITION DE l'UNION

4. L'Union se compose de :

4.a. Membres Nationaux (organisations),

4.b. Membres Individuels (personnes physiques).

IV. MEMBRES NATIONAUX

5. Une organisation représentant une communauté astronomique professionnelle nationale, désireuse de développer sa participation sur la scène de l'astronomie internationale et soutenant les objectifs de l'Union, peut adhérer à l'Union en qualité de Membre National.

6. Une organisation désireuse de rejoindre l'Union en qualité de Membre National tout en développant l'astronomie professionnelle dans la communauté qu'elle représente peut le faire :

6.a. De manière temporaire, selon les conditions précitées, pour une période maximale de neuf ans. Passé ce délai, elle doit demander à devenir Membre National de manière permanente, à défaut son adhésion á l'Union sera résiliée.

6.b. De manière prospective, pour une période maximale de six ans si sa communauté compte moins de six Membres Individuels. Passé ce délai, elle devra demander á devenir Membre National de manière temporaire ou permanente, à défaut de quoi son adhésion sera résiliée.

7. Un Membre National est admis dans l'Union de manière permanente, temporaire ou prospective par l'Assemblée Générale. Il peut se retirer de l'Union en informant le Secrétaire Général de son retrait par écrit.

8. Un Membre National peut être :

8.a. L'organisation par laquelle les scientifiques de la nation correspondante ou du territoire correspondant, adhérant à l'ICSU ou :

8.b. Une Société ou Comité National(e) compétent(e) d'Astronomie, ou :

8.c. Un établissement compétent d'enseignement supérieur.

9. L'adhésion d'un Membre National est automatiquement suspendue si ses cotisations annuelles, telles que définies aux Articles 23c et 23e ci-dessous n'ont pas été payées pendant cinq ans ; elle sera rétablie, sur approbation du Comité Exécutif, lorsque les arriérés relatifs à ses cotisations auront été payés en totalité. Aprés cinq ans de suspension d'un Membre National, le Comité Exécutif peut recommander á l'Assemblée Générale l'exclusion de ce Membre.

10. Un Membre National est admis dans l'Union au travers de l'une des catégories spécifiées dans le Réglement Intérieur.

V. MEMBRES INDIVIDUELS

11. Un scientifique professionnel exerçant son activité dans un des domaines de l'astronomie peut être admis dans l'Union par le Comité Exécutif en qualité de Membre Individuel. Un Membre Individuel peut quitter de l'Union en informant de son retrait le Secrétaire Général, par écrit.

VI. GOUVERNANCE

12. La gouvernance de l'Union est constituée par :

12.a. L'Assemblée Générale ;

12.b. Le Comité Exécutif ;

12.c. Les Membres du Bureau du Comité Executif.

VII. ASSEMBLEE GENERALE

13. L'Assemblée Générale se compose des Membres Nationaux et des Membres Individuels. L'Assemblée Générale détermine les grandes orientations de l'Union.

13.a. L'Assemblée Générale ratifie les Statuts de L'Union, y compris tous changements apportés à ces Statuts.

13.b. L'Assemblée Générale ratifie le Réglement Intérieur spécifiant les Règles de Procédure devant être suivies lors de l'application des Statuts.

13.c. L'Assemblée Générale élit un Comité Exécutif afin de mettre en œuvre ses décisions et le charger de diriger les affaires de l'Union entre d'une Assemblée Générale á l'autre. Le Comité Exécutif présente ses comptes-rendus á l'Assemblée Générale.

13.d. L'Assemblée Générale nomme un Comité des Finances pour conseiller le Comité Exécutif en son nom sur les questions budgétaires entre les Assemblées Générales et pour proposer ses recommandations á l'Assemblée Générale concernant l'approbation du budget et des comptes de l'Union. Le Comité des Finances se compose au plus de huit Membres de différentes représentations nationales et comprend un Président, proposé par les Membres Nationaux, qui conservera ses fonctions jusqu'à la fin de l'Assemblée Générale suivante.

13.e. L'Assemblée Générale nomme un Comité Spécial des Nominations afin de préparer un éventail de candidats en vue de l'élection du Comité Exécutif suivant.

13.f. L'Assemblée Générale nomme un Comité des Admissions pour permettre au Comité Exécutif de se prononcer sur l'admission de nouveaux Membres Individuels. Le Comité des Admissions se compose au plus de huit Membres de différentes représentations nationales et comprend un Président, proposé par les Membres Nationaux, qui conservera ses fonctions jusqu'à la fin de l'Assemblée Générale suivante.

14. Les votes lors de l'Assemblée Générale portant sur des questions de nature essentiellement scientifique, telles que déterminées par le Comité Exécutif, se font par les Membres Individuels. Les votes portant sur toutes les autres questions se font par les Membres Nationaux. Chaque Membre National nomme un représentant pour voter en son nom.

14.a. Concernant les questions ayant trait au budget de l'Union, le nombre de voix pour chaque Membre National est supérieur d'une unité au nombre définissant sa catégorie, telle que définie à l'Article 10 des Statuts. Les Membres Nationaux dont le statut est temporaire ou prospectif, ou qui ne sont pas à jour de leur cotisation au moment de l'Assemblée Générale, ne peuvent pas prendre part au vote.

14.b. Concernant les questions relatives à l'administration de l'Union, mais n'ayant pas trait au budget, chaque Membre National dispose d'une voix, selon les mêmes conditions relatives au paiement des cotisations mentionnées à l'Article 14.a.

14.c. Les Membres Nationaux peuvent voter par correspondance sur des questions mises à l'ordre du jour de l'Assemblée Générale.

14.d. Un vote n'est valide que si au moins deux tiers des Membres Nationaux ayant le droit de voter en vertu de l'Article 14.a. y participent, soit en votant, soit en se prononçant par une abstention. Une abstention n'est pas considérée comme l"expression d'un vote.

15. Les décisions de l'Assemblée Générale sont prises à la majorité absolue des voix. Cependant, la décision de modifier les Statuts exige l'approbation d'au moins deux tiers de tous les Membres Nationaux ayant droit de voter en vertu de l'Article 14.a. Lorsqu'il y a égalité des voix, il revient au Président de statuer sur l'issue du vote.

15.a. Afin de favoriser la plus large participation possible des Membres Individuels, le Comité Exécutif pourra décider que le vote portant sur certaines questions de nature essentiellement scientifique, tel que le déterminera le Comité Exécutif, pourront faire l'objet d'un vote électronique 31 jours au plus à partir de la clôture de l'Assemblée Générale à laquelle la question a été soulevée.

15.b. Le Comité Exécutif enverra un préavis d'au moins trois mois aux Membres avant l'ouverture de l'Assemblée Générale les informant des questions devant être soumises á un vote électronique après l'Assemblée Générale.

16. Les changements apportés aux Statuts ou au Règlement Intérieur ne peuvent être examinés par l'Assemblée Générale que si une proposition spécifique a été dûment soumise aux Membres Nationaux et mentionnée á l'ordre du jour de l'Assemblée Générale par la procédure et dans des délais spécifiés dans le Réglement Intérieur.

VIII. COMITE EXÉCUTIF

17. Le Comité Exécutif se compose du Président de l'Union, du Président désigné (son successeur), de six Vice-Présidents, du Secrétaire Général, et du Secrétaire Général Adjoint, élus par l'Assemblée Générale sur proposition du Comité Spécial des Nominations.

IX. MEMBRES DU BUREAU

18. Les Membres de l'Union sont le Président, le Secrétaire Général, le Président désigné et le Secrétaire Général Adjoint. Les Membres du Bureau prennent des décisions sur le court terme dans le cadre des orientations générales de l'Union telles que décidées par l'Assemblée Générale et appliquées par le Comité Exécutif.

X. DIVISIONS SCIENTIFIQUES

19. En tant que moyen efficace pour favoriser les progrès dans les principaux domaines de l'astronomie, le travail scientifique de l'Union est structuré autour de ses Divisions Scientifiques. Chaque Division recouvre un large domaine bien défini de l'astronomie, ou traite de sujets internationaux de nature interdisciplinaire. Autant que possible, les Divisions comprendront chacune un nombre comparable de Membres Individuels de l'Union.

20. Les Divisions sont créées ou supprimées par l'Assemblée Générale sur recommandation du Comité Exécutif. Les activités d'une Division sont organisées par un Comité d'Organisation présidé par le Président de Division. Le Président de Division et le Vice-Président sont élus par l'Assemblée Générale sur proposition du Comité Exécutif, et sont Membres ès-qualité du Comité d'Organisation.

XI. COMMISSIONS SCIENTIFIQUES

21. A l'intérieur des Divisions, les activités scientifiques réparties en disciplines bien définies peuvent être organisées á travers des Commissions Scientifiques. Dans certains cas spécifiques, une Commission peut recouvrir une discipline commune à une ou plusieurs Divisions et devient alors une Commission de l'ensemble de ces Divisions.

22. Les Commissions sont créées ou supprimées par le Comité Exécutif sur recommandation de la ou des Division(s) désirant les créer ou les supprimer. Les activités d'une Commission sont organisées par un Comité d'Organisation présidé par le Président de

Commission. Le Président de Commission et un Vice-Président sont nommés par le ou les Comité(s) d'Organisation de la ou des Division(s) correspondantes sur proposition du Comité d'Organisation de la Commission.

XII. BUDGET ET COTISATIONS

23. En vue de chaque Assemblée Générale, le Comité Exécutif prépare une proposition de budget couvrant la période jusqu'à l'Assemblée Générale suivante, accompagnée des comptes de l"Union pour la période précédente. Il les soumet au Comité des Finances en vue de leur examen avant leur soumission au vote de l'Assemblée Générale.

23.a. Le Comité des Finances examine les comptes de l'Union au regard des dépenses approuvées par l'Assemblée Générale précédente, selon l'exécution qu'en a fait le Comité Exécutif. Il se penche aussi sur l'adéquation ou non du budget proposé en vue de la mise en œuvre des orientations de l'Assemblée Générale. Il soumet son rapport à l'Assemblée Générale avant ses décisions concernant l'approbation des comptes et du budget.

23.b. Le montant de l' unité de cotisation est décidé par l'Assemblée Générale au titre du processus d'approbation du budget.

23.c. Chaque Membre National verse annuellement un nombre d'unités de cotisation correspondant à sa catégorie. Le nombre d'unités de cotisation pour chaque catégorie est spécifié dans le Règlement Intérieur.

23.d. Un vote sur les questions du ressort de l'Article 23 n'est valable que si les deux tiers des représentants des Membres Nationaux ayant le droit de vote en vertu de l'Article 14.a. prennnent part à ce vote. Dans tous les cas, une abstention ne constitue par un vote, mais une déclaration selon laquelle que le Membre National décide de ne pas voter.

23.e. Les Membres Nationaux ayant un statut d'intérim payent annuellement une demi-unité de cotisation.

23.f. Les Membres Nationaux ayant un statut prospectif ne payent aucune cotisation.

23.g. Le paiement des contributions relève de la responsabilité des Membres Nationaux. La responsabilité de chaque Membre National relative à l'Union se limite au montant des cotisations dues au cours de l'année.

XIII. PLEINS POUVOIRS

24. Si, en cas d'événements hors du contrôle de l'Union, les circonstances font qu'il est impossible de respecter pleinement les dispositions des Statuts et du Règlement Intérieur de l'Union, le Comité Exécutif et les Membres du Bureau, dans l'ordre spécifié ci-aprés, agiront comme ils l'estimeront nécessaire pour assurer la continuité du fonctionnement de l'Union. De telles décisions seront rapportées aux Membres Nationaux dès que cela

sera possible, jusqu'á ce qu'une Assemblée Générale Ordinaire ou Extraordinaire soit convoquée.

L'ordre hiérarchique est le suivant : le Comité Exécutif, en réunion ou par correspondance; le Président de l'Union; le Secrétaire Général; ou á défaut de possibilité ou de disponibilité de ce qui précéde, un des Vice-Présidents.

XIV. DISSOLUTION DE L'UNION

25. La décision de dissoudre l'Union n'est recevable que si elle est prise par l'Assemblée Générale avec l'approbation des trois quarts des Membres Nationaux ayant le droit de voter en vertu de l'Article 14.a. Une telle décision spécifiera une procédure visant à apurer les comptes et à disposer des actifs de l'Union.

XV. CLAUSES FINALES

26. Les présents Statuts entrent en vigueur le 21 août 2012.

27. Ils devront être publiés en français et en anglais. A toute fin juridique, la version française fera autorité.

3. IAU Bye-Laws

I. MEMBERSHIP

1. An application for admission to the Union as a National Member shall be submitted to the General Secretary by the proposing organization at least eight months before the next ordinary General Assembly.

2. The Executive Committee shall examine the application and resolve any outstanding issues concerning the nature of the proposed National Member and the category of membership (§VII.25). Subsequently, the Executive Committee shall forward the application to the General Assembly for decision, with its recommendation as to its approval or rejection.

3. The Executive Committee shall examine any proposal by a National Member to change its category of adherence to a more appropriate level. If the Executive Committee is unable to approve the request, either party may refer the matter to the next General Assembly.

4. Individual Members are admitted by the Executive Committee upon the nomination of a National Member or the President of a Division. The Executive Committee shall publish the criteria and procedures for membership, and shall consult the Membership Committee before admitting new Individual Members.

II. GENERAL ASSEMBLY

5. The ordinary General Assembly meets, as a rule, once every three years. Unless determined by the previous General Assembly, the place and date of the ordinary General Assembly shall be fixed by the Executive Committee and be communicated to the National Members at least one year in advance.

6. The President may summon an extraordinary General Assembly with the consent of the Executive Committee, and must do so at the request of at least one third of the National Members. The date, place, and agenda of business of an extraordinary General Assembly must be communicated to all National Members at least two months before the first day of the Assembly.

7. Matters to be decided upon by the General Assembly shall be submitted for consideration by those concerned as follows, counting from the first day of the General Assembly:

7.a. A motion to amend the Statutes or Bye-Laws may be submitted by a National Member or by the Executive Committee. Any such motion shall be submitted to the General Secretary at least nine months in advance and be forwarded, with the recommendation of the Executive Committee as to its adoption or rejection, to the National Members at least six months in advance.

7.b. The General Secretary shall distribute the draft budget prepared by the Executive Committee to the National Members at least eight months in advance. Any motion to modify this budget, or any other matters pertaining to it, shall be submitted to the General Secretary at least six months in advance. The Executive Committee shall consider whether or not to adopt any such motion in a modified budget, which shall be distributed to the National Members at least four months in advance. Should the Executive Committee decide to reject the motion it shall also be submitted to the General Assembly with the reasons for its rejection.

7.c. Any motion or proposal concerning the administration of the Union, and not affecting the budget, by a National Member, or by the Organizing Committee of a Scientific Division of the Union, shall be placed on the Agenda of the General Assembly, provided it is submitted to the General Secretary, in specific terms, at least six months in advance.

7.d. Any motion of a scientific character submitted by a National Member, a Scientific Division of the Union, or by an ICSU Scientific Committee or Program on which the Union is formally represented, shall be placed on the Agenda of the General Assembly, provided it is submitted to the General Secretary, in specific terms, at least six months in advance.

7.e. The complete agenda, including all such motions or proposals, shall be prepared by the Executive Committee and submitted to the National Members at least four months in advance.

8. The President may invite representatives of other organizations, scientists in related fields, and young astronomers to participate in the General Assembly. Subject to the

agreement of the Executive Committee, the President may authorise the General Secretary to invite representatives of other organizations, and the National Members or other appropriate IAU bodies to invite scientists in related fields and young astronomers.

III. SPECIAL NOMINATING COMMITTEE

9. The Special Nominating Committee consists of the President and past President of the Union, a member proposed by the retiring Executive Committee, and four members selected by the representatives of the National Members from up to twelve candidates proposed by Presidents of Divisions, with due regard to an appropriate distribution over the major branches of astronomy.

9.a. Except for the President and immediate past President, present and former members of the Executive Committee shall not serve on the Special Nominating Committee. No two members of the Special Nominating Committee shall belong to the same nation or National Member.

9.b. The General Secretary and the Assistant General Secretary participate in the work of the Special Nominating Committee in an advisory capacity, and the President-Elect may participate as an observer.

10. The Special Nominating Committee is appointed by the General Assembly, to which it reports directly. It assumes its duties immediately after the end of the General Assembly and remains in office until the end of the ordinary General Assembly next following that of its appointment, and it may fill any vacancy occurring among its members.

IV. OFFICERS AND EXECUTIVE COMMITTEE

11. Terms of office:

11.a. The President of the Union remains in office until the end of the ordinary General Assembly next following that of election. The President-Elect succeeds the President at that moment.

11.b. The General Secretary and the Assistant General Secretary remain in office until the end of the ordinary General Assembly next following that of their election. Normally the Assistant General Secretary succeeds the General Secretary, but both officers may be re-elected for another term.

11.c. The Vice-Presidents remain in office until the end of the ordinary General Assembly following that of their election. They may be immediately re-elected once to the same office.

11.d. The elections take place at the last session of the General Assembly, the names of the candidates proposed having been announced at a previous session.

12. The Executive Committee may fill any vacancy occurring among its members. Any person so appointed remains in office until the end of the next ordinary General Assembly.

13. The past President and General Secretary become advisers to the Executive Committee until the end of the next ordinary General Assembly. They participate in the work of the Executive Committee and attend its meetings without voting rights.

14. The Executive Committee shall formulate Working Rules to clarify the application of the Statutes and Bye-Laws. Such Working Rules shall include the criteria and procedures by which the Executive Committee will review applications for Individual Membership; standard Terms of Reference for the Scientific Commissions of the Union; rules for the administration of the Union's financial affairs by the General Secretary; and procedures by which the Executive Committee may conduct business by electronic or other means of correspondence. The Working Rules shall be published electronically and in the Transactions of the Union.

15. The Executive Committee appoints the Union's official representatives to other scientific organizations.

16. The Officers and members of the Executive Committee cannot be held individually or personally liable for any legal claims or charges that might be brought against the Union.

V. SCIENTIFIC DIVISIONS

17. The Divisions of the Union shall pursue the scientific objects of the Union within their respective fields of astronomy. Activities by which they do so include the encouragement and organization of collective investigations, and the discussion of questions relating to international agreements, cooperation, or standardization. They shall report to each General Assembly on the work they have accomplished and such new initiatives as they are undertaking.

18. Each Scientific Division shall consist of:

18.a. An Organizing Committee, normally of 6-12 persons, including the Division President and Vice-President, and a Division Secretary appointed by the Organizing Committee from among its members. The Committee is responsible for conducting the business of the Division.

18.b. Members of the Union accepted by the Organizing Committee in recognition of their special experience and interests.

19. Normally, the Division President is succeeded by the Vice-President at the end of the General Assembly following their election, but both may be re-elected for a second term. Before each General Assembly, the Organizing Committee shall organize an election from among the membership, by electronic or other means suited to its scientific structure, of a new Organizing Committee to take office for the following term. Election procedures should, as far as possible, be similar among the Divisions and require the

approval of the Executive Committee.

20. Each Scientific Division may structure its scientific activities by creating a number of Commissions. In order to monitor and further the progress of its field of astronomy, the Division shall consider, before each General Assembly, whether its Commission structure serves its purpose in an optimum manner. It shall subsequently present its proposals for the creation, continuation or discontinuation of Commissions to the Executive Committee for approval.

21. With the approval of the Executive Committee, a Division may establish Working Groups to study welldefined scientific issues and report to the Division. Unless specifically re-established by the same procedure, such Working Groups cease to exist at the next following General Assembly.

VI. SCIENTIFIC COMMISSIONS

22. A Scientific Commission shall consist of:

22.a. A President and an Organizing Committee consisting of 4-8 persons elected by the Commission membership, subject to the approval of the Organizing Committee of the Division;

22.b. Members of the Union, accepted by the Organizing Committee, in recognition of their special experience and interests, subject to confirmation by the Organizing Committee of the Division.

23. A Commission is initially created for a period of six years. The parent Division may recommend its continuation for additional periods of three years at a time, if sufficient justification for its continued activity is presented to the Division and the Executive Committee. The activities of a Commission are governed by Terms of Reference, which are based on a standard model published by the Executive Committee and are approved by the Division.

24. With the approval of the Division, a Commission may establish Working Groups to study well-defined scientific issues and report to the Commission. Unless specifically re-appointed by the same procedure, such Working Groups cease to exist at the next following General Assembly.

VII. ADMINISTRATION AND FINANCES

25. Each National Member pays annually to the Union a number of units of contribution corresponding to its category as specified below. National Members with interim status pay annually one half unit of contribution, and those with prospective status pay no dues.

Categories as defined in Statutes, §10:	I	II	III	IV	V	VI	VII	VIII	IX	X	XI	XII
Number of units of contribution:	1	2	4	6	10	14	20	27	35	45	60	80

26. The income of the Union is to be devoted to its objects, including:

26.a. the promotion of scientific initiatives requiring international co-operation;

26.b. the promotion of the education and development of astronomy world-wide;

26.c. the costs of the publications and administration of the Union.

27. Funds derived from donations are reserved for use in accordance with the instructions of the donor(s). Such donations and associated conditions require the approval of the Executive Committee.

28. The General Secretary is the legal representative of the Union. The General Secretary is responsible to the Executive Committee for not incurring expenditure in excess of the amount specified in the budget as approved by the General Assembly.

29. The General Secretary shall consult with the Finance Committee (cf. Statutes §13.d.) in preparing the accounts and budget proposals of the Union, and on any other matters of major importance for the financial health of the Union. The comments and advice of the Finance Committee shall be made available to the Officers and Executive Committee as specified in the Working Rules.

30. An Administrative office, under the direction of the General Secretary, conducts the correspondence, administers the funds, and preserves the archives of the Union.

31. The Union has copyright to all materials printed in its publications, unless otherwise arranged.

VIII. FINAL CLAUSES

32. These Bye-Laws enter into force on 21 August 2012.

33. The present Bye-Laws are published in French and English versions. For legal purposes, the French version is authoritative.

4. IAU Working Rules

INTRODUCTION AND RATIONALE

The Statutes of the International Astronomical Union (IAU) define the goals and organizational structure of the Union, while the Bye-Laws specify the main tasks of the various bodies of the Union in implementing the provisions of the Statutes. The Working Rules are designed to assist the membership and governing bodies of the Union in carrying out these tasks in an appropriate and effective manner. Each of the sections below is preceded by an introduction outlining the goals to be accomplished by the procedures specified in the succeeding paragraphs. The Executive Committee updates the Working

Rules as necessary to reflect current procedures and to optimize the services of the IAU to its membership.

I. NON-DISCRIMINATION

The International Astronomical Union (IAU) follows the regulations of the International Council for Science (ICSU) and concurs with the actions undertaken by their Standing Committee on Freedom in the Conduct of Science on non-discrimination and universality of science (cf. §22 below)

II. NATIONAL MEMBERSHIP

The aim of the rules for applications for National Membership is to ensure that the proposed National Member adequately represents an astronomical community not already represented by another Member, and that such membership will be of maximum benefit for the community concerned (cf. Statutes §IV).

1. Applications for National Membership should therefore clearly describe the following essential conditions:

1.a. the precise definition of the astronomical community to be represented by the proposed Member;

1.b. the present state and expected development of that astronomical community;

1.c. the manner in which the proposed National Member represents this community;

1.d. whether the application is for membership on a permanent, interim, or prospective basis; and

1.e. the category in which the prospective National Member wishes to be classified (cf. Bye-Laws §25).

1.f. the process by which the National Membership annual dues will be paid promptly and in full.

2. Applications for National Membership shall be submitted to the General Secretary, who will forward them to the Executive Committee for review as provided in the Statutes.

III. INDIVIDUAL MEMBERSHIP

Professional scientists whose research is directly relevant to some branch of astronomy are eligible for election as Individual Members of the Union (cf. Statutes §V). Individual Members are normally admitted by the Executive Committee on the proposal of a

National Member. However, Presidents of Divisions may also propose individuals for membership in cases when the normal procedure is not applicable or practicable (cf. Bye-Laws §4). The present rules are intended to ensure that all applications for membership are processed on a uniform basis, and that all members are fully integrated in and contributing to the activities of the Union.

3. The term "Professional Scientist" shall normally designate a person with a doctoral degree (Ph.D.) or equivalent experience in astronomy or a related science, and whose professional activities have a substantial component of work related to astronomy.

4. National Members and Division Presidents may propose Individual Members who fall outside the category of professional scientist but who have made major contributions to the science of astronomy, e.g., through education or research related to astronomy. Such proposals should be accompanied by a detailed motivation for what should be seen as exceptions to the rule.

5. Eight months before an ordinary General Assembly, National Members and Presidents of Divisions will be invited to propose new Individual Members; these proposals should reach the General Secretary no later than five months before the General Assembly. Late proposals will normally not be taken into consideration. Proposals from Presidents of Divisions will be communicated by the IAU 3 months before the General Assembly to the relevant National Members, if any, who may add the person(s) in question to their own list of proposals.

6. National Members shall promptly inform in writing the General Secretary of the death of any Individual Member represented by them. National Members are also urged to propose the deletion of Individual Members who are no longer active in astronomy by including a written agreement of the member concerned. Such proposals should be submitted to the General Secretary at the same time as proposals for new Individual Members.

7. Proposals for membership shall include the full name, date of birth, and nationality of the candidate, postal and electronic addresses, the University, year, and subject of the M.Sc./Ph.D. or equivalent degree, current affiliation and occupation, the proposing National Member or Division, the Division(s) and/or Commission(s) which the candidate wishes to join, and any further detail that might be relevant.

8. The standing Membership Committee advises the Executive Committee on its behalf on matters related to the admission of Individual Members (Statute 13f.).

8.a Three months before a General Assembly, the Membership Committee shall prepare a list of at least 10 Individual Members of the Union who accept to serve on the Committee for the next triennium if elected, including a nominee for Chair. The General Secretary shall forward this list to the National Members and invite additional nominations from them with a deadline of one month before the General Assembly. The Membership Committee shall verify that the resulting slate complies with the rules in Statutes 13 and with general principles of scientific, geographical and gender balance. Members shall not normally serve more than two consecutive terms, and it is desirable that roughly half of the members are replaced at each election. The Chair of the Membership Committee shall present the resulting slate of nominations to the National Members together with the report of the Committee on the previous triennium at the beginning of the General Assembly, for final election at its closing session.

9. The General Secretary shall submit all proposals for Individual Membership to the Membership Committee for review, consolidated into two lists:

9.a one containing all proposals by National Members; and

9.b one containing all proposals by Presidents of Divisions, in accordance with Bye-Law §4 .

10. The Membership Committee shall examine all proposals for individual membership and advise the Executive Committee on the proposals for individual membership.

11. In exceptional cases, the Executive Committee may, on the proposal of a Division, admit an Individual Member between General Assemblies. Such proposals shall be prepared as described above (cf. §2) and submitted with a justification of the request to bypass the normal procedure. The Executive Committee shall consult the Nominating Committee or relevant National Member before approving such exceptions to the normal procedure.

12. The General Secretary shall maintain updated lists of all National and Individual Members, and shall make these available to the membership in electronic form. The procedures for dissemination of these lists shall be set by the General Secretary in such a way that the membership directory be properly protected against unintended or inappropriate use.

IV. RESOLUTIONS OF THE UNION

Traditionally, the decisions and recommendations of the Union on scientific and organizational matters of general and significant importance are expressed in the Resolutions of the Union. In order for such Resolutions to carry appropriate weight in the international community, they should address astronomical matters of significant impact on the international society, or matters of international policy of significant importance for the international astronomical community as a whole.

Resolutions should be adopted by the Union only after thorough preparation by the relevant bodies of the Union. The proposed resolution text should be essentially complete before the beginning of the General Assembly, to allow Individual and National Members time to study them before discussion and debate by the General Assembly. The following procedures have been designed to accomplish this:

13. Proposals for Resolutions to be adopted by the Union may be submitted by a National Member, by the Executive Committee, a Division, a Commission or a Working Group. They should address specific issues of the nature described above, define the objectives to be achieved, and describe the action(s) to be taken by the Officers, Executive Committee, or Divisions to achieve these objectives.

14. Resolutions proposed for vote by the Union fall in three categories as set out in Article 14 of the Union's Statutes:

14.a. Resolutions with implications for the budget of the Union (Statute 14a); or

14.b. Resolutions affecting the administration of the Union but without financial implications (Statute 14b).

14.c. Resolutions of a primarily scientific nature (Statute 14).

Proposals for Resolutions should be submitted on standard forms appropriate for each type, which are available from the IAU Secretariat. They may be submitted in either English or French and will be discussed and voted upon in the original language. Upon submission each proposed Resolution is posted on the Union web site. When the approved Resolutions are published, a translation to the other language will be provided.

15. Resolutions with implications for the budget of the Union must be submitted to the General Secretary at least nine months before the General Assembly in order to be taken into account in the budget for the impending triennium. All other Resolutions must be submitted to the General Secretary six months (Bye-Laws 7c and 7d) before the beginning of the General Assembly. The Executive Committee may decide to accept late proposals in exceptional circumstances.

16. Before being submitted to the vote of the General Assembly, proposed Resolutions will be examined by the Executive Committee, Division Presidents, and by a Resolutions Committee, which is nominated by the Executive Committee. The Resolutions Committee consists of at least three members of the Union, one of whom should be a member of the Executive Committee, and one of whom should be a continuing member from the previous triennium. It is appointed by the General Assembly during its final session and remains in office until the end of the following General Assembly.

17. The Resolutions Committee will examine the content, wording, and implications of all proposed Resolutions promptly after their submission. In particular, it will address the following points:

i. suitability of the subject for an IAU Resolution;

ii. correct and unambiguous wording;

iii. consistency with previous IAU Resolutions.

The Resolutions Committee may refer a Resolution back to the proposers for revision or withdrawal if it perceives significant problems with the text, but can neither withdraw nor modify its substance on its own initiative. The Resolutions Committee advises the Executive Committee whether the subject of a proposed Resolution is primarily a matter of policy or primarily scientific. The Resolutions Committee will also notify the Executive Committee of any perceived problems with the substance of a proposed Resolution.

18. The Executive Committee will examine the substance and implications of all proposed Resolutions. Proposed Resolutions shall be published in the General Assembly

Newspaper before the final session, and shall state if the Resolution is open to electronic voting after the General Assembly. The Resolutions Committee will present the proposals during a plenary session of the General Assembly with its own recommendations, and those of the Executive Committee, if any, for their approval or rejection. A representative of the body proposing the Resolution will be given the opportunity to defend the Resolution in front of the General Assembly, after which a general discussion shall take place

19. Resolutions with implications for the budget of the Union are voted upon by the National Members during the final plenary session of the General Assembly. Other resolutions may be voted upon by the National Members or by Individual Members as appropriate according to the Statutes of the Union by correspondence after the General Assembly. The Union will facilitate electronic discussion of all Resolutions on the Union website in advance of a vote either at the General Assembly or electronically (Statutes 15a. and 15.b.).

V. EXTERNAL RELATIONS

Contacts with other international scientific organizations, national and international public bodies, the media, and the public are increasing in extent and importance. In order to maintain coherent overall policies in matters of international significance, clear delegation of authority is required. Part of this is accomplished by having the Union's representatives in other scientific organization appointed by the Executive Committee (cf. Bye-Laws §15). Supplementary rules are given in the following.

20. Representatives of the Union in other scientific organizations are appointed by the Executive Committee upon consultation with the Division(s) in the field(s) concerned.

21. In other international organizations, e.g. in the United Nations Organization, the Union is normally represented by the General Secretary or Assistant General Secretary, as decided by the Executive Committee.

22. The Union strongly supports the policies of the International Council for Science (ICSU) as regards the freedom and universality of science. Participants in IAU sponsored activities who feel that they may have been subjected to discrimination are urged, first, to seek clarification of the origin of the incident, which may have been due to misunderstandings or to the cultural differences encountered in an international environment. Should these attempts not prove successful, contact should be made with the General Secretary who will take steps to resolve the issue.

23. Public statements that are attributed to the Union as a whole can be made only by the President, the General Secretary, or the Executive Committee. The General Secretary may, in consultation with the relevant Division, appoint Individual Members of the Union with special expertise in questions that attract the attention of media and the general public as IAU spokespersons on specific matters.

VI. FINANCIAL MATTERS

The great majority of the Union's financial resources are provided by the National Members, as laid out in the Statutes §XII and Bye-Laws §VII. The purpose of the procedures described below is twofold: (i) to provide the best possible advice and guidance to the General Secretary and Executive Committee in planning and managing the Union?s financial affairs, and (ii) to provide National Members with a mechanism for continuing input to and oversight over these affairs between and in preparation for the General Assemblies. The procedures adopted to accomplish this are as follows:

24. At the end of each of its final sessions the General Assembly appoints a Finance Sub-Committee of 5-6 members, including a Chair.

24.a. Three months before a General Assembly, the Finance Committee shall prepare a list of at least 10 Individual Members of the Union who accept to serve on the Committee for the next triennium if elected, including a nominee for Chair. The General Secretary shall forward this list to the National Members and invite additional nominations from them with a deadline of one month before the General Assembly. The Finance Committee shall verify that the resulting slate complies with the rules in Statutes 13 and with general principles of scientific, geographical and gender balance. Members shall not normally serve more than two consecutive terms, and it is desirable that roughly half of the members are replaced at each election. The Chair of the Finance Committee shall present the resulting slate of nominations to the National Members together with the report of the Committee on the previous triennium at the beginning of the General Assembly, for final election at its closing session. The Finance Committee remains in office until the end of the next General Assembly (cf. Statutes §13.d.) and cooperates with the National Members, Executive Committee and General Secretary in the following manner:

24.b. After the end of each year the General Secretary will call for a legal audit of the accounts by a properly licensed, external auditor. The auditor will make a report addressed to the General Assembly. The General Secretary provides the Finance Sub-Committee with the auditor's report and summary reports covering the financial performance of the Union as compared to the approved budget, together with an analysis of any significant departures, and information on any Executive Committee approvals of budget changes. Upon receipt of the above reports from the General Secretary, the Finance Sub-Committee examines the accounts of the Union in the light of the corresponding budget and any relevant later decisions by the Executive Committee. It reports its findings and recommendations to the Executive Committee at its next meeting. The Finance Sub-Committee may at any time, at the request of the Executive Committee or the General Secretary, or on its own initiative, advise the General Secretary and/or the Executive Committee on any aspect of the Union's financial affairs.

24.c. Towards the end of the year preceding that of a General Assembly, the General Secretary shall submit a preliminary draft of the budget for the next triennium to the Finance Committee for review. The draft budget, updated as appropriate following the comments and advice of the Finance Committee, shall be submitted to the Executive Committee for approval together with the report of the Finance Committee and shall be sent to the National Members as a draft budget as stipulated in Bye-Law 7b. The final budget proposal as approved by the Executive Committee shall be submitted to the National Members with a statement of the views of the Finance Committee on the proposal.

24.d. Before the first session of a General Assembly, the Finance Sub-Committee shall submit a report, including the auditor's reports, to the Executive Committee on its findings and recommendations concerning the development of the Union's finances over the preceding triennium. The Finance Sub-Committee shall also prepare, in consultation with the National Members, a slate of candidates for the composition of the Finance Sub-Committee in the next triennium, preferably providing a balance between new and continuing members.

24.e. The report of the Finance Committee, together with the audited detailed accounts and the earlier comments on the proposed budget for the next triennium, will form a suitable basis for the discussions of the Finance Committee leading to its recommendations to the General Assembly concerning the approval of the accounts for the previous triennium and the budget for the next triennium, as well as the new Finance Sub-Committee to serve during that period.

25. The General Secretary is responsible for managing the Union's financial affairs according to the approved budget (cf. Bye-Laws §28).

25.a. In response to changing circumstances, the Executive Committee may approve such specific changes to the annual budgets as are consistent with the intentions of the General Assembly when the budget was approved.

25.b. Unless authorized by the Executive Committee, the General Secretary shall not approve expenses exceeding the approved budget by more than 10% of any corresponding major budget line or 2% of the total budget in a given year, whichever is larger. This restriction does not apply in cases when external funding has been provided for a specific purpose, e.g. travel grants to a General Assembly.

25.c. Unless specifically identified in the approved budget, contractual commitments in excess of €50,000, or with performance terms in excess of 3 years require the additional approval of the Union President.

26. The National Representatives, in approving the accounts for the preceding triennium, discharge the General Secretary and the Executive Committee of liability for the period in question.

VII. RULES OF PROCEDURE FOR THE EXECUTIVE COMMITTEE

The Executive Committee must respond quickly to events and thus it needs to be able to have discussions and take decisions on a relatively short timescale and without meeting in person. The following rules, as required by Bye-Law 14, are designed to facilitate EC action in a flexible manner, while giving such decisions the same legal status as those taken at actual physical meetings.

27. The Executive Committee should meet in person at least once per year. In years of a General Assembly it should meet in conjunction with and at the venue of the General Assembly. In other years, the Executive Committee decides on the date and venue of its regular meeting. The meetings of the Executive Committee are chaired by the President or, if the President is unavailable, by the President Elect or by one of the Vice-Presidents chosen by the Executive Committee to serve in this capacity.

28. The date and venue of the next regular meeting of the Executive Committee shall be communicated at least six months in advance to all its members and the Advisors, and

to all Presidents of Divisions. Any of these persons may then propose items for inclusion in the Draft Agenda of the meeting before the date posted on the IAU Deadlines page.

29. Outgoing and incoming Presidents of Divisions are invited to attend all non-confidential sessions of the outgoing and incoming Executive Committee, respectively, in the year of a General Assembly. The President will invite Presidents of Divisions to attend the meetings of the Executive Committee in the years preceding a General Assembly. Division Presidents attend these sessions with speaking right, but do not participate in any voting.

30. The Executive Committee may take official decisions if at least half of its members participate in the discussion and vote on an issue. Decisions are taken by a simple majority of the votes cast. In case of an equal division of votes, the Chair's vote decides the issue. Members who are unable to attend may, by written or electronic correspondence with the President before the meeting, authorize another member to vote on her/his behalf or submit valid votes on specific issues.

31. If events arise that require action from the Executive Committee between its regular meetings, the Committee may meet by teleconference or by such electronic or other means of correspondence as it may decide. In such cases, the Officers shall submit a clear description of the issue at hand, with a deadline for reactions. If the Officers propose a specific decision on the issue, the decision shall be considered as approved unless a majority of members vote against it by the specified deadline. In case of a delay in communication, or if the available information is considered insufficient for a decision, the deadline shall be extended or the decision deferred until a later meeting at the request of at least two members of the Executive Committee.

32. The Officers of the IAU should, as a rule, meet once a year at the IAU Secretariat in order to discuss all matters of importance to the Union. The other members of the Executive Committee and the Division Presidents shall be invited to submit items for discussion at the Officers' Meetings and shall receive brief minutes of these Meetings.

33. Should any member of the Executive Committee have a conflict of interest on a matter before the Executive Committee that might compromise their ability to act in the best interests of the Union, they shall declare their conflict of interest to the Executive Committee, and such conflict shall be recorded by the Secretariat. The remaining members of the Executive Committee determine the appropriate level of participation in such issues for members with a potential conflict of interest.

VIII. SCIENTIFIC MEETINGS AND PUBLICATIONS

Meetings and their proceedings remain a major part of the activities of the Union. The purpose of scientific meetings is to provide a forum for the development and dissemination of new ideas, and the proceedings are a written record of what transpired.

34. The General Secretary shall publish in the Transactions and on the IAU web site rules for scientific meetings organized or sponsored by the Union.

35. The proceedings of the General Assemblies and other scientific meetings organized or sponsored by the Union shall, as a rule, be published. To ensure prompt publication of Proceedings of IAU Symposia and Colloquia, the Assistant General Secretary is authorized to oversee the production of the material for the Proceedings. The Union shall publish an Information Bulletin at regular intervals to keep Members informed of current and future events in the Union. The Union shall also publish a more informal, periodic Newsletter which it distributes electronically to its members. The Executive Committee decides on the scope, format, and production policies for such publications, with due regard to the need for prompt publication of new scientific results and to the financial implications for the Union. At the present time, publications are in printed and in electronic form.

36. Divisions, Commissions, and Working Groups shall, with the approval of the Executive Committee, be encouraged to issue Newsletters or similar publications addressing issues within the scope of their activity.

IX. TERMS OF REFERENCE FOR DIVISIONS

The Divisions are the scientific backbone of the IAU. They have a main responsibility for monitoring the scientific and international development of astronomy within their subject areas, and for ensuring that the IAU will address the most significant issues of the time with maximum foresight, enterprising spirit, and scientific judgment. To fulfill this role IAU Divisions should maintain a balance between innovation and continuity. The following standard Terms of Reference have been drafted to facilitate that process, within the rules laid down in the Statutes §X and the Bye-Laws §V.

37. As specified in Bye-Law 18, the scientific affairs of the Division are conducted by an Organizing Committee of up to 12 members of the Division, headed by the Division President, Vice-President, and Secretary. Thus, all significant decisions of the Division require the approval of the Organizing Committee, and the President and Vice-President are responsible for organizing the work of the Committee so that its members are consulted in a timely manner. Contact information for the members of the Organizing Committee shall be maintained at the Division web site.

Unless agreed otherwise by the Executive Committee on a case by case basis, the President of a Division cannot be President of another Division or of a Commission, or be Chair of a Working Group.

38. Individual Members of the Union are admitted to membership in a Division by its Organizing Committee (cf. Bye-Laws §18). Individual Members active within the field of activity of the Division and interested in contributing to its development should contact the Division Secretary, who will consult the Organizing Committee on the admission of the candidates.

38.a. The Division Secretary shall maintain a list of Division members for ready consultation by the community, including their Commission memberships if any.

Updates to the list shall be provided to the IAU Secretariat on a running basis.

38.b. Members may resign from a Division by so informing the Division Secretary.

38.c. In the event of a Division being newly formed, Individual Members can themselves elect to join the Division. Before the General Assembly following that at which the new Division was created its Organising Committee shall scrutinise and confirm the Division membership.

39. The effectiveness of the Division relies strongly on the scientific stature and dedication of its President and Vice-President to the mission of the Division. The Executive Committee, in proposing new Division Presidents and Vice-Presidents for election by the General Assembly, will rely heavily on the recommendations of the Organizing Committee of the Division. In order to prepare a strong slate of candidates for these positions, and for the succession on the Organizing Committee itself, the following procedures shall normally apply:

39.a. Candidates are proposed and selected from the membership of the Division on the basis of their qualifications, experience, and stature in the fields covered by the Division. In addition, the Organizing Committees should have proper gender balance and broad geographical representation.

39.b. At least six months before a General Assembly, the Organizing Committee submits to the membership of the Division a list of candidates for President, Vice-President (for which there should be at least two persons willing to serve), Secretary, and the Organizing Committee for the next triennium. The Organizing Committee requests nominations from the entire membership in preparing this list, and then conducts a vote, normally electronically, among all Division members for the above offices, the results of which are reported to the General Secretary at least three months before the General Assembly. The Vice-President is normally nominated to succeed the President. The outgoing President participates in the deliberations of the new Organizing Committee in an advisory capacity.

39.c. If more names are proposed than there are positions to be filled on the new Organizing Committee, the outgoing Organizing Committee devises the procedure by which the requisite number of candidates is elected by the membership. The resulting list is communicated to the General Secretary at least two months before the General Assembly. The General Secretary may allow any outstanding issues to be resolved at the business meeting of the Division during the General Assembly. If for any reason the Organizing Committee has not been able to arrange for the election of new officers and an Organizing Committee by two months before the GA, the EC will nominate a VP and Organizing Committee at its first General Assembly meeting.

39.d. A member of the Organizing Committee normally serves a maximum of two terms, unless elected Vice-President of the Division in her/his second term. Presidents may serve for only one term.

39.e. The Organizing Committee decides on the procedures for designating the Division Secretary, who maintains the web site, records of the business and membership of the Division, and other rules for conducting its business by physical meetings or by correspondence.

39.f. In the event of a newly formed Division, paragraphs 39a - 39c do not apply. The Executive Committee shall consult the Organizing Committees of the relevant predecessor Divisions on possibl candidates for President and Vice-President of the new Division for the next triennium. The Executive Committee shall select the names to be proposed to the Generals Assembly for election.

39.g. As soon as possible after their election at a General Assembly, the President and Vice-President of the new Division shall request nominations to the Organising Committee from the membership of the Division and then conduct a vote among Division members, the results of which are reported to the General Secretary. The Organising Committee elects a Secretary from its membership.

40. A key responsibility of the Organizing Committee is to maintain an internal organization of Commissions and Working Groups in the Division which is conducive to the fulfillment of its mission. The Organizing Committee shall take the following steps to accomplish this task in a timely and effective manner:

40.a. Within the first year after a General Assembly - with the business meeting of the Commission at the General Assembly itself as a natural starting point - the Organizing Committee shall discuss with its Commissions, and within the Organizing Committee itself, if changes in its Commission and Working Group structure may enable it to accomplish its mission better in the future. As a rule, Working Groups should be created (following the rules in Bye-Law 21 and Bye-Law 23) for new activities that are either of a known, finite duration or are exploratory in nature. If experience, possibly from an existing Working Group, indicates that a major section of the Division's activities require a coordinating body for a longer period (a decade or more), the creation of a new Commission may be in order.

40.b. Whenever the Organizing Committee is satisfied that the creation of a new Working Group or Commission is well motivated, it may take immediate action as specified in Bye-Law 21 or Bye-Law 23. In any case, the Organizing Committee submits its complete proposal for the continuation, discontinuation, or merger of its Commissions and Working Groups to the General Secretary at least three months before the next General Assembly.

40.c. The President and Organizing Committee maintain frequent contacts with the other IAU Divisions to ensure that any newly emerging or interdisciplinary matters are addressed appropriately and effectively.

X. TERMS OF REFERENCE FOR COMMISSIONS

The role of the Commissions is to organize the work of the Union in specialized subsets of the fields of their parent Division(s), when the corresponding activity is judged to be of considerable significance over times of a decade or more. Thus, new Commissions may be created by the Executive Committee with the agreement of all the Divisions when fields emerge that are clearly in sustained long-term development and where the Union may play a significant role in promoting this development at the international level. Similarly,

Commissions may be discontinued by the Executive Committee upon the recommendation of the parent Division when their work can be accomplished effectively by the parent Division. In keeping with the many-sided activities of the Union, Commissions may have purely scientific as well as more organizational and/or interdisciplinary fields. They will normally belong and report to one of the IAU Divisions, but may be common to two or more Divisions. The following rules apply if a Division has more than one Commission.

41. The activities of a Commission are directed by an Organizing Committee of 4-8 members of the Commission, headed by a Commission President and Vice-President (cf. Bye-Laws §22). A member of the Organizing Committee normally serves a maximum of two terms, unless elected Vice-President of the Commission in her/his second term. Presidents may serve for only one term. All members of the Organizing Committee are expected to be active in this task, and are to be consulted on all significant actions of the Commission. The Organizing Committee appoints a Commission Secretary who maintains the records of the membership and activities of the Commission in co-operation with the Division Secretary and the IAU Secretariat. Contact information for the members of the Organizing Committee shall be maintained at the Commission web site.

Unless agreed otherwise by the Executive Committee on a case by case basis, the President of a Commission cannot be President of a Division or of another Commission, or be Chair of a Working Group.

42. Individual Members of the Union, who are active in the field of the Commission and wish to contribute to its progress, are admitted as members of the Commission by the Organizing Committee. Interested Members should contact the Commission Secretary, who will bring the request before the Organizing Committee for decision. Members may resign from the Commission by notifying the Commission Secretary. Before each General Assembly, the Organizing Committee may also decide to terminate the Commission membership of persons who have not been active in the work of the Commission; the individuals concerned shall be informed of such planned action before it is put into effect. The Commission Secretary will report all changes in the Commission membership to the Division Secretary and the IAU Secretariat.

43. At least six months before a General Assembly, the Organizing Committee submits to the membership of the Commission a list of candidates for President, Vice-President (for which there should be the names of two persons willing to serve), the Organizing Committee, and heads of Program Groups for the next triennium. The Organizing Committee requests nominations from the entire membership in preparing this list, and then conducts a vote, normally electronically, among all the members for the above offices, the results of which are reported to the General Secretary at least three months before the General Assembly. The Vice-President is normally nominated to succeed the President. The outgoing President participates in the deliberations of the new Organizing Committee in an advisory capacity. If more names are proposed than available elective positions, the outgoing Organizing Committee devises the procedure by which the requisite number of candidates is elected by the membership. The resulting list is submitted to the Organizing Committee of the parent Division(s) for approval before the end of the General Assembly. Members of the Organizing Committee normally serve a maximum of two terms, unless elected Vice-President of the Commission. Presidents may serve for only one term.

44. At least six months before each General Assembly, the Organizing Committee shall submit to the parent Division(s) a report on its activities during the past triennium, with its recommendation as to whether the Commission should be continued for another three years, or merged with one or more other Commissions, or discontinued. If a continuation is proposed, a plan for the activities of the next triennium should be presented, including those of any Working Groups which the Commission proposes to maintain during that period.

45. The Organizing Committee decides its own rules for the conduct of its business by physical meetings or (electronic) correspondence. Such rules require approval by the Organizing Committee of the parent Division(s).

46. The procedural rules applying to the establishment of a new Division shall also apply to the establishment of a new Commission. Where there is no 'relevant predecessor Commission(s)' the parent Division(s) shall submit to the potential membership of the new Commission a list of candidates for President, Vice-President and Organising Committee for the next triennium.

XI. SPECIAL NOMINATING COMMITTEE

47. Approximately six months before the start of the General Assembly the General Secretary shall invite Division Presidents and the members of the Executive Committee to nominate potential members of the SNC with a deadline of 3 months before the General Assembly. The Executive Committee shall prepare a list of candidates in consultation with the Membership Committee for appointment at the final Business Session of the General Assembly. The SNC, once appointed, shall elect its own Chair.

Transactions IAU, Volume XXVIIIB
Proc. XXVIII IAU General Assembly, August 2012
Thierry Montmerle, ed.

doi:10.1017/S1743921315005669

CHAPTER VIII

NEW MEMBERS AND DECEASED MEMBERS AT THE GENERAL ASSEMBY

1. New members admitted at the General Assembly

The following lists give the names of the 1008 new Individual Members admitted at the XXVIIIth General Assembly, ordered by National Member. New National Members are indicated by an asterisk (Ethiopia, Kazakhstan, Democratic People's Republic of Korea).

For a complete list of IAU members, please consult the IAU Directory:
`http://www.iau.org/administration/membership/individual/`.

For a complete list of IAU National Members (member countries), please visit:
`http://www.iau.org/administration/membership/national/`.

Argentina

Abrevaya, Ximena
Althaus, Leandro G.
Andruchow, Ileana
Cremades Fernandez, M. H.
De Rossi, Mara E.
Faifer, Favio R.
Hgele, Guillermo F.
Lopez, Alejandro M.
Parisi, Maria C.
Smith Castelli, Analia V.

Armenia

Hakopian, Susanna A.
Nikoghosyan, Elena H.

Australia

Bauer, Amanda E.
Bekki, Kenji
Brough, Sarah
Campbell, Simon W.
Collet, Remo
De Silva, Gayandhi M.
Edwards, Philip G.
Emonts, Bjorn
Farrell, Sean A.
Floyd, David J.
Kimball, Amy E.
Kudryavtseva, Nadezhda A.
Lenc, Emil
Mackey, Alasdair D.
Owers, Matthew S.
Price, Daniel J.
Rathborne, Jill M.
Reid, Warren A.
Rowell, Gavin P.
Seymour, Nicholas
Soria, Roberto
Stancliffe, Richard J.
Wayth, Randall B.
Wendt, Harry W.
Wicenec, Andreas J.
Wittenmyer, Robert A.
Zadnik, Marjan G.

Austria

Böhm, Johannes
Dannerbauer, Helmut
Dionatos, Odysseas
Kausch, Wolfgang
Khodachenko, Maxim L.
Lammer, Helmut
Maier, Christian
Marleau, Francine
Noll, Stefan
Nowotny-Schipper, Walter
Rank-Lüftinger, Theresa
Zhang, Tielong
Zwintz, Konstanze

Belgium

Absil, Olivier
De Becker, Michal D.
Gillon, Michal
Gentile, Gianfranco
Jacobs, Carla
Libert, Anne-Sophie
Morel, Thierry
Montalban, Josefina
Lobel, Alex J.
Parenti, Susanna
Noyelles, Benot B.
Nazé, Yael
Seaton, Daniel B.
Van Doorsselaere, Tom

Brazil

Alves, Virgnia M.
Bortoletto, Alexandre E.
Bretones, Paulo S.
Bueno De Camargo, Julio I.
Canto Martins, Bruno L.
Carciofi, Alex C.
Ferrari, Fabricio
Figueiredo, Newton
Krabbe, Angela C.
Lopes, Paulo A.
Lopes De Oliveira, R.
Ogando, Ricardo L.
Pilling, Sergio
Rojas, Gustavo D.
Sobreira, Paulo H.
Westera, Pieter W.

Bulgaria

Borisov, Borislav S.
Borisov, Galin B.
Borisova, Ana P.
Slavcheva-Mihova, Lyuba S.
Stoyanov, Kiril A.

Canada

Anderson, Jay
Bietenholz, Michael F.
Cannon, Kipp
Dufour, Patrick
Edwards, Louise O.
Ellison, Sara
Hanna, David S.
Heyl, Jeremy S.
Ivanova, Natalia
Marois, Christian
Patton, David R.
Reid, Michael
Rosolowsky, Erik W.
Sawicki, Marcin
Seyed-Mahmoud, Behnam
Shalchi, Andreas
Shelton, Ian K.
Sivakoff, Gregory R.
Strubbe, Linda E.
Willis, Jon P.

Chile

Anderson, Joseph P.
Berger, Jean-Philippe
Boquien, Mdric
Casassus, Simon
Cieza, Lucas A.
Clocchiatti, Alejandro
Cuadra, Jorge
Day Jones, Avril C.
Demarco, Ricardo J.
Gadotti, Dimitri A.
Girard, Julien H.
Kabath, Petr
Kameno, Seiji
Levenson, Nancy A.
Lo, Wing-Chi Nadia
Mardones, Diego
Martayan, Christophe D.
Martin, Sergio
Mawet, Dimitri P.
Monaco, Lorenzo
Nitschelm, Christian H.
Ohnaka, Keiichi
Pretorius, Magaretha L.
Puzia, Thomas H.
Reisenegger, Andreas
Siringo, Giorgio
Steenbrugge, Katrien C.
Tristram, Konrad R.
Unda-Sanzana, Eduardo
Van Der Bliek, Nicole S.
Vlahakis, Catherine E.
Wesson, Roger
Zoccali, Manuela

China Nanjing

An, Tao
Cai, Mingsheng
Cai, Yong
Chang, Hong
Chen, Dongni
Chen, Linfei
Chen, Xinyang
Chen, Zhijun
Dai, Zhibin
Dou, Jiangpei
Fan, Yufeng
Foucaud, Sebastien R.
Frew, David J.
Gao, Jian
Gong, Biping
Gu, Minfeng
Guo, Ji
Hao, Lei
He, Han
Hou, Xiyun
Ji, Li
Jia, Lei
Jiang, Zhibo
Jin, Liping
Kang, Xi
Kouwenhoven, M.B.N.
Lee, Jun
Li, Min
Li, Yuqiang
Liang, Guiyun
Lin, Chuang-Jia
Lin, Jun
Liu, Chengzhi
Liu, Guoqing
Liu, Jifeng
Liu, Siming
Liu, Tao
Liu, Xiaoqun
Liu, Yujuan
Lu, Youjun
Ma, Guanyi
Ning, Xiaoyu
Peng, Eric W.
Qu, Jinlu
Shen, Juntai
Shu, Fengchun
Tan, Baolin
Tian, Feng
Wang, Bo
Wang, Chen
Wang, Guangchao
Wang, Guangli
Wang, Junzhi
Wang, Min
Wang, Sen
Wang, Xiaoya
Wang, Yu
Wang, Yulin
Wei, Erhu
Wu, Xue-Feng
Wu, Zhen-Yu
Xiao, Dong
Xie, Yi
Xu, Ye
Yu, Cong
Yu, Qingjuan
Zhang, Chengmin
Zhang, Mian
Zhang, Shu
Zhang, Yang
Zhang, Yong
Zhang, Zhibin
Zheng, Xianzhong
Zheng, Xiaonian
Zhou, Xia
Zhu, Ming
Zhu, Qingfeng

China Taipei

Chou, Mei-Yin
Huang, Hui-Chun
Kemper, Francisca
Lin, Lihwai
Lin, Yen-Ting
Mizuno, Yosuke
Otsuka, Masaaki
Shang, Hsien
Tsai, An-Li
Tseng, Yao-Huan
Urata, Yuji
Wang, Shiang-Yu

Costa Rica

Barrantes, Marco N.
Frutos-Alfaro, Francisco

Croatia, the Republic of

Jelic, Vibor
Ruzdjak, Domagoj
Sudar, Davor

Czech Republic

Capek, David
Druckmueller, Miloslav
Gunar, Stanislav
Hledik, Stanislav
Jachym, Pavel
Janik, Jan
Kovar, Jiří
Scheirich, Peter
Shrbeny, Luk
Slany, Petr
Slechta, Miroslav
Stepan, Jiří
Svanda, Michal
Torok, Gabriel
Zasche, Petr
Zejda, Miloslav

Denmark

Bjaelde, Ole E.
Buchhave, Lars A.
Karoff, Christoffer
Leloudas, Georgios
Zirm, Andrew W.

Estonia

Aret, Anna
Hirv, Anti
Tempel, Elmo

Ethiopia*

Tessema, Solomon B.

Finland

Comeron, Sbastien
Granvik, Mikael
Hackman, Thomas
Koivisto, Tomi S.
Kostama, Veli-Petri
Neustroev, Vitaly V.
Rekola, Rami T.
Tammi, Joni P.
Tsygankov, Sergey
Valiviita, Jussi-Pekka

France

Augereau, Jean-Charles
Belkacem, Kevin
Benisty, Myriam
Bonfils, Xavier
Bot, Caroline L.
Botti, Thierry
Bourda, Géraldine
Cabanac, Remi A.
Cassan, Arnaud
Ceccarelli, Cecilia
Chauvin, Gaël
Chemin, Laurent
Copin, Yannick
Cottin, Hervé
Courtois, Hélène M.
Damé, Luc
Deleuil, Magali A.
Derrière, Sebastien
Dole, Hervé
Domiciano De Souza, A.
Dougados, Catherine L.
Falize, Emeric
Fouchard, Marc
Fromang, Sebastien
Gavras, Panagiotis
Gilles, Dominique
Hersant, Franck
Huertas-Company, Marc
Ibata, Rodrigo A.
Joblin, Christine
Jubier, Xavier M.
Kern, Pierre Y.
Kervella, Pierre
Kretzschmar, Matthieu
Langlois, Maud
Las Vergnas, Olivier
Le Bouquin, Jean-Baptiste
Le Poncin-Lafitte, C.
Lefloch, Bertrand
Lopez, Bruno
Mccracken, Henry J.
Millour, Florentin A.
Minazzoli, Olivier L.
Mousis, Olivier
Nardetto, Nicolas
Nesvadba, Nicole
Noterdaeme, Pasquier
Palacios, Ana
Perraut, Karine
Pinte, Christophe
Puech, Mathieu
Richard, Johan
Richard, Olivier A.
Rogister, Yves J.
Siebert, Arnaud N.
Taris, François
Thiébaut, Eric M.
Tresse, Laurence
Wakelam, Valentine

Georgia

Simonia, Irakli

Germany

Basu, Kaustuv
Braithwaite, Jonathan
Brandner, Wolfgang
Brunthaler, Andreas
Champion, David J.
Feulner, Georg
Haas, Martin F.
Hatziminaoglou, Evanthia
Hekker, Saskia
Horn, Martin E.
Horns, Dieter
Kaltenegger, Lisa
Koch, Andreas
Liermann, Adriane
Madjarska, Maria S.
Maguire, Kate L.
Müller, Jürgen E.
Nieva, Maria F.
Nilsson, Tobias J.
Nuza, Sebastian E.
Onel, Hakan
Oskinova, Lidia M.
Percheron, Isabelle
Schreiber, Karl U.
Seck, Friedrich
Semenov, Dmitry A.
Smida, Radomr
Steigenberger, Peter
Tepper Garcia, Thorsten
Thomas, Maik
Todt, Helge
Van Eymeren, Janine
Van Kampen, Eelco
Wendt, Martin
Weratschnig, Julia M.

Greece

Georgoulis, Manolis K.
Stergioulas, Nikolaos
Tsiganis, Kleomenis

Hungary

Csabai, Istvan
Derekas, Aliz
Gabanyi, Krisztina E.
Kiss, Csaba
Kospal, Ágnes
Kovacs, József
Moor, Attila
Mosoni, Laszlo
Pal, András
Sodor, dm
Suli, Áron L.
Szabo, Gyula M.
Szekely, Péter

India

Banerjee, Dipankar
Ch, Ishwara Chandra
Chelliah Subramonian, Stalin
Gangadhara, R.T.
Goswami, Aruna
Kasiviswanathan, S.
Mazumdar, Anwesh
Mookerjea, Bhaswati
Mukhopadhyay, Banibrata
Nandi, Dibyendu
Saripalli, Lakshmi V.
Sharma, Prateek
Srivastava, Nandita
Tripathi, Durgesh
Wadadekar, Yogesh G.

Indonesia

Arifyanto, Mochamad I.
Vierdayanti, Kiki
Wulandari, Hesti R.

Iran, Islamic Rep. of

Aghaee, Alireza
Bigdeli, Mohsen
Haghi, Hosein
Karami, Kayoomars
Pazhouhesh, Reza
Safari, Hossein
Saffari, Reza

Ireland

Aharonian, Felix A.
Chernyakova, Maria
Scholz, Alexander

Israel

Akashi, Muhammad S.
Bear, Ealeal
Beck, Sara C.
Behar, Ehud
Bromberg, Omer
Chelouche, Doron
Gal-Yam, Avishay
Helled, Ravit
Keshet, Uri
Laufer, Diana
Nakar, Ehud
Ofek, Eran O.
Pat-El, Igal O.
Perets, Hagai B.
Poznanski, Dovi
Pustilnik, Lev A.
Sari, Re'Em
Shimon, Meir
Wandel, Amri S.
Yair, Yoav Y.

Italy

Bianchi, Stefano
Campana, Riccardo
Cassano, Rossella
Cesetti, Mary
Cora, Alberto
Cracco, Valentina
Cristallo, Sergio
Crosta, Mariateresa
D'Ammando, Filippo
De Lucia, Gabriella
Del Popolo, Antonino
Fabiani, Sergio
Gallazzi, Anna R.
Gastaldello, Fabio
Guglielmino, Salvatore L.
Iorio, Lorenzo
La Mura, Giovanni
Lapi, Andrea
Magrini, Laura
Mancini, Dario
Mapelli, Michela
Marino, Antonietta
Masciadri, Elena
Massaro, Francesco
Morelli, Lorenzo
Orienti, Monica
Pareschi, Giovanni
Pentericci, Laura
Risaliti, Guido
Romano, Paolo
Schipani, Pietro
Sigismondi, Costantino
Vergani, Daniela
Zibetti, Stefano

Japan

Asano, Katsuaki
Baba, Junichi
Hanado, Yuko
Hasegawa, Sunao
Hazumi, Masashi
Hiramatsu, Masaaki
Hirose, Shigenobu
Honda, Mitsuhiko
Ikeda, Norio
Ishihara, Daisuke
Ishii, Miki
Itoh, Yoichi
Kajisawa, Masaru
Kamazaki, Takeshi
Kamegai, Kazuhisa
Kayo, Issha
Kimura, Hiroshi
Kobayashi, Masakazu
Kurayama, Tomoharu
Lykawka, Patryk S.
Matsumura, Tomotake
Matsunaga, Noriyuki
Matsushita, Kyoko
Morokuma, Tomoki
Murakami, Naoshi
Muraoka, Kazuyuki
Murayama, Hitoshi
Nagataki, Shigehiro
Narita, Norio
Niinuma, Kotaro
Niwa, Yoshito
Oka, Tomoharu
Okamoto, Takenori Joten
Ota, Naomi
Otsuki, Kaori
Sakamoto, Tsuyoshi
Sakon, Itsuki
Sato, Bunei
Seto, Naoki
Shimoikura, Tomomi
Shirahata, Mai
Shirasaki, Yuji
Stone, Jennifer M.
Sumi, Takahiro
Sumiyoshi, Kosuke
Susa, Hajime H.
Takada, Masahiro
Takahashi, Hidenori
Takahashi, Kunio
Takahashi, Rohta
Tanaka, Masayuki
Terada, Yukikatsu
Tominaga, Nozomu
Ueda, Haruhiko
Yamada, Shimako
Yamamoto, Hiroaki
Yokoyama, Takaaki
Yonetoku, Daisuke

Kazkhstan*

Denisyuk, Edvard
Tejfel, Victor
Vilkoviskij, Emmanuil

Korea, Rep. of

Ahn, Sang-Hyeon
Han, Inwoo
Hwang, Junga
Karouzos, Marios
Kim, Joo Hyeon
Kim, Minsun
Kim, Sang Chul
Kim, Sang Hyuk
Kim, Woong-Tae
Kusakabe, Motohiko
Lee, Chung-Uk
Lee, Jae Woo
Lee, Jeong-Eun J.
Lee, Joon Hyeop
Lee, Jung-Won
Lee, Ki-Won
Lee, Sang-Sung
Moon, Hong-Kyu
Oh, Suyeon Y.
Seo, Haingja
Song, Yong-Seon
Sung, Hyun-Il
Trippe, Sascha
Woo, Jong-Hak

Korea, D.P.R.*

Baek, Chang Ryong
Bang, Yong
Cha, Du
Cha, Gi Ung
Chio, Chol
Choe, Won Chol
Dong, Il
Hong, Hyon
Kang, Jin Sok
Kim, Jik
Kim, Yong
Kim, Yong
Kim, Yul
Kim, Zong
Li, Gi
Li, Gyong
Li, Hyok
Li, Sin
Ri, Son Jae

Latvia

Smirnova, Olesja

Lithuania

Kucinskas, Arunas
Laugalys, Vygandas
Stonkutė, Rima
Zdanavičius, Justas

Mexico

Ambrocio-Cruz, Silvia P.
Clark, David M.
Corral, Luis J.
Ferrusca, Daniel
Fox-Machado, Lester
Garcia-Segura, Guillermo
Ramos-Larios, Gerardo
Reyes-Ruiz, Mauricio
Richer, Michael G.
Roman-Zuniga, Carlos G.
Rubio-Herrera, Eduardo
Sabin, Laurence
Saucedo Morales, Julio C.
Torres-Papaqui, Juan P.
Vega, Olga

Netherlands

Baneke, David M.
Benkhoff, Johannes J.
Birkmann, Stephan M.
Bouwens, Rychard J.
Caputi, Karina I.
Cazaux, Stéphanie
Deller, Adam T.
Gandolfi, Davide
Kenworthy, Matthew A.
Koschny, Detlef V.
Labbe, Ivo
Mckean, John P.
Pilbratt, Göran L.
Rossi, Elena M.
Snik, Frans
Uttley, Philip
Van Den Berg, Maureen C.
Zender, Joe J.

New Zealand

Dunne, Loretta
Easther, Richard
Eldridge, John J.
Gordon, Chris
Johnston-Hollitt, Melanie

Nigeria

Opara, Fidelix E.

Norway

Hervik, Sigbjørn
Kristiansen, Jostein R.
Ortiz Carbonell, Ada

Oman

Ioannou, Zacharias

Panamá

Chung, Eduardo E.
Forero Villão, Vicente
Medina, Etelvina D.
Saenz, Eduardo

Philippines

Esguerra, José Perico H.
Sese, Rogel Mari D.

Poland

Bartczak, Przemyslaw P.
Borczyk, Wojciech M.
Gawronski, Marcin P.
Jamrozy, Marek
Kaminski, Krzysztof Z.
Kozlowski, Szymon
Lokas, Ewa L.
Maciesiak, Krzysztof
Marciniak, Anna
Melikidze, Giorgi I.
Migaszewski, Cezary
Pietrukowicz, Pawel
Rosinska, Dorota H.
Rozanska, Agata
Smolec, Radoslaw
Stachowski, Grzegorz S.

Romania

Nedelcu, Dan A.
Turcu, Vlad S.

Russian Federation

Antokhina, Eleonora A.
Busarev, Vladimir V.
Filippov, Boris P.
Gavrilov, Mikhail G.
Getling, Alexander V.
Ikonnikova, Natalia
Ilin, Gennadii N.
Ivanov, Pavel B.
Izmailov, Igor S.
Kaisin, Serafim S.
Kaminker, Alexander D.
Kolobov, Dmitri Y.
Kopatskaya, Evgenia N.
Malogolovets, Evgeny V.
Malov, Igor F.
Marshalov, Dmitriy
Melnik, Anna M.
Naroenkov, Sergey
Petrov, Sergey D.
Piotrovich, Mikhail Y.
Potapov, Vladimir A.
Pshirkov, Maxim S.
Rastegaev, Denis A.
Savanov, Igor S.
Shimansky, Vladislav V.
Skurikhina, Elena A.
Smirnova, Aleksandrina A.
Smirnova, Tatiana V.
Suleymanova, Svetlana A.
Surkis, Igor F.
Titov, Vladimir B.
Valeev, Azamat F.
Zhilkin, Andrey G.

Serbia, Republic of

Arbutina, Bojan R.
Bon, Edi
Bon, Natasa Z.
Borka Jovanovic, Vesna V.
Ilic, Dragana
Kovacevic, Jelena
Novakovic, Bojan S.
Simic, Sasa
Simic, Zoran J.
Vukotic, Branislav

Slovakia

Dobrotka, Andrej
Gomory, Peter
Husarik, Marek
Kanuchova, Zuzana
Kornos, Leonard

South Africa

Chiang, Hsin C.
Clarkson, Chris
Crause, Lisa A.
De Swardt, Bonita E.
Depagne, ric
Engelbrecht, Chris
Gilbank, David G.
Hess, Kelley M.
Letarte, Bruno
Loubser, Ilani S.
Lucero, Danielle M.
Mcbride, Vanessa A.
Middleton, Christopher T.
Miszalski, Brent
Monard, Libert A.
Moodley, Kavilan
Oozeer, Nadeem O.
Vaccari, Mattia
Venter, Christo
Wolleben, Maik
Worters, Hannah L.

Spain

Alonso Sobrino, Roi
Alonso-Herrero, Almudena
Arregui, Inigo
Bosch-Ramon, Valenti
Campo Bagatin, Adriano
Cristobal, David
De Ugarte Postigo, Antonio
Ederoclite, Alessandro
Garcia, Miriam
Garcia-Hernandez, Domingo A.
Goicoechea, Luis J.
Hernandez-Monteagudo, Carlos
Jones, David
Khomenko, Elena
Lammers, Uwe
Luque-Escamilla, Pedro L.
Maiz Apellaniz, Jesús
Marin-Franch, Antonio
Ortiz Gil, Amelia
Papitto, Alessandro
Perez-Garrido, Antonio
Perez-Gonzalez, Pablo G.
Perozzi, Ettore
Povic, Mirjana
Rea, Nanda
Sanchez-Blazquez, Patricia
Santos-Sanz, Pablo
Simon-Diaz, Sergio
Thoene, Christina C.
Ulla Miguel, Ana M.
Uytterhoeven, Katrien
Valdivielso, Luisa
Varela Lopez, Jesús
Vavrek, Roland D.
Viironen, Kerttu E.

Sweden

Axelsson, Magnus
Brandeker, Alexis
Haas, Rüdiger
Hayes, Matthew J.
Hjalmarsdotter, Linnea
Hobiger, Thomas
Johansen, Anders
Leenaarts, Jorrit
Mcmillan, Paul J.
Mitra, Dhrubaditya
Mortsell, Edvard
Nymark, Tanja K.
Vlemmings, Wouter H.
Wagner, Robert M.

Switzerland

Alibert, Yann
Ehrenreich, David
Ekstrom Garcia N., Sylvia
Faure, Cécile
Groh, Jose H.
Heng, Kevin
Jaeggi, Adrian
Lake, George
Lovis, Christophe
Meyer, Michael R.
Mordasini, Christoph
Panafidina, Natalia
Quanz, Sascha P.
Rassat, Anais M.
Read, Justin I.

Saha, Prasenjit
Stadel, Joachim G.
Teyssier, Romain
Thaller, Daniela
Vidotto, Aline A.

Tajikistan

Shoyoqubov, Shoayub

Thailand

Asanok, Kitiyanee
Burikham, Piyabut
Krittinatham, Watcharawuth
Kriwattanawong, Wichean
Nammahachak, Suwit
Wannawichian, Suwicha

Turkey

Ak, Serap
Ak, Tansel
Balman, Solen
Beklen, Elif
Kilcik, Ali
zeren, Ferhat F.
Ozisik, Tuncay

Ukraine

Balyshev, Marat A.
Bannikova, Elena Y.
Gorbaneva, Tatyana I.
Kaydash, Vadym G.
Korokhin, Viktor V.
Pulatova, Nadiia G.
Pushkarev, Alexander B.
Savanevich, Vadim Y.
Sergijenko, Olga
Sybiryakova, Yegeniya S.
Velikodsky, Yuri I.
Zhuk, Alexander I.

United Kingdom

Aigrain, Suzanne
Allan, Alasdair
Breton, René
Bushby, Paul J.
Chaplin, William J.
Croston, Judith H.
Dalton, Gavin B.
Davies, Benjamin
Davis, Timothy A.
Demory, Brice-Olivier
Ferreira, Pedro G.
Fletcher, Andrew F.
Fletcher, Leigh N.
Foullon, Claire
Georgy, Cyril
Gill, Michael J.
Gomez, Edward L.
Gomez, Haley L.
Hannah, Iain G.
Ho, Wynn
Hoenig, Sebastian F.
Iliev, Ilian T.
Izzard, Robert G.
Kaviraj, Sugata
Lintott, Chris J.
Masters, Karen L.
Matthews, Sarah A.
Miglio, Andrea
Peel, Michael W.
Pitkin, Matthew D.
Rushton, Anthony P.
Russell, Alexander J.
Salaris, Maurizio
Schure, Klara M.
Stott, John P.
Sullivan, Mark
Thurston, Mark R.
Vasta, Magda
Verma, Aprajita
Young, John S.

United States

Acton, Charles H.
Airapetian, Vladimir
Ajhar, Edward A.
Ammons, Stephen M.
Antoniou, Vallia
Aufdenberg, Jason P.
Baines, Ellyn K.
Barkhouse, Wayne A.
Bartlett, Jennifer L.
Benacquista, Matthew J.
Bernardini, Federico
Bertello, Luca
Bishop, Marsha
Bodaghee, Arash
Bodewits, Dennis
Bosken, Sarah M.
Bouton, Ellen N.
Boyajian, Tabetha S.
Brammer, Gabriel B.
Brenneman, Laura W.
Brown, Joanna M.

Brozovic, Marina
Brunner, Robert J.
Carlin, Jeffrey L.
Cenko, Stephen B.
Chakrabarti, Supriya
Cohen, David H.
Conti, Alberto
Creech-Eakman, Michelle J.
Croft, Steve
Cui, Wei
De Mink, Selma E.
De Val-Borro, Miguel
Denisco, Kenneth R.
Derosa, Marc L.
Diaz-Santos, Tanio
Dotter, Aaron L.
Espenak, Fred
Fernandez, Yanga R.
Figer, Donald F.
Finch, Charlie T.
Ford, Eric B.
Fraschetti, Federico
Fruscione, Antonella
Gammie, Charles F.
Gay, Pamela L.
Geller, Aaron M.
Gonzalez, Gabriela
Gordon, Karl D.
Grillmair, Carl J.
Grundstrom, Erika
Hallinan, Gregg
Hattori, Takashi
Hayashi, Keiji
Heinz, Sebastian
Hsieh, Henry H.
Jacobs, Christopher S.
Janches, Diego
Jha, Saurabh W.
Johns, Bethany R.
Johnson, Christian I.
Kafka, Styliani (Stella)
Keenan, Ryan C.
Keeney, Brian A.
Kent, Brian R.
Kim, Ji Hoon
Knight, Matthew M.
Korpela, Eric J.
Kotulla, Ralf C.
Kraft, Ralph P.
Kramer, William A.
Kraus, Stefan
Kuiper, Rolf
Lee, Hyun-Chul
Li, Jian-Yang
Livadiotis, George I.
Lowenthal, James D.
Magnier, Eugene A.
Mainzer, Amy
Masiero, Joseph R.
Mccall, Benjamin J.
Mcgehee, Peregrine M.
Meibom, Soren
Melbourne, Jason L.
Miesch, Mark
Miller, Eric D.
Minowa, Yosuke
Montgomery, Michele M.
Morris, Patrick W.
Morsony, Brian J.
Nakata, Fumiaki
Nitta, Atsuko
Nixon, Conor A.
Nota, Antonella
Oberst, Thomas E.
Odonoghue, Aileen A.
Olive, Don H.
Oluseyi, Hakeem M.
Oya, Shin
Page, Gary L.
Penny, Matthew T.
Pooley, David
Price, Charles A.
Prsa, Andrej
Puxley, Phil
Rauch, Michael
Reitze, David H.
Reyes, Reinabelle
Rhee, Jaehyon
Rho, Jeonghee
Rhodes, Jason D.
Richardson, Derek C.
Ricker, Paul M.
Riles, Keith
Rizzi, Luca
Robberto, Massimo
Roth, Ilan
Sanders, Gary H.
Sankrit, Ravi
Shoemaker, David H.
Simons, Douglas A.
Simunac, Kristin D.
Smolinski, Jason P.
Stacy, Athena R.
Stewart, Susan G.
Stewart-Mukhopadhyay, S.
Sundqvist, Jon O.
Tanaka, Ichi Makoto
Templeton, Matthew R.
Teng, Stacy H.
Thanjavur, Karunananth
Tremonti, Christy A.
Trenti, Michele
Tzanavaris, Panayiotis
Ubeda, Leonardo
Ud-Doula, Asif
Ueta, Toshiya
Usuda, Tomonori
Vakoch, Douglas A.
Valluri, Monica
Venters, Tonia M.
Viall, Nicholeen M.
Voit, Gerard M.
Weiss, John W.
Wilson, Gillian
Winter, Lisa M.
Womack, Maria P.
Wong, Tony H.
Wood, Brian E.
Wu, Chin-Chun
Yun, Min S.
Zaritsky, Dennis F.
Zeiler, Michael N.
Zhang, Jie

Uruguay

Sosa, Andrea

Vatican City State

Brown, David A.
Gabor, Pavel
Kikwaya Eluo, Jean-B.

Venezuela

Downes Wallace, Juan J.
Mateu, Cecilia

Viet Nam

Nguyen, Khanh V.
Nguyen, Lan Q.
Nguyen, Phuong T.
Pham, Diep N.
Phan, Dong V.
Tran, Ha Q.

2. Deceased members (2009-2012)

Akim, Efraim,
Albers, Henry,
Alksne, Zenta,
Anderson, Kinsey,
Andrienko, Dmitry,
Arnaud, Jean-Paul,
Arnold, James,
Axon, David,
Backer, Donald,
Baldwin, John,
Baldwin, Ralph,
Baum, William,
Bell, Roger,
Beurle, Kevin,
Bhattacharyya, J.,
Bidelman, William,
Birkle, Kurt,
Biswas, Sukumar,
Blaauw, Adriaan,
Blackwell, Donald,
Blanco, Victor,
Bone, Neil,
Bookmyer, Beverly,
Boulon, Jacques,
Bowen, George,
Boyd, Robert,
Buchler, J.,
Burgess, Alan,
Cahn, Julius,
Carr, Thomas,
Cayrel de Strobel, Giusa,
Ceplecha, Zdenek,
Chamberlain, Joseph,
CHOU, Kyong-Chol,
Christy, Robert,
Chubb, Talbot,
Clausen, Jens,
Coradini, Angioletta,
Cuisinier, Francois,
Danby, J.,
de Young, David,
Denishchik, Yurii,
Dewhirst, David,
Dollfus, Audouin,
Donn, Bertram,
Duerbeck, Hilmar,
Dyson, John,
Dzervitis, Uldis,
Eddy, John,
Efimov, Yuri,
Elliot, James,
FANG, Li-Zhi,
Fejes, Istvan,
Fiala, Alan,
Finkelstein, Andrej,
Finzi, Arrigo,
Firneis, Friedrich,
Fomin, Piotr,
Fridman, Aleksej,
Friedjung, Michael,
Galletto, Dionigi,
Galt, John,
Gascoigne, S.,
Gaska, Stanislaw,
Gelfreikh, Georgij,
Glushneva, Irina,
Goldsmith, S.,
Gontier, Anne-Marie,
Gorgolewski, Stanislaw,
Gould, Robert,
Guest, John,
Guseinov, O.,
Hansen, Carl,
HAYASHI , Chushiro,
Heeschen, David,
Hers, Jan,
Hingley, Peter,
Horsky, Jan,

Hovhannessian, Rafik,
HUANG, Yinn-Nien,
Huchra, John,
Hunten, Donald,
Iannini, Gualberto,
Idlis, Grigorij,
Jorgensen, Henning,
JUGAKU, Jun,
KAWABATA, Kinaki,
Kharin, Arkadiy,
Kiang, Tao,
KITAMURA, Masatoshi,
Koch, David,
Koch, Robert,
Kowal, Charles,
Krasinsky, George,
Krygier, Bernard,
Kundu, Mukul,
Lainela, Markku,
Lal, Devendra,
Lecar, Myron,
Legg, Thomas,
Leschiutta, Sigfrido,
Lindblad, Bertil,
Loucif, Mohammed,
Lovell, Bernard,
Major, John,
Mancuso, Santi,
Marsden, Brian,
May, Jorge,
Mead, Jaylee,
Milogradov-Turin, Jelena,
Minakov, Anatoliy,
MORIMOTO, Masaki,
MORITA, Koh-ichiro,
Moss, Christopher,
Nedbal, Dalibor,
Ogelman, Hakki,
Okoye, Samuel,
Oliver, John,
Olsen, Kenneth,
Pacini, Franco,
Page, Arthur,
Papushev, Pavel,
Peery, Benjamin,
Petford, Alfred,
Phillips, John,
Praderie, Franoise,
Price, Stephan,
Radhakrishnan, V.
Raimond, Ernst,
Rakos, Karl,
Rawlings, Steven,
Razin, Vladimir,
ReVelle, Douglas,
Richardson, Eric,
Riihimaa, Jorma,
Robinson, Leif,
Rood, Robert,
Rose, William,
Roy, Archie,
Rudzikas, Zenonas,
Sahade, Jorge,
Sandage, Allan,
Sargent, Wallace,
Schatzman, Evry,
Schoier, Fredrik,
Schwartz, Richard,
Searle, Leonard,
Segonds, Alain-Philippe,
Sehnal, Ladislav,
Semel, Meir,
Serrano, Alfonso,
Shakhbazian, Romelia,
Shakhovskoj, Nikolay,
Sheffer, Evgenij,
Sheridan, Kevin,
Shkodrov, Vladimir,
Soboleva, Natalja,
Spoelstra, T.,
Stavrev, Konstantin,
Steiger, W.,
Steinert, Klaus,
Stibbs, Douglas,
Sundman, Anita,
Svechnikov, Marij,
Szecsenyi-Nagy, Gbor,
Tahtinen, Leena,
TAKAKUBO, Keiya,
Tempesti, Piero,
Tinbergen, Jaap,
Tlamicha, Antonin,
Todoran, Ioan,
Toro, Tibor,
Torres, Carlos,
Tran-Minh, Franoise,
Tuominen, Ilkka,
Udal'tsov, Vyacheslav,
UENO, Sueo,
Vaidya, P.,
van Bueren, Hendrik,
Vargha, Magda,
WAKO, Kojiro,
Walter, Hans,
Weidemann, Volker,
Wesemael, Francois,
Westerhout, Gart,
Wilson, Albert,
Wlerick, Gerard,
Wolf, Bernhard,
Woszczyk, Andrzej,
Yoss, Kenneth,
Zimmermann, Helmut,
Zirin, Harold,

Transactions IAU, Volume XXVIIIB
Proc. XXVIII IAU General Assembly, August 2012
Thierry Montmerle, ed.

doi:10.1017/S1743921315005670

CHAPTER IX

DIVISIONS, COMMISSIONS, & WORKING GROUPS (until 31 August 2012)

NOTE. This chapter gives the main IAU scientific bodies (Division, Commissions and their Working Groups) in force until the end of the XVIIIth General Assembly. As a result of the adoption of Resolution B4 by this Assembly, a new Divisional structure was established (see Chapters II and IV of these *Transactions*), to take effect on 1 September 2012.

The Organizing Committee of each Division is given, followed by the list of Commissions affiliated to this Division and their respective Working Groups.

The members of the terminated Divisions and their Commissions are listed in the following Chapters (X and XI respectively).

Division I Fundamental Astronomy

President:
Dennis D. McCarthy
2432 Riviera Drive
22181, Vienna, Virginia
United States

Vice-President:
Sergei A. Klioner
Technische Universität Dresden
Lohrmann Observatory
Mommsenstr 13
01062, Dresden
Germany

Organizing Committee Members:

Dafydd Wyn Evans (United Kingdom), Catherine Y. Hohenkerk (United Kingdom), Mizuhiko HOSOKAWA (Japan), Cheng-Li Huang (China Nanjing), George H. Kaplan (United States), Zoran Knežević (Serbia, Republic of), Richard N. Manchester (Australia), Alessandro Morbidelli (France), Gérard Petit (France), Harald Schuh (Germany), Michael H. Soffel (Germany), Jan Vondrák (Czech Republic), Norbert Zacharias (United States)

Commissions:

Commission 4 Ephemerides
Commission 7 Celestial Mechanics & Dynamical Astronomy
Commission 8 Astrometry
Commission 19 Rotation of the Earth
Commission 31 Time
Commission 52 Relativity in Fundamental Astronomy

Division II Sun & Heliosphere

President:
Valentin Martínez Pillet
National Solar Observatory
3010 Coronal Loop
88349 , Sunspot, NM
United States

Vice-President:
James A. Klimchuk
NASA
Goddard Space Flight Center
Solar Physics Laboratory
Code 671
MD 20771, Greenbelt
United States

Organizing Committee Members:

Gianna Cauzzi (Italy), Natchimuthuk Gopalswamy (United States), Alexander Kosovichev (United States), Ingrid Mann (Sweden), Karel Schrijver (United States), Lidia van Driel-Gesztelyi (United Kingdom)

Commissions:

Commission 10 Solar Activity
Commission 12 Solar Radiation & Structure
Commission 49 Interplanetary Plasma & Heliosphere

Division III Planetary Systems Sciences

President:
Karen J. Meech
University of Hawaii Honolulu
Institute of Astronomy
2680 Woodlawn Drive
HI 96822, Honolulu
United States

Vice-President:
Giovanni B. Valsecchi
INAF
IAPS
Via Fosso del Cavaliere 100
Tor Vergata
00133, Roma
Italy

Organizing Committee Members:

Dominique Bockelée-Morvan (France), Alan Paul Boss (United States), Alberto Cellino (Italy), Guy Joseph Consolmagno (Vatican City State), Julio Angel Fernández (Uruguay), William M. Irvine (United States), Daniela Lazzaro (Brazil), Patrick Michel (France), Keith S. Noll (United States), Rita M. Schulz (Netherlands), Jun-ichi WATANABE (Japan), Makoto YOSHIKAWA (Japan), Jin ZHU (China Nanjing),

Commissions:

Commission 15 Physical Studies of Comets & Minor Planets
Commission 16 Physical Study of Planets & Satellites
Commisison 20 Positions & Motions of Minor Planets, Comets & Satellites
Commission 22 Meteors, Meteorites & Interplanetary Dust
Commission 51 Bio-Astronomy
Commission 53 Extrasolar Planets

Division IV Stars

President:
Christopher Corbally
Specola Vaticana
Steward Observatory-Vatican Observatory Research Group
00120, Città del Vaticano
Vatican City State

Vice-President:
Francesca D'Antona
INAF
Osservatorio Astronomico di Roma
Via Frascati 33
Monte Porzio Catone
00040, Roma
Italy

Organizing Committee Members:
Martin Asplund (Australia), Corinne Charbonnel (Switzerland), Jose-Angel Docobo (Spain), Richard O. Gray (United States), Nikolai E. Piskunov (Sweden)

Commissions:
Commission 26 Double & Multiple Stars
Commission 29 Stellar Spectra
Commission 35 Stellar Constitution
Commission 36 Theory of Stellar Atmospheres
Commission 45 Stellar Classification

Division V Variable Stars

President:
Steven D. Kawaler
Iowa State University
Department of Physics and Astronomy
A323 Zaffarano Hall
IA 50011-3160, Ames
United States

Vice-President:
Ignasi Ribas
Institut de Ciencies de l'Espai
IEEC-CSIC
Cami de Can Magrans, s/n
08193, Bellaterra, Barcelona
Spain

Organizing Committee Members:
Michel Breger (United States), Edward F. Guinan (United States), Gerald Handler (Poland), Slavek M. Rucinski (Canada)

Commissions:
Commission 27 Variable Stars
Commission 42 Close Binary Stars

Division VI Interstellar Matter

President:
You-Hua Chu
Academia Sinica
Institute of Astronomy and Asrophysics
11F of Astronomy-Mathematics Building
No.1, Sec. 4, Roosevelt Rd.
10617, Taipei
China Taipei

Vice-President:
Sun Kwok
University of Hong Kong
Faculty of Science
Chong Yuet Ming Physics Bldg
Pokfulam Rd
Hong Kong
China Nanjing

Organizing Committee Members:
Dieter Breitschwerdt (Germany), Michael G. Burton (Australia), Sylvie Cabrit (France), Paola Caselli (Italy), Elisabete M. de Gouveia Dal Pino (Brazil), Neal J. Evans (United States), Thomas Henning (Germany), Mika J. Juvela (Finland), Bon-Chul KOO (Korea, Rep of), Michal Różyczka (Poland), Laszlo Viktor Tóth (Hungary), Masato TSUBOI (Japan), Ji YANG (China Nanjing)

Commissions:
Commission 34 Interstellar Matter

Division VII Galactic System

President:
Despina Hatzidimitriou
University of Athens
Department of Physics
Panepistimiopolis
Zografos
157 84, Athens
Greece

Vice-President:
Rosemary F. Wyse
Johns Hopkins University
Physics-Astronomy Department CAS
Charles and 34th Str
MD 21218-2686, Baltimore
United States

Organizing Committee Members:
Giovanni Carraro (Chile), Bruce G. Elmegreen (United States), Birgitta Nordström (Denmark)

Commissions:
Commission 33 Structure & Dynamics of the Galactic System
Commission 37 Star Clusters & Associations

Division VIII Galaxies & the Universe

President:
Elaine M. Sadler
University of Sydney
School of Physics A28
NSW 2006, Sydney
Australia

Vice-President:
Françoise Combes
Observatoire de Paris
LERMA
61 Av de l'Observatoire
Bat. A
75014 Paris
France

Organizing Committee Members:

Roger L. Davies (United Kingdom), John S. Gallagher III (United States), Thanu Padmanabhan (India)

Commissions:

Commission 28 Galaxies
Commission 47 Cosmology

Division IX Optical & Infrared Techniques

President:
Andreas Quirrenbach
Universität Heidelberg
Landessternwarte
Königstuhl 12
69117, Heidelberg
Germany

Vice-President:
David Richard Silva
National Optical Astronomy Observatory
950 North Cherry Avenue
AZ 85719, Tucson
United States

Organizing Committee Members:

Michael G. Burton (Australia), Xiangqun Cui (China Nanjing), Ian S. McLean (United States), Eugene F. Milone (Canada), Jayant Murthy (India), Stephen T. Ridgway (United States), Gražina Tautvaišene (Lithuania), Andrei A. Tokovinin (Chile), Guillermo Torres (United States)

Commissions:

Commission 24 Galactic & Extragalactic Background Radiation
Commission 25 Stellar Photometry & Polarimetry
Commission 30 Radial Velocities
Commission 54 Optical & Infrared Interferometry

Division X Radio Astronomy

President:
A. Russell Taylor
University of Calgary
Department of Physics and Astronomy
2500 University Dr NW
AB T2N 1N4, Calgary
Canada

Vice-President:
Jessica Mary Chapman
CSIRO Astronomy and Space Science
PO Box 76
NSW 1710, Epping
Australia

Organizing Committee Members:

Christopher L. Carilli (United States), Gabriele Giovannini (Italy), Richard E. Hills (United Kingdom), Hisashi HIRABAYASHI (Japan), Justin L. Jonas (South Africa), Joseph Lazio (United States), Raffaella Morganti (Netherlands), Ren-Dong NAN (China Nanjing), Monica Rubio (Chile), Prajval Shastri (India)

Commissions:

Commission 40 Radio Astronomy

Division XI Space & High Energy Astrophysics

President:
Christine Jones
Harvard Smithsonian
Center for Astrophysics
High Energy Astrophysics Division
MS 2
60 Garden Str
MA 02138-1516, Cambridge
United States

Vice-President:
Noah Brosch
Tel Aviv University
Department of Physics & Astronomy
Wise Observatory
Ramat Aviv
PO Box 39040
69978, Tel Aviv
Israel

Organizing Committee Members:
Matthew G. Baring (United States), Martin Adrian Barstow (United Kingdom), João Braga (Brazil), Eugene M. Churazov (Russian Federation), Jean Eilek (United States), Hideyo KUNIEDA (Japan), Jayant Murthy (India), Isabella Pagano (Italy), Marco Salvati (Italy), Kulinder Pal Singh (India), Diana Mary Worrall (United Kingdom)

Commissions:
Commission 44 Space & High Energy Astrophysics

Division XII Union-Wide Activities

President:
Françoise Genova
Observatoire Astronomique de Strasbourg
Centre de Données astronomiques de Strasbourg (CDS)
11 rue de l'Université
67000, Strasbourg
France

Vice-President:
Raymond P. Norris
CSIRO/ATNF
PO Box 76
NSW 1710, Epping
Australia

Organizing Committee Members:
Dennis Crabtree (Canada), Olga B. Dluzhnevskaya (Russian Federation), Masatoshi OHISHI (Japan), Rosa M. Ros (Spain), Clive L.N. Ruggles (United Kingdom), Nikolay N. Samus (Russian Federation), Xiaochun Sun (China Nanjing), Virginia Trimble (United States), Wim van Driel (France), Glenn Michael Wahlgren (United States)

Commissions:
Commission 5 Documentation & Astronomical Data
Commission 6 Astronomical Telegrams
Commission 14 Atomic & Molecular Data
Commission 41 History of Astronomy
Commission 46 Astronomy Education & Development
Commission 50 Protection of Existing & Potential Observatory Sites
Commission 55 Communicating Astronomy with the Public

Division Working Groups

Division	*Working Group Name*	*Chair*
DI	Astrometry by Small Ground-Based Telescopes	M. Assafin
DI	Numerical Standards in Fundamental Astronomy	B. Luzum
DII	Comparative Solar Minima	S. Gibson
DII	International Collaboration on Space Weather	D. F. Webb
DII	International Data Access	R. D. Bentley
DII	Solar Eclipses	J. M. Pasachoff
DIII	Near Earth Objects	A. W. Harris
DIII	Planetary System Nomenclature (WGPSN)	R. M. Schulz
DIII	Small Bodies Nomenclature (SBN)	J. Tichá
DIV	Abundances in Red Giants	J. C. Lattanzio
DIV	Massive Stars	Artemio Herrero Davó
DVII	Galactic Center	J. Lazio
DVIII	Supernovae	W. Hillebrandt
DIX	Site Testing Instruments	A. A. Tokovinin
DIX	Sky Surveys	Q. A. Parker
DX	Interference Mitigation	W. A. Baan
DXI	Particle Astrophysics	R. Schlickeiser

Inter-Division Working Groups

Divisions	*Working Group Name*	*Chair*
DI-III	Cartographic Coordinates & Rotational Elements	B. A. Archinal
DIV-V	Active B Stars	C. E. Jones
DIV-V	Ap & Related Stars	G. Mathys
DIX-X	Encouraging the International Development of Antarctic Astronomy	M. G. Burton
DIX-X-XI	Astronomy from the Moon	H. D. Falcke
DX-XII	Historic Radio Astronpetitomy	K. I. Kellermann

Inter-Commission Working Group

Commissions	*Working Group Name*	*Chair*
C4, C7, C8, C16, C20	Natural Planetary Satellites	J.-E. Arlot

Commission Working Groups

Commission	*Working Group Name*	*Chair*
C5	TG On Preservation & Digitization of Photographic Plates	R. E. Griffin
C5	Astronomical Data	R. P. Norris
C5	Designations	M. Schmitz
C5	FITS	L. Chiappetti
C5	Libraries	M.Bishop; R. Hanisch
C5	Virtual Observatories, Data Centers & Networks	R. J. Hanisch
C12	Coordination of Synoptic Observations of the Sun	A. A. Pevtsov
C14	Atomic Data	G. Nave
C14	Collision Processes	M. S. Dimitrijevic
C14	Molecular Data	J. H. Black
C14	Solids & Their Surfaces	G. Vidali
C15	TG On Asteroid Magnitudes	R. A. Gil-Hutton
C15	TG On Asteroid Polarimetric Albedo Calibration	R. A. Gil-Hutton
C15	TG On Cometary Magnitudes	G. Tancredi
C15	Physical Studies of Asteroids	R. A. Gil-Hutton
C15	Physical Studies of Comets	D. C. Boice
C22	Meteor Shower Nomenclature	T. J. Jopek
C22	Professional-Amateur Cooperation in Meteors	D. V. Koschny
C26	Catolog of Orbital Elements of Spectroscopic Binary Systems	D. Pourbaix
C26	Maintenance of the Visual Double Star Database	W. I. Hartkopf
C30	Radial-Velocity Standard Stars	G. Jasniewicz
C30	Stellar Radial Velocity Bibliography	O. H. Levato
C34	Astrochemistry	T. J. Millar
C34	Planetary Nebulae	A. Machado
C40	Astrophysically Important Spectral Lines	M. OHISHI
C40	Historic Radio Astronomy	K. I. Kellermann
C41	Archives	I. Chinnici
C41	Astronomy and World Heritage	C. L. N. Ruggles
C41	Historical Instruments	L. Pigatto
C41	Johannes Kepler	T. J. Mahoney
C41	Transits of Venus	S. J. Dick
C50	Controlling Light Pollution	R. F. Green
C55	CAP Conferences	I. E. Robson
C55	CAP Journal	G. R. Bladon
C55	Communicating Heliophysics	C. Briand
C55	New Ways of CAP	M. J. West
C55	Washington Charter for CAP	Dennis Crabtree

Executive Committee Working Groups

Working Group Name	*Chair*
IAU General Assemblies	Daniela Lazzaro
Future Large Scale Facilities	Roger L. Davies
Women in Astronomy	Sarah Maddison

Transactions IAU, Volume XXVIIIB
Proc. XXVIII IAU General Assembly, August 2012
Thierry Montmerle, ed.

doi:10.1017/S1743921315005682

CHAPTER X

DIVISIONS MEMBERSHIP
(until 31 August 2012)

NOTE. This chapter gives the membership of the IAU Divisions in force until the end of the XVIIIth General Assembly. As a result of the adoption of Resolution B4 by this Assembly, a new Divisional structure was established (see Chapters II and IV of these *Transactions*), to take effect on 1 September 2012.

Division I Fundamental Astronomy

President: Dennis D. McCarthy
Vice-President: Sergei A. Klioner

Organizing Committee Members:

Dafydd Wyn Evans
Catherine Y. Hohenkerk
Mizuhiko HOSOKAWA
Cheng-Li Huang
George H. Kaplan
Zoran Knežević,
Richard N. Manchester
Alessandro Morbidelli
Gérard Petit
Harald Schuh
Michael H. Soffel
Jan Vondrák
Norbert Zacharias.

Members:

Abad Hiraldo, Carlos
Abad Medina, Alberto
Abalakin, Viktor
Abbas, Ummi
Abele, Maris
Ahmed, Abdel-aziz
AHN, Youngsook
Aksnes, Kaare
Alley, Carrol
Andrade, Manuel
Andrei, Alexandre
Anosova, Joanna
Antonacopoulos, Gregory
Antonelli, Lucio Angelo
Arabelos, Dimitrios
ARAKIDA, Hideyoshi
Archinal, Brent
Arenou, Frédéric
Argyle, Robert
Arias, Elisa
Arlot, Jean-Eudes
Ashby, Neil
Assafin, Marcelo
Athanassoula, Evangelie
Babusiaux, Carine
Badescu, Octavian
Bakhtigaraev, Nail
Ballabh, Goswami
Balmino, Georges
Bangert, John
Banni, Aldo
Barabanov, Sergey
Barberis, Bruno
Barbosu, Mihail
Barkin, Yuri
Bastian, Ulrich
Bazzano, Angela
Beauge, Cristian
Belizon, Fernando
Bell, Steven
Benedict, George
Benest, Daniel
Benevides Soares, Paulo
Beutler, Gerhard
Bhatnagar, K.
Bien, Reinhold
Boboltz, David
Bois, Eric

Bolotin, Sergei
Bolotina, Olga
Boss, Alan
Bouchard, Antoine
Boucher, Claude
Bougeard, Mireille
Bozis, George
Bradley, Arthur
Branham, Richard
Breakiron, Lee
Breiter, Slawomir
Brentjens, Michiel
Brieva, Eduardo
Brosche, Peter
Brouw, Willem
Brown, Anthony
Broz, Miroslav
Brumberg, Victor
Brunini, Adrian
Bruyninx, Carine
Bucciarelli, Beatrice
Bursa, Michal
CAI, Michael
Calabretta, Mark
Capitaine, Nicole
Caranicolas, Nicholas
Carpino, Mario
Carruba, Valerio
Carter, William
Casetti, Dana
Cazenave, Anny
Cefola, Paul
Celletti, Alessandra
CHAE, Kyu Hyun
Chambers, John
Chao, Benjamin
Chapanov, Yavor
Chapront, Jean
Chapront-Touze, Michelle
Charlot, Patrick
CHEN, Li
CHEN, Alfred
CHOI, Kyu Hong
CHOU, Yi
Christou, Apostolos
Cionco, Dr Rodolfo
Cioni, Maria-Rosa
Colin, Jacques
Conrad, Albert
Contopoulos, George
Cooper, Nicholas
Corbin, Thomas
Costa, Edgardo
Creze, Michel
Crifo, Francoise
Cudworth, Kyle
da Rocha-Poppe, Paulo
Dahn, Conard
Damljanovic, Goran
Danylevsky, Vassyl
De Biasi, Maria
de Bruijne, Jos
de Felice, Fernando
de Viron, Olivier
Debarbat, Suzanne
Dehant, Véronique
Dejaiffe, Rene
Del Santo, Melania
Deleflie, Florent
Delmas, Christian
Descamps, Pascal
Devyatkin, Aleksandr
Di Sisto, Romina
Dick, Wolfgang
Dick, Steven
Dickey, Jean
Dickman, Steven
Dikova, Smilyana
Dominguez, Mariano
DONG, Xiaojun
DONG, Shaowu
Douglas, R.
Dourneau, Gerard
Drozyner, Andrzej
DU, Lan
Ducourant, Christine
Duma, Dmitrij
Dunham, David
Duriez, Luc
Dvorak, Rudolf
Efroimsky, Michael
Elipe, Antonio
Emelianov, Nikolaj
Emelyanenko, Vacheslav
Emilio, Marcelo
Eppelbaum, Lev
Erdi, Bálint
Eroshkin, Georgij
Escapa, Alberto
Evans, Daniel
Fabricius, Claus
Fallon, Frederick
FAN, Yu
Fernandes-Martin, Vera
Fernandez, Silvia
Fernandez, Laura
Ferrandiz, Jose
Ferraz -Mello, Sylvio
Ferrer, Martinez
Fey, Alan
Fienga, Agnès
Firneis, Maria
Fliegel, Henry
Floria Peralta, Luis
Folgueira, Marta
Folkner, William
Fomin, Valery
Fominov, Aleksandr
FONG, Chugang
Fors, Octavi
Foschini, Luigi
Franz, Otto
Fredrick, Laurence
Fresneau, Alain
Froeschle, Michel
Froeschle, Claude
FU, Yanning
FUJIMOTO, Masa Katsu
FUJISHITA , Mitsumi
Gambis, Daniel
GAO, Yuping
GAO, Buxi
Gaposchkin, Edward
Gatewood, George
Gaume, Ralph
Gauss, Stephen
Gayazov, Iskander
Geffert, Michael
GENG, Lihong
Germain, Marvin
Getino Fernandez, Juan
Giacaglia, Giorgio
Giordano, Claudia
Giorgini, Jon
Giuliatti Winter, Silvia
Glebova, Nina
Goddi, Ciriaco
Goldreich, Peter
Gomes, Rodney
Gontcharov, George

GOUDA , Naoteru
Gozdziewski, Krzysztof
Gozhy, Adam
Gray, Norman
Greenberg, Richard
Gronchi, Giovanni
Gross, Richard
Guibert, Jean
Guinot, Bernard
Gumjudpai, Burin
Gusev, Alexander
Guseva, Irina
Hackman, Christine
Haghighipour, Nader
Hajian, Arsen
Hamilton, Douglas
HAN, Yanben
HAN, Tianqi
Hanslmeier, Arnold
Hanson, Robert
Harper, David
Hartkopf, William
Hau, George
Hefty, Jan
Heggie, Douglas
Helmer, Leif
Hemenway, Paul
Hering, Roland
Hestroffer, Daniel
Heudier, Jean-Louis
Hill, Graham
Hilton, James
Hobbs, David
Hobbs, George
Hoeg, Erik
Hohenkerk, Catherine
Holz, Daniel
HONG, Zhang
Horner, Jonathan
Howard, Sethanne
HSU, Rue-Ron
HU, Xiaogong
HU, Yonghui
HU, Hongbo
HUA, Yu
HUANG, Tianyi
HUANG, Cheng
Hugentobler, Urs
Hurley, Jarrod
Hut, Piet
Ianna, Philip
Ipatov, Sergei
Irwin, Michael
Ismail, Mohamed
ITO, Takashi
Ivanov, Dmitrii
Ivanova, Violeta
Ivantsov, Anatoliy
Jahreiss, Hartmut
Jakubik, Marian
Jefferys, William
JI, Jianghui
JIANG , Ing-Guey
JIN, WenJing
Johnson, Thomas
Johnston, Kenneth
Jones, Burton
Jones, Derek
Jordi, Carme
KAKUTA, Chuichi
Kalomeni, Belinda
Kalvouridis, Tilemachos
KAMEYA , Osamu
Kanayev, Ivan
Kazantseva , Liliya
Kharchenko, Nina
Khoda, Oleg
Kholshevnikov, Konstantin
Khumlumlert, Thiranee
KIM, Sungsoo
Kim, Yoo Jea
KING, Sun-Kun
KINOSHITA , Hiroshi
Kitiashvili, Irina
Klemola, Arnold
Klepczynski, William
Klock, Benny
Klocok, Lubomir
Klokocnik, Jaroslav
KOKUBO , Eiichiro
Kolaczek, Barbara
Kopeikin, Sergei
Korchagin, Vladimir
Korsun, Alla
Kosek, Wieslaw
Koshelyaevsky, Nikolay
Koshkin , Nikolay
Kostelecky, Jan
Kouba, Jan
Kovacevic, Andjelka
Kovalevsky, Jean
KOZAI, Yoshihide
Krivov, Alexander
Kuimov, Konstantin
Kurzynska, Krystyna
Kuznetsov, Eduard
Kwok, Sun
La Spina, Alessandra
Lala, Petr
Lara, Martin
Lattanzi, Mario
Lazorenko, Peter
Le Poole, Rudolf
Lecavelier des Etangs, Alain
Lega, Elena
Lehmann, Marek
Lemaitre, Anne
Lenhardt, Helmut
Lepine, Sebastien
Levine, Stephen
LI, Qi
LI, Jinling
LI, Zhigang
LI, xiaohui
LI, Yong
LIAO, Dechun
LIAO, Xinhao
Lieske, Jay
Lin, Douglas
Lindegren, Lennart
Lissauer, Jack
LIU, Ciyuan
Lopez Moratalla, Teodoro
Lu, Phillip
LU, Chunlin
LU, BenKui
LU, Xiaochun
Lucchesi, David
Luck, John
Lukashova, Marina
Lundquist, Charles
Luzum, Brian
MA, Wenzhang
MA, Jingyuan
MA, Lihua
MacConnell, Darrell
Maciejewski, Andrzej
Madsen, Claus
Maigurova, Nadiia
Majid, Abdul

Makarov, Valeri
Malhotra, Renu
Malkin, Zinovy
Mallamaci, Claudio
MANABE , Seiji
Mandel, Ilya
Marchal, Christian
Marranghello, Guilherme
Marschall, Laurence
Martinet, Louis
Martins, Roberto
MASAKI , Yoshimitsu
Matas, Vladimir
Matsakis, Demetrios
Mavraganis, Anastasios
McAlister, Harold
McLean, Brian
Melbourne, William
Melnyk, Olga
Mendes, Virgilio
Merriam, James
Metris, Gilles
Michel, Patrick
Mignard, François
Mikkola, Seppo
Milani Comparetti, Andrea
Millar, Thomas
Mink, Jessica
Monet, Alice
Monet, David
Morabito, David
Morbidelli, Roberto
Moreira Morais, Maria
Morgan, Peter
Morrison, Leslie
Mota, David
Mueller, Ivan
Mueller , Andreas
Muinos Haro, Jose
Muzzio, Juan
Mysen, Eirik
Nacozy, Paul
NAKAGAWA, Akiharu
NAKAJIMA, Koichi
Namouni, Fathi
Nastula, Jolanta
Navone, Hugo
Nefedyev, Yury
Newhall, X.
Nobili, Anna
Nothnagel, Axel
Nunez, Jorge
O'Handley, Douglas
OHNISHI , Kouji
Oja, Tarmo
Olivier, Enrico
Olsen, Hans
Orellana, Rosa
Orellana, Mariana
Orlov, Victor
Osborn, Wayne
Osorio, José
Osorio, Isabel
Pakvor, Ivan
Panessa, Francesca
Paquet, Paul
PARK , Pil Ho
Parv, Bazil
Pascu, Dan
Pauwels, Thierry
Pavlyuchenkov, Yaroslav
Pejovic, Nadezda
Penna, Jucira
Perryman, Michael
Pesek, Ivan
Petit, Jean-Marc
Petrovskaya, Margarita
Picca, Domenico
Pilat-Lohinger, Elke
Pilkington, John
Pineau des Forets, Guillaume
PING, Jinsong
Pinigin, Gennadiy
Pireaux, Sophie
Pitjeva, Elena
Platais, Imants
Podolsky, Jiri
Polyakhova, Elena
Poma, Angelo
Popescu, Petre
Pourbaix, Dimitri
Protsyuk, Yuri
Proverbio, Edoardo
Puetzfeld, Dirk
Rafferty, Theodore
Ray, James
Ray, Paul
Reasenberg, Robert
Reddy, Bacham
Reffert, Sabine
Reynolds, John
Richter, Bernd
Robertson, Douglas
Rodin, Alexander
Roemer, Elizabeth
Roeser, Siegfried
Roig, Fernando
Ron, Cyril
Rossello, Gaspar
Rossi, Alessandro
Rothacher, Markus
Ruder, Hanns
Russell, Jane
Ryabov, Yurij
Rykhlova, Lidiya
Saad, Abdel-naby
Salstein, David
Sanders, Walter
Sansaturio, Maria
Sarasso, Maria
SASAO , Tetsuo
SATO, Koichi
Schartel, Norbert
Scheeres, Daniel
Schilbach, Elena
Schildknecht, Thomas
Schillak, Stanislaw
Scholl, Hans
Scholz, Ralf-Dieter
Schubart, Joachim
Schutz, Bob
Schwekendiek, Peter
Segan, Stevo
Segransan, Damien
Seidelmann, P.
Sein-Echaluce, M.
Seitz, Florian
Shankland, Paul
Shapiro, Irwin
Sheikh, Suneel
Shelus, Peter
SHEN, Kaixian
SHEN, Zhiqiang
SHENG, Wan
Shevchenko, Ivan
SHI, Huli
Shiryaev, Alexander
Shulga, Oleksandr
Shuygina, Nadia

Sidlichovsky, Milos
Sidorenkov, Nikolay
Sima, Zdislav
Simo, Carles
Simon, Jean-Louis
Skripnichenko, Vladimir
Smart, Richard
Smylie, Douglas
Soderhjelm, Staffan
Sokolov, Leonid
Solaric, Nikola
SOMA, Mitsuru
Souchay, Jean
Sovers, Ojars
Sozzetti, Alessandro
Spoljaric, Drago
Standish, E.
Stappers, Benjamin
Stein, John
Steinmetz, Matthias
Stellmacher, Irène
Stephenson, F.
Steves, Bonnie
SUGANUMA , Masahiro
SUN, Yisui
SUN, Fuping
Sweatman, Winston
Szenkovits, Ferenc
TANG, Zheng-Hong
TAO, Jin-he
Tapley, Byron
Tarady, Vladimir
Tatevyan, Suriya
Taylor, Donald
Tedds, Jonathan
Teixeira, Ramachrisna
Teixeira, Paula
ten Brummelaar, Theo
Thomas, Claudine
Thuillot, William
Tiscareno, Matthew
Titov, Oleg
Tommei, Giacomo
Tremaine, Scott
Tsuchida, Masayoshi
TSUJIMOTO , Takuji
Tuckey, Philip
Turon, Catherine
Unwin, Stephen
Upgren, Arthur
Urban, Sean
Vallejo, Miguel
Valsecchi, Giovanni
Valtonen, Mauri
van Altena, William
van Leeuwen, Floor
van Leeuwen, Joeri
Varvoglis, Harry
Vashkovyak, Sofja
Vass, Gheorghe
Vassiliev, Nikolaj
Veillet, Christian
Vernotte, François
Vicente, Raimundo
Vieira Neto, Ernesto
Vienne, Alain
Vilhena de Moraes, R.
Vilinga, Jaime
Vilkki, Erkki
Vinet, Jean-Yves
Virtanen, Jenni
Vityazev, Veniamin
Vokrouhlicky, David
Volyanska, Margaryta
Walch, Jean-Jacques
Wallace, Patrick
WANG, Kemin
WANG, Jiaji
WANG, Zhengming
WANG, Xiao-bin
Wasserman, Lawrence
WATANABE , Noriaki
Watts, Anna
Weber, Robert
WEN, Linqing
Whipple, Arthur
White, Graeme
Wiegert, Paul
Wielen, Roland
Wilkins, George
Williams, Carol
Williams, James
Winkler, Gernot
Winter, Othon
Wnuk, Edwin
Wooden, William
WU, Haitao
WU, Bin
WU, Guichen
WU, Lianda
WU, Jiun-Huei
Wucknitz, Olaf
Wytrzyszczak, Iwona
XIA, Yifei
XIAO, Naiyuan
XIONG, Jianning
Yakut, Kadri
Yallop, Bernard
YAMADA , Yoshiyuki
YANG, Tinggao
YANG, Xuhai
YANO , Taihei
Yatsenko, Anatolij
Yatskiv, Yaroslav
YE, Shuhua
Yokoyama, Tadashi
YOSHIDA , Haruo
Yseboodt, Marie
YU, Nanhua
YUASA, Manabu
Zacchei, Andrea
Zafiropoulos, Basil
Zare, Khalil
ZHANG, Zhongping
ZHANG, Sheng-Pan
ZHANG, Shougang
ZHANG, Wei
ZHANG, Weiqun
ZHANG, Xiaoxiang
ZHANG, Wei
ZHANG, Haotong
ZHAO, Changyin
ZHAO, You
ZHAO, Haibin
Zharov, Vladimir
Zhdanov, Valery
ZHENG, Jia-Qing
ZHENG, Yong
ZHONG , Min
ZHOU, Ji-Lin
ZHOU, Yonghong
ZHOU, Li-Yong
ZHU, Yaozhong
ZHU, Wenyao
ZHU, Zi

Division II Sun & Heliosphere

President: Valentin Martínez Pillet
Vice-President: James A. Klimchuk

Organizing Committee Members:

Gianna Cauzzi
Natchimuthuk G.
Alexander Kosovichev
Ingrid Mann,
Karel Schrijver
Lidia van Driel-Gesztelyi

Members:

Abbett, William
Abdelatif, Toufik
Aboudarham, Jean
Abraham, Péter
Abramenko, Valentina
Acton, Loren
Afram, Nadine
Ahluwalia, Harjit
AI, Guoxiang
Aime, Claude
AKITA , Kyo
Alissandrakis, Costas
Altrock, Richard
Altschuler, Martin
Altyntsev, Alexandre
Aly, Jean-Jacques
Ambastha, Ashok
Ambroz, Pavel
Ananthakrishnan, S.
Anastasiadis, Anastasios
Andersen, Bo Nyborg
ANDO, Hiroyasu
Andretta, Vincenzo
Andries, Jesse
Ansari, S.M.
Antia, H.
Antiochos, Spiro
Antonucci, Ester
Anzer, Ulrich
Artzner, Guy
ASAI , Ayumi
Aschwanden, Markus
Asplund, Martin
Atac, Tamer
Ayres, Thomas
Babayev, Elchin
Babin, Arthur
Bagala, Liria
Bagare, S.
Balasubramaniam, K
Balikhin, Michael
Baliunas, Sallie
Ballester, Jose
Balthasar, Horst
BAO, Shudong
Baranovsky, Edward
Barnes, Aaron
Barrow, Colin
Barta, Miroslav
Basu, Sarbani
Batchelor, David
Baturin, Vladimir
Beckers, Jacques
Beckman, John
Bedding, Timothy
Beebe, Herbert
Beiersdorfer, Peter
Bell, Barbara
Bellot Rubio, Luis
Belvedere, Gaetano
Bemporad, Alessandro
Benevolenskaya, Elena
Benford, Gregory
Benz, Arnold
Berger, Mitchell
Bergeron, Jacqueline
Berghmans, David
Berrilli, Francesco
Bertaux, Jean-Loup
Bewsher, Danielle
Bhardwaj, Anil
BI, Shao
Bianda, Michele
Bingham, Robert
Blandford, Roger
Bobylev, Vadim
Bochsler, Peter
Bogdan, Thomas
Bommier, Veronique
Bonnet, Roger
Bornmann, Patricia
Borovik, Valerya
Botha, Gert
Bothmer, Volker
Bouchard, Antoine
Brajsa, Roman
Brandenburg, Axel
Brandt, Peter
Brandt, John
Braun, Douglas
Breckinridge, James
Brekke, Pål
Briand, Carine
Bromage, Barbara
Brooke, John
Brosius, Jeffrey
Brown, John
Browning, Philippa
Bruls, Jo
Brun, Allan
Bruner, Marilyn
Bruning, David
Bruno, Roberto
Bruns, Andrey
Brynildsen, Nils
Buccino, Andrea
Buchlin, Eric
Buechner, Joerg

Bumba, Vaclav
Burlaga, Leonard
Busa, Innocenza
Buti, Bimla
Cadez, Vladimir
Cairns, Iver
Cally, Paul
Cane, Hilary
Carbonell, Marc
Cargill, Peter
Cavallini, Fabio
Cecconi, Baptiste
Ceppatelli, Guido
CHAE, Jongchul
Chambe, Gilbert
CHAN, Kwing
Chandra, Suresh
CHANG, Heon-Young
Channok, Chanruangrit
Chapman, Gary
Chapman, Sandra
Charbonneau, Paul
Chashei, Igor
Chassefiere, Eric
CHEN, Peng-Fei
CHEN, Zhiyuan
Chernov, Gennadij
Chertok, Ilya
Chertoprud, Vadim
Chitre, Shashikumar
Chiuderi-Drago, Franca
CHIUEH , Tzihong
CHO, Kyung Suk
CHOE, Gwangson
CHOU, Chih-Kang
Choudhary, Debi Prasad
Choudhuri, Arnab
Christensen-Dalsgaard, J.
Clark, Thomas
Clette, Frederic
Cliver, Edward
Coffey, Helen
Collados, Manuel
Conway, Andrew
Correia, Emilia
Costa, Joaquim
Couturier, Pierre
Couvidat, Sebastien
Craig, Ian
Cram, Lawrence
Cramer, Neil
Crannell, Carol
Culhane, John
Cuperman, Sami
Curdt, Werner
Daglis, Ioannis
Dalla, Silvia
Dara, Helen
Dasso, Sergio
Datlowe, Dayton
Davila, Joseph
De Groof, Anik
de Jager, Cornelis
De Keyser, Johan
de Toma, Giuliana
Dechev, Momchil
Degenhardt, Detlev
Del Toro Iniesta, Jose
Del Zanna, Luca
Deliyannis, John
Demarque, Pierre
Deming, Leo
Demoulin, Pascal
DENG, YuanYong
Dennis, Brian
Dere, Kenneth
Deubner, Franz-Ludwig
Di Mauro, Maria Pia
Dialetis, Dimitris
DING, Mingde
Diver, Declan
Dobler, Wolfgang
Dobrzycka, Danuta
Donea, Alina
Dorch, Søren Bertil
Dorotovic, Ivan
Dravins, Dainis
Dryer, Murray
Dubau, Jacques
Dubois, Marc
Duchlev, Peter
Duldig, Marcus
Dumitrache, Cristiana
Durney, Bernard
Duvall Jr, Thomas
Dwivedi, Bhola
Efimenko, Volodymyr
Ehgamberdiev, Shuhrat
Einaudi, Giorgio
Emslie, Gordon
Engvold, Oddbjørn
Erdelyi, Robert
Ermolli, Ilaria
Esenoglu, Hasan
Esser, Ruth
Eviatar, Aharon
Fahr, Hans
Falewicz, Robert
FAN, Yuhong
FANG, Cheng
Farnik, Frantisek
Feldman, Uri
Fernandes, Francisco
Ferreira, Joao
Ferriz Mas, Antonio
Feynman, Joan
Fichtner, Horst
Field, George
Fisher, George
Fleck, Bernhard
Fletcher, Lyndsay
Fluri, Dominique
Foing, Bernard
Fomichev, Valerij
Fontenla, Juan
Forbes, Terry
Forgacs-Dajka, Emese
Fossat, Eric
Foukal, Peter
Fraenz, Markus
Froehlich, Claus
FU, Hsieh-Hai
Gabriel, Alan
Gaizauskas, Victor
Galsgaard, Klaus
Galvin, Antoinette
GAN, Weiqun
Garcia, Rafael
Garcia de la Rosa, Ignacio
Garcia-Berro, Enrique
Gary, Gilmer
Gedalin, Michael
Gergely, Tomas
Ghizaru, Mihai
Gill, Peter
Gilliland, Ronald
Gilman, Peter
Gimenez de Castro, Carlos
Gizon, Laurent
Glatzmaier, Gary

Gleisner, Hans
Goedbloed, Johan
Gokhale, Moreshwar
Goldman, Martin
Gomez, Daniel
Gomez, Maria
Gontikakis, Constantin
Goossens, Marcel
Gopasyuk, Olga
Gosling, John
Grandpierre, Attila
Gray, Norman
Grechnev, Victor
Gregorio, Anna
Grevesse, Nicolas
Grib, Sergey
Grzedzielski, Stanislaw
Gudiksen, Boris
Guhathakurta, Madhulika
Gupta, Surendra
Gurman, Joseph
Gyori, Lajos
Habbal, Shadia
Haberreiter, Margit
Hagyard, Mona
Hamedivafa, Hashem
Hammer, Reiner
Hanaoka, Yoichiro
Hanasz, Jan
Hanslmeier, Arnold
HARA , Hirohisa
Harra, Louise
Harrison, Richard
Harvey, John
Hasan, S. Sirajul
Hathaway, David
Haugan, Stein Vidar
Hayward, John
Heinzel, Petr
Hejna, Ladislav
Henoux, Jean-Claude
Herdiwijaya, Dhani
Heynderickx, Daniel
HIEI , Eijiro
Hildebrandt, Joachim
Hildner, Ernest
Hill, Frank
Hoang, Binh
Hochedez, Jean-François
Hoeksema, Jon
Hollweg, Joseph
Holman, Gordon
Holzer, Thomas
Hood, Alan
Howard, Robert
Hoyng, Peter
HUANG, Guangli
Huber, Martin
Hudson, Hugh
Humble, John
Hurford, Gordon
Illing, Rainer
INAGAKI, Shogo
ISHII, Takako
Ishitsuka, Mutsumi
Isliker, Heinz
Ivanchuk, Victor
Ivanov, Evgenij
Ivchenko, Vasily
Jackson, Bernard
Jain, Rajmal
Jakimiec, Jerzy
Janssen, Katja
Jardine, Moira
Jefferies, Stuart
JI, Haisheng
JIANG , Yun
JIANG, Aimin
Jimenez, Mancebo
JING, Hairong
Jokipii, Jack
Jones, Harrison
Jordan, Stuart
Jordan, Carole
Jurcak, Jan
KABURAKI , Osamu
Kahler, Stephen
KAKINUMA , Takakiyo
Kalkofen, Wolfgang
Kallenbach, Reinald
Kalman, Bela
Kaltman, Tatyana
Kapyla, Petri
Karlicky, Márian
Karpen, Judith
Kasparova, Jana
Katsova, Maria
Kaufmann, Pierre
Keil, Stephen
Keller, Horst
Khan, J
Khumlumlert, Thiranee
KIM, Yong Cheol
KIM, Kap sung
Kim, Iraida
Kiplinger, Alan
KITAI, Reizaburo
Kitchatinov, Leonid
Kitiashvili, Irina
Kjeldseth-Moe, Olav
Klein, Karl
Kliem, Bernhard
Klvana, Miroslav
Kneer, Franz
Knoelker, Michael
KO, Chung-Ming
KOJIMA , Masayoshi
Kolomanski, Sylwester
Kondrashova, Nina
Kontar, Eduard
Kopylova, Yulia
Kostik, Roman
Kotov, Valery
Kotrc, Pavel
Koutchmy, Serge
Koza, Julius
Kozak, Lyudmyla
Kozlovsky, Ben
Krimigis, Stamatios
Krivova, Natalie
Krucker, Sam
Kryshtal, Alexander
Kryvodubskyj, Valery
KUBOTA , Jun
Kucera, Ale"
Kulhanek, Petr
Kurochka, Evgenia
KUROKAWA , Hiroki
KUSANO , Kanya
Kuznetsov, Vladimir
Labrosse, Nicolas
Lai, Sebastiana
Lallement, Rosine
Landi, Simone
Landi Degl'Innocenti, E.
Landolfi, Marco
Lang, Kenneth
Lanzafame, Alessandro
Lapenta, Giovanni
Lario, David

Lawrence, John
Lazrek, Mohamed
Leibacher, John
Leiko, Uliana
Leka, K.D.
Leroy, Bernard
Leroy, Jean-Louis
Levy, Eugene
LI, Wei
LI, Hui
LI, Kejun
Li, Bo
LI, Zhi
Lie-Svendsen, Oystein
Lima, Joao
Lin, Yong
Linsky, Jeffrey
Liritzis, Ioannis
LIU, Yang
LIU, Yu
Livingston, William
Livshits, Moisey
Lopez Arroyo, M.
Lopez Fuentes, Marcelo
Lotova, Natalja
Loukitcheva, Maria
Low, Boon
Lozitskij, Vsevolod
Luest, Reimar
Lundstedt, Henrik
Machado, Marcos
Mackay, Duncan
MacKinnon, Alexander
MacQueen, Robert
MAKITA , Mitsugu
Malandraki, Olga
Malara, Francesco
Malherbe, Jean-Marie
Malitson, Harriet
Malville, J.
MANABE , Seiji
Mandrini, Cristina
Mangeney, André
Mann, Gottfried
Manoharan, P.
Marcu, Alexandru
Maricic, Darije
Marilena, Mierla
Marilli, Ettore
Maris, Georgeta
Mariska, John
Markova, Eva
Marmolino, Ciro
Marsch, Eckart
Martens, Petrus
Mason, Glenn
MASUDA , Satoshi
Matsuura, Oscar
Mattig, W.
Mavromichalaki, Helen
McAteer, R.T. James
McCabe, Marie
McIntosh, Patrick
McKenna Lawlor, Susan
Mein, Pierre
Meister, Claudia
Melnik, Valentin
Messerotti, Mauro
Messmer, Peter
Mestel, Leon
Meszarosova, Hana
Meyer, Friedrich
Michalek, Grzegorz
Miletsky, Eugeny
Milkey, Robert
Miralles, Mari Paz
Mohan, Anita
Moncuquet, Michel
Monteiro, Mario Joao
Moore, Ronald
Morabito, David
MORITA, Satoshi
MORIYAMA, Fumio
Motta, Santo
Mouradian, Zadig
Moussas, Xenophon
Muller, Richard
MUNETOSHI , Tokumaru
Munro, Richard
Musielak, Zdzislaw
NAKAJIMA, Hiroshi
Nakariakov, Valery
Neidig, Donald
Nesis, Anastasios
Neukirch, Thomas
Neupert, Werner
Nickeler, Dieter
Nicolas, Kenneth
NING, Zongjun
Nocera, Luigi
Noens, Jacques-Clair
Nordlund, Aake
Noyes, Robert
Nozawa, Satoshi
Nussbaumer, Harry
Obridko, Vladimir
Ofman, Leon
OHKI , Kenichiro
Oliver, Ramó n
Orlando, Salvatore
Ossendrijver, Mathieu
Owocki, Stanley
Ozguc, Atila
Padmanabhan, Janardhan
Paletou, Frédéric
Palle, Pere
Palle Bago, Enric
Palus, Pavel
PAN, Liande
Pandey, Birendra
Pap, Judit
Papathanasoglou, Dimitrios
Paresce, Francesco
Parfinenko, Leonid
Parhi, Shyamsundar
Pariat, Etienne
PARK , Young Deuk
Parker, Eugene
Parnell, Clare
Pasachoff, Jay
Paterno, Lucio
Pecker, Jean-Claude
Peres, Giovanni
Perkins, Francis
Peter, Hardi
Petrie, Gordon
Petrosian, Vahe
Petrov, Nikola
Petrovay, Kristof
Pevtsov, Alexei
Pflug, Klaus
Phillips, Kenneth
Picazzio, Enos
Pick, Monique
Pipin, Valery
Plainaki, Christina
Podesta, John
Poedts, Stefaan
Pohjolainen, Silja
Poland, Arthur

Poquerusse, Michel
Preka-Papadema, P.
Pres, Pawek
Priest, Eric
Proctor, Michael
QU, Zhongquan
Quemerais, Eric
Raadu, Michael
Radick, Richard
Ramelli, Renzo
Rao, A.
Raulin, Jean-Pierre
Readhead, Anthony
Reale, Fabio
Reardon, Kevin
Reeves, Hubert
Regnier, Stephane
Regulo, Clara
Reinard, Alysha
Rendtel, Juergen
Rengel, Miriam
Reshetnyk, Volodymyr
Riehokainen, Aleksandr
Riley, Pete
Ripken, Hartmut
Roca Cortes, Teodoro
Roddier, Francois
Rodriguez Hidalgo, Inés
Roemer, Max
Romoli, Marco
Rompolt, Bogdan
Rosa, Dragan
Rosa, Reinaldo
Rosner, Robert
Roudier, Thierry
Rouppe van der Voort, Luc
Rovira, Marta
Roxburgh, Ian
Rozelot, Jean-Pierre
Rudawy, Pawel
Ruediger, Guenther
Ruffolo, David
Ruiz Cobo, Basilio
Rusin, Vojtech
Russell, Christopher
Rust, David
Rutten, Robert
Ryabov, Boris
Rybak, Jan
Rybansky, Milan
Sagdeev, Roald
Sahal-Brechot, Sylvie
Saiz, Alejandro
SAKAI , Junichi
SAKAO, Taro
Samain, Denys
Sanahuja Parera, Blai
Sanchez Almeida, Jorge
Saniga, Metod
Sarris, Emmanuel
Sasso, Clementina
Sastri, Hanumath
Sattarov, Isroil
Sauval, A.
Sawyer, Constance
Scherb, Frank
Schindler, Karl
Schleicher, Helmold
Schlichenmaier, Rolf
Schmahl, Edward
Schmelz, Joan
Schmidt, Wolfgang
Schmieder, Brigitte
Schober, Hans
Schou, Jesper
Schreiber, Roman
Schuessler, Manfred
Schwartz, Pavol
Schwartz, Steven
Setti, Giancarlo
Severino, Giuseppe
Shchukina, Nataliia
Shea, Margaret
Sheeley, Neil
Sheminova, Valentina
SHIBASAKI , Kiyoto
SHIBATA , Kazunari
Shimizu, Toshifumi
Shimojo , Masumi
SHIN'ICHI , Nagata
Shine, Richard
Sigalotti, Leonardo
Sigwarth, Michael
Simnett, George
Simon, Guy
Simon, George
Singh, Jagdev
Sinha, Krishnanand
Sivaraman, Koduvayur
Skumanich, Andrew
Smaldone, Luigi
Smith, Dean
Smith, Peter
Smol'kov, Gennadij
Snegirev, Sergey
Sobotka, Michal
Socas-Navarro, Hector
Solanki, Sami
Soloviev, Alexandr
Somov, Boris
Spadaro, Daniele
Spicer, Daniel
Spruit, Hendrik
Stathopoulou, Maria
Staude, Juergen
Stebbins, Robin
Steffen, Matthias
Steiner, Oskar
Stellmacher, Götz
Stenflo, Jan
Stepanian, Natali
Stepanov, Alexander
Steshenko, N.
Stix, Michael
Stodilka, Myroslav
Straus, Thomas
Strong, Keith
Struminsky, Alexei
Sturrock, Peter
Subramanian, K.
Subramanian, Prasad
SUEMATSU , Yoshinori
Suess, Steven
SUZUKI, Takeru
Sylwester, Janusz
Sylwester, Barbara
Szalay, Alex
TAKANO, Toshiaki
Tapping, Kenneth
Tarashchuk, Vera
Teplitskaya, Raisa
Ternullo, Maurizio
Teske, Richard
Thomas, John
Tikhomolov, Evgeniy
Tlatov, Andrej
Tobias, Steven
Tomczak, Michal
Tripathy, Sushanta
Tritakis, Basil

Trottet, Gerard
Trujillo Bueno, Javier
Tsap, Yuri
Tsap, Teodor
Tsiklauri, David
Tsinganos, Kanaris
Tsiropoula, Georgia
TSUNETA , Saku
Tyul'bashev, Sergei
Uddin, Wahab
Usoskin, Ilya
Vainshtein, Leonid
Valio, Adriana
van den Oord, Bert
van der Heyden, Kurt
van der Linden, Ronald
Van Hoven, Gerard
Vandas, Marek
Vaughan, Arthur
Veck, Nicholas
Vekstein, Gregory
Velli, Marco
Venkatakrishnan, P.
Ventura, Paolo
Ventura, Rita
Verheest, Frank
Verma, V.
Verwichte, Erwin
Vial, Jean-Claude
Vieytes, Mariela
Vilinga, Jaime
Vilmer, Nicole
Vinod, S.
Voitenko, Yuriy
Voitsekhovska, Anna
von der Luehe, Oskar
Vrsnak, Bojan
Vucetich, Héctor
Walsh, Robert
WANG, Huaning
WANG, Jingxiu
WANG, Haimin
Wang, Yi-ming
WANG, Dongguang
WANG, Shujuan
WANG, Jingyu
WATANABE , Takashi
WATARI, Shinichi
Webb, David
Weiss, Nigel
Weller, Charles
White, Stephen
Wiehr, Eberhard
Wiik Toutain, Jun
Wikstol, Oivind
Winebarger, Amy
Wittmann, Axel
Woehl, Hubertus
Wolfson, Richard
Woltjer, Lodewijk
Worden, Simon
WU, De Jin
WU, Hsin-Heng
Wu, Shi
XIE, Xianchun
XU, Aoao
XU, Jun
XU, Zhi
YAN, Yihua
YANG, Zhiliang
YANG, Hong-Jin
YANG, Jing
YANG, Lei
Yesilyurt, Ibrahim
YI, Yu
YOICHIRO, Suzuki
YOSHIMURA, Hirokazu
YU, Dai
Yun, Hong-Sik
Zachariadis, Theodosios
Zampieri, Luca
Zappala, Rosario
Zarro, Dominic
Zelenka, Antoine
ZHANG, Mei
ZHANG, Jun
ZHAO, Junwei
ZHOU, Guiping
Zhugzhda, Yuzef
Zirker, Jack
Zlobec, Paolo

Division III Planetary Systems Sciences

President: Karen J. Meech
Vice-President: Giovanni B. Valsecchi

Organizing Committee Members:

D. Bockelée-Morvan
Alan Paul Boss,
Alberto Cellino
Guy Joseph Consolmagno
Julio Angel Fernández
William M. Irvine
Daniela Lazzaro
Patrick Michel
Keith S. Noll
Rita M. Schulz
Jun-ichi WATANABE
Makoto YOSHIKAWA
Jin ZHU

Members:

A'Hearn, Michael
Abalakin, Viktor
ABE, Shinsuke
AGATA, Hidehiko
Aikman, G.
Akimov, Leonid
Aksnes, Kaare
Al-Naimiy, Hamid
Alexandrov, Alexander
Alexandrov, Yuri
Allard, France
Allegre, Claude
Allison, Michael
Almar, Ivan
Alsabti, Abdul Athem
Altwegg, Kathrin
ANDO, Hiroyasu
Angione, Ronald
Apai, Daniel
Archinal, Brent
Ardila, David
Arlot, Jean-Eudes
Arpigny, Claude
Arthur, David
Asher, David
Atkinson, David
Atreya, Sushil
Babadzhanov, Pulat
Baggaley, William
Bailey, Mark
Balazs, Bela
Balbi, Amedeo
Balikhin, Michael
Ball, John
Bania, Thomas
Bar-Nun, Akiva
Barabanov, Sergey
Baran, Andrzej
Baransky, Olexander
Barber, Robert
Barbieri, Cesare
Barker, Edwin
Barkin, Yuri
Barlow, Nadine
Barriot, Jean-Pierre
Barrow, Colin
Barucci, Maria
Baryshev, Andrey
Battaner, Eduardo
Beckman, John
Beckwith, Steven
Beebe, Reta
Beer, Reinhard
Behrend, Raoul
Belkovich, Oleg
Bell III, James
Belskaya, Irina
Belton, Michael
Bemporad, Alessandro
Ben-Jaffel, Lofti
Bender, Peter
Bendjoya, Philippe
Benest, Daniel
Bennett, David
Berendzen, Richard
Bergstralh, Jay
Bernardi, Fabrizio
Bertaux, Jean-Loup
Berthier, Jerôme
Bezard, Bruno
Bhandari, N.
Bhardwaj, Anil
Biazzo, Katia
Bien, Reinhold
Billebaud, Francoise
Bingham, Robert
Binzel, Richard
Biraud, François
Birlan, Mirel
Biver, Nicolas
Blanco, Armando
Blanco, Carlo
Bless, Robert
Boerngen, Freimut
Boice, Daniel
Bond, Ian
Bondarenko, Lyudmila
Bonev, Tanyu
Borde, Pascal
Borovicka, Ji?í
Borysenko, Serhii
Bosma, Pieter
Bowyer, C.
Boyce, Peter
Brahic, André
Brandt, John
Branham, Richard
Brecher, Aviva
Britt, Daniel
Broadfoot, A.
Broderick, John
Brown, Peter

Brown, Robert
Brownlee, Donald
Buccino, Andrea
Buie, Marc
Buratti, Bonnie
Burba, George
Burke, Bernard
Burlaga, Leonard
Burns, Joseph
Butler, Paul
Butler, Bryan
CAI, Kai
Caldwell, John
Calvin, William
Campbell, Donald
Campbell-Brown, Margaret
Campins, Humberto
Campusano, Luis
Capaccioni, Fabrizio
Capria, Maria
Cardenas, Rolando
Cardoso Santos, Nuno
Carpino, Mario
Carruba, Valerio
Carruthers, George
Carsenty, Uri
Carusi, Andrea
Carvano, Jorge
Cecconi, Baptiste
Cellino, Alberto
Cerroni, Priscilla
Chaisson, Eric
Chandrasekhar, Th.
Chapman, Clark
Chapman, Robert
Chapront-Touze, Michelle
Chernetenko, Yulia
Chesley, Steven
Chevrel, Serge
Chodas, Paul
Chubko , Larysa
Cirkovic, Milan
Clairemidi, Jacques
Clayton, Donald
Clayton, Geoffrey
Cochran, William
Cochran, Anita
Colom, Pierre
Combi, Michael
Connes, Pierre
Connes, Janine
Connors, Martin
Conrad, Albert
Cooper, Timothy
Cooper, Nicholas
Cosmovici, Cristiano
Cottini, Valeria
Coude du Foresto, Vincent
Couper, Heather
Coustenis, Athena
Cremonese, Gabriele
Crovisier, Jacques
Cruikshank, Dale
Cuesta Crespo, Luis
Cunningham, Maria
Cuypers, Jan
d'Hendecourt, Louis
da Silveira, Enio
Daigne, Gerard
Dall , Thomas
Danks, Anthony
Darhmaoui, Hassane
Davidsson, Björn
Davies, John
Davies, Ashley
Davis, Michael
Davis, Gary
de Almeida, Amaury
de Bergh, Catherine
de Jager, Cornelis
de Loore, Camiel
de Pater, Imke
de Sanctis, Giovanni
de Sanctis, Maria
Deeg, Hans
Delbo, Marco
Dell'Oro, Aldo
Delplancke , Francoise
Delsanti, Audrey
Dermott, Stanley
Di Martino, Mario
Di Sisto, Romina
Dick, Steven
Dickel, John
Dickey, Jean
Dieleman, Pieter
Djorgovski, Stanislav
Dlugach, Zhanna
Dodonov, Sergej
Dominis Prester, Dijana
Donnison, John
Doressoundiram, Alain
Dorschner, Johann
Dotto, Elisabetta
Doubinskij, Boris
Dourneau, Gerard
Doval, Jorge M.
Downs, George
Drake, Frank
Drossart, Pierre
Dryer, Murray
Dubin, Maurice
Duffard, Rene
Dumont, Rene
Duncan, Martin
Dunham, David
Dunkin Beardsley, Sarah
Durech, Josef
Durrance, Samuel
Dutil, Yvan
Dvorak, Rudolf
Dwek, Eli
Dybczynski, Piotr
Dyson, Freeman
Ehrenfreund, Pascale
El-Baz, Farouk
Ellis, George
Elst, Eric
Emelianov, Nikolaj
Emelyanenko, Vacheslav
Encrenaz, Therese
Epishev, Vitali
Epstein, Eugene
Erard, Stéphane
Ershkovich, Alexander
Esenoglu, Hasan
Esposito, Larry
Evans, Neal
Evans, Michael
Eviatar, Aharon
Farnham, Tony
Fazio, Giovanni
Feldman, Paul
Feldman, Paul
Fernandez Lajus, Eduardo
Ferrari, Cécile
Ferraz-Mello, Sylvio
Ferreri, Walter
Ferrin, Ignacio
Field, George

Filacchione, Gianrico
Fink, Uwe
Firneis, Maria
Fisher, Philip
Fitzsimmons, Alan
Fornasier, Sonia
Fors, Octavi
Foryta, Dietmar
Franklin, Fred
Fraser, Brian
Fraser, Helen
Fredrick, Laurence
Freire Ferrero, Rubens
Freitas Mourao, Ronaldo
Froeschle, Christiane
Froeschle, Claude
FUJIMOTO, Masa Katsu
FUJIWARA , Akira
FUKAGAWA, Misato
Fulchignoni, Marcello
FURUSHO , Reiko
FUSE, Tetsuharu
Gajdos, Stefan
Galad, Adrián
Gammelgaard, Peter
Gatewood, George
Gautier, Daniel
Geiss, Johannes
Gerakines, Perry
Gerard, Jean-Claude
Gerard, Eric
Ghigo, Francis
Giani, Elisabetta
Gibson, James
Gierasch, Peter
Gil-Hutton, Ricardo
Gilmore, Alan
Giorgini, Jon
Giovane, Frank
Giovannelli, Franco
Golden, Aaron
Goldreich, Peter
Goldsmith, Donald
Goody, Richard
Gorbanev, Jury
Gorenstein, Paul
Gorkavyi, Nikolai
Gorshanov, Denis
Goswami, J.
Gott, J.
Goudis, Christos
Gounelle, Matthieu
Gradie, Jonathan
Grady, Monica
Grav, Tommy
Green, Daniel
Green, Jack
Green, Simon
Greenberg, Richard
Gregory, Philip
Grieger, Björn
Gronkowski, Piotr
Grossman, Lawrence
Gruen, Eberhard
Grundy, William
Gulbis, Amanda
Gulkis, Samuel
Gunn, James
Gurshtein, Alexander
Hadamcik, Edith
Hahn, Gerhard
Haisch, Bernard
Hajdukova, Maria
Hajdukova, Jr., Maria
Hale, Alan
Halliday, Ian
Hammel, Heidi
Hanner, Martha
Hanninen, Jyrki
Hapke, Bruce
Harper, David
Harris, Alan
Harris, Alan
Hart, Michael
Hartmann, William
Harvey, Gale
Harwit, Martin
HASEGAWA, Ichiro
Haupt, Hermann
Hauser, Michael
Hawkes, Robert
Hecht, James
Hemenway, Paul
Henry, Richard
Hershey, John
Hestroffer, Daniel
Heudier, Jean-Louis
HIRABAYASHI , Hisashi
Hoang, Binh
Hodge, Paul
Hofmann, Wilfried
Hogbom, Jan
Hol, Pedro
Holberg, Jay
Hollis, Jan
Holm, Nils
Homeier, Derek
HONG, Seung-Soo
Horedt, Georg
Horowitz, Paul
Hovenier, J.
Howard, Andrew
Howell, Ellen
HU, Zhong wen
Hubbard, William
Hudkova, Ludmila
Hunt, Garry
Hunter, James
Huntress, Wesley
Hurnik, Hieronim
Hurwitz, Mark
Hysom, Edmund
Ianna, Philip
Ibadinov, Khursand
Ibadov, Subhon
IP, Wing-Huen
Ireland, Michael
Iro, Nicolas
Irvine, William
Irwin, Patrick
Israel, Frank
Israelevich, Peter
Ivanov-Kholodny, Gor
Ivanova, Violeta
Ivanova, Oleksandra
Ivantsov, Anatoliy
Ivezic, Zeljko
IWASAKI, Kyosuke
Jackson, Bernard
Jackson, William
Jacobson, Robert
Jakubik, Marian
James, John
Jayawardhana, Ray
Jedicke, Robert
Jeffers, Sandra
Jenniskens, Petrus
Jockers, Klaus
Johnson, Torrence
Jones, James

Jopek, Tadeusz
Jorda, Laurent
Jordan, Andrés
Joubert, Martine
Jurgens, Raymond
Kaasalainen, Mikko
Kablak, Nataliya
Kaeufl, Hans Ulrich
Kafatos, Menas
Kalenichenko, Valentin
Kane, Stephen
Karakas, Amanda
Karatekin, Özgür
Kardashev, Nicolay
Kascheev, Rafael
Kaufmann, Pierre
Kavelaars, JJ.
Kawada, Mitsunobu
KAWAKITA, Hideyo
Kazantsev, Anatolii
Keay, Colin
Keheyan, Yeghis
Keil, Klaus
Keller, Hans-Ulrich
Keller, Horst
Kellermann, Kenneth
Khovritchev, Maxim
Kidger, Mark
Killen, Rosemary
Kilmartin, Pamela
Kilston, Steven
KIM, Yongha
KIM, Sang Joon
KIM, Yoo Jea
KING, Sun-Kun
KINOSHITA, Hiroshi
Kiselev, Nikolai
Kisseleva, Tamara
Kitiashvili, Irina
Klacka, Jozef
Klahr , Hubert
Klemola, Arnold
Kley, Wilhelm
Knacke, Roger
Knezevic, Zoran
Kocer, Dursun
Koeberl, Christian
Kohoutek, Lubos
Kokhirova, Gulchehra
Kolomiyets, Svitlana
Konacki, Maciej
Kopylov, Alexander
Korsun, Pavlo
KOSAI, Hiroki
Koshkin, Nikolay
Koten, Pavel
Koutchmy, Serge
KOZAI, Yoshihide
Kozak, Lyudmyla
Kozak, Pavlo
KOZASA, Takashi
Kramer, Busaba
Krimigis, Stamatios
Krishna, Swamy
Kristensen, Leif
Krolikowska-Soltan, M.
Kruchinenko, Vitaliy
Krugly, Yurij
Kryszczynska, Agnieszka
Ksanfomality, Leonid
KUAN, Yi-Jehng
Kueppers, Michael
Kuiper, Thomas
Kulikova, Nelly
Kulkarni, Prabhakar
Kumar, Shiv
Kurt, Vladimir
La Spina, Alessandra
Lacour, Sylvestre
Lagage, Pierre-Olivier
Lagerkvist, Claes-Ingvar
Lamy, Philippe
Lane, Arthur
Lara, Luisa
Larsen, Jeffrey
Larson, Harold
Larson, Stephen
Latham, David
Laurin, Denis
Lazzarin, Monica
Lazzaro, Daniela
Lebofsky, Larry
LEE, Thyphoon
LEE, Sang-Gak
Léger, Alain
Lellouch, Emmanuel
Lemaitre, Anne
Lemke, Dietrich
Lemmon, Mark
Lewis, John
LI, Guangyu
LI, Yong
Licandro, Javier
Lichtenegger, Herbert
Lieske, Jay
Liller, William
Lilley, Edward
Lillie, Charles
Lindsey, Charles
Lineweaver, Charles
Lippincott Zimmerman, S.
Lipschutz, Michael
Lissauer, Jack
Lisse, Carey
LIU, Sheng-Yuan
LIU, Michael
Lo Curto, Gaspare
Lockwood, G.
Lodders, Katharina
Lodieu, Nicolas
Lomb, Nicholas
Lopes, Rosaly
Lopez Gonzalez, Maria
Lopez Moreno, Jose
Lopez Puertas, Manuel
Lopez Valverde, M.
Lovas, Miklos
Lugaro, Maria
Lukyanyk, Igor
Lutz, Barry
Luu, Jane
Luz, David
Lyon, Ian
MA, Yuehua
Magee-Sauer, Karen
Magnusson, Per
MAIHARA, Toshinori
Makalkin, Andrei
Manara, Alessandro
Mann, Ingrid
Maran, Stephen
Marchi, Simone
Marcialis, Robert
Margot, Jean-Luc
Margrave Jr, Thomas
Maris, Michele
Martin, Maria
Martin, Donn
Martinez Fiorenzano, Aldo
Martinez-Frias, Jesus

Martins, Roberto
Marvin, Ursula
Marzari, Francesco
Mason, John
Matese, John
Mather, John
Matsakis, Demetrios
Matson, Dennis
MATSUDA, Takuya
MATSUI, Takafumi
MATSUMOTO, Toshio
Matsuura, Oscar
Mattila, Kalevi
Maucherat, Jean
Maury, Alain
Mazzotta Epifani, Elena
McAlister, Harold
McCord, Thomas
McCullough, Peter
McDonough, Thomas
McElroy, Michael
McFadden, Lucy
McGrath, Melissa
McKenna Lawlor, Susan
McKinnon, William
McMillan, Robert
McNaught, Robert
Meadows, A.
Medvedev, Yurij
Meisel, David
Melita, Mario
Mendillo, Michael
Mendoza, V.
Merin Martin, Bruno
Merline, William
Michalowski, Tadeusz
Michel, Patrick
Mickelson, Michael
Mignard, François
Milani Comparetti, Andrea
Millan Gabet, Rafael
Millis, Robert
Mills, Franklin
MINN, Young-Ki
Minniti, Dante
Mirabel, Igor
Misconi, Nebil
Moehlmann, Diedrich
Moerchen, Margaret
Mokhele, Khotso
Molina, Antonio
Moncuquet, Michel
Monet, Alice
Montmessin, Franck
Moore, Marla
Moore, Elliott
Moravec, Zdeněk
Moreno-Insertis, Fernando
Morozhenko, A.
Morris, Mark
Morris, Charles
Morrison, David
Mosser, Benoît
Mothe-Diniz, Thais
Mueller, Thomas
Muinonen, Karri
MUKAI, Tadashi
Muller, Richard
Mumma, Michael
Murphy, Robert
Murray, Carl
Murthy, Jayant
Nacozy, Paul
Naef, Dominique
NAGAHARA, Hiroko
NAKAGAWA, Yoshitsugu
NAKAMURA, Akiko
NAKAMURA, Takuji
NAKAMURA, Tsuko
NAKANO, Syuichi
NAKAZAWA, Kiyoshi
Napier, William
Nelson, Robert
Nelson, Richard
Neslusan, Lubos
Ness, Norman
Neuhaeuser, Ralph
Niarchos, Panagiotis
Niedner, Malcolm
Ninkov, Zoran
Nishimura, Tetsuo
Nobili, Anna
Nolan, Michael
Noll, Keith
Norris, Raymond
Nuth, Joseph
O'Dell, Charles
OHTSUKI, Keiji
Ortiz, Jose
Ostriker, Jeremiah
Owen, Tobias
Owen Jr, William
Palle Bago, Enric
Pandey, A.
Pang, Kevin
Paolicchi, Paolo
Paresce, Francesco
Parijskij, Yurij
PARK, Yong Sun
PARK, Byeong-Gon
Pascu, Dan
Pascucci, Ilaria
Patten, Brian
Pauwels, Thierry
Pecina, Petr
Peixinho, Nuno
Pellinen-Wannberg, Asta
Pendleton, Yvonne
Penny, Alan
Perek, Luboš
Perez, Mario
Perez de Tejada, Hector
Perrin, Jean-Marie
Perrin, Marshall
Persson, Carina
Petit, Jean-Marc
Pettengill, Gordon
Pfleiderer, Jorg
Picazzio, Enos
Pierce, David
Pilcher, Carl
PING, Jinsong
Pittich, Eduard
Pittichova, Jana
Plavchan, Jr., Peter
Politi, Romolo
Pollacco, Don
Polyakhova, Elena
Ponsonby, John
Poole, Graham
Porubcan, Vladimir
Potter, Andrew
Pozhalova, Zhanna
Pravec, Petr
Prialnik, Dina
Psaryov, Volodymyr
QIAO, Rongchuan
Quintana, Jose
Quirrenbach, Andreas
Raju, Vasundhara

Rao, M.
Rapaport, Michel
Rawlings, Mark
Reach, William
Reddy, Bacham
Rees, Martin
Reffert, Sabine
Reiners, Ansgar
Reitsema, Harold
Renard, Jean-Baptiste
Rendtel, Juergen
Rickman, Hans
Ripken, Hartmut
Rodionova, Zhanna
Rodriguez, Luis
Roemer, Elizabeth
Roeser, Siegfried
Roig, Fernando
Rojo, Patricio
Roos-Serote, Maarten
Roques, Françoise
Rosenbush, Vera
Rossi, Alessandro
Rousselot, Philippe
Rowan-Robinson, Michael
RUI, Qi
Ruskol, Evgeniya
Russel, Sara
Russell, Jane
Russo, Pedro
Ryabova, Galina
Saffe, Carlos
Sagdeev, Roald
SAITO, Takao
Samarasinha, Nalin
Sampson, Russell
Sanchez, Francisco
Sanchez Bejar, Victor
Sanchez-Lavega, Agustin
Sanchez-Saavedra, M.
Sancisi, Renzo
Sarre, Peter
Sasaki, Sho
SATO, Isao
Scargle, Jeffrey
Schaller, Emily
Schild, Rudolph
Schleicher, David
Schloerb, F.
Schmadel, Lutz
Schmidt, Maarten
Schneider, Jean
Schneider, Nicholas
Schober, Hans
Scholl, Hans
Schubart, Joachim
Schuch, Nelson
Schuh, Harald
Schuster, William
Seidelmann, P.
Seielstad, George
Sekanina, Zdenek
SEKIGUCHI, Tomohiko
Selam, Selim
Sen, Asoke
Sergis, Nick
Serra Ricart, Miquel
Shankland, Paul
Shanklin, Jonathan
Shapiro, Irwin
Sharma, A.
Shefov, Nikolaj
Shelus, Peter
SHEN, Kaixian
Shevchenko, Vasilij
Shevchenko, Vladislav
SHI, Huli
Shkuratov, Yurii
Shor, Viktor
Shostak, G.
Sicardy, Bruno
Sims, Mark
Singh, Harinder
Sivaram, C.
Sivaraman, Koduvayur
Sizonenko, Yuri
Smith, Bradford
Smith, Robert
Snellen, Ignas
Snyder, Lewis
Soberman, Robert
Soderblom, Larry
SOFUE , Yoshiaki
Solc, Martin
Solovaya, Nina
SOMA, Mitsuru
SONG, In Ok
Sozzetti, Alessandro
Spahr, Timothy
Spencer, John
Spinrad, Hyron
Sprague, Ann
Stallard, Thomas
Stam, Daphne
Standish, E.
Staude, Hans
Steel, Duncan
Stein, John
Stellmacher, Irène
Stern, S.
Sterzik, Michael
Stoev, Alexey
Stokes, Grant
Stone, Edward
Straizys, Vytautas
Strobel, Darrell
Strom, Robert
Sturrock, Peter
Sullivan, III, Woodruff
Surdej, Jean
Svestka, Jiri
Svoren, Jan
Swade, Daryl
Sykes, Mark
Synnott, Stephen
Szego, Karoly
Szutowicz, Slawomira
Tacconi-Garman, Lowell
TAKABA, Hiroshi
TAKADA-HIDAI, M.
TAKEDA, Hidenori
Tancredi, Gonzalo
Tanga, Paolo
TAO, Jun
Tarashchuk, Vera
Tarter, Jill
Tatum, Jeremy
Tavakol, Reza
Taylor, Fredric
Taylor, Donald
Tchouikova, Nadezhda
Tedds, Jonathan
Tedesco, Edward
Tejfel, Victor
Terrile, Richard
Terzian, Yervant
Thaddeus, Patrick
Tholen, David
Thomas, Nicolas
Thuillot, William

Tichá, Jana
Tiscareno, Matthew
Tolbert, Charles
Toller, Gary
Toshihiro, Kasuga
Tosi, Federico
Toth, Imre
Toth, Juraj
Tovmassian, Hrant
Tozzi, Gian
Trafton, Laurence
Trigo-Rodriguez, Josep
Trimble, Virginia
Trujillo, Chadwick
Tsuchida, Masayoshi
Tuccari, Gino
Turner, Edwin
Turner, Kenneth
Tyler Jr, G.
Tyson, John
Udry, Stephane
UENO, Munetaka
Ugolnikov, Oleg
Valdes-Sada, Pedro
Vallee, Jacques
van Houten-Groeneveld, I.
Varshalovich, Dmitrij
Vaubaillon, Jérémie
Vauclair, Gérard
Vazquez, Manuel
Vazquez, Roberto
Veeder, Glenn
Veiga, Carlos
Veillet, Christian
Venugopal, V.
Verschuur, Gerrit
Veverka, Joseph
Vidal-Madjar, Alfred
Vidmachenko, Anatoliy
Vienne, Alain
Vilas, Faith
Villaver, Eva
Virtanen, Jenni
Voelzke, Marcos
Vogt, Nikolaus
Voloschuk, Yuri
von Braun, Kaspar
Vrtilek, Jan
Walker, Alta
Wallace, Lloyd
Wallace, James
Wallis, Max
Walsh, Wilfred
Walsh, Andrew
WANG, Xiao-bin
Wasserman, Lawrence
Wasson, John
Watson, Frederick
Wdowiak, Thomas
Weaver, Harold
Webster, Alan
Weidenschilling, S.
Weinberg, Jerry
Weissman, Paul
Welch, William
Wesson, Paul
West, Richard
Wheatley, Peter
Whelan, Emma
Whipple, Arthur
Wielebinski, Richard
Williams, Gareth
Williams, James
Willson, Robert
Wilson, Thomas
Wolstencroft, Ramon
Wooden, Diane
Woolfson, Michael
Wright, Alan
Wright, Ian
Wu, Yanqin
Wurz, Peter
Wyckoff, Susan
XIONG, Jianning
XU, Weibiao
YABUSHITA, Shin
YAMAMOTO, Tetsuo
YAMAMOTO, Masayuki
YAMASHITA, Kojun
YANAGISAWA, Masahisa
Yanamandra-Fisher, P.
YANG, Jongmann
YANG, Xiaohu
YANO, Hajime
YE, Shuhua
Yeomans, Donald
Yesilyurt, Ibrahim
YI, Yu
YIM, Hong Suh
Yoder, Charles
YOSHIDA , Fumi
YOSHIKAWA, Makoto
Young, Andrew
YUASA, Manabu
Yuce, Kutluay
YUMIKO , Oasa
YUTAKA , Shiratori
Zagretdinov, Renat
Zapatero-Osorio, Maria R.
Zarka, Philippe
Zarnecki, John
ZHANG, You-Hong
ZHANG, Xiaoxiang
ZHANG, Wei
ZHANG, Jun
ZHAO, Haibin
Zharkov, Vladimir
ZHU, Jin
Ziolkowski, Krzysztof
Zuckerman, Benjamin

Division IV Stars

President: Christopher Corbally
Vice-President: Francesca D'Antona

Organizing Committee Members:

Martin Asplund
Corinne Charbonnel
Jose-Angel Docobo
Richard O. Gray
Nikolai E. Piskunov

Members:

Abia, Carlos
Abt, Helmut
Adams, Mark
Adelman, Saul
Afram, Nadine
Ahumada, Javier
Aikman, G.
Aizenman, Morris
Ake III, Thomas
Alcala, Juan Manuel
Alecian, Georges
Alencar, Silvia
Allard, France
Allende Prieto, Carlos
Altrock, Richard
Andrade, Manuel
Andretta, Vincenzo
Andreuzzi, Gloria
Angelov, Trajko
Annuk, Kalju
Anosova, Joanna
Antia, H.
AOKI, Wako
Appenzeller, Immo
ARAI, Kenzo
Ardeberg, Arne
Ardila, David
Arellano Ferro, Armando
Arenou, Frédéric
Arentoft, Torben
Argast, Dominik
Argyle, Robert
Arias, Maria
ARIMOTO, Nobuo
Arkharov, Arkadij
Armstrong, John
Arnett, W.
Arnould, Marcel
Arpigny, Claude
Asplund, Martin
Atac, Tamer
Atanackovic, Olga
Audard, Marc
Audouze, Jean
Aungwerojwit, Amornrat
Avrett, Eugene
Ayres, Thomas
Baade, Dietrich
Babu, G.S.D.
Baglin, Annie
Bagnulo, Stefano
Bagnuolo Jr, William
Bailer-Jones, Coryn
Bailyn, Charles
Baird, Scott
Bakker, Eric
Balega, Yurij
Baliunas, Sallie
Ballereau, Dominique
Balona, Luis
Banerjee, Dipankar
Barber, Robert
Barbuy, Beatriz
Barnes, Sydney
Baron, Edward
Bartkevicius, Antanas
Baschek, Bodo
Basri, Gibor
Basu, Sarbani
Batalha, Celso
Batten, Alan
Bauer, Wendy
Baym, Gordon
Bazot, Michael
Beavers, Willet
Becker, Stephen
Beckman, John
Beiersdorfer, Peter
Bellas-Velidis, Ioannis
Belmonte Aviles, Juan A.
Bennett, Philip
Bensby, Thomas
Benz, Willy
Berdyugina, Svetlana
Bergeron, Pierre
Bernat, Andrew
Bertelli, Gianpaolo
Berthomieu, Gabrielle
Bertone, Emanuele
Bertout, Claude
Bessell, Michael
Biazzo, Katia
Bikmaev, Ilfan
Bingham, Robert
Bisnovatyi-Kogan, G.
Blaga, Cristina
Blanco, Carlo
Bless, Robert
Blomme, Ronny
Bludman, Sidney
Boden, Andrew
Bodenheimer, Peter
Bodo, Gianluigi
Boehm, Torsten
Boesgaard, Ann
Boggess, Albert
Bohlender, David
Bombaci, Ignazio
Bond, Howard
Bonifacio, Piercarlo
Bonneau, Daniel

Bono, Giuseppe
Bopp, Bernard
Boss, Alan
Bouvier, Jerôme
Bragaglia, Angela
Brandi, Elisande
Brassard, Pierre
Bravo, Eduardo
Bressan, Alessandro
Breysacher, Jacques
Brickhouse, Nancy
Briot, Danielle
Brosche, Peter
Brown, Alexander
Brown, Douglas
Browning, Matthew
Brownlee, Robert
Bruenn, Stephen
Bruhweiler, Frederick
Brun, Allan
Bruning, David
Bruntt, Hans
Buchlin, Eric
Budaj, Jan
Bues, Irmela
Burkhart, Claude
Busa, Innocenza
Buser, Roland
Busso, Maurizio
Butkovskaya, Varvara
Butler, Keith
Callebaut, Dirk
Caloi, Vittoria
Cameron, Andrew
Canal, Ramon
Caputo, Filippina
Carbon, Duane
Carlsson, Mats
Carney, Bruce
Carpenter, Kenneth
Carretta, Eugenio
Carson, T.
Carter, Bradley
Cassinelli, Joseph
Castelli, Fiorella
Castor, John
Catala, Claude
Catalano, Franco
Catanzaro, Giovanni
Catchpole, Robin
Cayrel, Roger
Chaboyer, Brian
Chabrier, Gilles
Chadid, Merieme
Chamel, Nicolas
Chan, Roberto
CHAN, Kwing
Charbonnel, Corinne
Charpinet, Stéphane
Chavez-Dagostino, Miguel
Chechetkin, Valerij
CHEN, Peisheng
CHEN, Wen Ping
CHEN, Alfred
CHEN, Yuqin
Cherepashchuk, Anatolij
Chiosi, Cesare
Chitre, Shashikumar
Christensen-Dalsgaard, J.
Christlieb , Norbert
Christy, James
Chugai, Nikolaj
Cidale, Lydia
Claria, Juan
Claudi, Riccardo
Connolly, Leo
Conti, Peter
Cornide, Manuel
Corsico, Alejandro
Cottrell, Peter
Cowan, John
Cowley, Anne
Cowley, Charles
Cram, Lawrence
Crawford, David
Crowther, Paul
Cruzado, Alicia
Cugier, Henryk
CUI, Wenyuan
Culver, Roger
Cunha, Katia
Cuntz, Manfred
Cure, Michel
Cvetkovic, Zorica
da Silva, Licio
Dacic, Miodrag
Daflon, Simone
DAISAKU, Nogami
Dal Ri Barbosa, Cassio
Dall, Thomas
Damineli Neto, Augusto
Das, Mrinal
Daszynska-Daszkiewicz, J.
de Castro, Elisa
De Cat, Peter
de Greve, Jean-Pierre
de Jager, Cornelis
de Koter, Alex
de Laverny, Patrick
de Loore, Camiel
de Medeiros, Jose
Dearborn, David
Decin, Leen
Deinzer, W.
del Peloso, Eduardo
Deliyannis, Constantine
Demarque, Pierre
Denisenkov, Pavel
Deupree, Robert
Di Mauro, Maria Pia
Dimitrijevic, Milan
Dluzhnevskaya, Olga
Dominguez, Inma
Dominis Prester, Dijana
Donati, Jean-Francois
Doppmann, Gregory
Doyle MRIA, John
Dragunova, Alina
Drake, Natalia
Drake, Stephen
Dreizler, Stefan
Drilling, John
Duari, Debiprosad
Dufton, Philip
Dukes Jr., Robert
Duncan, Douglas
Dunham, David
Dupree, Andrea
Dupuis, Jean
Durisen, Richard
Dworetsky, Michael
Dziembowski, Wojciech
Edvardsson, Bengt
Edwards, Suzan
Eenens, Philippe
Eggenberger, Patrick
Eggleton, Peter
Eglitis, Ilgmars
Egret, Daniel
Elkin, Vladimir

Elmhamdi, Abouazza
ERIGUCHI, Yoshiharu
Eriksson, Kjell
Evangelidis, E.
Eyer, Laurent
Fadeyev, Yurij
Falceta-Goncalves, Diego
Faraggiana, Rosanna
Faulkner, John
Faurobert, Marianne
Feast, Michael
Feigelson, Eric
Fekel, Francis
Felenbok, Paul
Feltzing, Sofia
Fernandes, Joao
Fernandez-Figueroa, M.
Ferreira, João
Ferrer, Osvaldo
Fitzpatrick, Edward
Flannery, Brian
Fluri, Dominique
Foing, Bernard
Fontaine, Gilles
Fontenla, Juan
Forbes, J.
Fors, Octavi
Forveille, Thierry
Fossat, Eric
Foukal, Peter
Foy, Renaud
Franchini, Mariagrazia
Francois, Patrick
Frandsen, Soeren
Franz, Otto
Fredrick, Laurence
Freire Ferrero, Rubens
Freitas Mourao, Ronaldo
Fremat, Yves
Freytag, Bernd
Friel, Eileen
Frisch, Helene
Frisch, Uriel
Froeschle, Christiane
FUJIMOTO, Masayuki
FUKUDA, Ichiro
Fullerton, Alexander
Gabriel, Maurice
Gail, Hans-Peter
Gallino, Roberto
Gamen, Roberto
Garcia, Lopez
Garcia, Domingo
Garmany, Katy
Garrison, Robert
Gatewood, George
Gaudenzi, Silvia
Gautier, Daniel
Gautschy, Alfred
Gebbie, Katharine
Gehren, Thomas
Gerbaldi, Michele
Geroyannis, Vassilis
Gershberg, R.
Gesicki, Krzysztof
Geyer, Edward
Ghez, Andrea
Giampapa, Mark
Giannone, Pietro
Gimenez, Alvaro
Giorgi, Edgard
Giovannelli, Franco
Girardi, Leo
Gizis, John
Glagolevskij, Yurij
Glatzmaier, Gary
Glazunova, Ljudmila
Goebel, John
Goedhart, Sharmila
Golay, Marcel
Gomboc, Andreja
Goncalves, Denise
Gonzalez, Guillermo
Gonzalez, Jean-Francois
Gopka, Vera
Goriely, Stephane
Gough, Douglas
Goupil, Marie-José
Grady, Carol
Graham, Eric
Grant, Ian
Gratton, Raffaele
Gray, David
Gray, Richard
Greggio, Laura
Grenon, Michel
Grevesse, Nicolas
Griffin, R. Elizabeth
Griffin, Roger
Grinin, Vladimir
Grosso, Monica
GU, Sheng-hong
Guedel, Manuel
Guenther, David
Guetter, Harry
Gun, Gulnur
Gupta, Ranjan
Gussmann, Ernst-August
Gustafsson, Bengt
Guzik, Joyce
Haberreiter, Margit
HACHISU, Izumi
Haisch, Bernard
Hakkila, Jon
Halbwachs, Jean-Louis
Hamann, Wolf-Rainer
Hammond, Gordon
HAN, Zhanwen
Hanson, Margaret
Hanuschik, Reinhard
Harmer, Charles
Harmer, Dianne
Harper, Graham
Hartigan, Patrick
Hartkopf, William
Hartman, Henrik
Hartmann, Lee
Harutyunian, Haik
HASHIMOTO, Osamu
HASHIMOTO, Masahaki
Hauck, Bernard
Hauschildt, Peter
HE, Jinhua
Heacox, William
Hearnshaw, John
Heasley, James
Heber, Ulrich
Heger, Alexander
Heiter, Ulrike
Hempel, Marc
Henrichs, Hubertus
Henry, Richard
Hernanz, Margarita
Hershey, John
Hessman, Frederic
Hidayat, Bambang
Hill, Vanessa
Hill, Grant
Hill, Graham
Hillier, John

Hillwig, Todd
Hindsley, Robert
Hinkle, Kenneth
HIRAI, Masanori
HIRATA, Ryuko
Hirschi, Raphael
Hoare, Melvin
Hoeflich, Peter
Hoefner, Susanne
Hollowell, David
Holzer, Thomas
Homeier, Derek
Honda, Satoshi
HORAGUCHI, Toshihiro
Horch, Elliott
Houk, Nancy
Houziaux, Leo
Hron, Josef
HU, Zhong wen
Hubeny, Ivan
Hubert-Delplace, A.-M.
Hubrig, Swetlana
Huenemoerder, David
Huggins, Patrick
Hui bon Hoa, Alain
Hummel, Wolfgang
Hummel, Christian
Humphreys, Roberta
Hutchings, John
Hyland, Harry
Ianna, Philip
Iben Jr, Icko
Ignace, Richard
Ignjatovic, Ljubinko
Iliev, Ilian
Imbroane, Alexandru
Imshennik, Vladimir
Ireland, Michael
Irwin, Michael
Isern, Jordi
ISHIZUKA, Toshihisa
Israelian, Garik
ITOH, Naoki
Ivanov, Vsevolod
Ivans, Inese
IWAMOTO, Nobuyuki
IZUMIURA, Hideyuki
Jahn, Krzysztof
Jahreiss, Hartmut
James, Richard
Jankov, Slobodan
Jassur, Davoud
Jatenco-Pereira, Vera
Jehin, Emmanuel
JEON, Young Beom
Jevremovic, Darko
Johnson, Hollis
Johnson, Jennifer
Johnston, Helen
Jones, Carol
Jordan, Stefan
Jordan, Carole
Jorgensen, Jes
Jose, Jordi
Josselin, Eric
Judge, Philip
Jurdana-Sepic, Rajka
Kaehler, Helmuth
Kaeufl, Hans Ulrich
Kalkofen, Wolfgang
Kamp, Lucas
Kamp, Inga
Kandel, Robert
Kapyla, Petri
Karakas, Amanda
Karp, Alan
Kasparova, Jana
KATO, Ken-ichi
KATO, Mariko
Katsova, Maria
Kawka, Adela
Kazantseva, Liliya
KIGUCHI, Masayoshi
Kippenhahn, Rudolf
Kipper, Tonu
Kirkpatrick, Joseph
Kiselman, Dan
Kisseleva-Eggleton, L.
Kitsionas, Spyridon
Kiziloglu, Nilgun
Klein, Richard
Kley, Wilhelm
Klochkova, Valentina
Knoelker, Michael
Kochhar, Rajesh
Kochukhov, Oleg
KODAIRA, Keiichi
Koehler, Rainer
Koester, Detlev
KOGURE, Tomokazu
Kolesov, Aleksandr
Kolka, Indrek
Konar, Sushan
KONDO, Yoji
Kontizas, Evangelos
Korcakova, Daniela
Kordi, Ayman
Korn, Andreas
Korotin, Sergey
Kosovichev, Alexander
Kotnik-Karuza, Dubravka
Koubsky, Pavel
Kovachev, Bogomil
Kovetz, Attay
Kovtyukh, Valery
Kraus, Michaela
Krempec-Krygier, Janina
Krikorian, Ralph
Krishna, Swamy
Kroupa, Pavel
Krticka, Jiri
Kubat, Jiri
Kudritzki, Rolf-Peter
Kuhi, Leonard
Kumar, Shiv
Kupka, Friedrich
Kurtanidze, Omar
Kurtz, Michael
Kurtz, Donald
Kurucz, Robert
Kwok, Sun
Labay, Javier
Lago, Maria
Lagrange, Anne-Marie
Laird, John
Lamb, Susan
Lamb Jr, Donald
Lambert, David
Lamers, Henny
Lamontagne, Robert
Lampens, Patricia
Lamzin, Sergei
Langer, Norbert
Lanz, Thierry
Larson, Richard
Lasala Jr., Gerald
Laskarides, Paul
Lasota-Hirszowicz, J.-P.
Latham, David
Lattanzi, Mario

Lattanzio, John
Le Contel, Jean-Michel
Leao, Joao Rodrigo
Lebovitz, Norman
Lèbre, Agnès
Lebreton, Yveline
Leckrone, David
LEE, Thyphoon
LEE, Sang-Gak
LEE, Jae Woo
Lee, William
Leedjarv, Laurits
Leggett, Sandy
Leibacher, John
Leitherer, Claus
Lepine, Jacques
Lepine, Sebastien
Lester, John
Leushin, Valerij
Levato, Orlando
LI, Qingkang
LI, Ji
LI, Jinzeng
LIANG, Yanchun
Liebendoerfer, Matthias
Liebert, James
Lignières, François
LIM, Jeremy
Limongi, Marco
Ling, Josefina
Linnell, Albert
Linsky, Jeffrey
Lippincott Zimmerman, S.
Little-Marenin, Irene
Littleton, John
LIU, Michael
Livio, Mario
Lodders, Katharina
Loskutov, Viktor
Lu, Phillip
Lubowich, Donald
Lucatello, Sara
Luck, R.
Ludwig, Hans
Lugaro, Maria
Lundstrom, Ingemar
Luo, Qinghuan
LUO, Ali
Luri, Xavier
Luttermoser, Donald
Lutz, Julie
Lyubchik, Yuri
Lyubimkov, Leonid
MacConnell, Darrell
Machado, Maria
Maddison, Sarah
Madej, Jerzy
MAEDA, Keiichi
Maeder, André
MAEHARA , Hideo
Magain, Pierre
Magazzu, Antonio
Magnan, Christian
Maheswaran, M.
Maillard, Jean-Pierre
Maitzen, Hans
Malagnini, Maria
Malaroda, Stella
Manteiga Outeiro, Minia
Marilli, Ettore
Marley, Mark
Marsakova, Vladislava
Marsden, Stephen
Martin, Eduardo
Martinez Fiorenzano, Aldo
Martins, Fabrice
Mashonkina, Lyudmila
Mason, Brian
Massaglia, Silvano
Massey, Philip
Mathieu, Robert
Mathis, Stephane
Mathys, Gautier
Matsuura, Mikako
Matteucci, Francesca
Mauas, Pablo
Mazurek, Thaddeus
Mazzali, Paolo
Mazzitelli, Italo
McAlister, Harold
McDavid, David
McGregor, Peter
McSwain, Mary
Medupe, Rodney
Megessier, Claude
Melendez, Jorge
Melo, Claudio
Mendes, Luiz
Mendoza, V.
Mennickent, Ronald
Merlo, David
Mestel, Leon
Meyer-Hofmeister, Eva
Meynet, meynet
Michaud, Georges
Mickaelian, Areg
Mihajlov, Anatolij
Mikkola, Seppo
Mikolajewski, Maciej
Mikuláek, Zden?k
Minniti, Dante
Mitalas, Romas
MIYAJI , Shigeki
Moellenhoff, Claus
Moffat, Anthony
Mohan, Chander
Moiseenko, Sergey
Molaro, Paolo
Monaghan, Joseph
Monier, Richard
Monin, Dmitry
Monteiro, Mario Joao
Montes, David
Moore, Daniel
Moos, Henry
Morbidelli, Roberto
Morel, Pierre-Jacques
Morgan, John
Morossi, Carlo
Morrell, Nidia
Morrison, Nancy
Moskalik, Pawe?
Moss, David
Mowlavi, Nami
Mueller, Ewald
Muench, Guido
Musielak, Zdzislaw
Mutschlecner, Joseph
Nadyozhin, Dmitrij
Nagendra, K.
Nagirner, Dmitrij
Najarro de la Parra, F.
NAKAMURA, Takashi
NAKANO, Takenori
NAKAZAWA, Kiyoshi
Napiwotzki, Ralf
Narasimha, Delampady
NARIAI, Kyoji
NARITA, Shinji
Nazarenko, Victor

Negueruela, Ignacio
Neiner, Coralie
Nelemans, Gijs
Neuhaeuser, Ralph
Newman, Michael
Nicolet, Bernard
Niedzielski, Andrzej
Nielsen, Krister
Niemczura, Ewa
Nikoghossian, Arthur
Nilsson, Hampus
NISHIMURA, Shiro
NISHIMURA, Masayoshi
Noels, Arlette
NOMOTO, Ken'ichi
Nordlund, Aake
Nordström, Birgitta
Norris, John
North, Pierre
Notni, Peter
Nuernberger, Dieter
Nugis, Tiit
O'Neal, Douglas
O'Toole, Simon
Oblak, Edouard
Odell, Andrew
Oja, Tarmo
OKAMOTO, Isao
OKAZAKI, Atsuo
Oliveira, Joana
Olsen, Erik
Orlov, Victor
OSAKI, Yoji
Osborn, Wayne
Ostriker, Jeremiah
Oswalt, Terry
Oudmaijer, Rene
Owocki, Stanley
Pacharin-Tanakun, P.
Pakhomov, Yury
Pamyatnykh, Alexey
Pandey, Birendra
Panei, Jorge
Papaloizou, John
Parimucha, Stefan
Parsons, Sidney
Pauls, Thomas
Pavani, Daniela
Pavlenko, Yakov
Pearce, Gillian
Pecker, Jean-Claude
Peraiah, Annamaneni
Pereira, Claudio
Peters, Geraldine
Peterson, Deane
Peterson, Ruth
Petit, Pascal
Petr-Gotzens, Monika
Philip, A.G.
Phillips, Mark
Pilachowski, Catherine
Pinotsis, Antonis
Pinsonneault, Marc
Pintado, Olga
Pinto, Philip
Piskunov, Nikolai
Pizzichini, Graziella
Plez, Bertrand
Pogodin, Mikhail
Polcaro, V.
Polidan, Ronald
Pollacco, Don
Polosukhina-Chuvaeva, N.
Pongracic, Helen
Pontoppidan, Klaus
Popovic, Georgije
Porto de Mello, Gustavo
Pottasch, Stuart
Pourbaix, Dimitri
Poveda, Arcadio
Prentice, Andrew
Preston, George
Prialnik, Dina
Prieto, Cristina
Prieur, Jean-Louis
Primas, Francesca
Prinja, Raman
Prires Martins, Lucimara
Proffitt, Charles
Provost, Janine
Przybilla, Norbert
Pulone, Luigi
Puls, Joachim
Pustynski, Vladislav-V.
Querci, Monique
Raassen, Ion
Rachkovsky, D.
Raedler, K.
Ramadurai, Souriraja
Ramsey, Lawrence
Randich, Sofia
Rangarajan, K.
Rao, N.
Rashkovskij, Sergey
Rastogi, Shantanu
Rauch, Thomas
Rauscher, Thomas
Rauw, Gregor
Rawlings, Mark
Ray, Alak
Rayet, Marc
Reale, Fabio
Rebolo, Rafael
Reddy, Bacham
Reeves, Hubert
Rego, Fernandez
Reimers, Dieter
Reiners, Ansgar
Reipurth, Bo
Renzini, Alvio
Rettig, Terrence
Reyniers, Maarten
Ringuelet, Adela
Ritter, Hans
Rivinius, Thomas
Roberts Jr, Lewis
Roca Cortes, Teodoro
Rodrigues de Oliveira F., I.
Roman, Nancy
Romanyuk, Iosif
Rose, James
Rossi, Silvia
Rossi, Corinne
Rostas, François
Rountree, Janet
Rovira, Marta
Roxburgh, Ian
Rucinski, Slavek
Ruiz-Lapuente, María
Russell, Jane
Rutten, Robert
Ryabchikova, Tatiana
Ryan, Sean
Rybicki, George
Ryde, Nils
Sachkov, Mikhail
Sackmann, Inge
SADAKANE, Kozo
Saffe, Carlos
Sagar, Ram

SAIO, Hideyuki
Sakhibullin, Nail
Sanchez Almeida, Jorge
Santos, Filipe
Sanwal, Basant
Sapar, Arved
Sapar, Lili
Sareyan, Jean-Pierre
Sarna, Marek
Sarre, Peter
Sasselov, Dimitar
Sasso, Clementina
SATO, Katsuhiko
Sauty, Christophe
Savanov, Igor
Savedoff, Malcolm
Savonije, Gerrit
Sbordone, Luca
Scalo, John
Scardia, Marco
Scarfe, Colin
Schaerer, Daniel
Scharmer, Goeran
Schatten, Kenneth
Schild, Rudolph
Schmid-Burgk, J.
Schmidtke, Paul
Schmutz, Werner
Schoeller, Markus
Schoenberner, Detlef
Scholz, M.
Schrijver, Karel
Schroeder, Klaus
Schuh, Sonja
Schuler, Simon
Schutz, Bernard
Scuflaire, Richard
Sedlmayer, Erwin
Seggewiss, Wilhelm
Seidov, Zakir
Selam, Selim
Sengbusch, Kurt
Sengupta, Sujan
Shakht, Natalia
Shatsky, Nicolai
Shaviv, Giora
Shetrone, Matthew
SHI, Huoming
SHI, Jianrong
SHIBAHASHI , Hiromoto
SHIBATA, Yukio
Shine, Richard
Shipman, Harry
Sholukhova, Olga
Shore, Steven
Short, Christopher
Shustov, Boris
Shvelidze, Teimuraz
Siess, Lionel
Sigalotti, Leonardo
Signore, Monique
Sigut, T. A. Aaron
Sills, Alison
Silvestro, Giovanni
Simon, Michal
Simon, Theodore
Simonneau, Eduardo
Sinachopoulos, Dimitris
Singh, Mahendra
Sion, Edward
Skokos, Charalambos
Skumanich, Andrew
Smak, Jozef
Smalley, Barry
Smeyers, Paul
Smith, Robert
Smith, Graeme
Smith, Verne
Smith, J.
Smith, Myron
Sneden, Chris
Snow, Theodore
Sobouti, Yousef
Socas-Navarro, Hector
Soderblom, David
Soderhjelm, Staffan
Sofia, Sabatino
Sonneborn, George
Soubiran, Caroline
Sowell, James
Sparks, Warren
Spiegel, Edward
Spite, François
Spruit, Hendrik
Sreenivasan, S.
St-Louis, Nicole
Starrfield, Sumner
Stateva, Ivanka
Stauffer, John
Stawikowski, Antoni
Stecher, Theodore
Stee, Philippe
Steffen, Matthias
Stein, Robert
Stein, John
Steinlin, Uli
Stellingwerf, Robert
Stencel, Robert
Stepien, Kazimierz
Stern, Robert
Sterzik, Michael
Straizys, Vytautas
Stringfellow, Guy
Strittmatter, Peter
Strobel, Andrzej
Strom, Stephen
Stuik, Remko
Suda, Takuma
SUGIMOTO, Daiichiro
Suntzeff, Nicholas
Sweigart, Allen
Swings, Jean-Pierre
Szabados, Laszlo
Szeifert, Thomas
Taam, Ronald
TAKADA-HIDAI, M.
TAKAHARA, Mariko
TAKASHI, Hasegawa
TAKEDA, Yoichi
Talavera, Antonio
Tamazian, Vakhtang
Tango, William
Tantalo, Rosaria
Tarasov, Anatolii
Tautvaisiene, Graina
Teixeira, Paula
ten Brummelaar, Theo
Terquem, Caroline
Thejll, Peter
Thevenin, Frederic
Thielemann, Friedrich-Karl
Tohline, Joel
Tokovinin, Andrei
Tomasella, Lina
Tomov, Toma
Toomre, Juri
Tornambe, Amedeo
Torrejon, Jose Miguel
Torres, Guillermo
Townsend, Richard

Trimble, Virginia
TruranJr, James
Tscharnuter, Werner
TSUJI, Takashi
Turck-Chièze, Sylvaine
Turner, Nils
Tutukov, Aleksandr
UCHIDA, Juichi
Ulmschneider, Peter
Ulrich, Roger
Ulyanov, Oleg
UNNO, Wasaburo
Upgren, Arthur
Usenko, Igor
Utrobin, Victor
UTSUMI , Kazuhiko
Vakili, Farrokh
Valenti, Jeff
Valtier, Jean-Claude
Valtonen, Mauri
Valyavin, Gennady
van Altena, William
van den Heuvel, Edward
van der Hucht, Karel
van Dessel, Edwin
van Eck, Sophie
van Horn, Hugh
van Loon, Jacco
van Riper, Kenneth
Van Winckel, Hans
van't Veer-Menneret, C.
VandenBerg, Don
Vanko, Martin
Vardavas, Ilias
Vasu-Mallik, Sushma
Vauclair, Gérard
Vaughan, Arthur
Vaz, Luiz Paulo
Velusamy, T.
Vennes, Stéphane
Ventura, Paolo
Verdugo, Eva
Verheijen, Marc
Vieytes, Mariela
Viik, Tõnu
Vilhu, Osmi
Vilkoviskij, Emmanuil
Vink, Jorick
Viotti, Roberto
Vladilo, Giovanni
Vogt, Nikolaus
Vogt, Steven
von Hippel, Theodore
Vreux, Jean
Wade, Gregg
Wahlgren, Glenn
Walborn, Nolan
Walker, Gordon
Wallerstein, George
Walter, Frederick
WANG, Jiaji
WANG, Feilu
Ward, Richard
Warren Jr, Wayne
WATANABE, Tetsuya
Waters, Laurens
Weaver, Thomas
Weaver, William
Webbink, Ronald
Weber, Stephen
Wegner, Gary
Weis, Edward
Weiss, Nigel
Weiss, Werner
Weiss, Achim
WEN, Linqing
Werner, Klaus
Wesselius, Paul
Wheeler, J.
Whelan, Emma
White, Richard
Wickramasinghe, N.
Williams, John
Williams, Peredur
Willson, Lee Anne
Wilson, Robert
Wing, Robert
Winkler, Karl-Heinz
Woehl, Hubertus
Wolff, Sidney
Wood, Matthew
Wood, Peter
Woosley, Stanford
Wright, Nicholas
WU, Hsin-Heng
Wyckoff, Susan
XIONG, Da Run
YAMAOKA, Hitoshi
YAMASHITA, Yasumasa
Yanovitskij, Edgard
YI, Sukyoung
Yorke, Harold
YOSHIDA, Shin'ichirou
YOSHIDA, Takashi
YOSHIOKA, Kazuo
Yuce, Kutluay
YUMIKO, Oasa
Yungelson, Lev
Yushkin, Maxim
Zacs, Laimons
Zaggia, Simone
Zahn, Jean-Paul
Zapatero-Osorio, Maria R.
Zavala, Robert
Zdanavicius, Kazimeras
ZHANG, Huawei
ZHANG, Bo
ZHANG, Haotong
ZHANG, Yanxia
Zheleznyak, Alexander
ZHU, Zhenxi
ZHU, Liying
Zinnecker, Hans
Ziolkowski, Janusz
Zorec, Juan
Zverko, Juraj

Division V Variable Stars

President: Steven D. Kawaler
Vice-President: Ignasi Ribas

Organizing Committee Members:

Michel Breger
Edward F. Guinan
Gerald Handler
Slavek M. Rucinski

Members:

Aizenman, Morris
Al-Naimiy, Hamid
Albinson, James
Albrow, Michael
Alencar, Silvia
Alfaro, Emilio
Alpar, Mehmet
Amado Gonzalez, Pedro
Andersen, Johannes
ANDO, Hiroyasu
Andrievsky, Sergei
Antipin, Sergei
Antokhin, Igor
Antonello, Elio
Antonopoulou, Evgenia
Antonyuk, Kirill
Antov, Alexandar
Anupama, G.
Aquilano, Roberto
Arefiev, Vadim
Arellano Ferro, Armando
Arentoft, Torben
Arias, Maria
Arkhipova, Vera
Asteriadis, Georgios
Aungwerojwit, Amornrat
Avgoloupis, Stavros
Awadalla, Nabil
Baade, Dietrich
BABA, Hajime
Babkovskaia, Natalia
Baglin, Annie
Bailyn, Charles
Balona, Luis
Baptista, Raymundo
Baran, Andrzej
Baransky, Olexander
Barban, Caroline
Barkin, Yuri
Barnes III, Thomas
Barone, Fabrizio
Bartolini, Corrado
Barway, Sudhashu
Barwig, Heinz
Baskill, Darren
Bastien, Pierre
Batten, Alan
Bauer, Wendy
Bazot, Michael
Beaulieu, Jean-Philippe
Bedding, Timothy
Bedogni, Roberto
Bell, Steven
Belmonte Aviles, Juan A.
Belserene, Emilia
Belvedere, Gaetano
Benko, Jozsef
Berdnikov, Leonid
Bersier, David
Berthomieu, Gabrielle
Bessell, Michael
Bianchi, Luciana
Bianchini, Antonio
Bjorkman, Karen
Blair, William
Blundell, Katherine
Boffin, Henri
Bolton, Charles
Bonazzola, Silvano
Bond, Howard
Bopp, Bernard
Borisov, Nikolay
Boyd, David
Boyle, Stephen
Bozic, Hrvoje
Bradley, Paul
Bradstreet, David
Brandi, Elisande
Briquet, Maryline
Broglia, Pietro
Brown, Douglas
Brownlee, Robert
Bruch, Albert
Bruhweiler, Frederick
Bruntt, Hans
Buccino, Andrea
Budding, Edwin
Bunner, Alan
Burderi, Luciano
Burki, Gilbert
Burwitz, Vadim
Busa, Innocenza
Busko, Ivo
Busso, Maurizio
Butkovskaya, Varvara
Butler, Christopher
Buzasi, Derek
Cacciari, Carla
Caldwell, John
Callanan, Paul
Cameron, Andrew
Canalle, Joao
CAO, Huilai
Carrier, Fabien
Casares, Jorge
Catchpole, Robin
Catelan, Márcio
Chadid, Merieme
Chambliss, Carlson
Chapman, Robert
Chaty, Sylvain

Chaubey, Uma
CHEN, An-Le
CHEN, Alfred
CHEN, Xuefei
Cherchneff, Isabelle
Cherepashchuk, Anatolij
Chochol, Drahomir
CHOI, Kyu Hong
CHOI, Chul-Sung
CHOU, Yi
Christie, Grant
Ciardi, David
Cioni, Maria-Rosa
Claria, Juan
Clement, Christine
Clementini, Gisella
Cohen, Martin
Connolly, Leo
Contadakis, Michael
Cook, Kem
Cornelisse, Remon
Corradi, Romano
Corsico, Alejandro
Costa, Vitor
Cottrell, Peter
Coulson, Iain
Cowley, Anne
Cropper, Mark
CUI, Wenyuan
Cunha, Margarida
Cutispoto, Giuseppe
Cuypers, Jan
D'Amico, Nicolo'
D'Antona, Francesca
DAISAKU , Nogami
Dall , Thomas
Dall'Ora, Massimo
Danford, Stephen
Daszynska-Daszkiewicz, J.
De Cat, Peter
de Greve, Jean-Pierre
de Loore, Camiel
de Ridder, Joris
Del Santo, Melania
Delgado, Antonio
Demers, Serge
Demircan, Osman
DENG, LiCai
Deupree, Robert
Di Mauro, Maria Pia
Diaz, Marcos
Dobrzycka, Danuta
Donahue, Robert
Dorfi, Ernst
Dorokhova, Tetyana
Dougherty, Sean
Downes, Ronald
Drechsel, Horst
Dubus, Guillaume
Dukes Jr., Robert
Dunlop, Storm
Dupree, Andrea
Durisen, Richard
Duschl, Wolfgang
Dziembowski, Wojciech
Eaton, Joel
Edwards, Paul
Edwards, Suzan
Efremov, Yurij
Eggenberger, Patrick
Eggleton, Peter
Elias II, Nicholas
Elkin, Vladimir
Esenoglu, Hasan
Etzel, Paul
Evans, Aneurin
Evans, Nancy
Evren, Serdar
Eyer, Laurent
Eyres, Stewart
Fabrika, Sergei
Fadeyev, Yurij
Faulkner, John
Feast, Michael
Fekel, Francis
Ferland, Gary
Ferluga, Steno
Fernandez Lajus, Eduardo
Fernie, J.
Ferrario, Lilia
Ferrer, Osvaldo
Flannery, Brian
Fokin, Andrei
Formiggini, Lilliana
Fors, Octavi
Frank, Juhan
Fredrick, Laurence
FU, Jian-Ning
FU, Hsieh-Hai
FUJIWARA, Tomoko
Gaensicke, Boris
Gahm, Goesta
Galis, Rudolf
Gallagher III, John
Gameiro, Jorge
Gamen, Roberto
Garcia, Lia
García de María, Juan
Garcia-Lorenzo, Maria
Garmany, Katy
Garrido, Rafael
Gasiprong, Nipon
Geldzahler, Barry
Genet, Russell
Gershberg, R.
Geyer, Edward
Giannone, Pietro
Gieren, Wolfgang
Gies, Douglas
Gillet, Denis
Giovannelli, Franco
Glagolevskij, Yurij
Goldman, Itzhak
Gomboc, Andreja
Gondoin, Philippe
Gonzalez Martinez Pais, I.
Gosset, Eric
Gough, Douglas
Goupil, Marie-José
Graham, John
Grasberg, Ernest
Green, Daniel
Grinin, Vladimir
Groenewegen, Martin
Groot, Paul
Grygar, Jiri
GU, Wei-Min
Guerrero, Gianantonio
Gulliver, Austin
Gun, Gulnur
Gunn, Alastair
Gunthardt, Guillermo
GUO, Jianheng
Guzik, Joyce
Hackwell, John
Hadrava, Petr
Haefner, Reinhold
Haisch, Bernard
Hakala, Pasi
Halbwachs, Jean-Louis

Hamdy, M.
HANAWA, Tomoyuki
Handler, Gerald
Hantzios, Panayiotis
HAO, Jinxin
Harmanec, Petr
Hassall, Barbara
Haswell, Carole
Hawley, Suzanne
HAYASAKI, Kimitake
HE, Jinhua
Hegedues, Tibor
Heiser, Arnold
Hellier, Coel
Helt, Bodil
Hempelmann, Alexander
Henden, Arne
Hensler, Gerhard
Hesser, James
Hilditch, Ronald
Hill, Henry
Hill, Graham
Hills, Jack
Hillwig, Todd
Hintz, Eric
Hoard, Donald
Hojaev, Alisher
Holmgren, David
Holt, Stephen
Honeycutt, R.
HORIUCHI, Ritoku
Horner, Scott
Houdek, Gunter
Houk, Nancy
Howell, Steve
Hric, Ladislav
Hrivnak, Bruce
Hube, Douglas
Huenemoerder, David
Humphreys, Elizabeth
Hutchings, John
Ibanoglu, Cafer
Iben Jr, Icko
Iijima, Takashi
Ikhsanov, Nazar
Imamura, James
Imbert, Maurice
Ireland, Michael
Ishida, Toshihito
Ismailov, Nariman
ITA, Yoshifusa
Ivezic, Zeljko
Jablonski, Francisco
Jankov, Slobodan
Jasniewicz, Gerard
Jeffers, Sandra
Jeffery, Christopher
JEON , Young Beom
JEONG, Jang-Hae
Jerzykiewicz, Mikolaj
Jetsu, Lauri
Jewell, Philip
JIANG, Biwei
JIN, Zhenyu
Joner, Michael
Jonker, Peter
Joss, Paul
Jurcsik, Johanna
Kaeufl, Hans Ulrich
Kaitchuck, Ronald
Kalomeni, Belinda
Kaluzny, Janusz
KAMBE, Eiji
KANAMITSU, Osamu
Kanbur, Shashi
KANG, Young Woon
Karetnikov, Valentin
Karitskaya, Evgeniya
Karovska, Margarita
Karp, Alan
KATO, Taichi
Katsova, Maria
Kaufer, Andreas
KAWABATA, Shusaku
Kaye, Anthony
Kazarovets, Elena
Keller, Stefan
KENJI, Nakamura
Kenny, Harold
Kenyon, Scott
Kepler, S.
Kilkenny, David
KIM, Tu Whan
KIM, Ho-il
KIM, Chulhee
KIM, Seung-Lee
KIM, Chun-Hwey
KIM, Young-Soo
King, Andrew
Kiplinger, Alan
Kippenhahn, Rudolf
Kiss, Laszlo
Kjeldsen, Hans
Kjurkchieva, Diana
Kley, Wilhelm
Kochukhov, Oleg
Koen, Marthinus
Kolb, Ulrich
Kolesnikov, Sergey
Kollath, Zoltan
Komonjinda, Siramas
Komzik, Richard
Konacki, Maciej
KONDO, Yoji
Konstantinova-Antova, R.
Kopacki, Grzegorz
Korhonen, Heidi
Koubsky, Pavel
Kovari, Zsolt
Kraicheva, Zdravka
Krautter, Joachim
Kreiner, Jerzy
Kreykenbohm , Ingo
Krisciunas, Kevin
Kruchinenko, Vitaliy
Kruszewski, Andrzej
Krzesinski, Jerzy
Kubiak, Marcin
Kudashkina, Larisa
Kuhi, Leonard
Kunjaya, Chatief
Kurtz, Donald
Kwee, K.
Lacy, Claud
Lago, Maria
Lamb Jr, Donald
Lampens, Patricia
Laney, Clifton
Lanza, Antonino
Lapasset, Emilio
Larionov, Valeri
Larsson, Stefan
Larsson-Leander, Gunnar
Laskarides, Paul
Lawlor, Timothy
Lawson, Warrick
Lazaro, Carlos
Le Bertre, Thibaut
Lebzelter, Thomas
LEE, Jae Woo

LEE, Woo baik
LEE, Yong Sam
LEE, Myung Gyoon
Lee, William
Leedjarv, Laurits
Leung, Kam
LI, Yan
LI, Zhiping
LI, Ji
LI, Zhi
LI, Lifang
LIM, Jeremy
LIN, Yi-qing
Linnell, Albert
Linsky, Jeffrey
Little-Marenin, Irene
LIU, Qingzhong
Livio, Mario
Lloyd, Christopher
Lockwood, G.
Longmore, Andrew
Lopez, De
Lorenz-Martins, Silvia
Lub, Jan
Lucy, Leon
MacDonald, James
Maceroni, Carla
Machado Folha, Daniel
Macri, Lucas
Madore, Barry
Maeder, André
Malasan, Hakim
Mandel, Ilya
Manimanis, Vassilios
Mantegazza, Luciano
Marchev, Dragomir
Marconi, Marcella
Mardirossian, Fabio
Margrave Jr, Thomas
Marilli, Ettore
Markoff, Sera
Markworth, Norman
Marsakova, Vladislava
Marsh, Thomas
Martic, Milena
Martinez, Peter
Mason, Paul
Mathias, Philippe
Mathieu, Robert
MATSUMOTO, Katsura
Matthews, Jaymie
Mauche, Christopher
Mayer, Pavel
Mazeh, Tsevi
McCluskey Jr, George
McGraw, John
Meintjes, Petrus
Melia, Fulvio
Meliani, Mara
Melikian, Norair
Mennickent, Ronald
Mereghetti, Sandro
Messina, Sergio
Meyer-Hofmeister, Eva
Mezzetti, Marino
Michel, Eric
Mikolajewska, Joanna
Mikolajewski, Maciej
Mikuláěek, Zdeněk
Milano, Leopoldo
Milone, Luis
Milone, Eugene
MINESHIGE, Shin
Minikulov, Nasridin
MIYAJI, Shigeki
Mkrtichian, David
Mochnacki, Stefan
Moffett, Thomas
Mohan, Chander
Monteiro, Mario João
Morales Rueda, Luisa
Morgan, Thomas
Morrell, Nidia
Morrison, Nancy
Moskalik, Pawel
Mouchet, Martine
Mukai, Koji
Munari, Ulisse
Murdin, Paul
Mutel, Robert
NAKAMURA, Yasuhisa
NAKAO, Yasushi
NARIAI, Kyoji
Nather, R.
Naylor, Tim
Neff, James
Neiner, Coralie
Nelemans, Gijs
Nelson, Burt
Newsom, Gerald
NGEOW , Chow Choong
NHA, Il Seong
Niarchos, Panagiotis
Niemczura, Ewa
Nordstrom, Birgitta
Norton, Andrew
Nugis, Tiit
O'Donoghue, Darragh
O'Toole, Simon
Ogloza, Waldemar
OH, Kyu-Dong
OKAZAKI, Akira
Olah, Katalin
Oliveira, Alexandre
Olivier, Enrico
Olson, Edward
OSAKI, Yoji
Ostensen, Roy
Oswalt, Terry
Ozkan, Mustafa
Pandey, Uma
Panei, Jorge
Papaloizou, John
Paparo, Margit
Parimucha, Stefan
PARK, Hong-Seo
PARK, Byeong-Gon
Parsamyan, Elma
Parthasarathy, Mudumba
Patat, Ferdinando
Paterno, Lucio
Patkos, Laszlo
Pavlenko, Elena
Pavlovski, Kresimir
Pearson, Kevin
Percy, John
Perez Hernandez, Fernando
Peters, Geraldine
Petersen, J.
Petit, Pascal
Petrov, Peter
Pettersen, Bjørn
Piirola, Vilppu
Pijpers, Frank
Plachinda, Sergei
Plavchan, Jr., Peter
Pojma?ski, Grzegorz
Polidan, Ronald
Pollacco, Don
Pollard, Karen

Pont, Frédéric
Pop, Alexandru
Pop, Vasile
Popov, Sergey
Postnov, Konstantin
Potter, Stephen
Pribulla, Theodor
Pricopi, Dumitru
Pringle, James
Pritzl, Barton
Prokhorov, Mikhail
Provost, Janine
Pugach, Alexander
Pustynski, Vladislav-V.
QIAO, Guojun
Rafert, James
Rahunen, Timo
Ramsey, Lawrence
Ransom, Scott
Rao, Pasagada
Rao, N.
Rasio, Frederic
Ratcliff, Stephen
Reale, Fabio
Reglero Velasco, Victor
Reiners, Ansgar
Reinsch, Klaus
Renson, P.
Rey, Soo-Chang
Ribas, Ignasi
Richards, Mercedes
Ringwald, Frederick
Ritter, Hans
Rivinius, Thomas
Robb, Russell
Robertson, John
Robinson, Edward
Rodrigues, Claudia
Rodriguez, Eloy
Romanov, Yuri
Rosenbush, Alexander
Rountree, Janet
Rovithis-Livaniou, Helen
Roxburgh, Ian
Ruffert, Maximilian
Russev, Ruscho
Russo, Guido
Sachkov, Mikhail
Saha, Abhijit
SAIJO, Keiichi
Samec, Ronald
Samus, Nikolay
Sandmann, William
Sareyan, Jean-Pierre
Sarty, Gordon
Sasselov, Dimitar
Savonije, Gerrit
Scarfe, Colin
Schaefer, Bradley
Schartel, Norbert
Schiller, Stephen
Schlegel, Eric
Schmid, Hans
Schmidt, Edward
Schmidtke, Paul
Schmidtobreick, Linda
Schober, Hans
Schuh, Sonja
Schwarzenberg-Czerny, A.
Schwope, Axel
Scuflaire, Richard
Seeds, Michael
Seggewiss, Wilhelm
Selam, Selim
Semeniuk, Irena
Shafter, Allen
Shahbaz, Tariq
Shakhovskaya, Nadejda
Shakura, Nikolaj
Shaviv, Giora
Shenavrin, Victor
Sherwood, William
SHU, Frank
Silvotti, Roberto
Sima, Zdislav
Simmons, John
Sion, Edward
Sistero, Roberto
Skinner, Stephen
Skopal, Augustin
Smak, Jozef
Smeyers, Paul
Smith, Robert
Smith, Myron
Smith, Horace
Soderhjelm, Staffan
Solheim, Jan
Somasundaram, Seetha
SONG, Liming
Soszynski, Igor
Southworth, John
Sowell, James
Sparks, Warren
Stagg, Christopher
Stanishev, Vallery
Starrfield, Sumner
Steiman-Cameron, Thomas
Steiner, João
Stellingwerf, Robert
Stello, Dennis
Stencel, Robert
Stepien, Kazimierz
Sterken, Christiaan
Strassmeier, Klaus
Stringfellow, Guy
Strom, Stephen
SUGIMOTO, Daiichiro
Szabados, Laszlo
Szabo, Robert
Szatmary, Karoly
Szkody, Paula
Taam, Ronald
TAKATA, Masao
Tammann, Gustav
TAMURA, Shin'ichi
TAN, Huisong
Tarasova, Taya
Tas, Günay
Tauris, Thomas
Teays, Terry
Teixeira, Paula
Terrell, Dirk
Tomov, Toma
Torres, Guillermo
Tout, Christopher
Townsend, Richard
Traulsen, Iris
Tremko, Jozef
Trimble, Virginia
Tsvetkov, Milcho
Tsvetkova, Katja
Turner, David
Turolla, Roberto
Tutukov, Aleksandr
Tylenda, Romuald
Udovichenko, Sergei
UEMURA, Makoto
Ureche, Vasile
Usher, Peter
Uslenghi, Michela

Utrobin, Victor
Vaccaro, Todd
Valtier, Jean-Claude
van den Heuvel, Edward
van Genderen, Arnoud
van Hamme, Walter
Van Hoolst, Tim
Vaz, Luiz Paulo
Vennes, Stephane
Ventura, Rita
Verheest, Frank
Vilhu, Osmi
Viotti, Roberto
Vivas, Anna
Vogt, Nikolaus
Voloshina, Irina
von Braun, Kaspar
Votruba, Viktor
Wachter, Stefanie
Wade, Richard
Waelkens, Christoffel
Walder, Rolf
Walker, Merle
Walker, William
Walker, Edward
Wallerstein, George
WANG, Xunhao
Ward, Martin
Warner, Brian
Watson, Robert
Webbink, Ronald
Wehlau, Amelia
Weiler, Edward
Weis, Kerstin
Weiss, Werner
Welch, Douglas
Wheatley, Peter
Wheeler, J.
White II, James
Whitelock, Patricia
Williamon, Richard
Williams, Robert
Williams, Glen
Willson, Lee Anne
Wilson, Robert
Wing, Robert
Wittkowski, Markus
Wood, Peter
XIONG, Da Run
XUE, Li
Yakut, Kadri
YAMAOKA, Hitoshi
YAMASAKI, Atsuma
YOON, Tae-Seog
Yuce, Kutluay
YUJI, Ikeda
Zakirov, Mamnum
Zamanov, Radoslav
Zavala, Robert
Zeilik, Michael
ZHANG, Er-Ho
ZHANG, Bo
ZHANG, Xiaobin
Zharikov, Sergey
ZHU, Liying
Zijlstra, Albert
Ziolkowski, Janusz
Zola, Stanislaw
Zsoldos, Endre
Zuckerman, Benjamin
Zwitter, Tomaš

Division VI Interstellar Matter

President: You-Hua Chu
Vice-President: Sun Kwok

Organizing Committee Members:

Dieter Breitschwerdt
Michael G. Burton
Sylvie Cabrit
Paola Caselli
Elisabete M. de G. Dal Pino
Neal J. Evans
Thomas Henning
Mika J. Juvela
Bon-Chul KOO
Michal Ró życzka
Laszlo Viktor Tó th
Masato TSUBOI
Ji YANG

Members:

Aannestad, Per
Abgrall, Herve
Acker, Agnes
Adams, Fred
AIKAWA, Yuri
Al-Mostafa, Zaki
Alcolea, Javier
Altenhoff, Wilhelm
Alves, Joao
Andersen, Anja
Andersen, Morten
Andersson, B-G
Andronov, Ivan
Anglada, Guillem
Ardila, David
Arkhipova, Vera
Arny, Thomas
Arthur, Jane
Audard, Marc
Azcarate, Diana
Baars, Jacob
Babkovskaia, Natalia
Bachiller, Rafael
BAEK, Chang Hyun
Baker, Andrew
Ballesteros-Paredes, Javier
Balser, Dana
Baluteau, Jean-Paul
Bania, Thomas
Barlow, Michael
Barnes, Aaron
Baryshev, Andrey
Bash, Frank
Basu, Shantanu
Baudry, Alain
Bautista, Manuel
Bayet, Estelle
Becklin, Eric
Beckman, John
Beckwith, Steven
Bedogni, Roberto
Bergeron, Jacqueline
Bergin, Edwin
Bergman, Per
Bergstrom, Lars
Berkhuijsen, Elly
Bernat, Andrew
Bertout, Claude
Bhat, Ramesh
Bhatt, H.
Bianchi, Luciana
Bieging, John
Bignall, Hayley
Bignell, R.
Binette, Luc
Black, John
Blair, William
Blair, Guy
Bless, Robert
Blitz, Leo
Bloemen, Hans
Bobrowsky, Matthew
Bocchino, Fabrizio
Bochkarev, Nikolai
Bode, Michael
Bodenheimer, Peter
Boggess, Albert
Bohlin, Ralph
Boissé, Patrick
Boland, Wilfried
Bontemps, Sylvain
Bordbar, Gholam
Borkowski, Kazimierz
Boulanger, Francois
Boumis, Panayotis
Bourke, Tyler
Bouvier, Jerôme
Bowen, David
Brand, Peter
Brand, Jan
Briceño, Cesar
Bromage, Gordon
Brooks, Kate
Brouillet, Nathalie
Bruhweiler, Frederick
Bujarrabal, Valentin
Burke, Bernard
Burton, W.
Bychkov, Konstantin
Bykov, Andrei
Bzowski, Maciej
CAI, Kai
Cambrésy, Laurent
Cami, Jan
Caplan, James
Cappa de Nicolau, Cristina
Capriotti, Eugene
Capuzzo Dolcetta, Roberto
Carretti, Ettore
Carruthers, George
Casasola, Viviana
Castaneda, Héctor

Castelletti, Gabriela
Cattaneo , Andrea
Cecchi-Pestellini, Cesare
Centurion Martin, Miriam
Cernicharo, José
Cerruti Sola, Monica
Cersosimo, Juan
Cesarsky, Diego
Cesarsky, Catherine
CHA, Seung-Hoon
Chandra, Suresh
CHEN, Yang
CHEN, Yafeng
CHEN, Huei-Ru
CHEN, Xuefei
CHENG, Kwang
Cherchneff, Isabelle
Chevalier, Roger
CHIHARA, Hiroki
Chini, Rolf
Christopoulou, P.-E.
Churchwell, Edward
Ciardullo, Robin
Cichowolski, Silvina
Ciroi, Stefano
Clark, Frank
Clarke, David
Clegg, Robin
Codella, Claudio
Coffey, Deirdre
Colangeli, Luigi
Collin, Suzy
Combes, Françoise
Corbelli, Edvige
Corradi, Wagner
Corradi, Romano
Costantini, Elisa
Costero, Rafael
Cowie, Lennox
Cox, Donald
Cox, Pierre
Coyne, S.J, George
Crane, Philippe
Crawford, Ian
Crovisier, Jacques
Cuesta Crespo, Luis
Cunningham, Maria
d'Hendecourt, Louis
d'Odorico, Sandro
Dahn, Conard
Dale, James
Danks, Anthony
Danly, Laura
Davies, Rodney
Davis, Christopher
de Almeida, Amaury
De Avillez, Miguel
De Bernardis, Paolo
De Buizer, James
de Gregorio-Monsalvo, I.
de Jong, Teije
de La Noe, Jerome
De Marco, Orsola
Decourchelle, Anne
DEGUCHI, Shuji
Deharveng, Lise
Deiss, Bruno
Dennefeld, Michel
Dewdney, Peter
Dias da Costa, Roberto
Diaz, Ruben
Dib, Sami
Dickel, John
Dickel, Helene
Dickey, John
Dieleman, Pieter
Dinerstein, Harriet
Dinh, Trung
Disney, Michael
Djamaluddin, Thomas
Docenko, Dmitrijs
Dokuchaev, Vyacheslav
Dokuchaeva, Olga
Dominik, Carsten
Dopita, Michael
Dorschner, Johann
Dottori, Horacio
Downes, Dennis
Draine, Bruce
Dreher, John
Dubner, Gloria
Dubout, Renee
Dudorov, Aleksandr
Dufour, Reginald
Duley, Walter
Dupree, Andrea
Dutrey, Anne
Duvert, Gilles
Dwarkadas, Vikram
Dwek, Eli
Edwards, Suzan
Egan, Michael
Ehlerova, Soňa
Eisloeffel, Jochen
Elia, Davide
Elitzur, Moshe
Elliott, Kenneth
Elmegreen, Bruce
Elmegreen, Debra
Emerson, James
Encrenaz, Pierre
ESAMDIN, Ali
Escalante, Vladimir
ESIMBEK, Jarken
Esipov, Valentin
Esteban, César
Evans, Aneurin
Evans, Neal
Falceta-Goncalves, Diego
Falgarone, Edith
Falle, Samuel
Federman, Steven
Feitzinger, Johannes
Felli, Marcello
Fendt, Christian
Ferlet, Roger
Fernandes, Amadeu
Ferriere, Katia
Ferrini, Federico
Fesen, Robert
Fiebig, Dirk
Field, George
Field, David
Fierro, Julieta
Fischer, Jacqueline
Flannery, Brian
Fleck, Robert
Florido, Estrella
Flower, David
Folini, Doris
Ford, Holland
Forster, James
Franco, José
Franco, Gabriel Armando
Fraser, Helen
Freimanis, Juris
Fridlund, Malcolm
Frisch, Priscilla
Fuente, Asuncion
Fukuda, Naoya

FUKUI, Yasuo
Fuller, Gary
Furniss, Ian
Furuya, Ray
Gaensler, Bryan
Galli, Daniele
GAO, Yu
Garcia, Paulo
Garcia-Lario, Pedro
Garnett, Donald
Gathier, Roel
Gaume, Ralph
Gaustad, John
Gay, Jean
Geballe, Thomas
Genzel, Reinhard
Gerard, Eric
Gérin, Maryvonne
Gezari, Daniel
Ghanbari, Jamshid
Giacani, Elsa
Gibson, Steven
Giovanelli, Riccardo
Glover , Simon
Goddi, Ciriaco
Godfrey, Peter
Goebel, John
Goldes, Guillermo
Goldreich, Peter
Goldsmith, Donald
Golovatyj, Volodymyr
Gomez, Gonzalez
Goncalves, Denise
Gonzales-Alfonso, Eduardo
Goodman, Alyssa
Gordon, Mark
Gosachinskij, Igor
Goss, W. Miller
Graham, David
Granato, Gian Luigi
Gredel, Roland
Green, James
Gregorio-Hetem, Jane
Greisen, Eric
Grewing, Michael
Guelin, Michel
Guertler, Joachin
Guesten, Rolf
Guilloteau, Stéphane
Gull, Theodore
Gunthardt, Guillermo
GUO, Jianheng
Habing, Harm
Hackwell, John
Haisch Jr, Karl
HANAMI, Hitoshi
Harrington, J.
Harris, Alan
Harris-Law, Stella
Hartl, Herbert
Hartquist, Thomas
Harvey, Paul
Hatchell, Jennifer
Haverkorn, Marijke
Hayashi, Saeko
Haynes, Raymond
HE, Jinhua
Hebrard, Guillaume
Hecht, James
Heikkila, Arto
Heiles, Carl
Helfer, H.
Helmich, Frank
Helou, George
Henkel, Christian
Henney, William
Henning, Thomas
Hernandez, Jesús
Herpin, Fabrice
Heydari-Malayeri, M.
Heyer, Mark
Hidayat, Bambang
Higgs, Lloyd
Hildebrand, Roger
Hillenbrand, Lynne
Hippelein, Hans
HIRANO, Naomi
Hiriart, David
HIROMOTO, Norihisa
Hjalmarson, Ake
Hobbs, Lewis
Hollenbach, David
Hollis, Jan
HONG, Seung-Soo
Hora, Joseph
Horacek, Jiri
Houde, Martin
Houziaux, Leo
Hudson, Reggie
Huggins, Patrick
Hutchings, John
Hutsemekers, Damien
Hyung, Siek
Ilin, Vladimir
INOUE, Akio
INUTSUKA, Shu-ichiro
Irvine, William
Israel, Frank
Jackson, James
Jacoby, George
Jacq, Thierry
Jaffe, Daniel
Jahnke, Knud
Jenkins, Edward
Jimenez-Vicente, Jorge
JIN, Zhenyu
Johnson, Fred
Johnston, Kenneth
Johnstone, Douglas
Jones, Christine
Jorgensen, Jes
Jourdain de Muizon, Marie
Jura, Michael
Just, Andreas
Justtanont-Liseau, Kay
Kafatos, Menas
Kaftan, May
KAIFU, Norio
Kalenskii, Sergei
Kaler, James
KAMAYA, Hideyuki
Kamp, Inga
Kanekar, Nissim
Kantharia, Nimisha
Kassim, Namir
Kawada, Mitsunobu
Keene, Jocelyn
Kegel, Wilhelm
Keheyan, Yeghis
Kennicutt, Robert
KIM, Jongsoo
KIMURA, Toshiya
King, David
Kirkpatrick, Ronald
Kirshner, Robert
Klessen, Ralf
Knacke, Roger
Knapp, Gillian
Knezek, Patricia
Knude, Jens

KO, Chung-Ming
KOBAYASHI, Naoto
Kohoutek, Lubos
KOIKE , Chiyoe
KONDO, Yoji
KONG, Xu
Koornneef, Jan
KOZASA, Takashi
Krajnovic, Davor
Kramer, Busaba
Krautter, Joachim
Kravchuk, Sergei
Kreysa, Ernst
Krishna, Swamy
Krumholz, Mark
KUAN, Yi-Jehng
KUDOH, Takahiro
Kuiper, Thomas
Kulhanek, Petr
Kumar, C.
Kunth, Daniel
Kutner, Marc
Kwitter, Karen
Kylafis, Nikolaos
Lada, Charles
LAI, Shih-Ping
Laloum, Maurice
Langer, William
Latter, William
Laureijs, Rene
Laurent, Claudine
Lauroesch, James
Lazarian, Alexandre
Lazio, Joseph
Leao, Joao Rodrigo
Lebron, Mayra
LEE, Dae Hee
LEE, Hee Won
LEE, Myung Gyoon
Léger, Alain
Lehtinen, Kimmo
Leisawitz, David
Lépine, Jacques
Lequeux, James
Leto, Giuseppe
LI, Jinzeng
LIANG, Yanchun
Ligori, Sebastiano
Likkel, Lauren
Liller, William
Limongi, Marco
LIN, Weipeng
Linnartz, Harold
Lis, Dariusz
Liseau, René
Liszt, Harvey
LIU, Sheng-Yuan
LIU, Xiaowei
Lloyd, Myfanwy
Lo, Fred K. Y.
Lockman, Felix
Lodders, Katharina
Loinard, Laurent
Lopez Garcia, José
Louise, Raymond
Lovas, Francis
Lozinskaya, Tatjana
Lucas, Robert
Lynds, Beverly
Lyon, Ian
MA, Jun
Mac Low, Mordecai-Mark
Maciel, Walter
MacLeod, John
Madsen, Gregory
MAIHARA, Toshinori
MAKIUTI, Sin'itirou
Malbet, Fabien
Mampaso, Antonio
Manchado, Arturo
Manchester, Richard
Manfroid, Jean
Mantere, Maarit
Maret, Sébastien
Marston, Anthony
Martin, Peter
Martin, Christopher
Martin, Robert
Martin-Pintado, Jesus
Masson, Colin
Mather, John
Mathews, William
Mathis, John
MATSUHARA, Hideo
MATSUMOTO, Tomoaki
MATSUMURA, Masafumi
Mattila, Kalevi
Mauersberger, Rainer
McCall, Marshall
McClure-Griffiths, Naomi
Mccombie, June
McCray, Richard
McGregor, Peter
McKee, Christopher
McNally, Derek
Meaburn, John
Mebold, Ulrich
Meier, Robert
Meixner, Margaret
Mellema, Garrelt
Melnick, Gary
Mennella, Vito
Menon, T.
Menzies, John
Meszaros, Peter
Meyer, Martin
Mezger, Peter
Miller, Joseph
Milne, Douglas
Minier, Vincent
MINN , Young-Ki
Minter, Anthony
Mitchell, George
MIYAMA, Syoken
Mo, Jinger
Monin, Jean-Louis
Montmerle, Thierry
Moore, Marla
Moreno-Corral, Marco
Moriarty-Schieven, Gerald
Morris, Mark
Morton, Donald
Mouschovias, Telemachos
Muench, Guido
Mufson, Stuart
Mulas, Giacomo
Muller, Erik
Muller, Sebastien
Murthy, Jayant
Myers, Philip
NAGAHARA, Hiroko
NAGATA, Tetsuya
NAKADA , Yoshikazu
NAKAGAWA, Takao
NAKAMOTO, Taishi
NAKAMURA, Fumitaka
NAKANO, Takenori
NAKANO, Makoto
Natta, Antonella
Nguyen-Quang, Rieu

Nikolic, Silvana
NISHI, Ryoichi
NOMURA, Hideko
Nordh, Lennart
Norman, Colin
Nuernberger, Dieter
Nulsen, Paul
Nussbaumer, Harry
Nuth, Joseph
O'Dell, Charles
O'Dell, Stephen
Oey, Sally
OHTANI, Hiroshi
OKUDA, Haruyuki
OKUMURA, Shin-ichiro
Olofsson, Hans
Omont, Alain
OMUKAI, Kazuyuki
ONAKA, Takashi
Onello, Joseph
Orlando, Salvatore
Osborne, John
Ostriker, Eve
Ott, Juergen
Oudmaijer, Rene
Pagani, Laurent
Pagano, Isabella
PAK, Soojong
Palla, Francesco
Palmer, Patrick
Palumbo, Maria Elisabetta
Panagia, Nino
Pandey, Birendra
Pankonin, Vernon
PARK, Yong Sun
Parker, Eugene
Paron, Sergio
Parthasarathy, Mudumba
Pauls, Thomas
Pecker, Jean-Claude
Peeters, Els
Peimbert, Manuel
Pellegrini, Silvia
Pena, Miriam
Pendleton, Yvonne
PENG, Qingyu
Penzias, Arno
Pequignot, Daniel
Perault, Michel
Persi, Paolo
Persson, Carina
Peters, William
Petrosian, Vahe
Petuchowski, Samuel
Philipp, Sabine
Phillips, Thomas
Pihlström, Ylva
Pineau des Forêts, G.
Plume, René
Poeppel, Wolfgang
Pongracic, Helen
Pontoppidan, Klaus
Porceddu, Ignazio
Pottasch, Stuart
Pound, Marc
Pouquet, Annick
Prasad, Sheo
Preite Martinez, Andrea
Price, R.
Prochaska, Jason
Pronik, Iraida
Prusti, Timo
Puget, Jean-Loup
Ramirez, Jose
Ranalli, Piero
Rastogi, Shantanu
Ratag, Mezak
Rawlings, Jonathan
Rawlings, Mark
Raymond, John
Recchi, Simone
Redman, Matthew
Reipurth, Bo
Rengarajan, Thinniam
Rengel, Miriam
Reshetnyk, Volodymyr
Reyes, Rafael
Reynolds, Ronald
Reynolds, Cormac
Reynoso, Estela
Richter, Philipp
Rickard, Lee
Roberge, Wayne
Roberts, Douglas
Roberts Jr, William
Robinson, Garry
Roche, Patrick
Rodrigues, Claudia
Rodriguez, Luis
Rodriguez, Monica
Roediger, Elke
Roelfsema, Peter
Roeser, Hans-peter
Roger, Robert
Rogers, Alan
Rosa, Michael
Rosado, Margarita
Rouan, Daniel
Roxburgh, Ian
Ryabov, Michael
Sabbadin, Franco
Sahu, Kailash
SAIGO, Kazuya
Sakano, Masaaki
Salama, Farid
Salinari, Piero
Salomé, Philippe
Salter, Christopher
Samodurov, Vladimir
Sanchez Doreste, Néstor
Sanchez-Saavedra, M.
Sancisi, Renzo
Sandell, Göran
Sandqvist, Aage
Sarazin, Craig
Sargent, Annelia
Sarma, N.
Sarre, Peter
SATO, Shuji
SATO, Fumio
Savage, Blair
Savedoff, Malcolm
Scalo, John
Scherb, Frank
Schilke, Peter
Schlemmer, Stephan
Schmid-Burgk, J.
Schroder, Anja
Schwarz, Ulrich
Scoville, Nicholas
SEKI, Munezo
Sellgren, Kristen
Sembach, Kenneth
Sen, Asoke
SEON, Kwang il
Shadmehri, Mohsen
Shane, William
Shapiro, Stuart
Shaver, Peter
Shawl, Stephen

Shchekinov, Yuri
Shematovich, Valerij
Sherwood, William
Shields, Gregory
Shipman, Russell
Shmeld, Ivar
SHU, Frank
Shull, John
Shull, Peter
Shustov, Boris
Siebenmorgen, Ralf
Sigalotti, Leonardo
Silich, Sergey
Silk, Joseph
Silva, Laura
Silvestro, Giovanni
Sitko, Michael
Sivan, Jean-Pierre
Skilling, John
Skulskyj, Mychajlo
Slane, Patrick
Sloan, Gregory
Smith, Tracy
Smith, Peter
Smith, Michael
Smith, Craig
Smith, Randall
Smith, Robert
Snell, Ronald
Snow, Theodore
Sobolev, Andrey
Sofia, Ulysses
Sofia, Sabatino
SOFUE, Yoshiaki
Solc, Martin
Somerville, William
SONG, In Ok
Spaans, Marco
Stahler, Steven
Stanga, Ruggero
Stanghellini, Letizia
Stanimirovic, Snezana
Stapelfeldt, Karl
Stark, Ronald
Stasinska, Grazyna
Stecher, Theodore
Stecklum, Bringfried
Stenholm, Björn
Stone, James
Strom, Richard
SUH, Kyung Won
Sutherland, Ralph
SUZUKI, Tomoharu
Swade, Daryl
Sylvester, Roger
Szczerba, Ryszard
TACHIHARA, Kengo
Tafalla, Mario
TAKAHASHI, Junko
TAKANO, Toshiaki
TAMURA, Shin'ichi
TAMURA, Motohide
TANAKA, Masuo
Tantalo, Rosaria
Teixeira, Paula
Tenorio-Tagle, Guillermo
Terzian, Yervant
Testi, Leonardo
Thaddeus, Patrick
The, Pik-Sin
Thompson, A.
Thonnard, Norbert
Thronson Jr, Harley
Tilanus, Remo
Tokarev, Yurij
Torrelles, Jose-Maria
Torres-Peimbert, Silvia
Tosi, Monica
Tothill, Nicholas
Trammell, Susan
Treffers, Richard
Trinidad, Miguel
Turner, Kenneth
Tyul'bashev, Sergei
Ulrich, Marie-Hélène
Urosevic, Dejan
Urquhart, James
van de Steene, Griet
van den Ancker, Mario
van der Hulst, Jan
van der Laan, Harry
van der Tak, Floris
van Dishoeck, Ewine
van Gorkom, Jacqueline
van Loon, Jacco
van Woerden, Hugo
VandenBout, Paul
Varshalovich, Dmitrij
Vazquez, Roberto
Velazquez, Pablo
Verdoes Kleijn, Gijsbert
Verheijen, Marc
Verner, Ekaterina
Verschuur, Gerrit
Viala, Yves
Viallefond, Francois
Vidal-Madjar, Alfred
Viegas, Sueli
Vijh, Uma
Vilchez, Jose
Villaver, Eva
Vink, Jacco
Viti, Serena
Volk, Kevin
Vorobyov, Eduard
Voronkov, Maxim
Voshchinnikov, Nikolai
Vrba, Frederick
Wakker, Bastiaan
Walker, Gordon
Walmsley, C.
Walsh, Wilfred
Walsh, Andrew
Walton, Nicholas
WANG, Hongchi
WANG, Jun-Jie
Wang, Q. Daniel
WANG, Hong-Guang
Wannier, Peter
Ward-Thompson, Derek
Wardle, Mark
Watt, Graeme
Weaver, Harold
Weiler, Kurt
Weinberger, Ronald
Wesselius, Paul
Weymann, Ray
Whelan, Emma
White, Glenn
White, Richard
Whitelock, Patricia
Whiteoak, John
Whittet, Douglas
Whitworth, Anthony
Wickramasinghe, N.
Wiebe, Dmitri
Wild, Wolfgang
Wilkin, Francis
Williams, David
Williams, Robin

Williams, Robert
Willis, Allan
Willner, Steven
Wilson, Christine
Wilson, Robert
Wilson, Thomas
Winnberg, Anders
Witt, Adolf
Wolff, Michael
Wolfire, Mark
Wolstencroft, Ramon
Wolszczan, Alexander
Woltjer, Lodewijk
Woodward, Paul
Woolf, Neville
Wootten, Henry
Wouterloot, Jan
Wright, Edward
Wunsch, Richard
Wynn-Williams, Gareth
YABUSHITA, Shin
YAMADA, Masako
YAMAMOTO, Satoshi
YAMAMURA, Issei
YAMASHITA, Takuya
YAN, Jun
York, Donald
Yorke, Harold
YOSHIDA, Shigeomi
YUI, Yukari
Yun, João
Zavagno, Annie
Zealey, William
Zeilik, Michael
ZENG, Qin
ZHANG, JiangShui
ZHANG, Jingyi
ZHOU, Jianjun
ZHU, Wenbai
Zinchenko, Igor
Zuckerman, Benjamin

Division VII Galactic System

President: Despina Hatzidimitriou
Vice-President: Rosemary F. Wyse

Organizing Committee Members:

Giovanni Carraro
Bruce G. Elmegreen
Birgitta Nordström

Members:

Aarseth, Sverre
Acosta Pulido, Jose
Afanas'ev, Viktor
Aguilar, Luis
Ahumada, Javier
Ahumada, Andrea
Akeson, Rachel
Alcobe, Santiago
Alfaro, Emilio
Alksnis, Andrejs
Allen, Christine
Allen, Lori
Allende Prieto, Carlos
Altenhoff, Wilhelm
Ambastha, Ashok
Andersen, Johannes
Andreuzzi, Gloria
Aparicio, Antonio
Ardeberg, Arne
Ardi, Eliani
Armandroff, Taft
Arnold, Richard
Asteriadis, Georgios
Athanassoula, Evangelie
Aurière, Michel
Babusiaux, Carine
BAEK, Chang Hyun
Baier, Frank
Bailyn, Charles
Balazs, Bela
Balazs, Lajos
Balbus, Steven
Balcells, Marc
Banhatti, Dilip
Barberis, Bruno
Barmby, Pauline
Barrado y Navascues, David
Bartasiute, Stanislava
Bash, Frank
Bastian, Nathan
Baud, Boudewijn
Baume, Gustavo
Baumgardt, Holger
Bellazzini, Michele
Bensby, Thomas
Berkhuijsen, Elly
Biazzo, Katia
Bienayme, Olivier
Bijaoui, Albert
Binney, James
Blitz, Leo
Bloemen, Hans
Blommaert, Joris
Blum, Robert
Bobylev, Vadim
Boily, Christian
Bonatto, Charles
Bosch, Guillermo
Bragaglia, Angela
Brand, Jan
Bronfman, Leonardo
Brown, Warren
Brown, Anthony
Buonanno, Roberto
Burderi, Luciano
Burke, Bernard
Burkhead, Martin
Burton, W.
Butler, Ray
Buzzoni, Alberto
Byrd, Gene
Calamida, Annalisa
Caldwell, John
Callebaut, Dirk
Caloi, Vittoria
Cane, Hilary
Cannon, Russell
Cantiello, Michele
CAO, Zhen
Caputo, Filippina
Capuzzo Dolcetta, Roberto
Caretta, Cesar
Carney, Bruce
Carollo, Daniela
Carpintero, Daniel
Carrasco, Luis
Cesarsky, Diego
Cesarsky, Catherine
CHA, Seung-Hoon
Chaboyer, Brian
Chapman, Jessica
Chavarria-K, Carlos
CHEN, Li
CHEN, Huei-Ru
CHEN, Yuqin
CHENG, Kwang
Chiosi, Cesare
Christian, Carol
Christodoulou, Dimitris
CHUN, Mun-suk
Churchwell, Edward
Cincotta, Pablo
Cioni, Maria-Rosa
Claria, Juan
Clemens, Dan
Clementini, Gisella
Colin, Jacques
Comins, Neil
Contopoulos, George
Corradi, Romano
Costa, Edgardo

Covino, Elvira
Crampton, David
Crawford, David
Creze, Michel
Cropper, Mark
Croton, Darren
Cubarsi, Rafael
Cudworth, Kyle
Cuperman, Sami
D'Amico, Nicolo'
D'Antona, Francesca
Da Costa, Gary
Dale, James
Dalla Bonta, Elena
Dambis, Andrei
Danford, Stephen
Dapergolas, Anastasios
Daube-Kurzemniece, Ilga
Dauphole, Bertrand
Davies, Melvyn
Davies, Rodney
Dawson, Peter
de Grijs, Richard
de Jong, Teije
De Marchi, Guido
Dejonghe, Herwig
Dekel, Avishai
Demarque, Pierre
Demers, Serge
DENG, LiCai
Diaferio, Antonaldo
Diaz, Ruben
Dickel, Helene
Dickel, John
Dickman, Robert
Djorgovski, Stanislav
Djupvik, Anlaug Amanda
Dluzhnevskaya, Olga
do Nascimento, José
Downes, Dennis
Drilling, John
Drimmel, Ronald
Drissen, Laurent
Ducati, Jorge
Ducourant, Christine
Durrell, Patrick
Eastwood, Kathleen
Efremov, Yurij
Egret, Daniel
Einasto, Jaan
Elmegreen, Debra
ESAMDIN, Ali
ESIMBEK, Jarken
Evangelidis, E.
Faber, Sandra
Fall, S.
Fathi, Kambiz
Feast, Michael
Feinstein, Alejandro
Feitzinger, Johannes
Ferguson, Annette
Figueras, Francesca
Flynn, Chris
Forbes, Douglas
Forte, Juan
Foster, Tyler
Freeman, Kenneth
Friel, Eileen
FUJIMOTO, Masa Katsu
FUJIWARA, Takao
FUKUSHIGE, Toshiyuki
Fusi-Pecci, Flavio
Ganguly, Rajib
Garcia, Beatriz
Garzon, Francisco
Geffert, Michael
Geisler, Douglas
Genzel, Reinhard
Giersz, Miroslav
Gilmore, Gerard
Giorgi, Edgard
Glushkova, Elena
Golay, Marcel
Goldreich, Peter
Gomez, Ana
Gordon, Mark
Gottesman, Stephen
Gouliermis, Dimitrios
Gratton, Raffaele
Grayzeck, Edwin
Green, Anne
Green, Elizabeth
Green, James
Grenon, Michel
Grindlay, Jonathan
Grundahl, Frank
Guetter, Harry
Gupta, Sunil
HABE, Asao
Habing, Harm
Haisch Jr, Karl
Hakkila, Jon
HANAMI, Hitoshi
Hanes, David
Hanson, Margaret
Harris, Gretchen
Harris, Hugh
Hartkopf, William
Hawkins, Michael
Hayli, Abraham
Haywood, Misha
Heggie, Douglas
Heiles, Carl
Helmi, Amina
Herbst, William
Hernandez-Pajares, Manuel
Hesser, James
Hetem Jr., Annibal
Heudier, Jean-Louis
Hilker, Michael
Hillenbrand, Lynne
Hills, Jack
Hodapp, Klaus
HONMA, Mareki
HOZUMI, Shunsuke
Hron, Josef
HU, Hongbo
Huensch, Matthias
Humphreys, Roberta
Humphreys, Elizabeth
Hut, Piet
Iben Jr, Icko
IGUCHI, Osamu
IKEUCHI, Satoru
Illingworth, Garth
INAGAKI, Shogo
Israel, Frank
Ivezic, Zeljko
IYE, Masanori
Jablonka, Pascale
Jackson, Peter
Jahreiss, Hartmut
Jalali, Mir Abbas
Janes, Kenneth
Jasniewicz, Gerard
JEON, Young Beom
JIANG , Ing-Guey
JIANG, Dongrong
Jog, Chanda
Johansson, Peter

Jones, Derek
Joshi, Umesh
Kalirai, Jason
Kalnajs, Agris
Kamp, Lucas
KANG, Yong-Hee
Karakas, Amanda
KATO, Shoji
Khovritchev, Maxim
KIM, Sungsoo
KIM, Seung-Lee
King, Ivan
Kinman, Thomas
Kitsionas, Spyridon
Knapp, Gillian
KO, Chung-Ming
Kontizas, Evangelos
Kontizas, Mary
Korchagin, Vladimir
Kormendy, John
Krajnovic, Davor
Kroupa, Pavel
Krumholz, Mark
Kulsrud, Russell
Kun, Maria
Kundu, Arunav
Kurtev, Radostin
Kutuzov, Sergej
Laloum, Maurice
Landolt, Arlo
Lapasset, Emilio
Larson, Richard
Larsson-Leander, Gunnar
Latham, David
Laval, Annie
Lee, Young-Wook
LEE, Jae Woo
LEE, Kang Hwan
LEE, Hyung-Mok
LEE, Sang-Gak
Leisawitz, David
Leonard, Peter
Lepine, Sebastien
LI, Jinzeng
Liebert, James
LIN, Qing
Lindblad, Per
LIU, Michael
Lockman, Felix
Lodieu, Nicolas
Loktin, Alexhander
Lu, Phillip
Lucatello, Sara
LUO, Ali
Lynden-Bell, Donald
Maccarone, Thomas
MacConnell, Darrell
Maeder, Andre
Majumdar, Subhabrata
Makalkin, Andrei
MAKINO , Junichiro
Mamajek, Eric
Manchester, Richard
Mandel, Ilya
Marco, Amparo
Mardling, Rosemary
Markkanen, Tapio
Markov, Haralambi
Marochnik, Leonid
Marraco, Hugo
Marsden, Stephen
Martin, Christopher
Martinet, Louis
Martinez Delgado, David
Martinez Roger, Carlos
Martins, Donald
Martos, Marco
Matteucci, Francesca
Mayor, Michel
McClure-Griffiths, Naomi
McGregor, Peter
Mendez Bussard, Rene
Menon, T.
Menzies, John
Merrifield, Michael
Meylan, Georges
Mezger, Peter
Mikkola, Seppo
Miller, Richard
Milone, Eugene
Minniti, Dante
Mirabel, Igor
Mishurov, Yury
Moehler, Sabine
Moffat, Anthony
Mohammed, Ali
Mohan, Vijay
Moitinho, André
Monet, David
Morales Rueda, Luisa
Moreno Lupianez, Manuel
Morris, Rhys
Morris, Mark
Mould, Jeremy
Muench, Guido
Muminov, Muydinjon
Muzzio, Juan
NAKASATO, Naohito
Namboodiri, P.
Napolitano, Nicola
Navone, Hugo
Naylor, Tim
Nelemans, Gijs
Nelson, Alistair
Nemec, James
Nesci, Roberto
Neuhaeuser, Ralph
Newberg, Heidi
Nikiforov, Igor
Ninkov, Zoran
Ninkovic, Slobodan
Norman, Colin
Nuernberger, Dieter
Oblak, Edouard
Ocvirk, Pierre
Oey, Sally
OGURA, Katsuo
OH, Kap Soo
Oja, Tarmo
Ojha, Devendra
OKUDA, Haruyuki
Olano, Carlos
Oliveira, Joana
Origlia, Livia
Orlov, Victor
Ortiz, Roberto
Ortolani, Sergio
Ostorero, Luisa
Ostriker, Eve
Ostriker, Jeremiah
Oudmaijer, Rene
Palmer, Patrick
Palous, Jan
Pandey, A.
Pandey, Birendra
Papayannopoulos, Theodoros
PARK, Byeong-Gon
Parmentier, Geneviève
Parsamyan, Elma

Patsis, Panos
Patten, Brian
Pauls, Thomas
Paunzen, Ernst
Pavani, Daniela
Pedreros, Mario
Peimbert, Manuel
Penny, Alan
Perek, Luboš
Perez, Mario
Perryman, Michael
Pesch, Peter
Peterson, Charles
Petrovskaya, Margarita
Peykov, Zvezdelin
Phelps, Randy
Philip, A.G.
Piatti, Andrés
Pier, Jeffrey
Pilachowski, Catherine
Pirzkal, Norbert
Piskunov, Anatolij
Platais, Imants
Polyachenko, Evgeny
Porras Juárez, Bertha
Portegies Zwart, Simon
Portinari, Laura
Poveda, Arcadio
Price, R.
Pritchet, Christopher
Pulone, Luigi
Rabolli, Monica
Raharto, Moedji
Raimondo, Gabriella
Ratnatunga, Kavan
Ravindranath, Swara
Rebull, Luisa
Recio-Blanco, Alejandra
Reid, Iain
Reif, Klaus
Renzini, Alvio
Rey, Soo-Chang
Reylé, Céline
Rich, Robert
Richer, Harvey
Richter, Philipp
Richtler, Tom
Riegel, Kurt
Roberts, Morton
Roberts Jr, William
Robin, Annie
Rocha-Pinto, Hélio
Rodrigues de Oliveira F., I.
Rothberg, Barry
Rountree, Janet
Royer, Pierre
Rubin, Vera
Ruelas-Mayorga, R.
Ruiz, Maria Teresa
Russeva, Tatjana
Ruzicka, Adam
Rybicki, George
Saar, Enn
Sagar, Ram
Sakano, Masaaki
Sala, Ferran
Samus, Nikolay
Sanchez Bejar, Victor
Sanchez Doreste, Néstor
Sanchez-Saavedra, M.
Sanders, Walter
Sandqvist, Aage
Santiago, Basilio
Santillan, Alfredo
Santos Jr., Joao
Sanz, Jaume
Sargent, Annelia
Schechter, Paul
Schmidt, Maarten
Schoedel, Rainer
Schuler, Simon
Schweizer, François
Seggewiss, Wilhelm
Seimenis, John
Seitzer, Patrick
Sellwood, Jerry
Semkov, Evgeni
Serabyn, Eugene
SHAN, Hongguang
Shane, William
Shawl, Stephen
Sher, David
SHI, Huoming
SHU, Frank
SHU, Chenggang
Sigalotti, Leonardo
Simonson, S.
Skinner, Stephen
Smith, J.
Smith, Graeme
Sobouti, Yousef
Song, Inseok
SONG, Liming
SONG, Qian
Sotnikova, Natalia
Soubiran, Caroline
Southworth, John
Sparke, Linda
Spergel, David
Spiegel, Edward
Spurzem, Rainer
Stauffer, John
Stecker, Floyd
Steinlin, Uli
Stetson, Peter
Stoehr, Felix
Stringfellow, Guy
Strobel, Andrzej
Stuik, Remko
SU, Cheng-yue
Subramaniam, Annapurni
SUGIMOTO, Daiichiro
SUNG, Hwankyung
Suntzeff, Nicholas
Surdin, Vladimir
Sygnet, Jean-Francois
Tadross, Ashraf
TAKAHASHI, Koji
TAKASHI, Hasegawa
Tammann, Gustav
Tas, Günay
Terranegra, Luciano
The, Pik-Sin
Thomas, Claudine
Thoul, Anne
TIAN, Wenwu
Tikhonov, Nikolai
Tinney, Christopher
Tobin, William
TOMISAKA, Kohji
Toomre, Juri
Toomre, Alar
Tornambe, Amedeo
Torra, Jordi
TOSA, Makoto
Tosi, Monica
Tripicco, Michael
Trullols, I.
TSUJIMOTO, Takuji
Tsvetkov, Milcho

Tsvetkova, Katja
Turner, David
Turon, Catherine
Twarog, Bruce
Upgren, Arthur
Urquhart, James
Valtonen, Mauri
van Altena, William
van den Bergh, Sidney
van der Kruit, Pieter
van Woerden, Hugo
VandenBerg, Don
Vandervoort, Peter
Varela Perez, Antonia
Vazquez, Ruben
Vega, E.
Veltchev, Todor
Ventura, Paolo
Venugopal, V.
Vergne, María
Verschueren, Werner
Verschuur, Gerrit
Vesperini, Enrico
Villas da Rocha, Jaime
Vivas, Anna
Volkov, Evgeni
Volonteri, Marta
von Hippel, Theodore
Voroshilov, Volodymyr
Wachlin, Felipe
Wagner, Alexander
Walker, Merle
Walker, Gordon
Warren Jr, Wayne
Weaver, Harold
Wehlau, Amelia
Weistrop, Donna
Whiteoak, John
Whittet, Douglas
Wielebinski, Richard
Wielen, Roland
Woltjer, Lodewijk
Woodward, Paul
Wouterloot, Jan
Wramdemark, Stig
Wright, Nicholas
WU, Hsin-Heng
Wunsch, Richard
Yakut, Kadri
YAMAGATA, Tomohiko
YI, Sukyoung
YIM, Hong Suh
YOSHII , Yuzuru
YUMIKO, Oasa
Zachilas, Loukas
Zaggia, Simone
Zakharova, Polina
Zapatero-Osorio, Maria R.
ZHANG, Fenghui
ZHANG, Haotong
ZHAO, Jun Liang
ZHOU, Jianjun
Zinn, Robert

Division VIII Galaxies & the Universe

President: Elaine M. Sadler
Vice-President: Françoise Combes

Organizing Committee Members:

Roger L. Davies
John S. Gallagher III
Thanu Padmanabhan

Members:

Aalto, Susanne
Abbas, Ummi
Ables, Harold
Abu Kassim, Hasan
Adami, Christophe
Adams, Jenni
Adler, David
Afanas'ev, Viktor
Aguero, Estela
Aguilar, Luis
Ahmad, Farooq
AKIYAMA, Masayuki
Alard, Christophe
Alcaniz, Jailson
Aldaya, Victor
Alexander, Tal
Alimi, Jean-Michel
Allan, Peter
Allen, Ronald
Allington-Smith, Jeremy
Alloin, Danielle
Almaini, Omar
Aloisi, Alessandra
Alonso, Maria
Alonso, Maria
Amendola, Luca
Amram, Philippe
Andernach, Heinz
Andersen, Michael
Andreani, Paola
ANN, Hong-Bae
Anosova, Joanna
Anton, Sonia
Antonelli, Lucio Angelo
Aoki, Kentaro
Aparicio, Antonio
Aragon-Salamanca, Alfonso
Ardeberg, Arne
Aretxaga, Itziar
Argo, Megan
Argueso, Francisco
Arkhipova, Vera
Arnaboldi, Magda
Artamonov, Boris
Athanassoula, Evangelie
Atrio Barandela, Fernando
Audouze, Jean
Aussel, Hervé
Avelino, Pedro
Avila-Reese, Vladimir
AYANI, Kazuya
AZUMA, Takahiro
Azzopardi, Marc
Babul, Arif
Bachev, Rumen
Baddiley, Christopher
Baes, Maarten
Bagla, Jasjeet
Bahcall, Neta
BAI, Jinming
Bailey, Mark
Bajaja, Esteban
Bajtlik, Stanislaw
Baker, Andrew
Balbi, Amedeo
Baldwin, Jack
Balkowski-Mauger, Chantal
Ballabh, Goswami
Balland, Christophe
Balogh, Michael
Bamford, Steven
Banday, Anthony
Banhatti, Dilip
Barberis, Bruno
Barbon, Roberto
Barbuy, Beatriz
Barcons, Xavier
Bardeen, James
Bardelli, Sandro
Barger, Amy
Barkana, Rennan
Barnes, David
Barrientos, Luis
Barrow, John
Bartelmann, Matthias
Barth, Aaron
Barthel, Peter
Barton, Elizabeth
Barway, Sudhashu
Baryshev, Andrey
Basa, Stephane
Bassett, Bruce
Bassino, Lilia
Battaner, Eduardo
Battinelli, Paolo
Battye, Richard
Baum, Stefi
Bautista, Manuel
Bayet, Estelle
Beaulieu, Sylvie
Bechtold, Jill
Beck, Rainer
Beckman, John
Beckmann, Volker
Beesham, Aroonkumar
Begeman, Kor
Belinski, Vladimir
Bender, Ralf
Benedict, George
Benetti, Stefano
Benitez, Erika
Bennett, Charles
Bennett, David
Bensby, Thomas

Bentz, Misty
Berczik, Peter
Bergeron, Jacqueline
Bergvall, Nils
Berkhuijsen, Elly
Berta, Stefano
Bertschinger, Edmund
Betancor Rijo, Juan
Bettoni, Daniela
Bharadwaj, Somnath
Bhavsar, Suketu
BIAN, Yulin
Bianchi, Simone
Bicknell, Geoffrey
Biermann, Peter
Bignall, Hayley
Bignami, Giovanni
Bijaoui, Albert
Binetruy, Pierre
Binette, Luc
Binggeli, Bruno
Binney, James
Biretta, John
Birkinshaw, Mark
Biviano, Andrea
Bjornsson, Claes-Ingvar
Blakeslee, John
Blanchard, Alain
Bland-Hawthorn, Jonathan
Bleyer, Ulrich
Blitz, Leo
Block, David
Bludman, Sidney
Blumenthal, George
Blundell, Katherine
Boehringer, Hans
Boeker, Torsten
Boissier, Samuel
Boisson, Catherine
Boksenberg, Alec
Boles, Thomas
Bolzonella, Micol
Bomans, Dominik
Bond, John
Bongiovanni, Angel
Borgani, Stefano
Borne, Kirk
Boschin, Walter
Bosma, Albert
Bouchet, François
Bowen, David
Bower, Gary
Boyle, Brian
Braine, Jonathan
Branchesi, Marica
Braun, Robert
Bravo-Alfaro, Hector
Brecher, Kenneth
Bressan, Alessandro
Bridges, Terry
Bridle, Sarah
Briggs, Franklin
Brinchmann, Jarle
Brinks, Elias
Brodie, Jean
Brosch, Noah
Brouillet, Nathalie
Brown, Thomas
Brown, Michael
Bruzual, Gustavo
Bryant, Julia
Buat, Véronique
Bunker, Andrew
Buote, David
Burbidge, Eleanor
Bureau, Martin
Burgarella, Denis
Burkert, Andreas
Burns, Jack
Busarello, Giovanni
Buta, Ronald
Butcher, Harvey
Byrd, Gene
BYUN, Yong Ik
CAI, Michael
Calderon, Jesús
Calura, Francesco
Calvani, Massimo
Calzetti, Daniela
Campusano, Luis
Cannon, John
Cannon, Russell
Cantiello, Michele
Canzian, Blaise
CAO, Xinwu
CAO, Li
Caon, Nicola
Capaccioli, Massimo
Cappellari, Michele
Cappi, Alberto
Caproni, Anderson
Cardenas, Rolando
Caretta, Cesar
Carigi, Leticia
Carollo, Marcella
Carr, Bernard
Carretti, Ettore
Carrillo, Rene
Carswell, Robert
Carter, David
Casasola, Viviana
Casoli, Fabienne
Castagnino, Mario
Cattaneo , Andrea
Cavaliere, Alfonso
Cayatte, Veronique
Cellone, Sergio
Cepa, Jordi
Cesarsky, Diego
CHA, Seung-Hoon
CHAE, Kyu Hyun
Chakrabarti, Sandip
Chamaraux, Pierre
CHANG, Ruixiag
CHANG, Heon-Young
CHANG, Kyongae
Charlot, Stephane
Charmandaris, Vassilis
Chatterjee, Tapan
Chatzichristou, Eleni
Chavushyan, Vahram
CHEN, DaMing
CHEN, Yang
CHEN, Hsiao-Wen
CHEN, Jiansheng
CHEN, Lin-wen
CHEN, Pisin
CHEN, Xuelei
CHENG, Fuzhen
Chiappini , Cristina
CHIBA, Masashi
CHIBA, Takeshi
Chincarini, Guido
Chodorowski, Michal
CHOU, Chih-Kang
Choudhury, Tirthankar
Christensen, Lise
CHU, Yaoquan
Chugai, Nikolaj
CHUN, Sun

Cid Fernandes, Roberto
Ciliegi, Paolo
Cioni, Maria-Rosa
Ciotti, Luca
Ciroi, Stefano
Claeskens, Jean-François
Claria, Juan
Clarke, Tracy
Clavel, Jean
Clementini, Gisella
Clowe, Douglas
Clowes, Roger
Cocke, William
Cohen, Ross
Colafrancesco, Sergio
Colbert, Edward
Cole, Shaun
Coles, Peter
Colina, Luis
Colless, Matthew
Colombi, Stephane
Comte, Georges
Condon, James
Conselice, Christopher
Contopoulos, George
Cook, Kem
Cooray, Asantha
Cora, Sofia
Corbin, Michael
Corsini, Enrico
Corwin Jr, Harold
Cote, Stéphanie
Cote, Patrick
Couch, Warrick
Courbin, Frederic
Courteau, Stéphane
Courvoisier, Thierry
Couto da Silva, Telma
Covone, Giovanni
Cowsik, Ramanath
Coziol, Roger
Crane, Patrick
Crane, Philippe
Crawford, Carolin
Crawford, Steven
Cress, Catherine
Cristiani, Stefano
Croom, Scott
Croton, Darren
CUI, Wenyuan
Cunniffe, John
Curran, Stephen
Cypriano, Eduardo
D'Odorico, Valentina
D'Odorico, Sandro
D'Onofrio, Mauro
Da Costa, Gary
da Costa, Luiz
Da Rocha , Cristiano
Dadhich, Naresh
Dahle, Haakon
Daigne, Frédéric
DAISUKE, Iono
Dalla Bonta, Elena
Dallacasa, Daniele
Danese, Luigi
Danks, Anthony
Dantas, Christine
Das, P.
Dasyra, Kalliopi
Davidge, Timothy
Davidson, William
Davies, Rodney
Davies, Paul
Davis, Michael
Davis, Tamara
Davis, Marc
De Bernardis, Paolo
De Blok, Erwin
de Bruyn, A.
de Carvalho, Reinaldo
de Diego Onsurbe, Jose
de Grijs, Richard
de Jong, Roelof
de Lapparent, Valérie
de Lima, José
de Mello, Duilia
de Petris, Marco
de Propris, Roberto
de Rijcke, Sven
de Ruiter, Hans
de Silva, Lindamulage
de Zeeuw, Pieter
de Zotti, Gianfranco
Dejonghe, Herwig
Dekel, Avishai
Dell'Antonio, Ian
Demers, Serge
Demianski, Marek
DENG, Zugan
Dennefeld, Michel
Désert, François-Xavier
Dettmar, Ralf-Juergen
Deustua, Susana
Devost, Daniel
Dhurandhar, Sanjeev
Diaferio, Antonaldo
Diaz, Angeles
Diaz, Ruben
Dickey, John
Dietrich, Matthias
Dietrich , Jörg
Djorgovski, Stanislav
Dobbs, Matt
Dobrzycki, Adam
DOI, Mamoru
Dokuchaev, Vyacheslav
Dominguez, Mariano
Dominis Prester, Dijana
Donas, Jose
Donea, Alina
DONG, Xiao-Bo
Donner, Karl
Donzelli, Carlos
Dopita, Michael
Dottori, Horacio
Dovciak, Michal
Doyon, Rene
Dressel, Linda
Dressler, Alan
Drinkwater, Michael
Driver, Simon
Duc, Pierre-Alain
Dufour, Reginald
Dultzin-Hacyan, Deborah
Dumont, Anne-Marie
Dunlop, James
Dunsby, Peter
Durret, Florence
Duval, Marie-France
Dyer, Charles
Eales, Stephen
Edelson, Rick
Edmunds, Michael
Edsjo, Joakim
Efstathiou, George
Ehle, Matthias
Einasto, Jaan
Ekers, Ronald
Elgaroy, Oystein

Elizalde, Emilio
Ellis, George
Ellis, Simon
Ellis, Richard
Elmegreen, Debra
Elvis, Martin
Elyiv, Andrii
Emsellem, Eric
Enginol, Turan
English, Jayanne
ENOKI, Motohiro
Espey, Brian
Ettori, Stefano
Eungwanichayapant, Anant
Evans, Robert
Fabbiano, Giuseppina
Faber, Sandra
Fabricant, Daniel
Falceta-Goncalves, Diego
Falco, Emilio
Falcon Barroso, Jesus
Fall, S.
Famaey, Benoit
FAN, Junhui
FAN, Zuhui
Fasano, Giovanni
Fassnacht, Christopher
Fathi, Kambiz
Fatkhullin, Timur
Feain, Ilana
Feast, Michael
Fedeli, Cosimo
Fedorova, Elena
Feinstein, Carlos
Feitzinger, Johannes
FENG, Long Long
Ferguson, Annette
Ferland, Gary
Ferrarese, Laura
Ferreras, Ignacio
Ferrini, Federico
Field, George
Filippenko, Alexei
Flin, Piotr
Florides, Petros
Florido, Estrella
Focardi, Paola
Foltz, Craig
Fong, Richard
Forbes, Duncan
Ford, Holland
Ford Jr, W.
Forman, William
Foschini, Luigi
Fouque, Pascal
Fox, Andrew
Fraix-Burnet, Didier
Franceschini, Alberto
Francis, Paul
Franx, Marijn
Freedman, Wendy
Freeman, Kenneth
Frenk, Carlos
Friaca, Amancio
Fricke, Klaus
Fried, Josef
Frogel, Jay
FUJITA, Yutaka
FUKUGITA, Masataka
FUKUI, Takao
FUNATO, Yoko
Funes, José
Furlanetto, Steven
Fuzfa, Andre
Fynbo, Johan
Gaensler, Bryan
Gallagher, Sarah
Gallart, Carme
Gallego, Jesús
Galletta, Giuseppe
Gallimore, Jack
Gangui, Alejandro
Ganguly, Rajib
GAO, Yu
Garcia-Lorenzo, Maria
Gardner, Jonathan
Garilli, Bianca
Garrison, Robert
Gavignaud, Isabelle
Gelderman, Richard
Geller, Margaret
GENG, Lihong
Georgiev, Tsvetan
Gerhard, Ortwin
Ghigo, Francis
Ghirlanda, Giancarlo
Ghosh, P.
Giacani, Elsa
Giallongo, Emanuele
Giani, Elisabetta
Gibson, Brad
Gigoyan, Kamo
Gioia, Isabella
Giovanardi, Carlo
Giovanelli, Riccardo
Giroletti, Marcello
Gitti, Myriam
Glass, Ian
Glazebrook, Karl
Glover , Simon
Godlowski, Wlodzimierz
Goldsmith, Donald
Gonzalez Delgado, Rosa
Gonzalez Sanchez, A.
Gonzalez-Serrano, J. I.
Goobar, Ariel
Goodrich, Robert
Goret, Philippe
Gorgas, Garcia
Goss, W. Miller
Gosset, Eric
GOTO, Tomotsugu
Gottesman, Stephen
Gottloeber, Stefan
GOUDA, Naoteru
Govinder, Keshlan
Graham, Alister
Graham, John
Granato, Gian Luigi
Gray, Richard
Gray, Meghan
Grebel, Eva
Green, Anne
Gregg, Michael
Gregorio, Anna
Greve, Thomas
Greyber, Howard
Griest, Kim
Griffiths, Richard
Griv, Evgeny
Gronwall, Caryl
Grove, Lisbeth
Grupe, Dirk
GU, Qiusheng
Gudmundsson, Einar
Gumjudpai, Burin
Gunn, James
Gunthardt, Guillermo
Guseva, Natalia
Gutierrez, Carlos

Guzzo, Luigi
Gyulbudaghian, Armen
Haehnelt, Martin
Hagen, Hans-Juergen
Hagen-Thorn, Vladimir
Hall, Patrick
HAMABE, Masaru
Hambaryan, Valeri
Hamilton, Andrew
Hammer, François
HAN, Cheongho
HANAMI, Hitoshi
Hannestad, Steen
Hansen, Frode
HARA, Tetsuya
Hardy, Eduardo
Harms, Richard
Harnett, Julienne
Hasan, Hashima
Hashimoto, Yasuhiro
HATTORI, Makoto
Hau, George
Haugboelle, Troels
Hawking, Stephen
HE, XiangTao
Heald, George
Heavens, Alan
Heckman, Timothy
Heidt, Jochen
Heinamaki, Pekka
Held, Enrico
Hellaby, Charles
Heller, Michael
Helou, George
Hendry, Martin
Henning, Patricia
Henriksen, Mark
Henry, Richard
Hensler, Gerhard
Heraudeau, Philippe
Hernandez, Xavier
Hewett, Paul
Heyrovsky, David
Hicks, Amalia
Hickson, Paul
HIDEKI, Asada
Hintzen, Paul
HIRASHITA, Hiroyuki
Hjalmarson, Ake
Hjorth, Jens
Hnatyk, Bohdan
Ho, Luis
Hodge, Paul
Hoekstra, Hendrik
Holz, Daniel
Hopkins, Andrew
Hopp, Ulrich
Horellou, Cathy
Hornschemeier, Ann
Hornstrup, Allan
HOU, Jinliang
Houdashelt, Mark
Hough, James
Hu, Esther
HU, Fuxing
HU, Hongbo
HUANG, Keliang
Huang, Jiasheng
Huchtmeier, Walter
Hudson, Michael
Huettemeister, Susanne
Hughes, David
Humphreys, Roberta
Humphreys, Elizabeth
Hunstead, Richard
Hunt, Leslie
Hunter, James
Hutsi, Gert
Huynh, Minh
HWANG, Jai-chan
HWANG, Chorng-Yuan
ICHIKAWA, Takashi
ICHIKAWA, Shin-ichi
Icke, Vincent
Idiart, Thais
IKEUCHI, Satoru
Illingworth, Garth
IM, Myungshin
IMANISHI, Masatoshi
Impey, Christopher
INADA, Naohisa
Infante, Leopoldo
INOUE, Akio
Iovino, Angela
Irwin, Judith
ISHIHARA, Hideki
ISHIMARU, Yuhri
Israel, Frank
Ivezic, Zeljko
Ivison, Robert
IWAMURO, Fumihide
IWATA, Ikuru
IYE, Masanori
Iyer, Balasubramanian
Izotov, Yuri
Izotova, Iryna
Jablonka, Pascale
Jaffe, Walter
Jahnke, Knud
Jakobsson, Pall
JANG, Minwhan
Jannuzi, Buell
Jaroszynski, Michal
Jarrett, Thomas
Jauncey, David
Jaunsen, Andreas
Jedamzik, Karsten
Jensen, Brian
Jerjen, Helmut
Jetzer, Philippe
JIANG, Ing-Guey
Jimenez-Vicente, Jorge
JING, Yipeng
Jog, Chanda
Johansson, Peter
Johnston, Helen
Jones, Heath
Jones, Paul
Jones, Christine
Jones, Bernard
Jones, Thomas
Jordan, Andrés
Jorgensen, Inger
Joshi, Umesh
Jovanovic, Predrag
Joy, Marshall
Jungwiert, Bruno
Junkes, Norbert
Junkkarinen, Vesa
Junor, William
KAJINO, Toshitaka
Kalloglian, Arsen
Kandalyan, Rafik
Kanekar, Nissim
KANEKO, Noboru
KANG, Hyesung
Kapoor, Ramesh
Karachentsev, Igor
Karachentseva, Valentina
KAROJI , Hiroshi

KASHIKAWA, Nobunari
Kaspi, Shai
Kassim, Namir
Katgert, Peter
KATO, Shoji
Katsiyannis, Athanassios
Kauffmann, Guinevere
Kaufman, Michele
Kaul, Chaman
KAWABATA, Kiyoshi
Kawada, Mitsunobu
KAWAKATU, Nozomu
KAWASAKI, Masahiro
Keel, William
Kellermann, Kenneth
Kelly, Brandon
Kembhavi, Ajit
Kemp, Simon
Kennicutt, Robert
Khachikian, Edward
Khanna, Ramon
Khare, Pushpa
Khmil, Sergiy
Khosroshahi, Habib
Kilborn, Virginia
KIM, Dong Woo
KIM, Sungsoo
KIM, Jik
King, Lindsay
King, Ivan
Kinman, Thomas
Kirilova, Daniela
Kirshner, Robert
Kissler-Patig, Markus
KIYOTOMO, Ichiki
Klein, Ulrich
Knapen, Johan
Knapp, Gillian
Kneib, Jean-Paul
Knezek, Patricia
Kniazev, Alexei
Knudsen, Kirsten
KO, Chung-Ming
Kobayashi, Chiaki
Kochhar, Rajesh
KODAIRA, Keiichi
KODAMA, Tadayuki
KODAMA, Hideo
Koekemoer, Anton
Kogoshvili, Natela
Kokkotas, Konstantinos
Kolb, Edward
Kollatschny, Wolfram
KOMIYAMA, Yutaka
Kompaneets, Dmitrij
KONG, Xu
Kontizas, Evangelos
Kontorovich, Victor
Koo, David
Koopmans, Leon
Koratkar, Anuradha
Koribalski, Bärbel
Kormendy, John
Kotilainen, Jari
Kovalev, Yuri
Kovetz, Attay
KOZAI, Yoshihide
Kozlovsky, Ben
Kraan-Korteweg, Renée
Krajnovic, Davor
Krasinski, Andrzej
Krause, Marita
Krishna, Gopal
Kriss, Gerard
Kron, Richard
Krumholz, Mark
Kudrya, Yury
KUMAI, Yasuki
Kunchev, Peter
Kunert-Bajraszewska, M.
KUNO, Nario
Kunth, Daniel
Kuntschner, Harald
Kunz, Martin
Kuzio de Naray, Rachel
La Barbera, Francesco
La Franca, Fabio
Lacey, Cedric
Lachièze-Rey, Marc
Lagache, Guilaine
Lahav, Ofer
Lake, Kayll
Lal, Dharam
Lancon, Ariane
Lanfranchi, Gustavo
Larionov, Mikhail
Larsen, Søren
Larson, Richard
Lasota-Hirszowicz, J.-P.
Laurikainen, Eija
Layzer, David
Le Fèvre, Olivier
Leao, João Rodrigo
Lebron, Mayra
LEE, Myung Gyoon
LEE, Wo-Lung
Leeuw, Lerothodi
Lehnert, Matthew
Lehto, Harry
Leibundgut, Bruno
Lequeux, James
Leubner, Manfred
Levin, Yuri
Levine, Robyn
Lewis, Geraint
LI, Ji
LIAN, Luo
LIANG, Yanchun
Liddle, Andrew
Liebscher, Dierck-E.
Lilje, Per
Lilly, Simon
LIM, Jeremy
Lima Neto, Gastao
LIN, Weipeng
Lindblad, Per
Linden-Vørnle, Michael
LIOU, Guo Chin
LIU, Yongzhen
Lo, Fred K. Y.
Lobo, Catarina
Lombardi, Marco
Londrillo, Pasquale
Longair, Malcolm
Longo, Giuseppe
Lonsdale, Carol
Lopez, Ericson
Lopez, Sebastian
Lopez Aguerri, Jose Alfonso
Lopez Cruz, Omar
Lopez Hermoso, Maria
Lopez-Corredoira, Martin
Lopez-Sanchez, Angel
Lord, Steven
Loup, Cécile
Loveday, Jon
Lu, Limin
LU, Tan
Lubin, Lori
Lugger, Phyllis

Lukash, Vladimir
Luminet, Jean-Pierre
LUO, Ali
Lutz, Dieter
Lynden-Bell, Donald
Lynds, Beverly
Lynds, Roger
MA, Jun
Maartens, Roy
Macalpine, Gordon
Maccagni, Dario
MacCallum, Malcolm
Maccarone, Thomas
Macchetto, Ferdinando
Maciejewski, Witold
Mackie, Glen
Macquart, Jean-Pierre
Madden, Suzanne
Maddox, Stephen
Madore, Barry
MAEDA, Kei-ichi
Magorrian, Stephen
Magris, Gladis
Maharaj, Sunil
Mahtessian, Abraham
Maia, Marcio
Mainieri, Vincenzo
Maiolino, Roberto
Majumdar, Subhabrata
Makarov, Dmitry
Makarova, Lidia
Malagnini, Maria
Malesani, Daniele
Malhotra, Sageeta
Mamon, Gary
Mandolesi, Nazzareno
Mangalam, Arun
Mann, Robert
Mannucci, Filippo
Manrique, Alberto
Mansouri, Reza
Mao, Shude
Maoz, Dan
Marano, Bruno
Marcelin, Michel
Marco, Olivier
Marconi, Alessandro
Mardirossian, Fabio
Marek, John
Maris, Michele
Markoff, Sera
Marquez, Isabel
Marr, Jonathon
Marranghello, Guilherme
Marston, Anthony
Martin, Crystal
Martin, Rene
Martin, Maria
Martinet, Louis
Martinez , Vicent
Martinez-Gonzalez, E.
Martini, Paul
Martinis, Mladen
Martins, Carlos
Marziani, Paola
Masegosa, Josefa
Mather, John
MATSUMOTO, Toshio
Matthews, Lynn
Matzner, Richard
Mauersberger, Rainer
Maurice, Eric
Mayya, Divakara
Mazzarella, Joseph
McBreen, Brian
McGaugh, Stacy
McNeil, Stephen
Mediavilla, Evencio
Mehlert, Dörte
Meier, David
Meikle, William
Meisenheimer, Klaus
Mellier, Yannick
Melnyk, Olga
Melott, Adrian
Mendes de Oliveira, C.
Meneghetti, Massimo
Menendez-Delmestre, Karin
Menon, T.
Mercurio, Amata
Merighi, Roberto
Merluzzi, Paola
Merrifield, Michael
Meszaros, Peter
Meszaros, Attila
Metevier, Anne
Meusinger, Helmut
Meyer, David
Meyer, Martin
Meyer, Angela
Meylan, Georges
Meza, Andres
Mezzetti, Marino
Mihov, Boyko
Miley, George
Miller, Joseph
Miller, Neal
Miller, Hugh
Miller, Richard
Milvang-Jensen, Bo
Mirabel, Felix
Miralda-Escude, Jordi
Miralles, Joan-Marc
Miranda, Oswaldo
Miroshnichenko, Alla
Misawa, Toru
Misner, Charles
Miyazaki, Satoshi
Miyoshi, Shigeru
MIZUNO, Takao
Mo, Houjun
Mohr, Joseph
Moiseev, Alexei
Moles Villamate, Mariano
Molinari, Emilio
Molla, Mercedes
Monaco, Pierluigi
Moody, Joseph
Moore, Ben
Moreau, Olivier
MORI, Masao
Moscardini, Lauro
Mota, David
MOTOHARA, Kentaro
Motta, Veronica
Mould, Jeremy
Mourao, Ana Maria
Muanwong, Orrarujee
Muecket, Jan
Mueller, Volker
Mueller, Andreas
Mujica, Raul
Mulchaey, John
Muller, Erik
Muller, Richard
Muller, Sebastien
Munoz Tunon, Casiana
MURAKAMI, Izumi
MURAYAMA, Takashi
Murphy, Michael

Murphy, John
Murray, Stephen
Mushotzky, Richard
Muzzio, Juan
Nagao, Tohru
NAGASHIMA, Masahiro
Nair, Sunita
NAKAI, Naomasa
NAKAMICHI, Akika
NAKANISHI, Kouichiro
NAKANISHI, Hiroyuki
Nakos, Theodoros
Namboodiri, P.
NAMBU, Yasusada
Napolitano, Nicola
Narasimha, Delampady
Narlikar, Jayant
Naselsky, Pavel
Nasr-Esfahani, Bahram
Navarro, Julio
Nedialkov, Petko
Neves de Araujo, José
Nicoll, Jeffrey
Nikolajuk, Marek
Ninkovic, Slobodan
Nipoti, Carlo
Nishikawa, Ken-Ichi
Nityananda, Rajaram
Noerdlinger, Peter
NOGUCHI, Masafumi
NOH, Hyerim
Norman, Dara
Norman, Colin
Nottale, Laurent
Novikov, Igor
Novosyadlyj, Bohdan
Novotny, Jan
Nozari, Kourosh
Nucita, Achille
Nulsen, Paul
O'Connell, Robert
O'Dea, Christopher
Ocvirk, Pierre
Oemler Jr, Augustus
Oey, Sally
OHTA, Kouji
OKAMOTO, Takashi
Oliver, Sebastian
Olling, Robert
Olofsson, Kjell
Olowin, Ronald
Omizzolo, Alessandro
Oosterloo, Thomas
Origlia, Livia
Oscoz, Alejandro
Ostlin, Göran
Ostorero, Luisa
Ostriker, Eve
Ott, Juergen
OUCHI, Masami
Ovcharov, Evgeni
OYABU, Shinki
Ozsvath, Istvan
Page, Mathew
Page, Don
PAK, Soojong
Palmer, Philip
Palumbo, Giorgio
Panessa, Francesca
Pannuti, Thomas
Papayannopoulos, Th.
Paragi, Zsolt
PARK, Jang Hyun
Parker, Quentin
Parnovsky, Sergei
Partridge, Robert
Pastoriza, Miriani
Paturel, Georges
Pearce, Frazer
Pecker, Jean-Claude
Pedersen, Kristian
Pedrosa, Susana
Peebles, P.
Peimbert, Manuel
Peletier, Reynier
Pellegrini, Silvia
Pello, Roser
Pen, Ue-Li
PENG, Qingyu
Penzias, Arno
Perea-Duarte, Jaime
Perez, Fournon
Perez-Torres, Miguel
Peroux, Céline
Perryman, Michael
Persides, Sotirios
Persson, Carina
Peters, William
Peterson, Bruce
Peterson, Charles
Petit, Jean-Marc
Petitjean, Patrick
Petrosian, Vahe
Petrosian, Artaches
Petrov, Georgi
Petuchowski, Samuel
Pfenniger, Daniel
Philipp, Sabine
Phillipps, Steven
Phillips, Mark
Pihlström, Ylva
Pikichian, Hovhannes
Pimbblet, Kevin
Pipino, Antonio
Pirzkal, Norbert
Pisano, Daniel
Pizzella, Alessandro
Plana, Henri
Plionis, Manolis
Podolsky, Jiri
Pogge, Richard
Poggianti, Bianca
Polletta, Maria del Carmen
Polyachenko, Evgeny
Pompei, Emanuela
Popescu, Cristina
Popescu, Nedelia
Popović, Luka
Portinari, Laura
Poveda, Arcadio
Power, Chris
Prabhu, Tushar
Pracy, Michael
Prandoni, Isabella
Premadi, Premana
Press, William
Prieto, Almudena
Prires Martins, Lucimara
Pritchet, Christopher
Proctor, Robert
Pronik, Iraida
Pronik, Vladimir
Proust, Dominique
Prugniel, Philippe
Puerari, Ivânio
Puetzfeld, Dirk
Puget, Jean-Loup
Pustilnik, Simon
Puy, Denis
QIN, Bo

QIN, Yi-Ping
Quinn, Peter
Rafanelli, Piero
Rahvar, Sohrab
Raiteri, Claudia
Ramella, Massimo
Rampazzo, Roberto
Ranalli, Piero
Rand, Richard
Rasmussen, Jesper
Ravindranath, Swara
Raychaudhury, Somak
Read, Andrew
Rebolo, Rafael
Reboul, Henri
Recchi, Simone
Rector, Travis
Reddy, Bacham
Rees, Martin
Reeves, Hubert
Reiprich, Thomas
Rejkuba, Marina
Rephaeli, Yoel
Reshetnikov, Vladimir
Reunanen, Juha
Revaz, Yves
Revnivtsev, Mikhail
Rey, Soo-Chang
Reynolds, Cormac
Riazi, Nematollah
Riazuelo, Alain
Ribeiro, Marcelo
Ribeiro, André Luis
Richer, Harvey
Richstone, Douglas
Richter, Philipp
Ricotti, Massimo
Ridgway, Susan
Rindler, Wolfgang
Rix, Hans-Walter
Robert, Carmelle
Roberts, Morton
Roberts, Timothy
Roberts, David
Roberts Jr, William
Rocca-Volmerange, Brigitte
Rodrigues de Oliveira F. I.
Roeder, Robert
Roediger, Elke
Romano, Patrizia
Romano-Diaz, Emilio
Romeo, Alessandro
Romer, Anita
Romero-Colmenero, E.
Rosa, Michael
Rosa Gonzalez, Daniel
Rosado, Margarita
Rose, James
Rosquist, Kjell
Rothberg, Barry
Rots, Arnold
Rottgering, Huub
Rowan-Robinson, Michael
Roxburgh, Ian
Rozas, Maite
Rubin, Vera
Rubino-Martin, J. A.
Rubio, Monica
Rudnick, Lawrence
Rudnicki, Konrad
Ruffini, Remo
Ruszkowski, Mateusz
Ruzicka, Adam
Ryder, Stuart
Saar, Enn
Sackett, Penny
Sadun, Alberto
Sahlen, Martin
Sahni, Varun
SAITOH, Takayuki
Saiz, Alejandro
Sakai, Shoko
Sala, Ferran
Salvador-Sole, Eduardo
Salzer, John
Samurović, Srdjan
Sanahuja Parera, Blai
Sancisi, Renzo
Sanders, Robert
Sanders, David
Sanroma, Manuel
Sansom, Anne
Santiago, Basilio
Santos-Lleo, Maria
Sapar, Arved
Sapre, Ashok
Saracco, Paolo
Sarazin, Craig
SASAKI, Minoru
SASAKI, Shin
SASAKI, Misao
Sasaki, Toshiyuki
Saslaw, William
SATO, Shinji
SATO, Katsuhiko
SATO, Jun'ichi
SATO, Humitaka
Saviane, Ivo
SAWA, Takeyasu
Sazhin, Mikhail
Scaramella, Roberto
Schaerer, Daniel
Schartel, Norbert
Schaye, Joop
Schechter, Paul
Schindler, Sabine
Schmidt, Brian
Schmidt, Maarten
Schmitt, Henrique
Schmitz, Marion
Schneider, Donald
Schneider, Raffaella
Schneider, Jean
Schneider, Peter
Schramm, Thomas
Schroder, Anja
Schuch, Nelson
Schucking, Engelbert
Schwarz, Ulrich
Schweizer, François
Scodeggio, Marco
Scorza, Cecilia
Scott, Douglas
Scoville, Nicholas
Seielstad, George
Seigar, Marc
Sellwood, Jerry
Semelin, Benoit
Semerak, Oldrich
Sempere, Maria
SEON, Kwang il
Sergeev, Sergey
Serjeant, Stephen
Serote Roos, Margarida
Setti, Giancarlo
Severgnini, Paola
SHAN, Hongguang
Shandarin, Sergei
Shanks, Thomas
SHAO, Zhengyi

Shapovalova, Alla
Sharp, Nigel
Sharples, Ray
Shaver, Peter
Shaviv, Giora
Shaya, Edward
SHEN, Zhiqiang
Sherwood, William
SHIBATA, Masaru
Shields, Joseph
Shields, Gregory
SHIMASAKU, Kazuhiro
Shostak, G.
Shukurov, Anvar
Siebenmorgen, Ralf
Signore, Monique
Sigurdsson, Steinn
Sil'chenko, Olga
Silk, Joseph
Sillanpaa, Aimo
Silva, David
Silva, Laura
Simkin, Susan
Singh, Kulinder Pal
Siopis, Christos
Sistero, Roberto
Skillman, Evan
Slezak, Eric
Smail, Ian
Smecker-Hane, Tammy
Smette, Alain
Smith, Malcolm
Smith, Haywood
Smith, Nigel
Smith, Rodney
Smith, Eric
Smoot III, George
Soares, Domingos Savio
Sobouti, Yousef
SOHN, Young Jong
Sokolowski, Lech
Sollerman, Jesper
Soltan, Andrzej
SONG, Doo-Jong
SONG, Liming
SORAI, Kazuo
Souradeep, Tarun
Sparks, William
Spinoglio, Luigi
Spinrad, Hyron
Spyrou, Nicolaos
Squires, Gordon
Srianand, Raghunathan
Sridhar, Seshadri
Srinivasan, Ganesan
Statler, Thomas
Staveley-Smith, Lister
Stecker, Floyd
Steigman, Gary
Steiman-Cameron, Thomas
Steinbring, Eric
Stiavelli, Massimo
Stirpe, Giovanna
Stoehr, Felix
Stolyarov, Vladislav
Stone, Remington
Storchi-Bergmann, Thaisa
Storrie-Lombardi, Lisa
Straumann, Norbert
Strauss, Michael
Stritzinger, Maximilian
Strom, Richard
Strom, Robert
Struble, Mitchell
Strukov, Igor
Stuchlik, Zdenek
Stuik, Remko
SU, Cheng-yue
Subrahmanya, C.
Subramaniam, Annapurni
SUGAI , Hajime
SUGINOHARA, Tatsushi
SUGIYAMA, Naoshi
Suhhonenko, Ivan
Sulentic, Jack
Sullivan, III, Woodruff
Sundin, Maria
Sunyaev, Rashid
Surdej, Jean
Sutherland, Ralph
Sutherland, William
SUTO, Yasushi
Szalay, Alex
Szydlowski, Marek
Tacconi, Linda
Tacconi-Garman, Lowell
Tagger, Michel
TAGOSHI, Hideyuki
TAKAGI, Toshinobu
TAKAHARA, Fumio
TAKASHI, Hasegawa
TAKATA, Tadafumi
TAKATO, Naruhisa
TAKEUCHI, Tsutomu
TAKIZAWA, Motokazu
Tamm, Antti
Tammann, Gustav
TANABE, Kenji
TANIGUCHI, Yoshiaki
Tantalo, Rosaria
Tarter, Jill
TARUYA , Atsushi
TATEKAWA, Takayuki
Taylor, Angela
Taylor, James
Telles, Eduardo
Temporin, Sonia
Tenjes, Peeter
Terlevich, Roberto
Terzian, Yervant
Theis, Christian
Thomasson, Magnus
Thonnard, Norbert
Thornley, Michele
Thuan, Trinh
Tifft, William
Tikhonov, Nikolai
Tilanus, Remo
Tipler, Frank
Tissera, Patricia
Tisserand , Patrick
Toffolatti, Luigi
Toft, Sune
Tolstoy, Eline
TOMIMATSU, Akira
Tomita, Akihiko
TOMITA, Kenji
Tonry, John
Toomre, Alar
Tormen, Giuseppe
TOTANI, Tomonori
Tovmassian, Hrant
TOYAMA, Kiyotaka
Tozzi, Paolo
Traat, Peeter
Trager, Scott
Tremaine, Scott
Treu, Tommaso
Trevese, Dario
Trimble, Virginia

Trinchieri, Ginevra
Trotta, Roberto
Trujillo Cabrera, Ignacio
Tsamparlis, Michael
TSUCHIYA, Toshio
Tsvetkov, Dmitry
Tuffs, Richard
Tugay, Anatoliy
Tully, Richard
Turner, Edwin
Turner, Michael
Turnshek, David
Tyson, John
Tytler, David
Tyul'bashev, Sergei
Ugolnikov, Oleg
Ulrich, Marie-Hélène
UMEMURA, Masayuki
Urbanik, Marek
Uslenghi, Michela
Uson, Juan
Utrobin, Victor
Vagnetti, Fausto
Vaisanen, Petri
Valcheva, Antoniya
Valdes Parra, Jose
Valentijn, Edwin
Vallenari, Antonella
Valls-Gabaud, David
Valotto, Carlos
Valtchanov, Ivan
Valtonen, Mauri
van Albada, Tjeerd
van den Bergh, Sidney
van der Hulst, Jan
van der Kruit, Pieter
van der Laan, Harry
van der Marel, Roeland
van Driel, Wim
van Gorkom, Jacqueline
van Haarlem, Michiel
van Moorsel, Gustaaf
van Woerden, Hugo
Van Zee, Liese
Vansevicius, Vladas
Varma, Ram
Vaughan, Simon
Vauglin, Isabelle
Vavilova, Iryna
Vazdekis, Alexandre
Vedel, Henrik
Veilleux, Sylvain
Vercellone, Stefano
Verdes-Montenegro, L.
Verdoes Kleijn, Gijsbert
Vermeulen, Rene
Vestergaard, Marianne
Vettolani, Giampaolo
Viana, Pedro
Viel, Matteo
Vigroux, Laurent
Villata, Massimo
Vishniac, Ethan
Vishveshwara, C.
Vivas, Anna
Vlasyuk, Valerij
Vollmer, Bernd
Volonteri, Marta
von Borzeszkowski, H.
Vrtilek, Jan
WADA, Keiichi
Waddington, Ian
Wagner, Stefan
Wagner, Alexander
Wagoner, Robert
Wainwright, John
WAKAMATSU, Ken-Ichi
Walker, Mark
Walter, Fabian
Walterbos, Rene
Wambsganss, Joachim
WANAJO, Shinya
Wanas, Mamdouh
WANG, Yiping
WANG, Tinggui
WANG, Huiyuan
WANG, Hong-Guang
Ward, Martin
Watson, Darach
Webb, Tracy
Webster, Adrian
Weedman, Daniel
WEI, Jianyan
Weilbacher, Peter
Weiler, Kurt
Weinberg, Steven
Welch, Gary
Wesson, Paul
West, Michael
Westmeier, Tobias
White, Simon
Whiting, Alan
Whiting, Matthew
Whitmore, Bradley
Widrow, Lawrence
Wielebinski, Richard
Wielen, Roland
Wiita, Paul
Wilcots, Eric
Wild, Wolfgang
Will, Clifford
Williams, Robert
Williams, Theodore
Williams, Barbara
Wills, Beverley
Wills, Derek
Windhorst, Rogier
Winkler, Hartmut
Wise, Michael
Wisotzki, Lutz
Wofford, Aida
Wold, Margrethe
Woltjer, Lodewijk
Woosley, Stanford
Worrall, Diana
Woszczyna, Andrzej
Woudt, Patrick
Wozniak, Hervé
Wright, Edward
Wrobel, Joan
WU, Xue-bing
WU, Xiangping
WU, Hong
WU, Jiun-Huei
WU, Wentao
WU, Jianghua
Wu, Yanling
Wucknitz, Olaf
Wunsch, Richard
Wyithe, Stuart
Wynn-Williams, Gareth
Xanthopoulos, Emily
XIA, Xiao-Yang
XIANG, Shouping
Xilouris, Emmanouel
XU, Dawei
XUE, Suijian
YAGI, Masafumi
Yakovleva, Valerija
YAMADA, Yoshiyuki

YAMADA, Toru
YAMAGATA, Tomohiko
YAMAUCHI, Aya
YASUDA, Naoki
Yesilyurt, Ibrahim
YI, Sukyoung
YOICHIRO, Suzuki
YOKOYAMA, Jun'ichi
YONEHARA, Atsunori
YOSHIDA, Michitoshi
YOSHIDA, Hiroshi
YOSHII, Yuzuru
YOSHIKAWA, Kohji
YOSHIOKA, Satoshi
Yushchenko, Alexander
Zacchei, Andrea
Zaggia, Simone
Zamorani, Giovanni
Zamorano, Jaime
Zanichelli, Alessandra
Zannoni, Mario
Zaroubi, Saleem
Zasov, Anatoly
Zeilinger, Werner
Zepf, Stephen
Zezas, Andreas
ZHANG, Xiaolei
ZHANG, Tong-Jie
ZHANG, Yang
ZHANG, Jialu
ZHANG, JiangShui
ZHANG, Jingyi
ZHANG, Bo
ZHANG, Fenghui
ZHAO, Donghai
ZHOU, Xu
ZHOU, Youyuan
ZHOU, Hongyan
ZHOU, Jianjun
ZHU, Xingfeng
Zieba, Stanislaw
Ziegler, Bodo
Ziegler, Harald
Zinn, Robert
ZOU, Zhenlong
Zucca, Elena
Zwaan, Martin

Division IX Optical & Infrared Techniques

President: Andreas Quirrenbach
Vice-President: David Richard Silva

Organizing Committee Members:

Michael G. Burton
Xiangqun Cui
Ian S. McLean
Eugene F. Milone
Jayant Murthy
Stephen T. Ridgway
Gražina Tautvaišiene
Andrei A. Tokovinin
Guillermo Torres

Members:

Ables, Harold
Abt, Helmut
Acke, Bram
Adelman, Saul
Ahumada, Javier
AKITAYA, Hiroshi
Albrecht, Rudolf
Alecian, Evelyne
Anandaram, Mandayam
Andersen, Johannes
Andreuzzi, Gloria
Angel, J.
Angione, Ronald
Anthony-Twarog, Barbara
Arcidiacono, Carmelo
Arnold, Richard
Ashok, N.
Aspin, Colin
Aungwerojwit, Amornrat
Babkovskaia, Natalia
Baddiley, Christopher
Bakker, Eric
Baliyan, Kiran
Balona, Luis
Bamford, Steven
Baran, Andrzej
Barnes III, Thomas
Barrett, Paul
Barrientos, Luis
Bastien, Pierre
Batten, Alan
Baume, Gustavo
Beavers, Willet
Beers, Timothy
Bellazzini, Michele
Bendjoya, Philippe
Benson, James
Berdyugin, Andrei
Bessell, Michael
Beuzit, Jean-Luc
Bjorkman, Jon
Boonstra, Albert
Borde, Pascal
Borra, Ermanno
Breger, Michel
Brown, Thomas
Brown, Douglas
Bryant, Julia
Burki, Gilbert
Buscher, David
Buser, Roland
Butkovskaya, Varvara
Butler, Paul
Cantiello, Michele
Cardoso Santos, Nuno
Carney, Bruce
Carter, Brian
Castelaz, Micheal
Chadid, Merieme
CHEN, Wen Ping
CHEN, An-Le
CHEN, Yuqin
CHEN, Zhiyuan
Ciliegi, Paolo
Cioni, Maria-Rosa
Clem, James
Cochran, William
Coffey, Deirdre
Connolly, Leo
Couto da Silva, Telma
Coyne, S.J, George
Crampton, David
Crawford, David
Crawford, Steven
Crifo, Françoise
Cuby, Jean-Gabriel
Cuillandre, Jean-Charles
Cuypers, Jan
da Costa, Luiz
Dahn, Conard
DAISAKU, Nogami
Danchi, William
Danford, Stephen
Davis, Robert
Davis, Marc
de Lange, Gert
de Medeiros, Jose
De Souza Pellegrini, Paulo
Delplancke, Françoise
Dennefeld, Michel
Deshpande, M.
Dieleman, Pieter
Dolan, Joseph
Dravins, Dainis
Dubath, Pierre
Dubout, Renee
Ducati, Jorge
Ducourant, Christine
Dutrey, Anne
Duvert, Gilles
Edwards, Paul
Eisner, Josh
Elkin, Vladimir
Elmhamdi, Abouazza
Fabregat, Juan

Fabrika, Sergei
Feinstein, Alejandro
Fekel, Francis
Fernandez Lajus, Eduardo
Fernie, J.
Florido, Estrella
Fluri, Dominique
Foltz, Craig
Forte, Juan
Forveille, Thierry
Freeman, Kenneth
Galadi-Enriquez, David
Garcia, Beatriz
Gehrz, Robert
Genet, Russell
Genzel, Reinhard
Gerbaldi, Michèle
Ghosh, Swarna
Giani, Elisabetta
Gilliland, Ronald
Gilmore, Gerard
Giorgi, Edgard
Giovanelli, Riccardo
Glass, Ian
Glushkova, Elena
Gnedin, Yurij
Golay, Marcel
GONG, Xuefei
Gonzalez, Jorge
Graham, John
Grauer, Albert
Gray, David
Grenon, Michel
Grewing, Michael
Griffin, Roger
Grundahl, Frank
GU, Xuedong
Guetter, Harry
Gulbis, Amanda
Halbwachs, Jean-Louis
Haniff, Christopher
Hauck, Bernard
Hearnshaw, John
Hensberge, Herman
Hewett, Paul
Hilditch, Ronald
Hill, Graham
Hora, Joseph
Howard, Andrew
Hrivnak, Bruce
HU, Zhong wen
Hube, Douglas
Hubrig, Swetlana
Hummel, Christian
Huovelin, Juhani
Hyland, Harry
Imbert, Maurice
Ireland, Michael
Irwin, Alan
Ivezic, Zeljko
IWATA, Ikuru
Jeffers, Sandra
Jerzykiewicz, Mikolaj
Jordan, Andrés
Jordi, Carme
Jorgensen, Anders
Jorissen, Alain
Joshi, Umesh
Kalomeni, Belinda
Karachentsev, Igor
Katz, David
Kawada, Mitsunobu
Kazlauskas, Algirdas
Kebede, Legesse
Keller, Stefan
KENTARO, Matsuda
Kepler, S.
Kilkenny, David
KIM, Seung-Lee
KIM, Young-Soo
King, Ivan
Knude, Jens
Koehler, Rainer
Konacki, Maciej
Kornilov, Victor
Kulkarni, Prabhakar
Kurtz, Donald
Lacour, Sylvestre
Landstreet, John
Laskarides, Paul
Latham, David
Lawson, Peter
Lazauskaite, Romualda
Lebohec, Stephan
Lehnert, Matthew
Leisawitz, David
Lemaître, Gérard
Lemke, Michael
Lenzen, Rainer
Leroy, Jean-Louis
Levato, Orlando
Lewis, Brian
LI, Qingkang
Linde, Peter
Lindgren, Harri
Lo Curto, Gaspare
Lockwood, G.
Lub, Jan
Luna, Homero
Maitzen, Hans
Malbet, Fabien
Manfroid, Jean
Manset, Nadine
Marcy, Geoffrey
Markkanen, Tapio
Marraco, Hugo
Marschall, Laurence
Marsden, Stephen
Martinez Fiorenzano, Aldo
Martinez Roger, Carlos
Maslennikov, Kirill
Mason, Paul
Mathieu, Robert
Mathys, Gautier
Maurice, Eric
Mayer, Pavel
Mayor, Michel
Mazeh, Tsevi
McDavid, David
McMillan, Robert
Melnick, Gary
Mendoza, V.
Menzies, John
Metcalfe, Travis
Meylan, Georges
Millan Gabet, Rafael
Miller, Joseph
Mink, Jessica
Mironov, Aleksey
Mkrtichian, David
Moffett, Thomas
Moitinho, André
Monnier, John
Morrell, Nidia
Mourard, Denis
Munari, Ulisse
Naef, Dominique
Napolitano, Nicola
Naylor, Tim
Neiner, Coralie

Nicolet, Bernard
NOGUCHI, Kunio
Nordström, Birgitta
Notni, Peter
Oblak, Edouard
Oestreicher, Roland
Orsatti, Ana
PAK, Soojong
Parimucha, Stefan
Paumard, Thibaut
Pavani, Daniela
Pedreros, Mario
Pel, Jan
Penny, Alan
Pepe, Francesco
Perez-Torres, Miguel
Perrier-Bellet, Christian
Perrin, Marshall
Peterson, Ruth
Petit, Pascal
Pfeiffer, Raymond
Philip, A.G.
Piirola, Vilppu
Platais, Imants
Pokrzywka, Bartlomiej
Pott, Jörg-Uwe
Pourbaix, Dimitri
Preston, George
Pulone, Luigi
QIAN, Shengbang
Queloz, Didier
Rajagopal, Jayadev
Rao, Pasagada
Rastorguev, Alexey
Ratnatunga, Kavan
Raveendran, A.
Rawlings, Mark
Reglero Velasco, Victor
Reshetnyk, Volodymyr
Rivinius, Thomas
Robb, Russell
Robinson, Edward
Rodrigues, Claudia
Romanov, Yuri
Romanyuk, Yaroslav
Rostopchina, Alla
Rousset, Gérard
Royer, Frédéric
Rubenstein, Eric
Rubin, Vera
Sachkov, Mikhail
Samus, Nikolay
Santos Agostinho, Rui
Scarfe, Colin
Schinckel, Antony
Schmidt, Edward
Schoedel, Rainer
Schroder, Anja
Schuller, Peter
Schuster, William
SEKIGUCHI, Kazuhiro
Sen, Asoke
Shankland, Paul
Shawl, Stephen
Sivan, Jean-Pierre
Smith, J.
Smith, Myron
Snowden, Michael
Solivella, Gladys
Stefanik, Robert
Steinlin, Uli
Steinmetz, Matthias
Stetson, Peter
Stickland, David
Stockman Jr, Hervey
Stone, Remington
Straizys, Vytautas
Strauss, Michael
Stritzinger, Maximilian
Subramaniam, Annapurni
Sudzius, Jokubas
Sullivan, Denis
Suntzeff, Nicholas
Szabados, Laszlo
Szkody, Paula
Szymanski, Michal
Tallon, Michel
Tallon-Bosc, Isabelle
Tandon, S.
Taranova, Olga
Tas, Günay
Tedds, Jonathan
Tokunaga, Alan
Tolbert, Charles
Tomasella, Lina
Tonry, John
Townsend, Richard
Turon, Catherine
Tuthill, Peter
Tycner, Christopher
UMEDA, Hideyuki
Ureche, Vasile
Urquhart, James
Uslenghi, Michela
Vakili, Farrokh
van Belle, Gerard
van Dessel, Edwin
Vanko, Martin
Vaughan, Arthur
Verhoelst, Tijl
Verma, R.
Verschueren, Werner
Vinko, Jozsef
Voloshina, Irina
von Braun, Kaspar
Vrba, Frederick
Walker, William
Walker, Alistair
Walker, Gordon
Wallace, James
WANG, Guomin
WANG, Shen
WANG, Jingyu
Warren Jr, Wayne
Wegner, Gary
Weiss, Werner
Weistrop, Donna
Wesselius, Paul
Wheatley, Peter
White, Nathaniel
Wielebinski, Richard
Willstrop, Roderick
Winiarski, Maciej
Woillez, Julien
Wramdemark, Stig
YAMASHITA, Yasumasa
YANG, Stephenson
YANG, Dehua
YAO, Yongqiang
Young, Andrew
YUAN, Xiangyan
Yudin, Ruslan
YUJI , Ikeda
Zaggia, Simone
Zavala, Robert
ZHANG, You-Hong
ZHANG, Haotong
ZHOU, Jianfeng
ZHU, Liying
Zwitter, Tomaž

Division X Radio Astronomy

President: A. Russell Taylor
Vice-President: Jessica Mary Chapman

Organizing Committee Members:

Christopher L. Carilli
Gabriele Giovannini
Richard E. Hills
Hisashi HIRABAYASHI
Justin L. Jonas
Joseph Lazio
Raffaella Morganti
Ren-Dong NAN
Monica Rubio
Prajval Shastri

Members:

Abraham, Péter
Ade, Peter
Akujor, Chidi
Alberdi, Antonio
Alexander, Paul
Alexander, Joseph
Allen, Ronald
Aller, Margo
Aller, Hugh
Altenhoff, Wilhelm
Altunin, Valery
Ambrosini, Roberto
Andernach, Heinz
Anglada, Guillem
Antonova, Antoaneta
Argo, Megan
Arnal, Edmundo
Asareh, Habibolah
Aschwanden, Markus
Aubier, Monique
Augusto, Pedro
Baan, Willem
Baars, Jacob
Baath, Lars
Babkovskaia, Natalia
Bachiller, Rafael
Bailes, Matthew
Bajaja, Esteban
Bajkova, Anisa
Baker, Andrew
Baker, Joanne
Balasubramanian, V.
Balasubramanyam, Ramesh
Ball, Lewis
Bally, John
Balonek, Thomas
Banhatti, Dilip
Barrow, Colin
Bartel, Norbert
Barthel, Peter
Bartkiewicz, Anna
Barvainis, Richard
Baryshev, Andrey
Bash, Frank
Baudry, Alain
Baum, Stefi
Bayet, Estelle
Beasley, Anthony
Beck, Rainer
Benaglia, Paula
Benn, Chris
Bennett, Charles
Benz, Arnold
Berkhuijsen, Elly
Bhandari, Rajendra
Bhat, Ramesh
Bieging, John
Biermann, Peter
Biggs, James
Bignall, Hayley
Bignell, R.
Biraud, François
Biretta, John
Birkinshaw, Mark
Blair, David
Blandford, Roger
Bloemhof, Eric
Blundell, Katherine
Boboltz, David
Bock, Douglas
Bockelée-Morvan, D.
Bolatto, Alberto
Bondi, Marco
Boonstra, Albert
Bower, Geoffrey
Branchesi, Marica
Bregman, Jacob
Brentjens, Michiel
Bridle, Alan
Brinks, Elias
Britzen, Silke
Broderick, John
Bronfman, Leonardo
Brooks, Kate
Brouw, Willem
Brown, Jo-Anne
Browne, Ian
Brunetti, Gianfranco
Bryant, Julia
Bujarrabal, Valentin
Burderi, Luciano
Burke, Bernard
Campbell, Robert
Campbell-Wilson, Duncan
Caproni, Anderson
Carlqvist, Per
Caroubalos, Constantinos
Carretti, Ettore
Carvalho, Joel
Casasola, Viviana
Casoli, Fabienne
Castelletti, Gabriela
Castets, Alain
Cecconi, Baptiste
Celotti, Anna Lisa

Cernicharo, Jose
CHAN, Kwing
Chandler, Claire
Charlot, Patrick
CHEN, Yongjun
CHEN, Huei-Ru
CHEN, Xuefei
CHEN, Zhiyuan
Chengalur, Jayaram
CHIKADA, Yoshihiro
CHIN, Yi-nan
Chini, Rolf
CHO, Se Hyung
Choudhury, Tirthankar
CHUNG, Hyun-Soo
Chyzy, Krzysztof
Cichowolski, Silvina
Ciliegi, Paolo
Clark, David
Clark, Frank
Clark, Barry
Clegg, Andrew
Clemens, Dan
Cohen, Marshall
Coleman, Paul
Colomer, Francisco
Combes, Françoise
Combi, Jorge
Condon, James
Conklin, Edward
Contreras, Maria
Conway, John
Corbel, Stéphane
Cordes, James
Cotton Jr, William
Crane, Patrick
Crawford, Fronefield
Crovisier, Jacques
Crutcher, Richard
Cunningham, Maria
D'Amico, Nicolo'
Dagkesamansky, Rustam
DAISHIDO, Tsuneaki
DAISUKE, Iono
Dallacasa, Daniele
Davies, Rodney
Davis, Robert
Davis, Richard
Davis, Michael
de Bergh, Catherine
De Bernardis, Paolo
de Gregorio-Monsalvo, I.
de Jager, Cornelis
de La Noë, Jerome
de Lange, Gert
de Ruiter, Hans
de Vicente, Pablo
Deshpande, Avinash
Despois, Didier
Dewdney, Peter
Dhawan, Vivek
Diamond, Philip
Dickel, Helene
Dickel, John
Dickey, John
Dickman, Robert
DOBASHI, Kazuhito
Dodson, Richard
Doubinskij, Boris
Dougherty, Sean
Downes, Dennis
Downs, George
Drake, Frank
Drake, Stephen
Dreher, John
Duffett-Smith, Peter
Dutrey, Anne
Dwarakanath, K.
Dyson, Freeman
Eales, Stephen
Edelson, Rick
Ehle, Matthias
Ekers, Ronald
Elia, Davide
Ellingsen, Simon
Emerson, Darrel
Epstein, Eugene
Erickson, William
ESAMDIN, Ali
ESIMBEK, Jarken
Ewing, Martin
EZAWA, Hajime
Facondi, Silvia
Falcke, Heino
Fanaroff, Bernard
Fanti, Roberto
Faulkner, Andrew
Feain, Ilana
Fedotov, Leonid
Feigelson, Eric
Feldman, Paul
Felli, Marcello
Feretti, Luigina
Fernandes, Francisco
Ferrari, Attilio
Fey, Alan
Field, George
Filipovic, Miroslav
Florkowski, David
Foley, Anthony
Fomalont, Edward
Fort, David
Forveille, Thierry
Fouque, Pascal
Frail, Dale
Frater, Robert
Frey, Sandor
Friberg, Per
Fuerst, Ernst
FUKUI, Yasuo
Gabuzda, Denise
Gaensler, Bryan
Gallego, Juan Daniel
Gallimore, Jack
GAO, Yu
Garay, Guido
Garrington, Simon
Gasiprong, Nipon
Gaume, Ralph
Geldzahler, Barry
GENG, Lihong
Genzel, Reinhard
Gerard, Eric
Gergely, Tomas
Gervasi, Massimo
Ghigo, Francis
Ghosh, Tapasi
Gil, Janusz
Gimenez, Alvaro
Gioia, Isabella
Giroletti, Marcello
Gitti, Myriam
Goddi, Ciriaco
Goedhart, Sharmila
Gomez, Gonzalez
Gomez Fernandez, Jose
Gopalswamy, Natchimuthuk
Gordon, Mark
Gorschkov, Aleksandr
Gosachinskij, Igor

Gottesman, Stephen
Gower, Ann
Graham, David
Green, Anne
Green, David
Green, James
Gregorini, Loretta
Gregorio-Hetem, Jane
Gregory, Philip
Grewing, Michael
GU, Xuedong
Gubchenko, Vladimir
Guelin, Michel
Guesten, Rolf
Guilloteau, Stéphane
Gulkis, Samuel
Gull, Stephen
GUO, Jianheng
Gupta, Yashwant
Gurvits, Leonid
Gwinn, Carl
Hall, Peter
HAN, JinLin
Hanasz, Jan
HANDA, Toshihiro
Hanisch, Robert
Hankins, Timothy
Hardee, Philip
Harnett, Julienne
Harris, Daniel
HASEGAWA, Tetsuo
Haverkorn, Marijke
Hayashi, Masahiko
Haynes, Raymond
Haynes, Martha
HE, Jinhua
Heald, George
Heeralall-Issur, Nalini
Heiles, Carl
Helou, George
Henkel, Christian
Herpin, Fabrice
Hewish, Antony
Hibbard, John
Higgs, Lloyd
HIROTA, Tomoya
Hjalmarson, Ake
Ho, Paul
Hoang, Binh
Hobbs, George
Hofner, Peter
Hogbom, Jan
Hogg, David
Hollis, Jan
HONG, Xiaoyu
Hopkins, Andrew
Horiuchi, Shinji
Hotan, Aidan
Howard III, William
Huchtmeier, Walter
Hughes, Philip
Hughes, David
Humphreys, Elizabeth
Hunstead, Richard
Huynh, Minh
HWANG, Chorng-Yuan
Ibrahim, Zainol Abidin
IGUCHI, Satoru
Ikhsanov, Robert
IMAI, Hiroshi
INATANI, Junji
INOUE, Makoto
Ipatov, Aleksandr
Irvine, William
ISHIGURO, Masato
Israel, Frank
Ivanov, Dmitrii
IWATA, Takahiro
Jackson, Carole
Jackson, Neal
Jacq, Thierry
Jaffe, Walter
Janssen, Michael
Jauncey, David
Jewell, Philip
JIN, Zhenyu
Johnston, Kenneth
Johnston, Helen
Jones, Paul
Jones, Dayton
Josselin, Eric
JUNG, Jae-Hoon
Kaftan, May
Kaidanovski, Mikhail
KAIFU, Norio
KAKINUMA, Takakiyo
Kalberla, Peter
Kaltman, Tatyana
KAMEYA, Osamu
Kandalyan, Rafik
Kanekar, Nissim
Kardashev, Nicolay
Kassim, Namir
KASUGA, Takashi
Kaufmann, Pierre
KAWABE, Ryohei
KAWAGUCHI, Kentarou
KAWAMURA, Akiko
Kedziora-Chudczer, Lucyna
Kellermann, Kenneth
Kesteven, Michael
Khaikin, Vladimir
Kijak, Jaroslaw
Kilborn, Virginia
Killeen, Neil
KIM, Tu Whan
KIM, Hyun-Goo
KIM, Kwang tae
KIM, SANG JOON
Kislyakov, Albert
Kitaeff, Vyacheslav
Klein, Ulrich
Klein, Karl
Knudsen, Kirsten
KOBAYASHI, Hideyuki
Kocharovsky, Vitaly
KODA, Jin
KOHNO, Kotaro
KOJIMA, Masayoshi
Kolomiyets, Svitlana
Kondratiev, Vladislav
Konovalenko, Alexander
Kopylova, Yulia
Korzhavin, Anatoly
Kovalev, Yuri
Kovalev, Yuri
KOYAMA, Yasuhiro
Kramer, Michael
Kramer, Busaba
Kreysa, Ernst
Krichbaum, Thomas
Krishna, Gopal
Krishnan, Thiruvenkata
Kronberg, Philipp
Krugel, Endrik
KUAN, Yi-Jehng
Kuijpers, H.
Kuiper, Thomas
Kulkarni, Vasant
Kulkarni, Shrinivas

Kulkarni, Prabhakar
Kumkova, Irina
Kundt, Wolfgang
Kunert-Bajraszewska, M.
Kus, Andrzej
Kutner, Marc
Kwok, Sun
La Franca, Fabio
Lada, Charles
LAI, Shih-Ping
Laing, Robert
Lal, Dharam
Landecker, Thomas
Lang, Kenneth
Langer, William
Langston, Glen
LaRosa, Theodore
Lasenby, Anthony
Lawrence, Charles
Leahy, J.
Lebron, Mayra
LEE, Youngung
LEE, Chang Won
LEE, Yong Bok
Lehnert, Matthew
Lépine, Jacques
Lequeux, James
Lesch, Harald
Lestrade, Jean-François
Li, Hong-Wei
LI, Zhi
LIANG , Shiguang
Likkel, Lauren
Lilley, Edward
LIM, Jeremy
Lindqvist, Michael
Lis, Dariusz
Liseau, René
Lister, Matthew
LIU, Xiang
LIU, Sheng-Yuan
Lo, Fred K. Y.
Lockman, Felix
Loiseau, Nora
Longair, Malcolm
Loukitcheva, Maria
Lovell, James
Lozinskaya, Tatjana
Lubowich, Donald
Luks, Thomas
Lyne, Andrew
Lytvynenko, Leonid
Macchetto, Ferdinando
MacDonald, Geoffrey
MacDonald, James
Machalski, Jerzy
Mack, Karl-Heinz
MacLeod, John
MAEHARA, Hideo
Malofeev, Valery
Manchester, Richard
Mandolesi, Nazzareno
Mantovani, Franco
MAO, Rui-Qing
Maran, Stephen
Marcaide, Juan-Maria
Mardyshkin, Vyacheslav
Marecki, Andrzej
Markoff, Sera
Marscher, Alan
Marti, Josep
Martin, Christopher
Martin, Robert
Martin-Pintado, Jesus
Marvel, Kevin
Masheder, Michael
Maslowski, Jozef
Mason, Paul
Masson, Colin
Matsakis, Demetrios
MATSUO, Hiroshi
MATSUSHITA, Satoki
Matthews, Brenda
Mattila, Kalevi
Matveenko, Leonid
Mauersberger, Rainer
McConnell, David
McCulloch, Peter
McKenna Lawlor, Susan
McMullin, Joseph
Mebold, Ulrich
Meier, David
Menon, T.
Menten, Karl
Meyer, Martin
Mezger, Peter
Michalec, Adam
Mikhailov, Andrey
Miley, George
Miller, Neal
Milne, Douglas
Mirabel, Igor
Miroshnichenko , Alla
Mitchell, Kenneth
Miyawaki, Ryosuke
MIYAZAKI, Atsushi
MIYOSHI, Makoto
MIZUNO, Akira
MIZUNO, Norikazu
Moellenbrock III, George
Moffett, David
Momjian, Emmanuel
MOMOSE, Munetake
Montmerle, Thierry
Morabito, David
Moran, James
Morison, Ian
MORIYAMA, Fumio
Morras, Ricardo
Morris, David
Morris, Mark
Moscadelli, Luca
Muller, Erik
Muller, Sebastien
Mundy, Lee
MURATA, Yasuhiro
Murphy, Tara
Mutel, Robert
Muxlow, Thomas
Myers, Philip
Nadeau, Daniel
Nagnibeda, Valerij
NAKANO, Takenori
NAKASHIMA, Jun-ichi
Neeser, Mark
Nguyen-Quang, Rieu
Nicastro, Luciano
Nice, David
Nicolson, George
Nikolic, Silvana
NISHIO , Masanori
Norris, Raymond
Nuernberger, Dieter
O'Dea, Christopher
O'Sullivan, John
OGAWA, Hideo
Ohashi, Nagayoshi
OHISHI, Masatoshi
Ojha, Roopesh
OKUMURA, Sachiko

Olberg, Michael
ONISHI, Toshikazu
Orchiston, Wayne
Otmianowska-Mazur, Katarzyna
Ott, Juergen
Owen, Frazer
Özel, Mehmet
Padman, Rachael
Palmer, Patrick
Panessa, Francesca
Pankonin, Vernon
Paragi, Zsolt
Paredes Poy, Josep
Parijskij, Yurij
PARK, Yong Sun
Parma, Paola
Paron, Sergio
Parrish, Allan
Pasachoff, Jay
Pashchenko, Mikhail
Patel, Nimesh
Pauls, Thomas
Pearson, Timothy
Peck, Alison
Pedersen, Holger
Pedlar, Alan
PENG, Bo
PENG, Qingyu
Penzias, Arno
Perez, Fournon
Perez-Torres, Miguel
Perley, Richard
Persson, Carina
Peters, William
Petrova, Svetlana
Pettengill, Gordon
Philipp, Sabine
Phillips, Christopher
Phillips, Thomas
Pick, Monique
PING, Jinsong
Pisano, Daniel
Planesas, Pere
Pogrebenko, Sergei
Polatidis, Antonios
Pompei, Emanuela
Ponsonby, John
Pooley, Guy
Porcas, Richard
Porras Juárez, Bertha
Prandoni, Isabella
Preston, Robert
Preuss, Eugen
Price, R.
Puschell, Jeffery
Radford, Simon
Rahimov, Ismail
Ransom, Scott
Rao, A.
Ray, Tom
Ray, Paul
Readhead, Anthony
Redman, Matthew
Reich, Wolfgang
Reid, Mark
Reif, Klaus
Reyes, Francisco
Reynolds, John
Reynolds, Cormac
RHEE , Myung Hyun
Ribo, Marc
Richer, John
Rickard, Lee
Ridgway, Susan
Rioja, Maria
Rizzo, Jose
Roberts, Morton
Roberts, David
Robertson, Douglas
Robertson, James
Roeder, Robert
Roelfsema, Peter
Roennaeng, Bernt
Roeser, Hans-peter
Roger, Robert
Rogers, Alan
Rogstad, David
Romanov, Andrey
Romero, Gustavo
Romney, Jonathan
Rosa Gonzalez, Daniel
Rudnick, Lawrence
Rudnitskij, Georgij
Russell, Jane
Rydbeck, Gustaf
Rys, Stanislaw
Sadler, Elaine
Saikia, Dhruba
SAKAMOTO, Seiichi
Salomé, Philippe
Salter, Christopher
Samodurov, Vladimir
Sandell, Göran
Sanders, David
Sargent, Annelia
Sarma, N.
Sarma, Anuj
Sastry, Ch.
SATO, Fumio
Savolainen, Tuomas
Sawada, Tsuyoshi
SAWADA-SATOH, Satoko
Sawant, Hanumant
Scalise Jr, Eugenio
Schilizzi, Richard
Schilke, Peter
Schlickeiser, Reinhard
Schmidt, Maarten
Schroder, Anja
Schuch, Nelson
Schwarz, Ulrich
Scott, Paul
Seaquist, Ernest
Seielstad, George
SEKIDO, Mamoru
SEKIMOTO, Yutaro
SETA, Masumichi
Setti, Giancarlo
Shaffer, David
Shaposhnikov, Vladimir
Shaver, Peter
SHEN, Zhiqiang
Shepherd, Debra
Shevgaonkar, R.
SHIBATA, Katsunori
Shinnaga, Hiroko
Shmeld, Ivar
Shone, David
Shulga, Valerii
Sieber, Wolfgang
Singal, Ashok
Sinha, Rameshwar
Skillman, Evan
Slade, Martin
Slee, O.
Smith, Francis
Smith, Niall
Smith, Dean
Smolentsev, Sergej

Smol?kov, Gennadij
Snellen, Ignas
Sobolev, Yakov
Sodin, Leonid
SOFUE, Yoshiaki
Somanah, Radhakhrishna
SONG, Qian
Sorochenko, Roman
Spencer, Ralph
Spencer, John
Sramek, Richard
Sridharan, Tirupati
Stairs, Ingrid
Stanghellini, Carlo
Stappers, Benjamin
Steffen, Matthias
Stewart, Paul
Stil, Jeroen
Storey, Michelle
Strom, Richard
Strukov, Igor
Subrahmanya, C.
Subrahmanyan, Ravi
SUGITANI, Koji
Sullivan, III, Woodruff
SUNADA, Kazuyoshi
Swarup, Govind
Swenson Jr, George
Szymczak, Marian
TAKABA, Hiroshi
TAKANO, Shuro
TAKANO, Toshiaki
Tapping, Kenneth
Tarter, Jill
TATEMATSU, Ken'ichi
te Lintel Hekkert, Peter
Terzian, Yervant
Theureau, Gilles
Thomasson, Peter
Thompson, A.
Thum, Clemens
TIAN, Wenwu
Tingay, Steven
Tiplady, Adrian
Tofani, Gianni
Tolbert, Charles
Tornikoski, Merja
TOSAKI, Tomoka
Tovmassian, Hrant
Trigilio, Corrado
Trinidad, Miguel
Tritton, Keith
Troland, Thomas
Trushkin, Sergey
TSUBOI, Masato
Tsutsumi, Takahiro
Tuccari, Gino
Turner, Jean
Turner, Kenneth
Tyul'bashev, Sergei
Tzioumis, Anastasios
Udaya, Shankar
Ulrich, Marie-Helene
Ulvestad, James
Ulyanov , Oleg
Umana, Grazia
UMEMOTO, Tomofumi
Unwin, Stephen
Urama, Johnson
Urosevic, Dejan
Urquhart, James
Uson, Juan
Val'tts, Irina
Vallee, Jacques
Valtaoja, Esko
Valtonen, Mauri
van der Hulst, Jan
van der Kruit, Pieter
van der Laan, Harry
van der Tak, Floris
van Driel, Wim
van Gorkom, Jacqueline
van Langevelde, Huib
van Leeuwen, Joeri
van Woerden, Hugo
VandenBout, Paul
Vats, Hari
Vaughan, Alan
Velusamy, T.
Venturi, Tiziana
Venugopal, V.
Verheijen, Marc
Verkhodanov, Oleg
Vermeulen, Rene
Verschuur, Gerrit
Verter, Frances
Vestergaard, Marianne
Vilas, Faith
Vilas-Boas, José
Vivekanand, M.
Vogel, Stuart
Volvach, Alexander
Voronkov, Maxim
WAJIMA, Kiyoaki
Walker, Robert
Wall, Jasper
Wall, William
Walmsley, C.
Walsh, Wilfred
Walsh, Andrew
WANG, Shouguan
WANG, Na
WANG, Hong-Guang
WANG, Shujuan
WANG, Jingyu
Wannier, Peter
Ward-Thompson, Derek
Wardle, John
Warmels, Rein
Warner, Peter
Watson, Robert
Wehrle, Ann
Wei, Mingzhi
Weigelt, Gerd
Weiler, Kurt
Weiler, Edward
Welch, William
WEN, Linqing
WENLEI, Shan
Westmeier, Tobias
Whiteoak, John
Whiting, Matthew
Wickramasinghe, N.
Wielebinski, Richard
Wiik, Kaj
Wiklind, Tommy
Wild, Wolfgang
Wilkinson, Peter
Willis, Anthony
Wills, Derek
Wills, Beverley
Willson, Robert
Wilner, David
Wilson, Robert
Wilson, Thomas
Wilson, William
Windhorst, Rogier
Winnberg, Anders
Wise, Michael
Witzel, Arno

Wolszczan, Alexander
Woltjer, Lodewijk
Woodsworth, Andrew
Wootten, Henry
Wright, Alan
Wrobel, Joan
WU, Yuefang
WU, Xinji
Wucknitz, Olaf
YANG, Zhigen
YANG, Ji
YAO, Qijun
YE, Shuhua
Yin, Qi-Feng
YONEKURA, Yoshinori
Yusef-Zadeh, Farhad
Zainal Abidin, Zamri
Zaitsev, Valerij
Zanichelli, Alessandra
Zannoni, Mario
Zarka, Philippe
Zavala, Robert
Zensus, J-Anton
ZHANG, Jian
ZHANG, Xizhen
ZHANG, Hongbo
ZHANG, Qizhou
ZHANG, JiangShui
ZHANG, Jingyi
ZHANG, Haiyan
ZHAO, Jun-Hui
Zheleznyak, Alexander
Zheleznyakov, Vladimir
ZHENG, Xinwu
ZHOU, Jianfeng
ZHOU, Jianjun
ZHU, LiChun
ZHU, Wenbai
Zieba, Stanislaw
Zinchenko, Igor
Zlobec, Paolo
Zlotnik, Elena
Zuckerman, Benjamin
Zwaan, Martin
Zylka, Robert

Division XI Space & High Energy Astrophysics

President: Christine Jones
Vice-President: Noah Brosch

Organizing Committee Members:

Matthew G. Baring
Martin Adrian Barstow
João Braga
Eugene M. Churazov
Jean Eilek
Hideyo KUNIEDA
Jayant Murthy
Isabella Pagano
Marco Salvati
Kulinder Pal Singh
Diana Mary Worrall

Members:

Abramowicz, Marek
Acharya, Bannanje
Acton, Loren
Agrawal, P.
Aguiar, Odylio
Ahluwalia, Harjit
Ahmad, Imad
Alexander, Joseph
Allington-Smith, Jeremy
Amati, Lorenzo
Andersen, Bo Nyborg
Antonelli, Lucio Angelo
Apparao, K.
ARAFUNE, Jiro
Arefiev, Vadim
Arnaud, Monique
Arnould, Marcel
Arons, Jonathan
Aschenbach, Bernd
Asvarov, Abdul
Audard, Marc
Audley, Michael
Audouze, Jean
AWAKI, Hisamitsu
AYA , Bamba
Ayres, Thomas
Baan, Willem
Bailyn, Charles
Balikhin, Michael
Baliunas, Sallie
Barret, Didier
Baskill, Darren
Baym, Gordon
Bazzano, Angela
Becker, Werner
Becker, Robert
Beckmann, Volker
Begelman, Mitchell
Beiersdorfer, Peter
Belloni, Tomaso
Bender, Peter
Benedict, George
Benford, Gregory
Bennett, Charles
Bennett, Kevin
Benvenuto, Omar
Bergeron, Jacqueline
Berta, Stefano
Beskin, Gregory
Beskin, Vasily
Bhattacharjee, Pijush
Bhattacharya, Dipankar
Bhattacharyya, Sudip
Bianchi, Luciana
Bicknell, Geoffrey
Biermann, Peter
Bignami, Giovanni
Bingham, Robert
Blandford, Roger
Bleeker, Johan
Bless, Robert
Blinnikov, Sergey
Bloemen, Hans
Blondin, John
Bludman, Sidney
Bocchino, Fabrizio
Boer, Michel
Boggess, Nancy
Boggess, Albert
Bohlin, Ralph
Boksenberg, Alec
Bonazzola, Silvano
Bonnet, Roger
Bonnet-Bidaud, Jean-Marc
Bonometto, Silvio
Borozdin, Konstantin
Bougeret, Jean-Louis
Bowyer, C.
Bradley, Arthur
Branchesi, Marica
Brandt, Soeren
Brandt, William
Brandt, John
Brecher, Kenneth
Breslin, Ann
Brinkman, Bert
Brown, Alexander
Bruhweiler, Frederick
Bruner, Marilyn
Brunetti, Gianfranco
Bumba, Vaclav
Bunner, Alan
Buote, David
Burderi, Luciano
Burenin, Rodion
Burke, Bernard
Burrows, David
Burrows, Adam
Bursa, Michal
Butler, Christopher
Caccianiga, Alessandro
CAI, Michael
Camenzind, Max
Campbell, Murray
CAO, Li

Cappi, Massimo
Caraveo, Patrizia
Cardenas, Rolando
Cardini, Daniela
Carlson, Per
Carpenter, Kenneth
Casandjian, Jean-Marc
Cash Jr, Webster
Cassé, Michel
Castro-Tirado, Alberto
Cavaliere, Alfonso
Celotti, Anna Lisa
Cesarsky, Catherine
Chakrabarti, Sandip
Chakraborty, Deo
CHANG, Hsiang-Kuang
CHANG, Heon-Young
Channok, Chanruangrit
Chapman, Sandra
Chapman, Robert
Charles, Philip
Chartas, George
Chechetkin, Valerij
CHEN, Lin-wen
Chenevez, Jérôme
CHENG, Kwongsang
Cheung, Cynthia
Chian, Abraham
Chiappetti, Lucio
CHIKAWA, Michiyuki
Chitre, Shashikumar
Chochol, Drahomir
CHOE, Gwangson
CHOI, Chul-Sung
CHOU, Yi
Ciotti, Luca
Clark, George
Clark, Thomas
Clay, Roger
Collin, Suzy
Comastri, Andrea
Condon, James
Contopoulos, Ioannis
Corbel, Stéphane
Corbet, Robin
Corbett, Ian
Corcoran, Michael
Cordova, France
Cornelisse, Remon
Costantini, Elisa
Courvoisier, Thierry
Cowie, Lennox
Cowsik, Ramanath
Crannell, Carol
Crocker, Roland
Cropper, Mark
Croton, Darren
Cruise, Adrian
Culhane, John
Cunniffe, John
Curir, Anna
Cusumano, Giancarlo
da Costa, Antonio
da Silveira, Enio
Dadhich, Naresh
DAI, Zigao
Dalla Bonta, Elena
DAmico, Flavio
Darriulat, Pierre
Davidson, William
Davis, Robert
Davis, Michael
Dawson, Bruce
de Felice, Fernando
de Jager, Cornelis
de Martino, Domitilla
Del Santo, Melania
Del Zanna, Luca
Della Ceca, Roberto
Dempsey, Robert
den Herder, Jan-Willem
Dennerl, Konrad
Dennis, Brian
Dermer, Charles
Di Cocco, Guido
Diaz Trigo, Maria
Digel, Seth
Disney, Michael
Dokuchaev, Vyacheslav
Dolan, Joseph
Domingo, Vicente
Dominis Prester, Dijana
Donea, Alina
DONG, Xiao-Bo
DOTANI, Tadayasu
Dovciak, Michal
Downes, Turlough
Drake, Frank
Drury, Luke
Dubus, Guillaume
Duorah, Hira
Dupree, Andrea
Durouchoux, Philippe
Edelson, Rick
Edwards, Paul
Ehle, Matthias
Eichler, David
Elvis, Martin
Elyiv, Andrii
Emanuele, Alessandro
Ensslin, Torsten
ESAMDIN, Ali
ESIMBEK, Jarken
Ettori, Stefano
Eungwanichayapant, Anant
Evans, Daniel
Fabian, Andrew
Fabricant, Daniel
Faraggiana, Rosanna
Fatkhullin, Timur
Fazio, Giovanni
Feldman, Paul
Fender, Robert
Fendt, Christian
Ferrari, Attilio
Field, George
Fisher, Philip
Fishman, Gerald
Florido, Estrella
Foing, Bernard
Fomin, Valery
Fonseca Gonzalez, Maria
Forman, William
Foschini, Luigi
Franceschini, Alberto
Frandsen, Soeren
Frank, Juhan
Fransson, Claes
Fredga, Kerstin
FUJIMOTO, Shin-ichiro
FUJITA, Mitsutaka
Furniss, Ian
Fyfe, Duncan
Gabriel, Alan
Gaensler, Bryan
Gaisser, Thomas
Galeotti, Piero
Galloway, Duncan
GAO, Yu
Garmire, Gordon

Gaskell, C.
Gathier, Roel
Gehrels, Neil
Gendre, Bruce
Georgantopoulos, Ioannis
Gezari, Daniel
Ghia, Piera Luisa
Ghirlanda, Giancarlo
Ghisellini, Gabriele
Giacconi, Riccardo
Gioia, Isabella
Giroletti, Marcello
Gitti, Myriam
Goldsmith, Donald
Goldwurm, Andrea
Gomboc, Andreja
Gomez de Castro, Ana
Gotthelf, Eric
Gotz, Diego
Grebenev, Sergei
Greenhill, John
Gregorio, Anna
Grenier, Isabelle
Grewing, Michael
Greyber, Howard
Griffiths, Richard
Grindlay, Jonathan
Grosso, Nicolas
Gull, Theodore
Gumjudpai, Burin
Gun, Gulnur
Gunn, James
Gutierrez, Carlos
Guziy , Sergiy
Hakkila, Jon
Halevin, Alexandros
Hameury, Jean-Marie
Hannikainen, Diana
Hardcastle, Martin
Harms, Richard
Harris, Daniel
Harvey, Paul
Harwit, Martin
Hasan, Hashima
HATSUKADE, Isamu
Haubold, Hans
Haugboelle, Troels
Hauser, Michael
Hawkes, Robert
Hawking, Stephen
HAYAMA, Kazuhiro
Haymes, Robert
Heger, Alexander
Heise, John
Helfand, David
Hempel, Marc
Henoux, Jean-Claude
Henriksen, Richard
Henry, Richard
Hensberge, Herman
Hicks, Amalia
Hill, Adam
Hoffman, Jeffrey
Holberg, Jay
Holloway, Nigel
Holt, Stephen
Holz, Daniel
Hora, Joseph
Horandel, Jörg
Hornschemeier, Ann
Hornstrup, Allan
Houziaux, Leo
Hoyng, Peter
HSU, Rue-Ron
HUANG, YongFeng
Huang, Jiasheng
Huber, Martin
Hulth, Per
Hurley, Kevin
Hutchings, John
HWANG, Chorng-Yuan
Ibrahim, Alaa
ICHIMARU, Setsuo
Ikhsanov, Nazar
Illarionov, Andrei
Imamura, James
Imhoff, Catherine
in't Zand, Johannes
INOUE, Hajime
INOUE, Makoto
IOKA, Kunihito
Ipser, James
ISHIDA, Manabu
Israel, Werner
ITOH, Masayuki
Jackson, John
Jaffe, Walter
Jakobsson, Pall
Jamar, Claude
Janka, Hans
Jaranowski, Piotr
Jenkins, Edward
Jokipii, Jack
Jones, Thomas
Jonker, Peter
Jordan, Stuart
Jordan, Carole
Joss, Paul
Kafatos, Menas
Kalemci, Emrah
KANEDA, Hidehiro
Kaper, Lex
Kapoor, Ramesh
Karakas, Amanda
Karpov, Sergey
Kaspi, Victoria
Kasturirangan, K.
Katarzynski, Krzysztof
KATO, Tsunehiko
KATO, Yoshiaki
Katsova, Maria
Katz, Jonathan
KAWAI, Nobuyuki
Kellermann, Kenneth
Kellogg, Edwin
Kelly, Brandon
Kembhavi, Ajit
KENJI, Nakamura
Kessler, Martin
Khumlumlert, Thiranee
Killeen, Neil
KIM, Yonggi
Kimble, Randy
KINUGASA, Kenzo
Kirk, John
KIYOSHI, Hayashida
Klinkhamer, Frans
Klose, Sylvio
Knapp, j.knapp
KO, Chung-Ming
Kobayashi, Shiho
Koch-Miramond, Lydie
Kohmura, Takayoshi
KOIDE, Shinji
KOJIMA, Yasufumi
Kokubun, Motohide
Kolb, Edward
KONDO, Yoji
KONDO, Masaaki
Kong, Albert

KOSHIBA, Masatoshi
KOSUGI, George
Koupelis, Theo
Kouveliotou, Chryssa
KOYAMA, Katsuji
Kozma, Cecilia
Kretschmar, Peter
Kreykenbohm, Ingo
Kryvdyk, Volodymyr
Kuiper, Lucien
Kulsrud, Russell
KUMAGAI, Shiomi
Kuncic, Zdenka
Kundt, Wolfgang
Kunz, Martin
Kurt, Vladimir
KUSUNOSE, Masaaki
La Franca, Fabio
Lagache, Guilaine
Lal, Dharam
Lamb, Frederick
Lamb, Susan
Lamb Jr, Donald
Lamers, Henny
Lampton, Michael
Lapington, Jonathan
Lattimer, James
Lea, Susan
Leckrone, David
LEE, Wo-Lung
Lee, William
Leighly, Karen
Lemaire, Philippe
Levin, Yuri
Levine, Robyn
Lewin, Walter
LI, Xiangdong
LI, Tipei
Liang, Edison
LIN, Xuan-bin
Linsky, Jeffrey
Liu, Bifang
Loaring, Nicola
Lochner, James
Long, Knox
Longair, Malcolm
Lovelace, Richard
LU, Tan
LU, Jufu
LU, Fangjun
LU, Ye
Luest, Reimar
Luminet, Jean-Pierre
Luo, Qinghuan
Lutovinov, Alexander
Lynden-Bell, Donald
Lyubarsky, Yury
MA, YuQian
Maccacaro, Tommaso
Maccarone, Thomas
Macchetto, Ferdinando
MACHIDA, Mami
Maggio, Antonio
Mainieri, Vincenzo
Majumdar, Subhabrata
Makarov, Valeri
Malesani, Daniele
Malitson, Harriet
Malkan, Matthew
Manara, Alessandro
Mandolesi, Nazzareno
Mangano, Vanessa
Maran, Stephen
Marar, T.
Maricic, Darije
Markoff, Sera
Marov, Mikhail
Marranghello, Guilherme
Martinez-Bravo, Oscar
Martinis, Mladen
MASAI, Kuniaki
Masnou, Jean-Louis
Mason, Glenn
Mather, John
MATSUMOTO, Hironori
MATSUMOTO, Ryoji
MATSUOKA, Masaru
Matt, Giorgio
Matz, Steven
Mazurek, Thaddeus
McBreen, Brian
McCluskey Jr, George
McCray, Richard
McWhirter, R.
Medina, Jose
Meier, David
Meiksin, Avery
Melatos, Andrew
Melia, Fulvio
Melnick, Gary
Melnyk, Olga
Melrose, Donald
Mendez, Mariano
Mereghetti, Sandro
Merlo, David
Mestel, Leon
Meszaros, Peter
Meyer, Jean-Paul
Meyer, Friedrich
Micela, Giuseppina
Miller, John
Miller, Guy
Miller, Michael
Mineo, Teresa
Miroshnichenko , Alla
Miyaji, Takamitsu
MIYAJI, Shigeki
Miyata, Emi
MIZUMOTO, Yoshihiko
MIZUTANI, Kohei
Moderski, Rafal
Molla, Mercedes
Monet, David
MOON, Shin-Haeng
Moos, Henry
Morgan, Thomas
MORI, Koji
MORI, Masaki
Morton, Donald
Mota, David
Motch, Christian
MOTIZUKI, Yuko
Mourao, Ana Maria
Mulchaey, John
MURAKAMI, Toshio
MURAKAMI, Hiroshi
Murdock, Thomas
Murtagh, Fionn
NAGATAKI, Shigehiro
NAKAYAMA, Kunji
Neff, Susan
Ness, Norman
Neuhaeuser, Ralph
Neupert, Werner
Nichols, Joy
Nicollier, Claude
Nielsen, Krister
Nikolajuk, Marek
NISHIMURA, Osamu
NITTA, Shin-ya

Nityananda, Rajaram
NOMOTO, Ken'ichi
Norci, Laura
Nordh, Lennart
Norman, Colin
Noyes, Robert
Nulsen, Paul
O'Brien, Paul
O'Connell, Robert
O'Sullivan, Denis
OGAWARA, Yoshiaki
Okeke, Pius
OKUDA, Toru
Olthof, Henk
Onken, Christopher
OOHARA, Ken-ichi
Orellana, Mariana
Orford, Keith
Orio, Marina
Orlandini, Mauro
Orlando, Salvatore
Osborne, Julian
Osten, Rachel
Ostriker, Jeremiah
Ostrowski, Michal
Ott, Juergen
Owen, Tobias
OZAKI, Masanobu
Özel, Mehmet
Paciesas, William
Page, Clive
Page, Mathew
PAK, Soojong
Paltani, Stéphane
Palumbo, Giorgio
Pandey, Uma
Panessa, Francesca
Papadakis, Iossif
Paragi, Zsolt
PARK, Myeong Gu
Parker, Eugene
Patten, Brian
Paul, Biswajit
Pavlov, George
Peacock, Anthony
Pearce, Mark
Pearson, Kevin
Pellegrini, Silvia
Pellizza, Leonardo
PENG, Qiuhe
PENG, Qingyu
Perez, Mario
Perola, Giuseppe
Perry, Peter
Peters, Geraldine
Peterson, Laurence
Peterson, Bruce
Pethick, Christopher
Petkaki, Panagiota
Petro, Larry
Petrosian, Vahe
Phillips, Kenneth
Pian, Elena
Pinkau, K.
Pinto, Philip
Pipher, Judith
Piran, Tsvi
Piro, Luigi
Polidan, Ronald
Polletta, Maria del Carmen
Popov, Sergey
Porquet, Delphine
Pottschmidt, Katja
Pounds, Kenneth
Poutanen, Juri
Pozanenko, Alexei
Prasanna, A.
Preuss, Eugen
Produit, Nicolas
Protheroe, Raymond
Prouza, Michael
Prusti, Timo
Raiteri, Claudia
Ramadurai, Souriraja
Ramirez, Jose
Ranalli, Piero
Rao, Arikkala
Rasmussen, Ib
Rasmussen, Jesper
Ray, Paul
Raychaudhury, Somak
Reale, Fabio
Rees, Martin
Reeves, Hubert
Reiprich, Thomas
Rengarajan, Thinniam
Revnivtsev, Mikhail
Rhoads, James
Robba, Natale
Roberts, Timothy
Roman, Nancy
Romano, Patrizia
Roming, Peter
Rosendhal, Jeffrey
Rosner, Robert
Rovero, Adrián
Rubino-Martin, J. A.
Ruder, Hanns
Ruffini, Remo
Ruffolo, David
Ruszkowski, Mateusz
Rutledge, Robert
Sabau-Graziati, Lola
Safi-Harb, Samar
Sagdeev, Roald
Sahlen, Martin
Saiz, Alejandro
Sakano, Masaaki
Sakelliou, Irini
Sanchez, Norma
Sanders III, Wilton
Santos-Lleo, Maria
Saslaw, William
SATO, Katsuhiko
Savage, Blair
Savedoff, Malcolm
Sazonov, Sergey
Sbarufatti, Boris
Scargle, Jeffrey
Schaefer, Gerhard
Schartel, Norbert
Schatten, Kenneth
Schilizzi, Richard
Schmitt, Juergen
Schnopper, Herbert
Schreier, Ethan
Schulz, Norbert
Schwartz, Daniel
Schwartz, Steven
Sciortino, Salvatore
Seielstad, George
Selvelli, Pierluigi
Semerak, Oldrich
SEON, Kwang il
Sequeiros, Juan
Setti, Giancarlo
Severgnini, Paola
Seward, Frederick
Shahbaz, Tariq
Shakhov, Boris

Shakura, Nikolaj
Shaver, Peter
Shaviv, Giora
SHEN, Zhiqiang
SHIBAI, Hiroshi
Shibanov, Yuri
SHIBAZAKI, Noriaki
Shields, Gregory
SHIGEYAMA, Toshikazu
SHIMURA, Toshiya
SHIN, Watanabe
Shukre, C.
Shustov, Boris
Signore, Monique
Sikora, Marek
Silvestro, Giovanni
Simon, Paul
Simon, Vojtech
Sims, Mark
Skilling, John
Skinner, Stephen
Skjaeraasen, Olaf
Smale, Alan
Smith, Linda
Smith, Bradford
Smith, Peter
Smith, Nigel
Snow, Theodore
Sofia, Sabatino
Sokolov, Vladimir
Somasundaram, Seetha
SONG, Qian
Sonneborn, George
Sood, Ravi
Spallicci di Filottrano, A.
Sreekumar, Parameswaran
Srinivasan, Ganesan
Srivastava, Dhruwa
Staubert, Rüdiger
Stecher, Theodore
Stecker, Floyd
Steigman, Gary
Steiner, Joao
Stencel, Robert
Stephens, S.
Stern, Robert
Stevens, Ian
Stier, Mark
Still, Martin
Stockman Jr, Hervey
Stoehr, Felix
Straumann, Norbert
Stringfellow, Guy
Strohmayer, Tod
Strong, Ian
Struminsky, Alexei
Stuchlik, Zdenek
Sturrock, Peter
SU, Cheng-yue
Subr, Ladislav
Suleimanov, Valery
SUN, Wei-Hsin
Sunyaev, Rashid
SUZUKI, Hideyuki
Swank, Jean
Tagliaferri, Gianpiero
TAKAHARA, Fumio
TAKAHASHI, Masaaki
TAKAHASHI, Tadayuki
TAKEI, Yoh
Tanaka, Yasuo
TASHIRO, Makoto
TATEHIRO, Mihara
Tavecchio, Fabrizio
TERASHIMA , Yuichi
Terrell, James
Thorne, Kip
Thronson Jr, Harley
TIAN, Wenwu
TOMIMATSU, Akira
Torres, Carlos Alberto
Torres, Diego
Tovmassian, Hrant
Traub, Wesley
Trimble, Virginia
Truemper, Joachim
TruranJr, James
Trussoni, Edoardo
TSUGUYA, Naito
Tsujimoto, Masahiro
TSUNEMI, Hiroshi
TSURU, Takeshi
Tsuruta, Sachiko
Tsygan, Anatolij
Tuerler, Marc
Tylka, Allan
UEDA, Yoshihiro
Ulyanov, Oleg
Uslenghi, Michela
Usov, Vladimir
Vahia, Mayank
Valtonen, Mauri
van den Heuvel, Edward
van der Hucht, Karel
van der Walt, Diederick
van Duinen, R.
van Putten, Maurice
van Riper, Kenneth
Vaughan, Simon
Vercellone, Stefano
Vestergaard, Marianne
Vial, Jean-Claude
Vidal-Madjar, Alfred
Vignali, Cristian
Vikhlinin, Alexey
Vilhu, Osmi
Villata, Massimo
Vink, Jacco
Viollier, Raoul
Viotti, Roberto
Voelk, Heinrich
Volonteri, Marta
Vrtilek, Saeqa
Wagner, Alexander
Walker, Helen
Wanas, Mamdouh
WANG, Shouguan
WANG, jiancheng
WANG, Ding-Xiong
WANG, Yi-ming
WANG, Zhenru
WANG, Hong-Guang
WANG, Feilu
WANG, Shujuan
Watanabe, Ken
WATARAI, Kenya
Watts, Anna
Waxman, Eli
Weaver, Kimberly
Weaver, Thomas
Webster, Adrian
Wehrle, Ann
WEI, Daming
Weiler, Kurt
Weiler, Edward
Weinberg, Jerry
Weisskopf, Martin
Wells, Donald
WEN, Linqing
Wesselius, Paul

Wheatley, Peter
Wheeler, J.
Whitcomb, Stanley
White, Nicholas
Wijers, Ralph
Wijnands, Rudy
Will, Clifford
Willis, Allan
Willner, Steven
Wilms, Jörn
Winkler, Christoph
Wise, Michael
Wolfendale FRS, Sir Arnold
Wolstencroft, Ramon
Wolter, Anna
Woltjer, Lodewijk
WU, Shaoping
WU, Jiun-Huei
Wunner, Guenter
XU, Renxin
XU, Dawei
Yadav, Jagdish
Yakut, Kadri
Yamada, Shoichi
Yamasaki, Tatsuya
YAMASAKI, Noriko
YAMASHITA, Kojun
YAMAUCHI, Makoto
YAMAUCHI, Shigeo
Yock, Philip
YOICHIRO, Suzuki
YOSHIDA, Atsumasa
YU, Wang
YU, Wenfei
YUAN, Ye-fei
YUAN, Weimin
YUAN, Feng
Zacchei, Andrea
Zamorani, Giovanni
Zane, Silvia
Zannoni, Mario
Zarnecki, John
Zdziarski, Andrzej
Zezas, Andreas
ZHANG, Jialu
ZHANG, William
ZHANG, Shuang Nan
ZHANG, Li
ZHANG, JiangShui
ZHANG, Jingyi
ZHANG, You-Hong
ZHANG, Yanxia
ZHENG, Xiaoping
ZHENG, Wei
ZHOU, Jianfeng
Zombeck, Martin

Division XII Union-Wide Activities

President: Françoise Genova
Vice-President: Raymond P. Norris

Organizing Committee Members:

Dennis Crabtree
Olga B. Dluzhnevskaya
Masatoshi OHISHI
Rosa M. Ros
Clive L.N. Ruggles
Nikolay N. Samus
Xiaochun Sun
Virginia Trimble
Wim van Driel
Glenn Michael Wahlgren

Members:

A'Hearn, Michael
Abalakin, Viktor
Abt, Helmut
Accomazzi, Alberto
Acker, Agnès
Adelman, Saul
Afram, Nadine
Aggarwal, Kanti
Aguilar, Maria
AHN, Youngsook
Aizenman, Morris
Aksnes, Kaare
Al-Naimiy, Hamid
Albanese, Lara
Alexandrov, Yuri
Allard, Nicole
Allard, France
Allen Jr, John
Allende Prieto, Carlos
Alsabti, Abdul Athem
Alvarez, Rodrigo
Alvarez, Pedro
Alvarez del Castillo, E.
Alvarez-Pomares, Oscar
Anandaram, Mandayam
Andernach, Heinz
Andrews, Frank
Ansari, S.M.
Antonelli, Lucio Angelo
Apai, Daniel
Apostolovska, Gordana
Arcidiacono, Carmelo
Ardeberg, Arne
Arellano Ferro, Armando
Argo, Megan
Arion, Douglas
Aslan, Zeki
Aubier, Monique
Baan, Willem
Babul, Arif
Baddiley, Christopher
Badescu, Octavian
Badolati, Ennio
BAEK, Chang Hyun
Bailey, Katherine
Bajaja, Esteban
Balanca, Christian
Balbi, Amedeo
Bamford, Steven
Banhatti, Dilip
Baransky, Olexander
Barber, Robert
Barclay, Charles
BARET, Bruny
Barklem, Paul
Barlow, Nadine
Barnbaum, Cecilia
Barthel, Peter
Baskill, Darren
Batten, Alan
Bautista, Manuel
Bayet, Estelle
Bazzano, Angela
Beckmann, Volker
Beiersdorfer, Peter
Belmonte Aviles, J. A.
Bely-Dubau, Francoise
Benacchio, Leopoldo
Benkhaldoun, Zouhair
Benn, Chris
Bennett, Jim
Bensammar, Slimane
Berendzen, Richard
Bernabeu, Guillermo
Berthier, Jerôme
Bessell, Michael
Bhardwaj, Anil
Biemont, Emile
Birlan, Mirel
Bishop, Roy
Black, John
Blackwell-Whitehead, Richard
Blanco, Carlo
Bobrowsky, Matthew
Boechat-Roberty, Heloisa
Bojurova, Eva
Bommier, Véronique
Bond, Ian
Bonoli, Fabrizio
Boonstra, Albert
Booth, Roy
Borde, Suzanne
Borysow, Aleksandra
Botez, Elvira
Bouchard, Antoine
Bowen, David
Boyce, Peter
Branscomb, L.
Briand, Carine
Brieva, Eduardo
Brinchmann, Jarle
Bromage, Gordon
Brooks, Randall
Brosch, Noah

Brosche, Peter
Brouw, Willem
Brown, Robert
Buchlin, Eric
Budding, Edwin
Burman, Ronald
CAI, Michael
CAI, Kai
Calabretta, Mark
Calvet, Nuria
Cannon, Wayne
Capaccioli, Massimo
Carbon, Duane
Caretta, Cesar
Carraminana, Alberto
Carrasco, Bertha
Carter, Brian
Casasola, Viviana
Cattaneo , Andrea
Cayrel, Roger
Celebre, Cynthia
Chamcham, Khalil
Chance, Kelly
CHANG, Hsiang-Kuang
CHEN, Huei-Ru
CHEN, An-Le
CHEN, Alfred
CHEN, Lin-wen
CHEN, Guoming
Cheung, Cynthia
Chiappetti, Lucio
CHIN, Yi-nan
Chinnici, Ileana
Christlieb , Norbert
CHU, Yaoquan
Cichowolski, Silvina
Ciroi, Stefano
Clarke, David
Clegg, Andrew
Clifton, Gloria
Coffey, Deirdre
Colafrancesco, Sergio
Colas, François
Corbally, Christopher
Corbin, Brenda
Cornille, Marguerite
Costero, Rafael
Cottrell, Peter
Couper, Heather
Couto da Silva, Telma
Covone, Giovanni
Coyne, S.J, George
Crawford, David
Crézé, Michel
Cuesta Crespo, Luis
CUI, Shizhu
CUI, Zhenhua
CUI, Chenzhou
Cunniffe, John
Cunningham, Maria
d'Hendecourt, Louis
Dadic, Zarko
DAISUKE, Iono
Dall'Ora, Massimo
Dalla, Silvia
Damineli Neto, Augusto
Danezis, Emmanuel
Daniel, Jean-Yves
Danner, Rolf
Darhmaoui, Hassane
Darriulat, Pierre
Davis, Robert
Davis, Donald
Davis, A. E. L.
de Frees, Douglas
de Greve, Jean-Pierre
de Grijs, Richard
de Jong, Teije
de Kertanguy, Amaury
de Lange, Gert
Débarbat, Suzanne
Del Santo, Melania
Delsanti, Audrey
Demircan, Osman
Devaney, Martin
DeVorkin, David
Dick, Wolfgang
Dick, Steven
Dickel, Helene
Diego, Francisco
Diercksen, Geerd
Dimitrijevic, Milan
Dobrzycki, Adam
Dominguez, Mariano
Donahue, Megan
Doran, Rosa
Dorokhova, Tetyana
Dubau, Jacques
Dubois, Pascal
Ducati, Jorge
Duffard, Rene
Dukes Jr., Robert
Dulieu, Francois
Durand, Daniel
Duval, Marie-France
Dworetsky, Michael
Eastwood, Kathleen
Edwards, Paul
Egret, Daniel
Ehgamberdiev, Shuhrat
Ehle, Matthias
Eidelsberg, Michele
Elia, Davide
Elyiv, Andrii
Esenoglu, Hasan
Esteban, César
Eze, Romanus
Falceta-Goncalves, Diego
Feain, Ilana
Feautrier, Nicole
Federici, Luciana
Fernandez, Julio
Fernandez-Figueroa, M.
Fernie, J.
Field, J. V.
Fienberg, Richard
Fierro, Julieta
Filacchione, Gianrico
Filippenko, Alexei
Fillion, Jean-Hugues
Fink, Uwe
Firneis, Maria
Fleck, Robert
Flin, Piotr
Florides, Petros
Flower, David
Fluke, Christopher
Fluri, Dominique
Forbes, Douglas
Fraser, Helen
Freeman, Kenneth
Freitas Mourao, Ronaldo
FU, Hsieh-Hai
Fuhr, Jeffrey
Fyfe, Duncan
Gabriel, Alan
Gabriel, Carlos
Gallagher III, John
Gallino, Roberto
Gangui, Alejandro

Ganguly, Rajib
Garcia, Beatriz
Garcia-Lorenzo, Maria
Gargaud, Muriel
Gasiprong, Nipon
George, Martin
Gerbaldi, Michele
Gergely, Tomas
Gill, Peter
Gills, Martins
Gimenez, Alvaro
Gingerich, Owen
Glagolevskij, Yurij
Glass, Ian
Glinski, Robert
Glover , Simon
Goddi, Ciriaco
Goldbach, Claudine
Gomez, Monique
Govender, Kevindran
Grant, Ian
Gray, Richard
Green, Anne
Green, Daniel
Green, Richard
Green, David
Gregorio, Anna
Gregorio-Hetem, Jane
Greisen, Eric
Grevesse, Nicolas
Griffin, R. Elizabeth
Griffin, Roger
Grindlay, Jonathan
Grosbol, Preben
Grothkopf, Uta
Guibert, Jean
Guinan, Edward
Gumjudpai, Burin
Gunthardt, Guillermo
GUO, Hongfang
Gurshtein, Alexander
Haenel, Andreus
HAN, Wonyong
Hanisch, Robert
Haque, Shirin
Hartman, Henrik
HASEGAWA, Ichiro
Hau, George
Haubold, Hans
Hauck, Bernard
Haupt, Hermann
Haverkorn, Marijke
Havlen, Robert
Hayli, Abraham
Haynes, Roslynn
Haynes, Raymond
Hearnshaw, John
Hefele, Herbert
Helmer, Leif
Helou, George
Hemenway, Mary
Hempel, Marc
Herrmann, Dieter
Hesser, James
Heudier, Jean-Louis
Hicks, Amalia
Hidayat, Bambang
Hillier, John
HIRAI, Masanori
Hoang, Binh
Hobbs, George
Hockey, Thomas
Hollow, Robert
Holmberg, Gustav
Homeier, Derek
Hopkins, Andrew
Horacek, Jiri
Horandel, Jörg
Hotan, Aidan
Houziaux, Leo
HSU, Rue-Ron
Huan, Nguyen
HUANG, Tianyi
Huber, Martin
Hudkova, Ludmila
Huebner, Walter
Huettemeister, Susanne
Hughes, Stephen
HWANG, Chorng-Yuan
Hysom, Edmund
Hyung, Siek
Ibrahim, Alaa
Ignjatovic, Ljubinko
Iliev, Ilian
Ilyasov, Sabit
Impey, Christopher
Inglis, Michael
Irwin, Alan
Irwin, Patrick
ISHIZAKA, Chiharu
Jafelice, Luiz
Jahnke, Knud
Jamar, Claude
Jauncey, David
Jenkner, Helmut
JEON, Young Beom
JEONG, Jang-Hae
JIANG, Xiaoyuan
Jimenez-Vicente, Jorge
Johnson, Fred
Johnston, Helen
Jordan, Carole
Jorgensen, Uffe
Jorissen, Alain
Kablak, Nataliya
Kalberla, Peter
Kalemci, Emrah
Kamp, Inga
Kanekar, Nissim
Kaplan, George
Karetnikov, Valentin
Karttunen, Hannu
Kastel, Galina
Kay, Laura
Keller, Hans-Ulrich
Kelly, Brandon
Kembhavi, Ajit
Kennedy, Eugene
Kerber, Florian
Khan, J
Kiasatpour, Ahmad
Kielkopf, John
KIM, Yonggi
KIM, Chun-Hwey
KIM, Sang Joon
KIM, Young-Soo
KIM, Yoo Jea
Kingston, Arthur
Kipper, Tonu
Kirby, Kate
Klinglesmith III, Daniel
Kochhar, Rajesh
Koechlin, Laurent
Kohl, John
Kolenberg, Katrien
Kolka, Indrek
Kollerstrom, Nicholas
Kolomanski, Sylwester
Kolomiyets, Svitlana
Komonjinda, Siramas

KONG, Xu
Kontizas, Evangelos
Kontizas, Mary
Kovaleva, Dana
Kovalevsky, Jean
KOZAI, Yoshihide
Krajnovic, Davor
Kramer, Busaba
Kramida, Alexander
Kreiner, Jerzy
Krisciunas, Kevin
Krishna, Gopal
Kroto, Harold
Krupp, Edwin
KUAN, Yi-Jehng
Kuin, Paul
Kunz, Martin
Kupka, Friedrich
Kurucz, Robert
Lago, Maria
Lai, Sebastiana
Lambert, David
Lanciano, Nicoletta
Lanfranchi, Gustavo
Lang, Kenneth
Langhoff, Stephanie
Launay, Françoise
Launay, Jean-Michel
Layzer, David
Le Bourlot, Jacques
Le Floch, André
Le Guet Tully, Françoise
Leach, Sydney
LEE, Eun Hee
LEE, Kang Hwan
LEE, Woo baik
LEE, Yong Sam
LEE, Yong Bok
Léger, Alain
Leibowitz, Elia
Lemaire, Jean-louis
Lequeux, James
Lerner, Michel-Pierre
Lesteven, Soizick
Leung, Kam
Levy, Eugene
Lewis, Brian
LI, Yong
LIN, Weipeng
Linde, Peter
Linden-Vørnle, Michael
Linnartz, Harold
Little-Marenin, Irene
LIU, Ciyuan
LIU, Sheng-Yuan
Loaring, Nicola
Locher, Kurt
Lomb, Nicholas
Longo, Giuseppe
Lonsdale, Carol
Lopes, Rosaly
Lopez-Sanchez, Angel
Lovas, Francis
LU, Fangjun
Luck, John
Lutz, Barry
Maciel, Walter
Maddison, Ronald
Madore, Barry
Madsen, Claus
Mahoney, Terence
Maillard, Jean-Pierre
Majumdar, Subhabrata
Malasan, Hakim
Malin, David
Malkov, Oleg
Mallamaci, Claudio
Mann, Robert
Marchi, Simone
Marco, Olivier
Maricic, Darije
Markkanen, Tapio
Martinet, Louis
Martinez, Peter
Martinez Delgado, David
Martinez-Bravo, Oscar
Mashonkina, Lyudmila
Mason, Helen
Massey, Robert
Mattig, W.
Matz, Steven
Maza, José
McKenna Lawlor, Susan
McKinnon, David
McLean, Brian
McNally, Derek
McWhirter, R.
Meadows, A.
Meidav, Meir
Mein, Pierre
Mendillo, Michael
Mendoza-Torress, J.-E.
Menzies, John
Merin Martin, Bruno
Merlo, David
Metaxa, Margarita
Michel, Laurent
Mickelson, Michael
Mihajlov, Anatolij
Minier, Vincent
Mink, Jessica
Mitton, Simon
Mitton, Jacqueline
MIZUNO, Takao
Moreels, Guy
Morrell, Nidia
Morris, Rhys
Morton, Donald
Mourao, Ana Maria
Mueller , Andreas
Mujica, Raul
Mumma, Michael
Murdin, Paul
Murphy, Tara
Murphy, John
Murtagh, Fionn
Nadal, Robert
Nahar, Sultana
Najid, Nour-Eddine
NAKAJIMA, Koichi
NAKAMURA, Tsuko
NAKANO, Syuichi
Narlikar, Jayant
Nave, Gillian
Navone, Hugo
Nayar, S.R.Prabhakaran
Nefedyev, Yury
Nelson, Burt
Newsom, Gerald
Nguyen-Quang, Rieu
Nicolaidis, Efthymios
Nicolson, Iain
Nielsen, Krister
Niemczura, Ewa
Nilsson, Hampus
Ninkovic, Slobodan
NISHIMURA, Shiro
Noels, Arlette
Nollez, Gerard
Nordström, Birgitta

Norton, Andrew
Nussbaumer, Harry
O'Brian, Thomas
Ochsenbein, François
Ocvirk, Pierre
Odman, Carolina
OH, Kyu-Dong
Ohashi, Nagayoshi
Oja, Heikki
Oka, Takeshi
Okeke, Pius
Olivier, Enrico
Olsen, Hans
Omont, Alain
Oproiu, Tiberiu
Orchiston, Wayne
Orton, Glenn
Osborn, Wayne
Osorio, José
Oswalt, Terry
Owen, Frazer
OZEKI, Hiroyuki
Özel, Mehmet
Pakhomov, Yury
Palmeri, Patrick
Pamyatnykh, Alexey
Pandey, Uma
Pankonin, Vernon
Pantoja, Carmen
Papathanasoglou, Dimitrios
Paragi, Zsolt
PARK, Yong Sun
Paron, Sergio
Pasachoff, Jay
Pasian, Fabio
Paturel, Georges
Pavani, Daniela
Peach, Gillian
Pecker, Jean-Claude
Pence, William
Penston, Margaret
Percy, John
Perez-Torres, Miguel
Perrin, Marshall
Peterson, Charles
Petrini, Daniel
Pettersen, Bjørn
Pettini, Marco
Philip, A.G.
Phillips, Mark
Piacentini, Ruben
Picazzio, Enos
Pigatto, Luisa
Piskunov, Anatolij
Pizzichini, Graziella
Politi, Romolo
Polyakhova, Elena
Pompea, Stephen
Popov, Sergey
Porras Juárez, Bertha
Pound, Marc
Pozhalova, Zhanna
Pradhan, Anil
Proverbio, Edoardo
Pucillo, Mauro
Quinet, Pascal
Radeva, Veselka
Rafferty, Theodore
Ralchenko, Yuri
Ramadurai, Souriraja
Ramirez, Jose
Rastogi, Shantanu
Ratnatunga, Kavan
Ravindranath, Swara
Reardon, Kevin
Reboul, Henri
Redman, Matthew
Renson, P.
Reynolds, Cormac
Rijsdijk, Case
Roberts, Morton
Roberts, Douglas
Roca Cortes, Teodoro
Roemer, Elizabeth
Rogers, Forrest
Roman, Nancy
Rosa Gonzalez, Daniel
Rosenzweig-Levy, Patrica
Rostas, François
Rots, Arnold
Roueff, Evelyne
Routly, Paul
Ruder, Hanns
Russo, Guido
Russo, Pedro
Ryabchikova, Tatiana
Sabra, Bassem
Safko, John
Sahal-Brechot, Sylvie
Salama, Farid
Samodurov, Vladimir
Sampson, Russell
Sanahuja Parera, Blai
Sanchez, Francisco
Sandqvist, Aage
Sandrelli, Stefano
Santos-Lleo, Maria
Saraiva, Maria de Fatima
Sarasso, Maria
Sarre, Peter
Sattarov, Isroil
Savanov, Igor
Savin, Daniel
Schade, David
Schaefer, Bradley
Schilbach, Elena
Schilizzi, Richard
Schleicher, David
Schmadel, Lutz
Schmitz, Marion
Schneider, Jean
Schrijver, Johannes
Schroder, Anja
Schroeder, Daniel
Schultz, David
Seaman, Rob
Seeds, Michael
SEKIGUCHI, Kazuhiro
Shank, Michael
Shankland, Paul
Sharp, Nigel
Shaw, Richard
Shetrone, Matthew
SHI, Jianrong
Shingareva, Kira
Shipman, Harry
Shore, Bruce
Siebenmorgen, Ralf
Signore, Monique
SIHER, El Arbi
Sima, Zdislav
Sinha, Krishnanand
Slater, Timothy
Smail, Ian
Smith, Francis
Smith, Peter
Smith, William
Smith, Robert
Sobouti, Yousef
Solc, Martin

Solheim, Jan
Somerville, William
SONG , In Ok
SONG, Liming
Soonthornthum, B.
Soriano, Bernardo
Spielfiedel, Annie
Spite, François
Stam, Daphne
Stancil, Philip
Stark, Glenn
Stathopoulou, Maria
Steele, John
Stefl, Vladimir
Stehle, Chantal
Steinle, Helmut
Stencel, Robert
Stenholm, Björn
Stephenson, F.
Sterken, Christiaan
Stoev, Alexey
Storey, Michelle
Strachan, Leonard
Straizys, Vytautas
Strelnitski, Vladimir
SU, Cheng-yue
Sullivan, III, Woodruff
Summers, Hugh
Sun, Xiaochun
Suntzeff, Nicholas
Sutherland, Ralph
Svestka, Jiri
Swarup, Govind
Swerdlow, Noel
Swings, Jean-Pierre
TAKAYANAGI, Kazuo
Tatum, Jeremy
Taub, Liba
Tayal, Swaraj
Tchang-Brillet, Lydia
Tedds, Jonathan
Tennyson, Jonathan
Teuben, Peter
Theodossiou, Efstratios
Tholen, David
Ticha, Jana
Tignalli, Horacio
Tobin, William
Tody, Douglas
Tolbert, Charles
Tomasella, Lina
Torres, Jesus Rodrigo
Torres-Peimbert, Silvia
Touma, Jihad
Tozzi, Gian
Tremko, Jozef
Trinidad, Miguel
Tsvetkov, Milcho
Tugay, Anatoliy
Turner, Kenneth
Tzioumis, Anastasios
Ugolnikov, Oleg
Ulyanov, Oleg
Upgren, Arthur
Urama, Johnson
Valdes Parra, Jose
Valeev, Sultan
van den Bergh, Sidney
van den Heuvel, Edward
van Dishoeck, Ewine
van Driel, Wim
van Gent, Robert
van Rensbergen, Walter
van Santvoort, Jacques
Varshalovich, Dmitrij
Vass, Gheorghe
Vauclair, Sylvie
Vavilova, Iryna
Verdoes Kleijn, Gijsbert
Verdun, Andreas
Vernin, Jean
Videira, Antonio
Vilinga, Jaime
Vilks, Ilgonis
Villar Martin, Montserrat
Vinuales Gavin, Ederlinda
Voelk, Heinrich
Voelzke, Marcos
Volyanska, Margaryta
Vujnovic, Vladis
Walker, Merle
Walker, Constance
Wallace, Patrick
Walsh, Wilfred
WANG, Shouguan
WANG, Junxian
WANG, Jian-Min
WANG, Xunhao
WANG, Feilu
Ward, Richard
Warner, Brian
Warren Jr, Wayne
Weilbacher, Peter
Wells, Donald
Wenger, Marc
West, Richard
Whelan, Emma
White, Graeme
White II, James
Whitelock, Patricia
Whiteoak, John
Whiting, Matthew
Wielen, Roland
Wiese, Wolfgang
Wilkins, George
Williamon, Richard
Williams, Gareth
Williams, Thomas
Willmore, A.
Wilson, Curtis
Wise, Michael
Wolfschmidt, Gudrun
Woolf, Neville
Wright, Alan
WU, Jiun-Huei
Wunner, Guenter
XIE, Xianchun
XIONG, Jianning
YAMAOKA, Hitoshi
YANG, Hong-Jin
YANG, Changgen
YANG, Xiaohu
YANO, Hajime
Yau, Kevin
YE, Shuhua
Yeomans, Donald
YIM, Hong Suh
Yoshino, Kouichi
YUMIKO, Oasa
Zacchei, Andrea
Zakirov, Mamnum
Zanini, Valeria
Zealey, William
Zeilik, Michael
Zeippen, Claude
ZENG, Qin
ZHANG, You-Hong
ZHANG, Haiyan
ZHANG, Yanxia
ZHAO, Gang

ZHAO, Yongheng
ZHAO, Jun Liang
ZHOU, Yonghong
ZHOU, Xu
ZHOU, Jianfeng
ZHU, Jin
Zsoldos, Endre

Transactions IAU, Volume XXVIIIB
Proc. XXVIII IAU General Assembly, August 2012
Thierry Montmerle, ed.

doi:10.1017/S1743921315005694

CHAPTER XI

COMMISSIONS MEMBERSHIP
(until 31 August 2012)

NOTE. This chapter gives the membership of the Commissions (listed by Commission number), as they were affiliated to the Divisions in force until the end of the XVIIIth General Assembly. As a result of the adoption of Resolution B4 by this Assembly, a new Divisional structure was established (see Chapters II and IV of these *Transactions*), to take effect on 1 September 2012. This new structure will affect the way Commissions are affiliated to Divisions in the next triennium (2012–2015).

Division I Commission 4 Ephemerides

President: George H. Kaplan
Vice-President: Catherine Y. Hohenkerk

Organizing Committee Members:

Arlot, Jean-Eudes
Bangert, John
Bell, Steven
Folkner, William
Kaplan, George
Lara, Martin
Pitjeva, Elena
Urban, Sean

Members:

Abalakin, Viktor
Acton, Charles
AHN, Youngsook
ARAKIDA, Hideyoshi
Bartlett, Jennifer
Brozovic, Marina
Brumberg, Victor
Bueno de Camargo, Julio
Capitaine, Nicole
Chapront, Jean
Chapront-Touze, Michelle
Cooper, Nicholas
Dickey, Jean
Dunham, David
Eroshkin, Georgij
Espenak, Fred
Fominov, Aleksandr
FU, Yanning
Giorgini, Jon
Glebova, Nina
Harper, David
Hilton, James
HOU, Xiyun
Howard, Sethanne
Husarik, Marek
Iorio, Lorenzo
Ivantsov, Anatoliy
Johnston, Kenneth
Jubier, Xavier
Kaplan, George
KINOSHITA , Hiroshi
Klepczynski, William
Kolaczek, Barbara
Lara, Martin
Laskar, Jacques
Lehmann, Marek
Lenhardt, Helmut
Lieske, Jay
Lopez Moratalla, Teodoro
Madsen, Claus
Majid, Abdul
Mallamaci, Claudio
MASAKI, Yoshimitsu
Morrison, Leslie
Mueller, Ivan
Newhall, X.
Noyelles, Benoît
O'Handley, Douglas

Ofek, Eran
Olivier, Enrico
Page, Gary
Pavlyuchenkov, Yaroslav
Reasenberg, Robert
Rodin, Alexander
Romero Perez, Maria
Rossello, Gaspar
Seidelmann, P.
Shapiro, Irwin
Shiryaev, Alexander
SHU, Fengchun
Shuygina, Nadia
Simon, Jean-Louis
Skripnichenko, Vladimir
Standish, E.
Stewart, Susan
Suli, Áron
Vilinga, Jaime
Vondrak, Jan
Wallace, Patrick
WANG, Xiao-bin
Weratschnig, Julia
Wielen, Roland
Wilkins, George
Williams, Carol
Williams, James
Winkler, Gernot
Wytrzyszczak, Iwona
XIE, Yi
Yallop, Bernard

Division XII Commission 5 Documentation & Astronomical Data

President: Masatoshi OHISHI
Vice-President: Robert J. Hanisch

Organizing Committee Members:

Andernach, Heinz
Bishop, Marsha
Griffin, R. Elizabeth
Kembhavi, Ajit
Murphy, Tara
OHISHI , Masatoshi
Pasian, Fabio

Members:

A'Hearn, Michael
Abalakin, Viktor
Abt, Helmut
Accomazzi, Alberto
Acharya, Bannanje
Adelman, Saul
Aerts, Conny
Agueros, Marcel
Aizenman, Morris
AK, Serap
Alexander, Paul
Allan, Alasdair
Allen, Lori
Alvarez, Pedro
Andernach, Heinz
Andreon, Stefano
Antonelli, Lucio Angelo
Anupama, G.
Arenou, Frédéric
Argyle, Robert
Arlot, Jean-Eudes
Armstrong, John
Aspin, Colin
Baffa, Carlo
Bagla, Jasjeet
Banhatti, Dilip
Barbieri, Cesare
Barbuy, Beatriz
Bartczak, Przemyslaw
Beasley, Anthony
Beckmann, Volker
Bell Burnell, Jocelyn
Benacchio, Leopoldo
Benetti, Stefano
Benn, Chris
Bentley, Robert
Bersier, David
Berthier, Jerôme
Bertout, Claude
Bessell, Michael
Bhat, Ramesh
Bignall, Hayley
Bishop, Marsha
Bolatto, Alberto
Bond, Howard
Bond, Ian
Borde, Suzanne
Borisova, Ana
Borne, Kirk
Bosken, Sarah
Bouton, Ellen
Boyce, Peter
Brammer, Gabriel
Brandt, William
Brescia, Massimo
Brinchmann, Jarle
Brosch, Noah
Brouw, Willem
Brown, Michael
Brunner, Robert
Bucciarelli, Beatrice
Calabretta, Mark
Campana, Riccardo
Cappellaro, Enrico
Caretta, Cesar
Catelan, Márcio
Cenko, Stephen
CHANG, Hsiang-Kuang
CHANG, Hong
CHEN, Xuelei
Cheung, Cynthia
Chiappetti, Lucio
Christlieb, Norbert
CHU, Yaoquan
Ciardi, David
Clayton, Geoffrey
Conti, Alberto
Corbally, Christopher
Corbin, Brenda
Cordes, James
Creze, Michel
Cristiani, Stefano
Cristobal, David
Csabai, Istvan
CUI, Chenzhou
Cunniffe, John
Dalla, Silvia
Davis, Robert
de Carvalho, Reinaldo
De Cuyper, Jean-Pierre
De Rossi, María
Depagne, Éric
Derriere, Sebastien
DeVorkin, David
Dickel, Helene
Dimitrijevic, Milan
Djorgovski, Stanislav
Dluzhnevskaya, Olga
Dobrzycki, Adam
Downes Wallace, Juan
Drimmel, Ronald
Dubois, Pascal
Ducati, Jorge
Ducourant, Christine
Durand, Daniel

Ederoclite, Alessandro
Ekstrom Garcia N., Sylvia
Elia, Davide
Ellingsen, Simon
Elyiv, Andrii
FAN, Yufeng
Feigelson, Eric
Ferrari, Fabricio
Folgueira, Marta
Forveille, Thierry
Foucaud, Sébastien
Fox-Machado, Lester
Fraix-Burnet, Didier
FUKUSHIMA, Toshio
Fyfe, Duncan
Gabriel, Carlos
Gallagher, Sarah
Gallagher III, John
Gastaldello, Fabio
Gehrels, Neil
Genova, Françoise
Gezari, Suvi
Golev, Valeri
Gomez, Monique
Goodman, Alyssa
Graham, Eric
Green, Daniel
Green, David
Gregory, Philip
Greisen, Eric
Griffin, R. Elizabeth
Griffin, Roger
Grindlay, Jonathan
Groot, Paul
Grosbol, Preben
Guibert, Jean
Guinan, Edward
GUO, Jianheng
GUO, Hongfang
Hamuy, Mario
Harmer, Dianne
Hauck, Bernard
Hefele, Herbert
Heiser, Arnold
Helou, George
Henden, Arne
Hessman, Frederic
Hestroffer, Daniel
Hledik, Stanislav
Hodge, Paul
Hopkins, Andrew
Horne, Keith
Howell, Steve
Hudec, Rene
Hudkova, Ludmila
Hunstead, Richard
Ioannou, Zacharias
Ivezic, Zeljko
Jacoby, George
Jauncey, David
Jenkner, Helmut
JIN, WenJing
Jones, Derek
Jordan, Andrés
Kalberla, Peter
Kaplan, George
Kazantseva, Liliya
Kedziora-Chudczer, Lucyna
Kelly, Brandon
Kent, Brian
Kimball, Amy
Kitaeff, Vyacheslav
Koen, Marthinus
Kolenberg, Katrien
Kolobov, Dmitri
KONG, Xu
Kopatskaya, Evgenia
Koribalski, Bärbel
Kovalev, Yuri
Kovaleva, Dana
Krishna, Gopal
Kubat, Jiri
Kudryavtseva, Nadezhda
Kuin, Paul
Kulkarni, Shrinivas
Kunz, Martin
Labbe, Ivo
Larson, Stephen
Lazio, Joseph
Lee, William
Leibundgut, Bruno
Lequeux, James
Lesteven, Soizick
Linde, Peter
Lintott, Chris
Lister, Matthew
LIU, Xiaoqun
LIU, Siming
Long, Knox
Longo, Giuseppe
Lonsdale, Carol
Lopes, Paulo
Loup, Cécile
Louys, Mireille
LU, Fangjun
Lubowich, Donald
Macquart, Jean-Pierre
Madore, Barry
MAEDA, Keiichi
Magnier, Eugene
Malkov, Oleg
Mann, Robert
Mannucci, Filippo
Martinez, Vicent
Mason, Brian
MATSUNAGA, Noriyuki
Matz, Steven
McAteer, R.T. James
McLean, Brian
McMahon, Richard
McNally, Derek
Meadows, A.
Mein, Pierre
Merin Martin, Bruno
Michel, Laurent
Mickaelian, Areg
Mink, Jessica
Minniti, Dante
Mitton, Simon
Moitinho, André
Monet, David
Montes, David
Mookerjea, Bhaswati
Morbidelli, Roberto
Morris, Rhys
Morrison, Nancy
Muinos Haro, Jose
Murphy, Tara
Murtagh, Fionn
Murtagh, Fionn
Murtagh, Fionn
NAGATA, Tetsuya
NAKAJIMA, Koichi
Nefedyev, Yury
Nesci, Roberto
NISHIMURA, Shiro
Ochsenbein, François
Ogando, Ricardo
OHISHI , Masatoshi
Ojha, Roopesh

Oluseyi, Hakeem
Orellana, Rosa
Ott, Juergen
Pakhomov, Yury
Pamyatnykh, Alexey
Panessa, Francesca
Paturel, Georges
Pauwels, Thierry
Peck, Alison
Pecker, Jean-Claude
Pence, William
Perez-Gonzalez, Pablo
PHAM, Diep
Philip, A.G.
Piskunov, Anatolij
Pitkin, Matthew
Pizzichini, Graziella
Plavchan, Jr., Peter
Potter, Stephen
Protsyuk, Yuri
Prsa, Andrej
Pucillo, Mauro
Pushkarev, Alexander
Puxley, Phil
QU, Jinlu
Quinn, Peter
Ratnatunga, Kavan
Ray, Alak
Reardon, Kevin
Renson, P.
Reyes-Ruiz, Mauricio
Richards, Mercedes
Rickard, Lee
Ridgway, Stephen
Rizzi, Luca
Robinson, Edward
Rocha-Pinto, Hélio
Rodrigues, Claudia
Roman, Nancy
Romaniello, Martino
Rossi, Corinne
Rots, Arnold
Rudnick, Lawrence
Russo, Guido
Saar, Enn
Saha, Abhijit
Sahal-Brechot, Sylvie
Samodurov, Vladimir
Sarasso, Maria
Savanevich, Vadim
Schade, David
Schilbach, Elena
Schmadel, Lutz
Schmitz, Marion
Schneider, Jean
Schreiber, Roman
Seaman, Rob
SEKIGUCHI, Kazuhiro
Seymour, Nicholas
Shafter, Allen
Sharp, Nigel
Shastri, Prajval
Shaw, Richard
Shelton, Ian
SHIRASAKI, Yuji
Silva, David
Smart, Richard
Smith, Robert
Smith, Randall
SONG, Liming
Soszynski, Igor
Spite, François
Srianand, Raghunathan
Sterken, Christiaan
Stickland, David
Stil, Jeroen
Stoehr, Felix
Strelnitski, Vladimir
Strom, Stephen
SU, Cheng-yue
Sullivan, Mark
SUNG, Hyun-Il
Surkis, Igor
Sutton, Edmund
Szalay, Alex
Szkody, Paula
TANAKA, Ichi Makoto
Taylor, A.
Tedds, Jonathan
Templeton, Matthew
Tessema, Solomon
Teuben, Peter
Titov, Vladimir
Torres-Papaqui, Juan
Trimble, Virginia
Tsvetkov, Milcho
Turatto, Massimo
Turner, Kenneth
Tyson, John
Ulrich, Roger
Urban, Sean
Valeev, Sultan
Valls-Gabaud, David
Vandenbussche, Bart
Varela Lopez, Jesús
Vavrek, Roland
Veillet, Christian
Velikodsky, Yuri
Verkhodanov, Oleg
Viotti, Roberto
Vishniac, Ethan
Vollmer, Bernd
Wallace, Patrick
WANG, Shiang-Yu
WANG, Jian-Min
Warren Jr, Wayne
Weilbacher, Peter
Weiss, Werner
Weller, Charles
Wells, Donald
Wenger, Marc
Whitelock, Patricia
Whiting, Matthew
Wicenec, Andreas
Wielen, Roland
Wilkins, George
Williams, Robert
Willis, Anthony
Woudt, Patrick
Wright, Alan
YAMADA, Shimako
YANG, Hong-Jin
YANG, Xiaohu
Zacchei, Andrea
Zacharias, Norbert
Zender, Joe
ZHANG, Yanxia
ZHANG, Shu
ZHANG, Zhibin
ZHAO, Jun Liang
ZHAO, Yongheng
ZHOU, Jianfeng

Division XII Commission 6 Astronomical Telegrams

President: Nikolay N. Samus
Vice-President: Hitoshi YAMAOKA

Organizing Committee Members:

Aksnes, Kaare
Green, Daniel
NAKANO, Syuichi
Spahr, Timothy
Williams, Gareth

Members:

Allan, Alasdair
Apostolovska, Gordana
Baransky, Olexander
Bazzano, Angela
Bouchard, Antoine
CHEN, Xinyang
Corbin, Brenda
Cracco, Valentina
D'Ammando, Filippo
Esenoglu, Hasan
Filippenko, Alexei
Gal-Yam, Avishay
Grindlay, Jonathan
Kaminker, Alexander
Kastel, Galina
Kouveliotou, Chryssa
LIU, Guoqing
MOROKUMA, Tomoki
NAKAMURA, Tsuko
Ofek, Eran
Paragi, Zsolt
Phillips, Mark
Poznanski, Dovi
Roemer, Elizabeth
Rushton, Anthony
Seaman, Rob
Sivakoff, Gregory
SOMA, Mitsuru
Sullivan, Mark
Tholen, David
Tsvetkov, Milcho
URATA, Yuji
Valeev, Azamat
West, Richard
Williams, Gareth

Division I Commission 7 Celestial Mechanics & Dynamical Astronomy

President: Zoran Knežević
Vice-President: Alessandro Morbidelli

Organizing Committee Members:

Athanassoula, Evangelie
Laskar, Jacques
Malhotra, Renu
Mikkola, Seppo
Roig, Fernando

Members:

Abad Medina, Alberto
Abalakin, Viktor
Aksnes, Kaare
Andrade, Manuel
Anosova, Joanna
Antonacopoulos, Gregory
ARAKIDA, Hideyoshi
Archinal, Brent
Athanassoula, Evangelie
Augereau, Jean-Charles
Balmino, Georges
Barabanov, Sergey
Barberis, Bruno
Barbosu, Mihail
Barkin, Yuri
Bartczak, Przemyslaw
Benest, Daniel
Beutler, Gerhard
Bhatnagar, K.
Bois, Eric
Borczyk, Wojciech
Borisov, Borislav
Boss, Alan
Bouchard, Antoine
Bozis, George
Branham, Richard
Breiter, Slawomir
Brieva, Eduardo
Broz, Miroslav
Brumberg, Victor
Brunini, Adrian
CAI, Michael
Caranicolas, Nicholas
Carlin, Jeffrey
Carpino, Mario
Carruba, Valerio
Cefola, Paul
Chambers, John
Chapanov, Yavor
Chapront, Jean
Chapront-Touze, Michelle
CHOI, Kyu Hong
Christou, Apostolos
Cionco, Dr Rodolfo
Colin, Jacques
Conrad, Albert
Contopoulos, George
Cooper, Nicholas
Deleflie, Florent
Descamps, Pascal
Di Sisto, Romina
Dikova, Smilyana
DONG, Xiaojun
Dourneau, Gérard
Drozyner, Andrzej
DU, Lan
Duriez, Luc
Dvorak, Rudolf
Efroimsky, Michael
Elipe, Antonio
Emelianov, Nikolaj
Emelyanenko, Vacheslav
Esguerra, Jose Perico
Fernandez, Silvia
Ferrari, Fabricio
Ferraz -Mello, Sylvio
Ferrer, Martinez
Finch, Charlie
Floria Peralta, Luis
FONG, Chugang
Fouchard, Marc
Froeschle, Claude
FUKUSHIMA, Toshio
Gaposchkin, Edward
GENG, Lihong
Giacaglia, Giorgio
Giordano, Claudia
Giuliatti Winter, Silvia
Goldreich, Peter
Gomes, Rodney
Gozdziewski, Krzysztof
Granvik, Mikael
Greenberg, Richard
Gronchi, Giovanni
Gusev, Alexander
Hamilton, Douglas
Hanslmeier, Arnold
Hau, George
Heggie, Douglas
Hersant, Franck
Heyl, Jeremy
Horner, Jonathan
HOU, Xiyun
Hsieh, Henry
HU, Xiaogong
HUANG, Tianyi
HUANG, Cheng
Hurley, Jarrod
Iorio, Lorenzo
Ipatov, Sergei
Ismail, Mohamed
ITO, Takashi
Ivanov, Pavel
Ivanova, Violeta
Ivantsov, Anatoliy
Izmailov, Igor
Jaeggi, Adrian
Jakubik, Marian

Jefferys, William
JI, Jianghui
JIANG, Ing-Guey
Kalvouridis, Tilemachos
Kent, Brian
Kenworthy, Matthew
Kholshevnikov, Konstantin
KIM, Sungsoo
KIM, Yoo Jea
KING, Sun-Kun
KINOSHITA , Hiroshi
Kitiashvili, Irina
Klioner, Sergei
Klocok, Lubomir
Klokocnik, Jaroslav
Knezevic, Zoran
KOKUBO, Eiichiro
Korchagin, Vladimir
Kornos, Leonard
Koshkin , Nikolay
Kouwenhoven, M.B.N.
Kovacevic, Andjelka
Kovalevsky, Jean
KOZAI, Yoshihide
Krivov, Alexander
Kuznetsov, Eduard
La Spina, Alessandra
Lala, Petr
Lammers, Uwe
Le Poncin-Lafitte, Ch.
Lecavelier des Etangs, Alain
Lega, Elena
Levine, Stephen
LI, Yuqiang
LIAO, Xinhao
Libert, Anne-Sophie
Lieske, Jay
Lin, Douglas
Lissauer, Jack
LIU, Chengzhi
Livadiotis, George
LU, BenKui
Lucchesi, David
Lundquist, Charles
LYKAWKA, Patryk
MA, Jingyuan
MA, Lihua
Malhotra, Renu
Mapelli, Michela
Marchal, Christian
Martinet, Louis
Martins, Roberto
MASAKI, Yoshimitsu
Matas, Vladimir
Mavraganis, Anastasios
Melbourne, William
Metris, Gilles
Michel, Patrick
Migaszewski, Cezary
Mignard, François
Milani Comparetti, Andrea
Minazzoli, Olivier
Montgomery, Michele
Moreira Morais, Maria
Muzzio, Juan
Mysen, Eirik
Nacozy, Paul
Namouni, Fathi
Naroenkov, Sergey
Navone, Hugo
Nefedyev, Yury
Ninkovic, Slobodan
Nobili, Anna
Novakovic, Bojan
Noyelles, Benoît
O'Handley, Douglas
Orellana, Rosa
Orlov, Victor
Osorio, Isabel
Osorio, José
Page, Gary
Pal, András
Parv, Bazil
Pauwels, Thierry
Perets, Hagai
Perozzi, Ettore
Petit, Jean-Marc
Petrov, Sergey
Petrovskaya, Margarita
Pilat-Lohinger, Elke
Polyakhova, Elena
Puetzfeld, Dirk
Reyes-Ruiz, Mauricio
Richardson, Derek
Rodin, Alexander
Rossi, Alessandro
Ryabov, Yurij
Saad, Abdel-naby
Sansaturio, Maria
Sari, Re'em
Scheeres, Daniel
Scholl, Hans
Schubart, Joachim
Segan, Stevo
Seidelmann, P.
Sein-Echaluce, M.
Shakht, Natalia
Shankland, Paul
Shapiro, Irwin
SHENG, Wan
Shevchenko, Ivan
Sidlichovsky, Milos
Sima, Zdislav
Simo, Carles
Simon, Jean-Louis
Skripnichenko, Vladimir
Soffel, Michael
Sokolov, Leonid
Souchay, Jean
Stadel, Joachim
Standish, E.
Stellmacher, Irène
Steves, Bonnie
Stewart, Susan
Suli, Áron
SUN, Yisui
SUN, Fuping
Sweatman, Winston
Szenkovits, Ferenc
TAO, Jin-he
Tatevyan, Suriya
Taylor, Donald
Thuillot, William
Tiscareno, Matthew
Titov, Vladimir
Tommei, Giacomo
Tremaine, Scott
Trenti, Michele
Tsiganis, Kleomenis
Tsuchida, Masayoshi
Valsecchi, Giovanni
Valtonen, Mauri
Varvoglis, Harry
Vashkovyak, Sofja
Vassiliev, Nikolaj
Veillet, Christian
Vieira Neto, Ernesto
Vienne, Alain
Vilhena de Moraes, R.
Vinet, Jean-Yves

Virtanen, Jenni
Vokrouhlicky, David
Vondrak, Jan
Walch, Jean-Jacques
WANG, Xiaoya
WANG, Guangli
WATANABE, Noriaki
WEI, Erhu
Weiss, John
Whipple, Arthur
Wiegert, Paul
Williams, Carol
Winter, Othon
Wnuk, Edwin
WU, Lianda
Wytrzyszczak, Iwona
XIONG, Jianning
YANG, Xuhai
YANO, Taihei
Yokoyama, Tadashi
YOSHIDA, Haruo
Yseboodt, Marie
YUASA, Manabu
Zafiropoulos, Basil
Zare, Khalil
ZHANG, Sheng-Pan
ZHANG, Wei
ZHANG, Xiaoxiang
ZHANG, Wei
ZHANG, Yang
ZHAO, Changyin
ZHAO, You
ZHAO, Haibin
Zhdanov, Valery
ZHENG, Jia-Qing
ZHENG, Yong
ZHOU, Li-Yong
ZHOU, Ji-Lin
ZHU, Wenyao

Division I Commission 8 Astrometry

President: Dafydd Wyn Evans
Vice-President: Norbert Zacharias

Organizing Committee Members:

Andrei, Alexandre
Brown, Anthony
Evans, Dafydd Wyn
GOUDA, Naoteru
Kumkova, Irina
Popescu, Petre
Souchay, Jean
Unwin, Stephen
ZHU, Zi

Members:

Abad Hiraldo, Carlos
Abbas, Ummi
Ahmed, Abdel-aziz
Ammons, Stephen
Andrei, Alexandre
Arenou, Frédéric
Argyle, Robert
Arias, Elisa
Arlot, Jean-Eudes
Assafin, Marcelo
BABA, Junichi
Babusiaux, Carine
Badescu, Octavian
Bakhtigaraev, Nail
Ballabh, Goswami
Bangert, John
Barkin, Yuri
Bartczak, Przemyslaw
Bartlett, Jennifer
Bastian, Ulrich
Belizon, Fernando
Benedict, George
Benevides Soares, Paulo
Bien, Reinhold
Boboltz, David
Bouchard, Antoine
Bougeard, Mireille
Bradley, Arthur
Branham, Richard
Brosche, Peter
Brouw, Willem
Bucciarelli, Beatrice
Bueno de Camargo, Julio
Capitaine, Nicole
Carlin, Jeffrey
Casetti, Dana
Chapanov, Yavor
Chemin, Laurent
CHEN, Li
CHEN, Alfred
CHEN, Linfei
Cioni, Maria-Rosa
Cooper, Nicholas
Corbin, Thomas
Costa, Edgardo
Crézé, Michel
Crifo, Françoise
Crosta, Mariateresa
Cudworth, Kyle
da Rocha-Poppe, Paulo
Dahn, Conard
Damljanovic, Goran
Danylevsky, Vassyl
Day Jones, Avril
de Bruijne, Jos
Dejaiffe, Rene
Del Santo, Melania
Deller, Adam
Delmas, Christian
Devyatkin, Aleksandr
Dick, Wolfgang
Dick, Steven
DU, Lan
Ducourant, Christine
Duma, Dmitrij
Emilio, Marcelo
Fabricius, Claus
FAN, Yu
Fernandes-Martin, Vera
Fey, Alan
Finch, Charlie
Firneis, Maria
Fomin, Valery
Fors, Octavi
Franz, Otto
Fredrick, Laurence
Fresneau, Alain
Froeschle, Michel
Frutos-Alfaro, Francisco
FUJISHITA, Mitsumi
FUKUSHIMA, Toshio
Gatewood, George
Gaume, Ralph
Gauss, Stephen
Gavras, Panagiotis
Geffert, Michael
Germain, Marvin
Gilles, Dominique
Goddi, Ciriaco
Gontcharov, George
Guibert, Jean
Guseva, Irina
Hajian, Arsen
HAN, Inwoo
Hanson, Robert
Hartkopf, William
Helmer, Leif
Hemenway, Paul
Hering, Roland
Heudier, Jean-Louis
Hill, Graham
Hobbs, David
Hoeg, Erik
HONG, Zhang
Ianna, Philip

Iorio, Lorenzo
Irwin, Michael
Ivantsov, Anatoliy
Izmailov, Igor
Jacobs, Christopher
Jahreiss, Hartmut
Jefferys, William
JIA, Lei
JIN, WenJing
Johnston, Kenneth
Jones, Burton
Jones, Derek
Jordi, Carme
Kalomeni, Belinda
Kanayev, Ivan
Kaplan, George
Kazantseva, Liliya
Kharchenko, Nina
King, Ivan
Klemola, Arnold
Klioner, Sergei
Klock, Benny
Kovalevsky, Jean
Kuimov, Konstantin
Kumkova, Irina
KURAYAMA, Tomoharu
Kurzynska, Krystyna
Lammers, Uwe
Lattanzi, Mario
Lazorenko, Peter
Le Poncin-Lafitte, Ch.
Le Poole, Rudolf
Lenhardt, Helmut
Lepine, Sebastien
LI, Zhigang
LI, Qi
Lindegren, Lennart
LIU, Chengzhi
Lu, Phillip
LU, Chunlin
MA, Wenzhang
MacConnell, Darrell
Magnier, Eugene
Maigurova, Nadiia
Mallamaci, Claudio
Marschall, Laurence
Marshalov, Dmitriy
Mason, Brian
McAlister, Harold
McLean, Brian
Mignard, François
Mink, Jessica
Monet, David
Morbidelli, Roberto
Muinos Haro, Jose
NAKAGAWA, Akiharu
NAKAJIMA, Koichi
Naroenkov, Sergey
Nefedyev, Yury
NIINUMA, Kotaro
NIWA, Yoshito
Noyelles, Benoît
Nunez, Jorge
Ofek, Eran
OHNISHI, Kouji
Oja, Tarmo
Olive, Don
Olsen, Hans
Osborn, Wayne
Osorio, José
Page, Gary
Pakvor, Ivan
Pascu, Dan
Pauwels, Thierry
Penna, Jucira
Perryman, Michael
Petrov, Sergey
PING, Jinsong
Pinigin, Gennadiy
Platais, Imants
Poma, Angelo
Popescu, Petre
Pourbaix, Dimitri
Protsyuk, Yuri
Proverbio, Edoardo
Prusti, Timo
Pugliano, Antonio
Rafferty, Theodore
Reddy, Bacham
Reffert, Sabine
Reynolds, John
Rodin, Alexander
Roemer, Elizabeth
Roeser, Siegfried
Russell, Jane
SAKAMOTO, Tsuyoshi
Sanders, Walter
Sarasso, Maria
SATO, Koichi
Schilbach, Elena
Schildknecht, Thomas
Scholz, Ralf-Dieter
Schreiber, Karl
Schwekendiek, Peter
Segransan, Damien
Seidelmann, P.
Shelus, Peter
SHEN, Kaixian
SHEN, Zhiqiang
SHU, Fengchun
Sivakoff, Gregory
Smart, Richard
Soderhjelm, Staffan
Solaric, Nikola
SOMA, Mitsuru
Sovers, Ojars
Sozzetti, Alessandro
Spoljaric, Drago
Standish, E.
Stein, John
Steinmetz, Matthias
Stewart, Susan
SUGANUMA, Masahiro
SUN, Fuping
SUNG, Hyun-Il
Surkis, Igor
TANG, Zheng-Hong
Taris, François
Tedds, Jonathan
Teixeira, Paula
ten Brummelaar, Theo
Thuillot, William
TSUJIMOTO, Takuji
Turon, Catherine
UEDA, Haruhiko
Upgren, Arthur
Urban, Sean
Vallejo, Miguel
van Altena, William
van Leeuwen, Floor
Vass, Gheorghe
Vilkki, Erkki
Volyanska, Margaryta
Wallace, Patrick
WANG, Jiaji
WANG, Zhengming
WANG, Xiaoya
WANG, Guangli
Wasserman, Lawrence
WEI, Erhu

White, Graeme
Wicenec, Andreas
Wielen, Roland
WU, Zhen-Yu
XIA, Yifei
XIE, Yi
YAMADA, Yoshiyuki
YANG, Tinggao
YANO, Taihei
Yatsenko, Anatolij
Yatskiv, Yaroslav
YE, Shuhua
Zacchei, Andrea
ZHANG, Wei
ZHANG, Yong
ZHENG, Yong
ZHU, Zi

Division II Commission 10 Solar Activity

President: Lidia van Driel-Gesztelyi
Vice-President: Karel Schrijver

Organizing Committee Members:

Charbonneau, Paul
Fletcher, Lyndsay
Hasan, S. Sirajul
Hudson, Hugh
KUSANO , Kanya
Mandrini, Cristina
Peter, Hardi
Vrsnak, Bojan
YAN, Yihua

Members:

Abbett, William
Abdelatif, Toufik
Aboudarham, Jean
Abraham, Péter
Abramenko, Valentina
Afram, Nadine
Ahluwalia, Harjit
AI, Guoxiang
Airapetian, Vladimir
AKITA, Kyo
Alissandrakis, Costas
Altrock, Richard
Altschuler, Martin
Altyntsev, Alexandre
Aly, Jean-Jacques
Ambastha, Ashok
Ambroz, Pavel
Anastasiadis, Anastasios
Andersen, Bo Nyborg
Andretta, Vincenzo
Andries, Jesse
Antiochos, Spiro
Antonucci, Ester
Anzer, Ulrich
Arregui, Inigo
Aschwanden, Markus
Atac, Tamer
Babin, Arthur
Bagala, Liria
Bagare, S.
Balasubramaniam, K.
Balikhin, Michael
Ballester, Jose
Banerjee, Dipankar
BAO, Shudong
Baranovsky, Edward
Barrantes, Marco
Barrow, Colin
Barta, Miroslav
Basu, Sarbani
Batchelor, David
Beckers, Jacques
Bedding, Timothy
Beebe, Herbert
Bell, Barbara
Bellot Rubio, Luis
Belvedere, Gaetano
Bemporad, Alessandro
Benevolenskaya, Elena
Benz, Arnold
Berger, Mitchell
Berghmans, David
Berrilli, Francesco
Bertello, Luca
Bewsher, Danielle
Bianda, Michele
Bingham, Robert
Bobylev, Vadim
Bogdan, Thomas
Bommier, Véronique
Bornmann, Patricia
Botha, Gert
Bothmer, Volker
Bouchard, Antoine
Bougeret, Jean-Louis
Brajsa, Roman
Brandenburg, Axel
Brandt, Peter
Braun, Douglas
Brekke, Pål
Bromage, Barbara
Brooke, John
Brosius, Jeffrey
Brown, John
Browning, Philippa
Brun, Allan
Bruner, Marilyn
Bruno, Roberto
Bruns, Andrey
Brynildsen, Nils
Buccino, Andrea
Buchlin, Eric
Buechner, Joerg
Bumba, Vaclav
Busa, Innocenza
Bushby, Paul
Cadez, Vladimir
CAI, Mingsheng
Cane, Hilary
Carbonell, Marc
Cargill, Peter
Cauzzi, Gianna
CHAE, Jongchul
Chambe, Gilbert
Chandra, Suresh
CHANG, Heon-Young
Channok, Chanruangrit
Chaplin, William
Chapman, Gary
CHEN, Peng-Fei
CHEN, Zhiyuan
Chernov, Gennadij
Chertok, Ilya
Chertoprud, Vadim
Chiuderi-Drago, Franca

CHIUEH, Tzihong
CHO, Kyung Suk
CHOE, Gwangson
Choudhary, Debi Prasad
Choudhuri, Arnab
Cliver, Edward
Coffey, Helen
Collados, Manuel
Conway, Andrew
Cora, Alberto
Correia, Emilia
Costa, Joaquim
Craig, Ian
Cramer, Neil
Crannell, Carol
Culhane, John
Curdt, Werner
Dalla, Silvia
Damé, Luc
Dasso, Sergio
Datlowe, Dayton
Davila, Joseph
De Groof, Anik
de Jager, Cornelis
de Toma, Giuliana
Dechev, Momchil
Del Toro Iniesta, Jose
Demoulin, Pascal
DENG, YuanYong
Dennis, Brian
Dere, Kenneth
DeRosa, Marc
Deubner, Franz-Ludwig
Dialetis, Dimitris
DING, Mingde
Dobler, Wolfgang
Dobrzycka, Danuta
Dorch, Søren Bertil
Dorotovic, Ivan
Druckmueller, Miloslav
Dryer, Murray
Dubau, Jacques
Dubois, Marc
Duchlev, Peter
Duldig, Marcus
Dumitrache, Cristiana
Dwivedi, Bhola
Efimenko, Volodymyr
Emslie, Gordon
Engvold, Oddbjørn
Erdelyi, Robert
Ermolli, Ilaria
Esenoglu, Hasan
Falewicz, Robert
FAN, Yuhong
FANG, Cheng
Farnik, Frantisek
Fernandes, Francisco
Ferreira, Joao
Ferriz Mas, Antonio
Feulner, Georg
Filippov, Boris
Fisher, George
Fludra, Andrzej
Fluri, Dominique
Foing, Bernard
Fontenla, Juan
Forbes, Terry
Forgacs-Dajka, Emese
Fossat, Eric
Foullon, Claire
FU, Hsieh-Hai
Gabriel, Alan
Gaizauskas, Victor
Galsgaard, Klaus
GAN, Weiqun
Garcia de la Rosa, Ignacio
Gary, Gilmer
Georgoulis, Manolis
Gergely, Tomas
Getling, Alexander
Ghizaru, Mihai
Gill, Peter
Gilliland, Ronald
Gilman, Peter
Gimenez de Castro, Carlos
Glatzmaier, Gary
Gleisner, Hans
Goedbloed, Johan
Gokhale, Moreshwar
Gomory, Peter
Gontikakis, Constantin
Goossens, Marcel
Gopasyuk, Olga
Grandpierre, Attila
Grechnev, Victor
Gregorio, Anna
Grib, Sergey
Gudiksen, Boris
Guglielmino, Salvatore
Guhathakurta, Madhulika
Gunar, Stanislav
Gupta, Surendra
Gurman, Joseph
Gyori, Lajos
Haberreiter, Margit
Hagyard, Mona
Hammer, Reiner
Hanaoka, Yoichiro
Hanasz, Jan
Hannah, Iain
Hanslmeier, Arnold
HARA, Hirohisa
Harra, Louise
Harvey, John
Hathaway, David
Haugan, Stein Vidar
HAYASHI, Keiji
Hayward, John
He, Han
Heinzel, Petr
Henoux, Jean-Claude
Herdiwijaya, Dhani
HIEI, Eijiro
Hildebrandt, Joachim
Hildner, Ernest
Hochedez, Jean-François
Hoeksema, Jon
Hohenkerk, Catherine
Hollweg, Joseph
Holman, Gordon
Holzer, Thomas
Hood, Alan
Howard, Robert
Hoyng, Peter
Hudson, Hugh
Hurford, Gordon
HWANG, Junga
ISHII, Takako
Ishitsuka, Mutsumi
Isliker, Heinz
Ivanchuk, Victor
Ivanov, Evgenij
Ivchenko, Vasily
Jackson, Bernard
Jacobs, Carla
Jain, Rajmal
Jakimiec, Jerzy
Janssen, Katja
Jardine, Moira

JI, Haisheng
JIANG , Yun
JIANG, Aimin
Jimenez, Mancebo
JING, Hairong
Jones, Harrison
Jordan, Stuart
Jubier, Xavier
Jurcak, Jan
KABURAKI, Osamu
Kahler, Stephen
Kallenbach, Reinald
Kalman, Bela
Kaltman, Tatyana
Kapyla, Petri
Karami , Kayoomars
Karlicky, Márian
Karoff, Christoffer
Karpen, Judith
Kasiviswanathan, S.
Kasparova, Jana
Katsova, Maria
Kaufmann, Pierre
Keppens, Rony
Khan, J
Khodachenko, Maxim
Khomenko, Elena
Khumlumlert, Thiranee
KILCIK, Ali
Kim, Iraida
KIM, Kap sung
Kiplinger, Alan
KITAI, Reizaburo
Kitchatinov, Leonid
Kitiashvili, Irina
Kjeldseth-Moe, Olav
Klein, Karl
Kliem, Bernhard
Klvana, Miroslav
Kolobov, Dmitri
Kolomanski, Sylwester
Kondrashova, Nina
Kontar, Eduard
Kopylova, Yulia
Kostik, Roman
Kotrc, Pavel
Koutchmy, Serge
Koza, Julius
Kozlovsky, Ben
Kramer, William
Kretzschmar, Matthieu
Krimigis, Stamatios
Krittinatham, Watcharawuth
Krucker, Sam
Kryshtal, Alexander
Kryvodubskyj, Valery
KUBOTA, Jun
Kucera, Aleš
Kulhanek, Petr
Kurochka, Evgenia
KUROKAWA, Hiroki
KUSANO, Kanya
Kuznetsov, Vladimir
Labrosse, Nicolas
Landi, Simone
Lang, Kenneth
Langlois, Maud
Lawrence, John
Lazrek, Mohamed
Leibacher, John
Leiko, Uliana
Leka, K.D.
Leroy, Jean-Louis
Leroy, Bernard
LI, Wei
LI, Hui
LI, Kejun
LI, Zhi
Lie-Svendsen, Oystein
Lima, Joao
Lin, Yong
LIN, Jun
Liritzis, Ioannis
LIU, Yang
LIU, Yu
LIU, Siming
Livadiotis, George
Livshits, Moisey
Lopez Fuentes, Marcelo
Loukitcheva, Maria
Low, Boon
Lozitskij, Vsevolod
Lundstedt, Henrik
MA, Guanyi
Machado, Marcos
Mackay, Duncan
MacKinnon, Alexander
MacQueen, Robert
Madjarska, Maria
MAKITA, Mitsugu
Malara, Francesco
Malherbe, Jean-Marie
Malitson, Harriet
Malville, J.
MANABE, Seiji
Mandrini, Cristina
Mann, Gottfried
Marcu, Alexandru
Maricic, Darije
Marilena, Mierla
Maris, Georgeta
Mariska, John
Markova, Eva
Martens, Petrus
Mason, Glenn
MASUDA, Satoshi
Matsuura, Oscar
Matthews, Sarah
Mattig, W.
McAteer, R.T. James
McCabe, Marie
McIntosh, Patrick
McKenna Lawlor, Susan
Mein, Pierre
Melnik, Valentin
Messerotti, Mauro
Messmer, Peter
Meszarosova, Hana
Michalek, Grzegorz
Miesch, Mark
Miletsky, Eugeny
Miralles, Mari Paz
MITRA, Dhrubaditya
Mohan, Anita
Moreno-Insertis, Fernando
MORITA, Satoshi
MORIYAMA, Fumio
Motta, Santo
Muller, Richard
Musielak, Zdzislaw
NAKAJIMA, Hiroshi
Nakariakov, Valery
Nandi, Dibyendu
Neidig, Donald
Neukirch, Thomas
Neupert, Werner
Nickeler, Dieter
NING, Zongjun
Nocera, Luigi
Noens, Jacques-Clair

Noyes, Robert
Nozawa, Satoshi
Nussbaumer, Harry
Obridko, Vladimir
Ofman, Leon
OH, Suyeon
OHKI, Kenichiro
OKAMOTO, Takenori J.
Oliver, Ramón
Oluseyi, Hakeem
Onel, Hakan
Opara, Fidelix
Orlando, Salvatore
Ortiz Carbonell, Ada
Ozguc, Atila
Ozisik, Tuncay
Padmanabhan, Janardhan
Paletou, Frédéric
Palle, Pere
Palle Bago, Enric
Palus, Pavel
PAN, Liande
Pap, Judit
Parenti, Susanna
Parfinenko, Leonid
Pariat, Etienne
PARK, Young Deuk
Parnell, Clare
Pasachoff, Jay
Paterno, Lucio
Peres, Giovanni
Peter, Hardi
Petrie, Gordon
Petrosian, Vahe
Petrov, Nikola
Petrovay, Kristof
Pevtsov, Alexei
Pflug, Klaus
PHAN, Dong
Phillips, Kenneth
Pick, Monique
Pipin, Valery
Plainaki, Christina
Podesta, John
Poedts, Stefaan
Pohjolainen, Silja
Poland, Arthur
Poquerusse, Michel
Preka-Papadema, P.
Pres, Pawek
Priest, Eric
Proctor, Michael
Pustilnik, Lev
Raadu, Michael
Ramelli, Renzo
Rao, A.
Raulin, Jean-Pierre
Reale, Fabio
Reeves, Hubert
Regnier, Stephane
Regulo, Clara
Reinard, Alysha
Rendtel, Juergen
Rengel, Miriam
Reshetnyk, Volodymyr
Riehokainen, Aleksandr
Riley, Pete
Roca Cortes, Teodoro
Roemer, Max
Romano, Paolo
Romoli, Marco
Rompolt, Bogdan
Rosa, Dragan
Roudier, Thierry
Rovira, Marta
Roxburgh, Ian
Rozelot, Jean-Pierre
Rudawy, Pawel
Ruediger, Guenther
Ruffolo, David
Ruiz Cobo, Basilio
Rusin, Vojtech
Russell, Alexander
Rust, David
Ruzdjak, Domagoj
Rybak, Jan
Rybansky, Milan
Safari, Hossein
Sahal-Brechot, Sylvie
Saiz, Alejandro
SAKAO, Taro
SAKURAI, Takashi
Sanchez Almeida, Jorge
Saniga, Metod
Sasso, Clementina
Sattarov, Isroil
Sawyer, Constance
Schindler, Karl
Schlichenmaier, Rolf
Schmahl, Edward
Schmelz, Joan
Schmieder, Brigitte
Schober, Hans
Schuessler, Manfred
Schwartz, Pavol
Seaton, Daniel
Shea, Margaret
Sheeley, Neil
SHIBASAKI, Kiyoto
Shimizu, Toshifumi
Shimojo, Masumi
SHIN'ICHI, Nagata
Shine, Richard
Sigalotti, Leonardo
Simnett, George
Simon, Guy
Simunac, Kristin
Sinha, Krishnanand
Smaldone, Luigi
Smith, Dean
Smol?kov, Gennadij
Snegirev, Sergey
Snik, Frans
Sobotka, Michal
Socas-Navarro, Hector
Solanki, Sami
Soloviev, Alexandr
Somov, Boris
Spadaro, Daniele
Spicer, Daniel
Spruit, Hendrik
Srivastava, Nandita
Steiner, Oskar
Stellmacher, Götz
Stenflo, Jan
Stepanian, Natali
Stepanov, Alexander
Steshenko, N.
Stix, Michael
Strong, Keith
Struminsky, Alexei
Sturrock, Peter
Subramanian, K.
Subramanian, Prasad
Sudar, Davor
SUZUKI, Takeru
Svanda, Michal
Sylwester, Janusz
Sylwester, Barbara
Szalay, Alex

TAKAHASHI, Kunio
TAKANO, Toshiaki
TAN, Baolin
Tapping, Kenneth
Tarashchuk, Vera
Ternullo, Maurizio
Teske, Richard
Thomas, John
Tikhomolov, Evgeniy
Tlatov, Andrej
Tobias, Steven
Tomczak, Michal
TRAN, Ha
Tripathi, Durgesh
Tripathy, Sushanta
Tritakis, Basil
Trottet, Gerard
Tsap, Yuri
Tsinganos, Kanaris
Uddin, Wahab
Usoskin, Ilya
Valio, Adriana
van den Oord, Bert
van der Heyden, Kurt
van der Linden, Ronald
Van Doorsselaere, Tom
Van Hoven, Gerard
Vaughan, Arthur
Veck, Nicholas
Vekstein, Gregory
Velli, Marco
Venkatakrishnan, P.
Ventura, Rita
Verheest, Frank
Verma, V.
Verwichte, Erwin
Vial, Jean-Claude
Viall, Nicholeen
Vieytes, Mariela
Vilinga, Jaime
Vilmer, Nicole
Vinod, S.
Voitenko, Yuriy
Vrsnak, Bojan
Walsh, Robert
WANG, Huaning
WANG, Yi-ming
WANG, Haimin
WANG, Dongguang
WANG, Shujuan
WANG, Jingyu
Webb, David
White, Stephen
Wiehr, Eberhard
Wikstol, Oivind
Winebarger, Amy
Wittmann, Axel
Woehl, Hubertus
Wolfson, Richard
Woltjer, Lodewijk
Wood, Brian
WU, De Jin
Wu, Shi
XIE, Xianchun
XU, Aoao
XU, Jun
XU, Zhi
YANG, Zhiliang
YANG, Hong-Jin
YANG, Lei
Yesilyurt, Ibrahim
YI, Yu
YOKOYAMA, Takaaki
YOSHIMURA, Hirokazu
YU, Dai
YU, Cong
Yun, Hong-Sik
Zachariadis, Theodosios
Zappala, Rosario
Zelenka, Antoine
Zender, Joe
ZHANG, Mei
ZHANG, Jun
Zhang, Jie
Zhang, Tielong
Zharkova, Valentina
ZHOU, Guiping
Zhugzhda, Yuzef
Zlobec, Paolo

Division II Commission 12 Solar Radiation & Structure

President: Alexander Kosovichev
Vice-President: Gianna Cauzzi

Organizing Committee Members:

Brandenburg, Axel
Kuznetsov, Vladimir
Shchukina, Nataliia
Venkatakrishnan, P.
Warren Jr, Wayne

Members:

Abbett, William
Aboudarham, Jean
Acton, Loren
AI, Guoxiang
Aime, Claude
Alissandrakis, Costas
Altrock, Richard
Altschuler, Martin
Andersen, Bo Nyborg
Anderson, Jay
ANDO, Hiroyasu
Andretta, Vincenzo
Ansari, S.M.
Antia, H.
Artzner, Guy
ASAI, Ayumi
Asplund, Martin
Avrett, Eugene
Ayres, Thomas
Babayev, Elchin
Baliunas, Sallie
Balthasar, Horst
Banerjee, Dipankar
Barta, Miroslav
Basu, Sarbani
Baturin, Vladimir
Beckers, Jacques
Beckman, John
Beebe, Herbert
Beiersdorfer, Peter
Bemporad, Alessandro
Benford, Gregory
Bertello, Luca
Bhardwaj, Anil
BI, Shao
Bingham, Robert
Boehm-Vitense, Erika
Bommier, Veronique
Bonnet, Roger
Bornmann, Patricia
Borovik, Valerya
Bougeret, Jean-Louis
Brandt, Peter
Breckinridge, James
Brosius, Jeffrey
Bruls, Jo
Brun, Allan
Bruner, Marilyn
Bruning, David
Buchlin, Eric
Bumba, Vaclav
Cadez, Vladimir
Cavallini, Fabio
Ceppatelli, Guido
Chambe, Gilbert
CHAN, Kwing
Chapman, Gary
Chertok, Ilya
CHOE, Gwangson
Christensen-Dalsgaard, J.
Clark, Thomas
Clette, Frederic
Collados, Manuel
Collet, Remo
Couvidat, Sebastien
Craig, Ian
Cramer, Neil
Damé, Luc
Dara, Helen
de Jager, Cornelis
de Toma, Giuliana
Dechev, Momchil
Degenhardt, Detlev
Del Toro Iniesta, Jose
Deliyannis, John
Demarque, Pierre
Deming, Leo
DeRosa, Marc
Deubner, Franz-Ludwig
Di Mauro, Maria Pia
Diver, Declan
Donea, Alina
Doyle MRIA, John
Dravins, Dainis
Druckmueller, Miloslav
Duvall Jr, Thomas
Ehgamberdiev, Shuhrat
Einaudi, Giorgio
Ermolli, Ilaria
Esser, Ruth
Fabiani, Sergio
Falewicz, Robert
FANG, Cheng
Feldman, Uri
Fernandes, Francisco
Feulner, Georg
Fisher, George
Fleck, Bernhard
Fluri, Dominique
Fomichev, Valerij
Fontenla, Juan
Forgacs-Dajka, Emese
Fossat, Eric
Foukal, Peter
Foullon, Claire
Froehlich, Claus
Gabriel, Alan
Gaizauskas, Victor

GAN, Weiqun
Garcia, Rafael
Garcia-Berro, Enrique
Georgoulis, Manolis
Getling, Alexander
Glatzmaier, Gary
Goldman, Martin
Gomez, Maria
Gomory, Peter
Gopalswamy, Natchimuthuk
Grevesse, Nicolas
Guhathakurta, Madhulika
Gunar, Stanislav
Hagyard, Mona
Hamedivafa, Hashem
Hammer, Reiner
Hannah, Iain
Harvey, John
Hejna, Ladislav
HIEI, Eijiro
Hildner, Ernest
Hill, Frank
Hoang, Binh
Hoeksema, Jon
Howard, Robert
Hoyng, Peter
HUANG, Guangli
HWANG, Junga
Illing, Rainer
Ivanov, Evgenij
Jackson, Bernard
Jacobs, Carla
Janssen, Katja
Jones, Harrison
Jordan, Carole
Jordan, Stuart
Jurcak, Jan
Kalkofen, Wolfgang
Kalman, Bela
Kaltman, Tatyana
Kapyla, Petri
Karlicky, Márian
Karoff, Christoffer
Karpen, Judith
Kasiviswanathan, S.
Kaufmann, Pierre
Keil, Stephen
Khan, J
Khomenko, Elena
Khumlumlert, Thiranee
KILCIK, Ali
KIM, Yong Cheol
Kim, Iraida
Kitiashvili, Irina
Klein, Karl
Kneer, Franz
Knoelker, Michael
Kolobov, Dmitri
Kolomanski, Sylwester
Kopylova, Yulia
Kosovichev, Alexander
Kostik, Roman
Kotov, Valery
Kotrc, Pavel
Koutchmy, Serge
Koza, Julius
Kramer, William
Krittinatham, Watcharawuth
Kryvodubskyj, Valery
Kucera, Aleš
Labrosse, Nicolas
Landi Degl'Innocenti, E.
Landolfi, Marco
Lanzafame, Alessandro
Leenaarts, Jorrit
Leibacher, John
Leka, K.D.
Leroy, Jean-Louis
LIN, Jun
Linsky, Jeffrey
Livadiotis, George
Livingston, William
Lopez Arroyo, M.
Lopez Fuentes, Marcelo
Loukitcheva, Maria
Luest, Reimar
Madjarska, Maria
MAKITA, Mitsugu
Mandrini, Cristina
Maricic, Darije
Marilli, Ettore
Marmolino, Ciro
Matthews, Sarah
Mattig, W.
McAteer, R.T. James
McKenna Lawlor, Susan
Mein, Pierre
Melrose, Donald
Meszarosova, Hana
Meyer, Friedrich
Miesch, Mark
Milkey, Robert
Monteiro, Mario Joao
Moore, Ronald
Morabito, David
MORITA, Satoshi
MORIYAMA, Fumio
Mouradian, Zadig
Muller, Richard
Munro, Richard
Nandi, Dibyendu
Nesis, Anastasios
Nicolas, Kenneth
Nordlund, Aake
Noyes, Robert
Nozawa, Satoshi
Obridko, Vladimir
OH, Suyeon
Onel, Hakan
Ortiz Carbonell, Ada
Ossendrijver, Mathieu
Owocki, Stanley
Padmanabhan, Janardhan
Paletou, Frédéric
Palle, Pere
Palus, Pavel
Papathanasoglou, Dimitrios
Parenti, Susanna
Pasachoff, Jay
Pecker, Jean-Claude
Petrie, Gordon
Petrov, Nikola
Petrovay, Kristof
Pevtsov, Alexei
Pflug, Klaus
Phillips, Kenneth
Picazzio, Enos
Poquerusse, Michel
Priest, Eric
QU, Zhongquan
Radick, Richard
Ramelli, Renzo
Reardon, Kevin
Regulo, Clara
Rengel, Miriam
Riehokainen, Aleksandr
Roca Cortes, Teodoro
Roddier, Francois
Roth, Markus
Roudier, Thierry

Rouppe van der Voort, Luc
Rovira, Marta
Rozelot, Jean-Pierre
Rusin, Vojtech
Russell, Alexander
Rutten, Robert
Ryabov, Boris
Rybak, Jan
Rybansky, Milan
Safari, Hossein
SAKAI, Junichi
Samain, Denys
Sanchez Almeida, Jorge
Saniga, Metod
Sasso, Clementina
Sauval, A.
Scherrer, Philip
Schleicher, Helmold
Schmahl, Edward
Schmidt, Wolfgang
Schmieder, Brigitte
Schober, Hans
Schou, Jesper
Schuessler, Manfred
Schwartz, Steven
Schwartz, Pavol
Seaton, Daniel
Severino, Giuseppe
Shchukina, Nataliia
Sheeley, Neil
Sheminova, Valentina
SHIBASAKI, Kiyoto
Shine, Richard
Sigalotti, Leonardo
Sigismondi, Costantino
Sigwarth, Michael
Simon, George
Simon, Guy
Simunac, Kristin
Singh, Jagdev
Sinha, Krishnanand
Sivaraman, Koduvayur
Skumanich, Andrew
Smith, Peter
Solanki, Sami
Soloviev, Alexandr
Spicer, Daniel
Srivastava, Nandita
Stathopoulou, Maria
Staude, Juergen
Stebbins, Robin
Steffen, Matthias
Steiner, Oskar
Stenflo, Jan
Stepan, Jiří
Stepanian, Natali
Stix, Michael
Stodilka, Myroslav
Straus, Thomas
SUEMATSU, Yoshinori
Svalgaard, Leif
Svanda, Michal
TAN, Baolin
Tarashchuk, Vera
Teplitskaya, Raisa
Thomas, John
Tlatov, Andrej
TRAN, Ha
Tripathi, Durgesh
Trujillo Bueno, Javier
Tsap, Teodor
Tsiklauri, David
Tsiropoula, Georgia
Usoskin, Ilya
Van Doorsselaere, Tom
Van Hoven, Gerard
Vaughan, Arthur
Venkatakrishnan, P.
Ventura, Paolo
Vial, Jean-Claude
Vilmer, Nicole
Voitsekhovska, Anna
von der Luehe, Oskar
WANG, Haimin
WANG, Jingxiu
WANG, Shujuan
Weiss, Nigel
Wiik Toutain, Jun
Wittmann, Axel
Woehl, Hubertus
Worden, Simon
WU, Hsin-Heng
XIE, Xianchun
XU, Zhi
YOICHIRO, Suzuki
YOSHIMURA, Hirokazu
Yun, Hong-Sik
Zampieri, Luca
Zarro, Dominic
Zelenka, Antoine
ZHANG, Mei
ZHANG, Jun
Zhang, Jie
ZHAO, Junwei
Zharkova, Valentina
Zhugzhda, Yuzef
Zirker, Jack
Zuccarello, Francesca

Division XII Commission 14 Atomic & Molecular Data

President: Glenn Michael Wahlgren
Vice-President: Ewine F. van Dishoeck

Organizing Committee Members:

Beiersdorfer, Peter
Dimitrijevic, Milan
Jorissen, Alain
Mashonkina, Lyudmila
Nilsson, Hampus
Salama, Farid
Tennyson, Jonathan

Members:

Adelman, Saul
Afram, Nadine
Aggarwal, Kanti
Allard, Nicole
Allen Jr, John
Allende Prieto, Carlos
Arion, Douglas
Balanca, Christian
Banerjee, Dipankar
Barber, Robert
Barnbaum, Cecilia
Bautista, Manuel
Bayet, Estelle
Behar, Ehud
Bely-Dubau, Francoise
Bhardwaj, Anil
Biemont, Emile
Black, John
Blackwell-Whitehead, Richard
Bodewits, Dennis
Boechat-Roberty, Heloisa
Bommier, Véronique
Borysow, Aleksandra
Branscomb, L.
Bromage, Gordon
Buchlin, Eric
Carbon, Duane
Casasola, Viviana
Cazaux, Stephanie
Chance, Kelly
CHEN, Huei-Ru
CHEN, Guoming
Christlieb, Norbert
Cichowolski, Silvina
Cornille, Marguerite
d'Hendecourt, Louis
de Frees, Douglas
de Kertanguy, Amaury
Depagne, Éric
Diercksen, Geerd
Dimitrijevic, Milan
Dubau, Jacques
Dulieu, Francois
Eidelsberg, Michele
Feautrier, Nicole
Federici, Luciana
Federman, Steven
Filacchione, Gianrico
Fillion, Jean-Hugues
Fink, Uwe
Flower, David
Fluri, Dominique
Fraser, Helen
Fuhr, Jeffrey
Gabriel, Alan
Gallagher III, John
Gargaud, Muriel
Glagolevskij, Yurij
Glinski, Robert
Glover , Simon
Goddi, Ciriaco
Goldbach, Claudine
Grant, Ian
Grevesse, Nicolas
Hartman, Henrik
Heiter, Ulrike
Henning, Thomas
Hesser, James
Hillier, John
Hoang, Binh
Homeier, Derek
Horacek, Jiri
Horandel, Jörg
Huber, Martin
Huebner, Walter
Ignjatovic, Ljubinko
Iliev, Ilian
Irwin, Alan
Irwin, Patrick
Jamar, Claude
Joblin, Christine
Johnson, Fred
Jordan, Carole
Jorgensen, Uffe
Jorissen, Alain
Kamp, Inga
Kanekar, Nissim
Kanuchova, Zuzana
Kennedy, Eugene
Kerber, Florian
Kielkopf, John
KIM, Sang Joon
KIM, Joo Hyeon
Kingston, Arthur
Kipper, Tonu
Kirby, Kate
Kohl, John
Kramida, Alexander
Kroto, Harold
KUAN, Yi-Jehng
Kupka, Friedrich
Kurucz, Robert
Lambert, David
Langhoff, Stephanie

Launay, Jean-Michel
Launay, Françoise
Layzer, David
Le Bourlot, Jacques
Le Floch, André
Leach, Sydney
Leger, Alain
Lemaire, Jean-louis
LIANG, Guiyun
Linnartz, Harold
LIU, Sheng-Yuan
Lo, Wing-Chi Nadia
Lobel, Alex
Lovas, Francis
Lucero, Danielle
Lutz, Barry
Maillard, Jean-Pierre
Mardones, Diego
Martin, Sergio
Mason, Helen
McCall, Benjamin
McWhirter, R.
Mickelson, Michael
Mihajlov, Anatolij
Mookerjea, Bhaswati
Morris, Patrick
Morton, Donald
Mumma, Michael
Nahar, Sultana
Nave, Gillian
Newsom, Gerald
NGUYEN, Phuong
Nielsen, Krister
Nollez, Gérard
Nussbaumer, Harry
O'Brian, Thomas
Oka, Takeshi
Omont, Alain
Orton, Glenn
OTSUKA, Masaaki
OZEKI, Hiroyuki
Palmeri, Patrick
PARK, Yong Sun
Paron, Sergio
Peach, Gillian
Petrini, Daniel
Pettini, Marco
Piacentini, Ruben
Pilling, Sergio
Pradhan, Anil
Quinet, Pascal
Ralchenko, Yuri
Ramirez, Jose
Rastogi, Shantanu
Redman, Matthew
Rogers, Forrest
Rostas, François
Roueff, Evelyne
Ruder, Hanns
Ryabchikova, Tatiana
Sahal-Brechot, Sylvie
Sarre, Peter
Savanov, Igor
Savin, Daniel
Schrijver, Johannes
Schultz, David
Semenov, Dmitry
SHI, Jianrong
Shore, Bruce
Simic, Zoran
Sinha, Krishnanand
Smith, William
Smith, Peter
Somerville, William
SONG, In Ok
Spielfiedel, Annie
Stancil, Philip
Stark, Glenn
Stehle, Chantal
Strachan, Leonard
Strelnitski, Vladimir
Summers, Hugh
Sutherland, Ralph
Swings, Jean-Pierre
TAKAYANAGI, Kazuo
Tatum, Jeremy
Tayal, Swaraj
Tchang-Brillet, Lydia
Tennyson, Jonathan
Tozzi, Gian
Tripathi, Durgesh
Ulyanov, Oleg
van Rensbergen, Walter
Varshalovich, Dmitrij
Vasta, Magda
Vavrek, Roland
Voelk, Heinrich
Vujnovic, Vladis
Wakelam, Valentine
WANG, Junxian
WANG, Feilu
WANG, Junzhi
Wiese, Wolfgang
Wunner, Guenter
XIAO, Dong
YANG, Changgen
Yoshino, Kouichi
Zeippen, Claude
ZENG, Qin
ZHAO, Gang

Division III Commission 15 Physical Study of Comets & Minor Planets

President: Alberto Cellino
Vice-President: Dominique Bockelée-Morvan

Organizing Committee Members:

Davidsson, Björn
Dotto, Elisabetta
Fitzsimmons, Alan
Jenniskens, Petrus
Mothe-Diniz, Thais
Tancredi, Gonzalo
Wooden, Diane

Members:

A'Hearn, Michael
ABE, Shinsuke
AGATA, Hidehiko
Alibert, Yann
Allègre, Claude
Altwegg, Kathrin
Archinal, Brent
Arpigny, Claude
Babadzhanov, Pulat
Bailey, Mark
Bar-Nun, Akiva
Barabanov, Sergey
Baransky, Olexander
Barber, Robert
Barker, Edwin
Barriot, Jean-Pierre
Barucci, Maria
Bear, Ealeal
Belton, Michael
Bemporad, Alessandro
Bendjoya, Philippe
Benkhoff, Johannes
Bhardwaj, Anil
Bingham, Robert
Binzel, Richard
Birlan, Mirel
Biver, Nicolas
Blanco, Armando
Bodewits, Dennis
Boice, Daniel
Bonev, Tanyu
Borisov, Galin
Borisov, Borislav
Borysenko, Serhii
Bowell, Edward
Brandt, John
Brecher, Aviva
Britt, Daniel
Brown, Robert
Brownlee, Donald
Brozovic, Marina
Bueno de Camargo, Julio
Buie, Marc
Buratti, Bonnie
Burlaga, Leonard
Burns, Joseph
Busarev, Vladimir
Butler, Bryan
Campins, Humberto
Campo Bagatin, Adriano
Capaccioni, Fabrizio
Capria, Maria
Carruba, Valerio
Carruthers, George
Carsenty, Uri
Carusi, Andrea
Carvano, Jorge
Cellino, Alberto
Cerroni, Priscilla
Chandrasekhar, Th.
Chapman, Clark
Chapman, Robert
Chubko , Larysa
Clairemidi, Jacques
Clayton, Geoffrey
Clayton, Donald
Cochran, William
Cochran, Anita
Colom, Pierre
Combi, Michael
Connors, Martin
Conrad, Albert
Consolmagno, Guy
Cosmovici, Cristiano
Cottin, Hervé
Cremonese, Gabriele
Crovisier, Jacques
Cruikshank, Dale
Cuypers, Jan
da Silveira, Enio
Danks, Anthony
Davidsson, Björn
Davies, John
de Almeida, Amaury
de Pater, Imke
de Sanctis, Maria
de Sanctis, Giovanni
de Val-Borro, Miguel
Delbo, Marco
Dell'Oro, Aldo
Delsanti, Audrey
Dermott, Stanley
Di Martino, Mario
Doressoundiram, Alain
Dotto, Elisabetta
Dryer, Murray
Duffard, Rene
Duncan, Martin
Durech, Josef
Encrenaz, Therese
Erard, Stéphane
Ershkovich, Alexander
Eviatar, Aharon
Farnham, Tony
Feldman, Paul

Fernandez, Julio
Fernandez, Yanga
Ferrin, Ignacio
Filacchione, Gianrico
Fitzsimmons, Alan
Fletcher, Leigh
Fornasier, Sonia
Foryta, Dietmar
Fouchard, Marc
Fraser, Helen
Froeschle, Christiane
FUJIWARA, Akira
Fulchignoni, Marcello
FURUSHO, Reiko
Gajdos, Stefan
Galad, Adrián
Gammelgaard, Peter
GAO, Jian
Geiss, Johannes
Gerakines, Perry
Gerard, Eric
Gibson, James
Giovane, Frank
Gounelle, Matthieu
Gradie, Jonathan
Grady, Monica
Granvik, Mikael
Green, Simon
Green, Daniel
Greenberg, Richard
Gronkowski, Piotr
Grossman, Lawrence
Gruen, Eberhard
Grundy, William
Gulbis, Amanda
Gustafson, Bo
Hadamcik, Edith
Halliday, Ian
Hanner, Martha
Hapke, Bruce
Harris, Alan
Hartmann, William
Harwit, Martin
HASEGAWA, Sunao
Haupt, Hermann
Howell, Ellen
Hsieh, Henry
Huebner, Walter
Huntress, Wesley
Husarik, Marek
Ibadinov, Khursand
Ibadov, Subhon
IP, Wing-Huen
Irvine, William
Irwin, Patrick
Israelevich, Peter
Ivanova, Violeta
Ivanova, Oleksandra
Ivezic, Zeljko
Jackson, William
Jakubik, Marian
Jedicke, Robert
Jockers, Klaus
Johnson, Torrence
Jorda, Laurent
Kaasalainen, Mikko
Kaeufl, Hans Ulrich
Kanuchova, Zuzana
Karatekin, Özgür
Kavelaars, JJ.
Kaydash, Vadym
Keay, Colin
Keil, Klaus
Keller, Horst
Kidger, Mark
Kim, Joo Hyeon
KING, Sun-Kun
Kiselev, Nikolai
Kiss, Csaba
Klacka, Jozef
Knacke, Roger
Knezevic, Zoran
Knight, Matthew
Koeberl, Christian
Kohoutek, Lubos
Kornos, Leonard
Korokhin, Viktor
Korsun, Pavlo
Koschny, Detlef
Koshkin, Nikolay
KOZASA, Takashi
Krimigis, Stamatios
Krishna, Swamy
Kristensen, Leif
Krugly, Yurij
Kryszczynska, Agnieszka
KUAN, Yi-Jehng
Kueppers, Michael
La Spina, Alessandra
Lagerkvist, Claes-Ingvar
Lamy, Philippe
Lane, Arthur
Lara, Luisa
Larson, Harold
Larson, Stephen
Laufer, Diana
Lazzarin, Monica
Lazzaro, Daniela
Lebofsky, Larry
LEE, Thyphoon
Levasseur-Regourd, A.-Ch.
Li, Jian-Yang
Liller, William
Lillie, Charles
Lindsey, Charles
Lipschutz, Michael
Lissauer, Jack
Lisse, Carey
LIU, Xiaoqun
Lo Curto, Gaspare
Lodders, Katharina
Lopes, Rosaly
Lukyanyk, Igor
Lutz, Barry
Luu, Jane
LYKAWKA, Patryk
Lyon, Ian
Magee-Sauer, Karen
Magnusson, Per
Mainzer, Amy
Makalkin, Andrei
Maran, Stephen
Marchi, Simone
Marcialis, Robert
Marciniak, Anna
Maris, Michele
Marzari, Francesco
Masiero, Joseph
Matson, Dennis
Matsuura, Oscar
Mazzotta Epifani, Elena
McCord, Thomas
McFadden, Lucy
McKenna Lawlor, Susan
Meech, Karen
Meisel, David
Mendillo, Michael
Merline, William
Michalowski, Tadeusz
Michel, Patrick

Milani Comparetti, Andrea
Millis, Robert
Moehlmann, Diedrich
MOON, Hong-Kyu
Moore, Elliott
Morrison, David
Mothe-Diniz, Thais
Mousis, Olivier
Mueller, Thomas
Muinonen, Karri
MUKAI, Tadashi
Mumma, Michael
NAKAMURA, Tsuko
NAKAMURA, Akiko
Napier, William
Nedelcu, Dan
Niedner, Malcolm
Ninkov, Zoran
Nolan, Michael
Noll, Keith
Novakovic, Bojan
O'Dell, Charles
Ortiz, Jose
Pal, András
Paolicchi, Paolo
Peixinho, Nuno
Pendleton, Yvonne
Perez de Tejada, Hector
Picazzio, Enos
Pilcher, Carl
Pittich, Eduard
Pittichova, Jana
Prialnik, Dina
Reyes-Ruiz, Mauricio
Richardson, Derek
Rickman, Hans
Roemer, Elizabeth
Roig, Fernando
Rosenbush, Vera
Rossi, Alessandro
Rousselot, Philippe
Russel, Sara
Sagdeev, Roald
SAITO, Takao
Samarasinha, Nalin
Santos-Sanz, Pablo
Sasaki, Sho
SATO, Isao
Schaller, Emily
Scheirich, Peter
Schleicher, David
Schloerb, F.
Schmidt, Maarten
Schober, Hans
Scholl, Hans
Sekanina, Zdenek
SEKIGUCHI, Tomohiko
Sen, Asoke
SEO, Haingja
Serra Ricart, Miquel
Shanklin, Jonathan
Sharma, A.
Shevchenko, Vasilij
Shor, Viktor
Shoyoqubov, Shoayub
Simonia, Irakli
Sims, Mark
Sivaraman, Koduvayur
Sizonenko, Yuri
Smith, Bradford
Snyder, Lewis
Solc, Martin
Sosa, Andrea
Spinrad, Hyron
Steel, Duncan
Stern, S.
Stewart-Mukhopadhyay, S.
Surdej, Jean
Svoren, Jan
Swade, Daryl
Sykes, Mark
Szego, Karoly
Szekely, Péter
Szutowicz, Slawomira
Tacconi-Garman, Lowell
TAKEDA, Hidenori
Tancredi, Gonzalo
Tanga, Paolo
TAO, Jun
Tarashchuk, Vera
Tatum, Jeremy
Tedesco, Edward
Tholen, David
Thomas, Nicolas
TIAN, Feng
Tiscareno, Matthew
Tosi, Federico
Toth, Imre
Tozzi, Gian
Trujillo, Chadwick
Ugolnikov, Oleg
Valdes-Sada, Pedro
Vazquez, Roberto
Veeder, Glenn
Velikodsky, Yuri
Veverka, Joseph
Vilas, Faith
Voelzke, Marcos
Wallis, Max
WANG, Xiao-bin
WANG, Shiang-Yu
Wasson, John
WATANABE, Jun-ichi
Wdowiak, Thomas
Weaver, Harold
Weidenschilling, S.
Weissman, Paul
West, Richard
Williams, Iwan
Wilson, Lionel
Womack, Maria
Wooden, Diane
Woolfson, Michael
Wyckoff, Susan
YABUSHITA, Shin
YANAGISAWA, Masahisa
YANG, Jongmann
YANG, Xiaohu
Yeomans, Donald
YI, Yu
YOSHIDA , Fumi
Zarnecki, John
ZHANG, Xiaoxiang
ZHANG, Jun
ZHAO, Haibin
ZHU, Jin

Division III Commission 16 Physical Study of Planets & Satellites

President: Melissa Ann McGrath
Vice-President: Mark T. Lemmon

Organizing Committee Members:

KIM, Sang Joon
Ksanfomality, Leonid
Lara, Luisa
Morrison, David
Tejfel, Victor
Yanamandra-Fisher, P.

Members:

Akimov, Leonid
Alexandrov, Yuri
Alibert, Yann
Allison, Michael
Archinal, Brent
Arthur, David
Atkinson, David
Atreya, Sushil
Balikhin, Michael
Barkin, Yuri
Barlow, Nadine
Barrow, Colin
Bartczak, Przemyslaw
Battaner, Eduardo
Beebe, Reta
Beer, Reinhard
Bell III, James
Belton, Michael
Ben-Jaffel, Lofti
Bender, Peter
Benkhoff, Johannes
Bergstralh, Jay
Bertaux, Jean-Loup
Bezard, Bruno
Bhardwaj, Anil
Billebaud, Francoise
Binzel, Richard
Blanco, Armando
Bodewits, Dennis
Bondarenko, Lyudmila
Borisov, Borislav
Bosma, Pieter
Boss, Alan
Boyce, Peter
Brahic, André
Brecher, Aviva
Broadfoot, A.
Brown, Robert
Buie, Marc
Buratti, Bonnie
Burba, George
Burns, Joseph
Busarev, Vladimir
Caldwell, John
Campbell, Donald
Campo Bagatin, Adriano
Capria, Maria
Cecconi, Baptiste
Chakrabarti, Supriya
Chapman, Clark
Chevrel, Serge
Chung, Eduardo
Clairemidi, Jacques
Cochran, Anita
Combi, Michael
Connes, Janine
Cottin, Hervé
Cottini, Valeria
Courtin, Régis
Coustenis, Athena
Cruikshank, Dale
Davies, Ashley
Davis, Gary
de Bergh, Catherine
de Pater, Imke
de Val-Borro, Miguel
Deleuil, Magali
Demory, Brice-Olivier
DeNisco, Kenneth
Dermott, Stanley
Dickel, John
Dickey, Jean
Dlugach, Zhanna
Doressoundiram, Alain
Drake, Frank
Drossart, Pierre
Duffard, Rene
Dunkin Beardsley, Sarah
Durrance, Samuel
Ehrenreich, David
El-Baz, Farouk
Encrenaz, Thérèse
Epishev, Vitali
Esposito, Larry
Evans, Michael
Ferrari, Cécile
Ferrusca, Daniel
Feulner, Georg
Filacchione, Gianrico
Fink, Uwe
Fletcher, Leigh
FUJIWARA, Akira
Gautier, Daniel
Geiss, Johannes
Gérard, Jean-Claude
Gierasch, Peter
Gillon, Michaël
Goldreich, Peter
Goody, Richard
Gorenstein, Paul
Gorkavyi, Nikolai
Gounelle, Matthieu
Grav, Tommy
Green, Jack
Grieger, Bjoern

Grossman, Lawrence
Gulkis, Samuel
Gurshtein, Alexander
Halliday, Ian
Hammel, Heidi
Hanninen, Jyrki
Harris, Alan
HASEGAWA, Ichiro
Helled, Ravit
Heng, Kevin
Hersant, Franck
Holberg, Jay
Horedt, Georg
Hovenier, J.
Hubbard, William
Hunt, Garry
HWANG, Junga
Irvine, William
Irwin, Patrick
IWASAKI, Kyosuke
Jin, Liping
Johnson, Torrence
Jordan, Andrés
Jurgens, Raymond
Kaeufl, Hans Ulrich
Kascheev, Rafael
Killen, Rosemary
KIM, Yongha
KIM, Sang Joon
KIM, Yoo Jea
KIM, Joo Hyeon
Kley, Wilhelm
Kostama, Veli-Petri
Kozak, Lyudmyla
Kraft, Ralph
Krimigis, Stamatios
Kumar, Shiv
Kurt, Vladimir
Lane, Arthur
Larson, Harold
Larson, Stephen
Laufer, Diana
LEE, Jun
Lellouch, Emmanuel
Lemmon, Mark
Lewis, John
Li, Jian-Yang
Licandro, Javier
Lichtenegger, Herbert
Lineweaver, Charles
Lissauer, Jack
LIU, Xiaoqun
LIU, Chengzhi
Lo Curto, Gaspare
Lockwood, G.
Lodders, Katharina
Lopes, Rosaly
Lopez Moreno, Jose
Lopez Puertas, Manuel
Lopez Valverde, M.
Lutz, Barry
Luz, David
LYKAWKA, Patryk
Lyon, Ian
Mainzer, Amy
Makalkin, Andrei
Marchi, Simone
Marcialis, Robert
Margot, Jean-Luc
Marov, Mikhail
Martinez-Frias, Jesus
Matson, Dennis
MATSUI, Takafumi
Mazzotta Epifani, Elena
McCord, Thomas
McCullough, Peter
McElroy, Michael
McGrath, Melissa
McKinnon, William
Meadows, A.
Mendillo, Michael
Mickelson, Michael
Millis, Robert
Mills, Franklin
Moehlmann, Diedrich
Molina, Antonio
Moncuquet, Michel
Montmessin, Franck
Moreno-Insertis, Fernando
Morozhenko, A.
Mosser, Benoît
Mousis, Olivier
Mumma, Michael
Murphy, Robert
NAKAGAWA, Yoshitsugu
Nelson, Richard
Ness, Norman
Nixon, Conor
Noll, Keith
OHTSUKI , Keiji
Ortiz, Jose
Owen, Tobias
Pang, Kevin
Paolicchi, Paolo
PARK, Yong Sun
Pascu, Dan
Peixinho, Nuno
Perez, Mario
Petit, Jean-Marc
Pettengill, Gordon
PING, Jinsong
Politi, Romolo
Potter, Andrew
Psaryov, Volodymyr
Quanz, Sascha
Rao, M.
Rodionova, Zhanna
Roos-Serote, Maarten
Roques, Françoise
Rosenbush, Vera
Rossi, Alessandro
Ruskol, Evgeniya
Russo, Pedro
Sampson, Russell
Sanchez-Lavega, Agustin
Sasaki, Sho
Schaller, Emily
Scheirich, Peter
Schleicher, David
Schloerb, F.
Schneider, Nicholas
SEO, Haingja
Sergis, Nick
Shapiro, Irwin
Shevchenko, Vladislav
SHI, Huli
Shkuratov, Yurii
Sicardy, Bruno
Sims, Mark
Smith, Bradford
Snellen, Ignas
Soderblom, Larry
Sprague, Ann
Stallard, Thomas
Stam, Daphne
Stern, S.
Sterzik, Michael
Stewart-Mukhopadhyay, S.
Stoev, Alexey
Stone, Edward

Strobel, Darrell
Strom, Robert
Synnott, Stephen
Tanga, Paolo
Taylor, Fredric
Tchouikova, Nadezhda
Tedds, Jonathan
Terrile, Richard
Tholen, David
Thomas, Nicolas
TIAN, Feng
Tiscareno, Matthew
Tosi, Federico
Trafton, Laurence
Trujillo, Chadwick
Tyler Jr, G.
Veiga, Carlos
Veverka, Joseph
Vidmachenko, Anatoliy
Walker, Alta
Wallace, Lloyd
Wannawichian, Suwicha
Wasserman, Lawrence
Wasson, John
Weidenschilling, S.
Williams, Iwan
Williams, James
Wilson, Lionel
Woolfson, Michael
Wu, Yanqin
Wurz, Peter
Yair, Yoav
Yanamandra-Fisher, P.
YANG, Xiaohu
YI, Yu
Yoder, Charles
Young, Andrew
Zarka, Philippe
ZHANG, Jun
ZHANG, Mian
Zhang, Tielong
Zharkov, Vladimir

Division I Commission 19 Rotation of the Earth

President: Harald Schuh
Vice-President: Cheng-Li Huang

Organizing Committee Members:

Chao, Benjamin
Gross, Richard
Huang, Cheng-Li
Kosek, Wieslaw
Malkin, Zinovy
Richter, Bernd
Salstein, David
Titov, Oleg

Members:

Arabelos, Dimitrios
Archinal, Brent
Arias, Elisa
Banni, Aldo
Barkin, Yuri
Bartlett, Jennifer
Beutler, Gerhard
Boehm, Johannes
Bolotin, Sergei
Bolotina, Olga
Boucher, Claude
Bougeard, Mireille
Bourda, Geraldine
Brentjens, Michiel
Brosche, Peter
Brzezinski, Aleksander
Capitaine, Nicole
Cazenave, Anny
Chao, Benjamin
Chapanov, Yavor
CHEN, Wen Ping
Damljanovic, Goran
De Biasi, Maria
de Viron, Olivier
Debarbat, Suzanne
Defraigne, Pascale
Dehant, Véronique
Dejaiffe, Rene
Deleflie, Florent
Dick, Wolfgang
Dickey, Jean
Dickman, Steven
DU, Lan
Eppelbaum, Lev
Escapa, Alberto
Fernandez, Laura
Ferrandiz, Jose
Fliegel, Henry
Folgueira, Marta
FONG, Chugang
FUJISHITA, Mitsumi
FUKUSHIMA, Toshio
Gambis, Daniel
GAO, Buxi
Gaposchkin, Edward
Gayazov, Iskander
Getino Fernandez, Juan
Gozhy, Adam
Guinot, Bernard
Gusev, Alexander
Haas, Rüdiger
HAN, Tianqi
HAN, Yanben
Hefty, Jan
Hobiger, Thomas
HUANG, Cheng
Hugentobler, Urs
Jaeggi, Adrian
JIN, WenJing
Johnson, Thomas
Jubier, Xavier
KAKUTA, Chuichi
KAMEYA, Osamu
Khoda, Oleg
Klepczynski, William
Kolaczek, Barbara
Korsun, Alla
Kosek, Wieslaw
Kostelecky, Jan
Kouba, Jan
LEE, Jun
Lehmann, Marek
LI, Jinling
LI, Yong
LIAO, Dechun
Lieske, Jay
LIU, Ciyuan
LIU, Chengzhi
Luzum, Brian
Ma, Chopo
MA, Lihua
Malkin, Zinovy
MANABE , Seiji
Marshalov, Dmitriy
MASAKI, Yoshimitsu
McCarthy, Dennis
Melbourne, William
Merriam, James
Minazzoli, Olivier
Monet, Alice
Morgan, Peter
Morrison, Leslie
Mueller, Ivan
Müller, Jürgen
Mysen, Eirik
Nastula, Jolanta
Navarro, Julio
Newhall, X.
Nilsson, Tobias
Nothnagel, Axel
Panafidina, Natalia
Paquet, Paul
PARK, Pil Ho
Pejovic, Nadezda
Pesek, Ivan
Petit, Gérard
Petrov, Sergey
Picca, Domenico

Pilkington, John
Poma, Angelo
Proverbio, Edoardo
Pugliano, Antonio
Ray, James
Richter, Bernd
Robertson, Douglas
Rogister, Yves
Ron, Cyril
Rothacher, Markus
Ruder, Hanns
Rykhlova, Lidiya
Salstein, David
Sansaturio, Maria
SASAO, Tetsuo
SATO, Koichi
Schillak, Stanislaw
Schreiber, Karl
Schuh, Harald
Schutz, Bob
Seitz, Florian
Sevilla, Miguel
Seyed-Mahmoud, Behnam
Shapiro, Irwin
Shelus, Peter
SHI, Huli
SHU, Fengchun
Shuygina, Nadia
Sidorenkov, Nikolay
Skurikhina, Elena
Soffel, Michael
Souchay, Jean
Steigenberger, Peter
Stephenson, F.
SUN, Fuping
Surkis, Igor
Tapley, Byron
Tarady, Vladimir
Thaller, Daniela
Thomas, Maik
Titov, Oleg
Veillet, Christian
Vicente, Raimundo
Wallace, Patrick
WANG, Zhengming
WANG, Kemin
WANG, Qi-jie
WANG, Xiaoya
WANG, Guangli
Weber, Robert
WEI, Erhu
Williams, James
Wooden, William
WU, Bin
XIAO, Naiyuan
Yatskiv, Yaroslav
YE, Shuhua
YU, Nanhua
ZHANG, Zhongping
ZHENG, Yong
ZHONG , Min
ZHOU, Yonghong
ZHU, Yaozhong

Division III Commission 20 Positions & Motions of Minor Planets, Comets & Satellites

President: Makoto YOSHIKAWA
Vice-President: Steven R. Chesley

Organizing Committee Members:

Gilmore, Alan
Pravec, Petr
Spahr, Timothy
Ticha, Jana
ZHU, Jin

Members:

A'Hearn, Michael
Abalakin, Viktor
Aikman, G.
Aksnes, Kaare
Arlot, Jean-Eudes
Babadzhanov, Pulat
Baggaley, William
Bailey, Mark
Baransky, Olexander
Behrend, Raoul
Benest, Daniel
Benkhoff, Johannes
Bernardi, Fabrizio
Berthier, Jérôme
Bien, Reinhold
Blanco, Carlo
Boerngen, Freimut
Borisov, Galin
Borisov, Borislav
Bowell, Edward
Branham, Richard
Bueno de Camargo, Julio
Burns, Joseph
Carpino, Mario
Carusi, Andrea
Chapront-Touze, Michelle
Chodas, Paul
Churyumov, Klim
Cooper, Nicholas
Darhmaouil, Hassane
de Sanctis, Giovanni
Delbo, Marco
Di Sisto, Romina
Donnison, John
Dourneau, Gerard
Doval, Jorge M.
Dunham, David
Dvorak, Rudolf
Dybczynski, Piotr
Elst, Eric
Emelianov, Nikolaj
Emelyanenko, Vacheslav
Epishev, Vitali
Evans, Michael
Fernandez, Julio
Fernandez, Yanga
Ferraz -Mello, Sylvio
Ferreri, Walter
Fors, Octavi
Franklin, Fred
Fraser, Brian
Freitas Mourao, Ronaldo
Froeschle, Claude
FUSE, Tetsuharu
Gibson, James
Giorgini, Jon
Gomez, Edward
Gorshanov, Denis
Green, Daniel
Greenberg, Richard
Hahn, Gerhard
Hainaut, Olivier
Harper, David
Harris, Alan
HASEGAWA, Ichiro
Haupt, Hermann
Hemenway, Paul
Heudier, Jean-Louis
Hoenig, Sebastian
Hol, Pedro
HOU, Xiyun
Hsieh, Henry
Hudkova, Ludmila
Hurnik, Hieronim
Husarik, Marek
Ianna, Philip
Ivanova, Violeta
Ivantsov, Anatoliy
Izmailov, Igor
Jacobson, Robert
Jakubik, Marian
Kablak, Nataliya
Kazantsev, Anatolii
Kilmartin, Pamela
KIM, Sang Joon
KINOSHITA, Hiroshi
Kisseleva, Tamara
Klemola, Arnold
Knezevic, Zoran
Knight, Matthew
Kohoutek, Lubos
KOSAI, Hiroki
Koschny, Detlef
Koshkin, Nikolay
KOZAI, Yoshihide
Kristensen, Leif
Krolikowska-Soltan, M.
Krugly, Yurij
Kulikova, Nelly
Lagerkvist, Claes-Ingvar
Larsen, Jeffrey
Laurin, Denis
Lemaître, Anne
LI, Guangyu
LI, Yong
Li, Jian-Yang
Lieske, Jay
Lomb, Nicholas

Lovas, Miklos
LYKAWKA, Patryk
MA, Guanyi
Mainzer, Amy
Manara, Alessandro
Mancini, Dario
Martins, Roberto
Matese, John
Maury, Alain
McMillan, Robert
McNaught, Robert
Medvedev, Yurij
Melita, Mario
Millis, Robert
Monet, Alice
Moravec, Zden?k
Morris, Charles
Murray, Carl
Nacozy, Paul
NAKAMURA, Tsuko
NAKANO, Syuichi
Nedelcu, Dan
Neslusan, Lubos
NGUYEN, Phuong
Nobili, Anna
Novakovic, Bojan
Noyelles, Benoît
Olive, Don
Owen Jr, William
Ozisik, Tuncay
Page, Gary
Pandey, A.
Pascu, Dan
Pat-El, Igal
Pauwels, Thierry
Perozzi, Ettore
Pierce, David
PING, Jinsong
Pittich, Eduard
Polyakhova, Elena
Porubcan, Vladimir
Pozhalova, Zhanna
QIAO, Rongchuan
Raju, Vasundhara
Rapaport, Michel
Reitsema, Harold
Rekola, Rami
Richardson, Derek
Rickman, Hans
Roemer, Elizabeth
Roeser, Siegfried
Rossi, Alessandro
RUI, Qi
Santos-Sanz, Pablo
SATO, Isao
Savanevich, Vadim
Scheirich, Peter
Schmadel, Lutz
Schober, Hans
Scholl, Hans
Schubart, Joachim
Schuster, William
Seidelmann, P.
Sekanina, Zdenek
SEO, Haingja
Shanklin, Jonathan
Shelus, Peter
SHEN, Kaixian
SHI, Huli
Shor, Viktor
Solovaya, Nina
SOMA, Mitsuru
Sosa, Andrea
Standish, E.
Steel, Duncan
Stellmacher, Irène
Stokes, Grant
Suli, Áron
Svoren, Jan
Sybiryakova, Yegeniya
Synnott, Stephen
Szabo, Gyula
Szutowicz, Slawomira
Tancredi, Gonzalo
Tatum, Jeremy
Taylor, Donald
Thuillot, William
Tiscareno, Matthew
Titov, Vladimir
Trujillo, Chadwick
Tsiganis, Kleomenis
Tsuchida, Masayoshi
Tuccari, Gino
van Houten-Groeneveld, I.
Veillet, Christian
Vienne, Alain
Virtanen, Jenni
Voelzke, Marcos
WANG, Xiaoya
Wasserman, Lawrence
Weiss, John
Weissman, Paul
West, Richard
Whipple, Arthur
Williams, Gareth
Williams, Iwan
Williams, James
XIONG, Jianning
YABUSHITA, Shin
Yeomans, Donald
YIM, Hong Suh
YOSHIKAWA, Makoto
YUASA, Manabu
Zadnik, Marjan
Zagretdinov, Renat
ZHANG, Xiaoxiang
ZHANG, Wei
ZHANG, Yang
ZHAO, Haibin
Ziolkowski, Krzysztof

Division IX Commission 21 Galactic & Extragalactic Background Radiation

President: Jayant Murthy
Vice-President: Jayant Murthy

Organizing Committee Members:

Baggaley, William
Dwek, Eli
Gustafson, Bo
Levasseur-Regourd, A.-Ch.
Mann, Ingrid
Mattila, Kalevi
WATANABE, Jun-ichi

Members:

Angione, Ronald
ASANO, Katsuaki
Behar, Ehud
Belkovich, Oleg
Boquien, Médéric
Bot, Caroline
Bouwens, Rychard
Bowyer, C.
Broadfoot, A.
Caputi, Karina
Chernyakova, Maria
Chiang, Hsin
Clairemidi, Jacques
d'Hendecourt, Louis
de Petris, Marco
Dermott, Stanley
Dodonov, Sergej
Dole, Herve
Dubin, Maurice
Dumont, René
Feldman, Paul
FUJIWARA, Akira
Giovane, Frank
Gounelle, Matthieu
Gruen, Eberhard
Hanner, Martha
HAO, Lei
Harwit, Martin
Hauser, Michael
Hecht, James
Henry, Richard
Hernandez-Monteagudo, C.
Hofmann, Wilfried
HONG, Seung-Soo
Horns, Dieter
Hurwitz, Mark
Ivanov-Kholodny, Gor
Jackson, Bernard
James, John
Jelic, Vibor
Johns, Bethany
Joubert, Martine
Keshet, Uri
KOBAYASHI, Masakazu
Kopylov, Alexander
Korpela, Eric
Koutchmy, Serge
Kozak, Lyudmyla
Kramer, Busaba
Kulkarni, Prabhakar
Lamy, Philippe
Lapi, Andrea
Léger, Alain
Leinert, Christoph
Lemke, Dietrich
Lenc, Emil
LI, Li-Xin
Lillie, Charles
Lopez Gonzalez, Maria
Lopez Moreno, José
Lopez Puertas, Manuel
MAIHARA, Toshinori
Martin, Donn
Massaro, Francesco
Mather, John
MATSUMOTO, Toshio
Maucherat, Jean
Misconi, Nebil
Moodley, Kavilan
Muinonen, Karri
MUKAI, Tadashi
NAKAMURA, Akiko
Nishimura, Tetsuo
Paresce, Francesco
Perez-Gonzalez, Pablo
Perrin, Jean-Marie
Pfleiderer, Jorg
Rauch, Michael
Reach, William
Renard, Jean-Baptiste
Risaliti, Guido
Sanchez, Francisco
Sanchez-Saavedra, M.
Schuh, Harald
Seymour, Nicholas
Shefov, Nikolaj
SHIRAHATA, Mai
Siringo, Giorgio
Smith, Robert
Soberman, Robert
Staude, Hans
Sykes, Mark
Tepper Garcia, Thorsten
Toller, Gary
Tsygankov, Sergey
Tyson, John
UENO, Munetaka
Ugolnikov, Oleg
Vaccari, Mattia
Venters, Tonia
Vrtilek, Jan
Wagner, Robert
Wallis, Max
Weinberg, Jerry
Wendt, Martin

Wesson, Paul
Wheatley, Peter
Witt, Adolf
Wolleben, Maik
Wolstencroft, Ramon
Woolfson, Michael
YAMAMOTO, Tetsuo
YAMASHITA, Kojun
YANG, Xiaohu
ZHANG, Yuying

Division III Commission 22 Meteors, Meteorites & Interplanetary Dust

President: Jun-ichi WATANABE
Vice-President: Petrus Matheus Marie Jenniskens

Organizing Committee Members:

Borovicka, Jiří
Campbell-Brown, Margaret
Consolmagno, Guy
Jenniskens, Petrus
Jopek, Tadeusz
Vaubaillon, Jérémie
Williams, Iwan

Members:

Alexandrov, Alexander
Apai, Daniel
Asher, David
Babadzhanov, Pulat
Baggaley, William
Belkovich, Oleg
Benkhoff, Johannes
Bhandari, N.
Brown, Peter
Brownlee, Donald
Busarev, Vladimir
Campbell-Brown, Margaret
Capek, David
Carusi, Andrea
Cooper, Timothy
Dieleman, Pieter
Djorgovski, Stanislav
Dubin, Maurice
Duffard, Rene
Fromang, Sébastien
Gajdos, Stefan
Gorbanev, Jury
Goswami, J.
Gounelle, Matthieu
Grady, Monica
Granvik, Mikael
Gruen, Eberhard
Gustafson, Bo
Hajdukova, Maria
Hajdukova, Jr., Maria
Halliday, Ian
Hanner, Martha
Harvey, Gale
HASEGAWA, Ichiro
HASEGAWA, Sunao
Hawkes, Robert
Hodge, Paul
HONG, Seung-Soo
Husarik, Marek
Jakubik, Marian
Jones, James
Jopek, Tadeusz
Kalenichenko, Valentin
Kanuchova, Zuzana
Karakas, Amanda
Keay, Colin
Khovritchev, Maxim
Kikwaya Eluo, J.-B.
KIMURA, Hiroshi
Koeberl, Christian
Kokhirova, Gulchehra
Kolomiyets, Svitlana
Kornos, Leonard
Koschny, Detlef
Kostama, Veli-Petri
Koten, Pavel
Kozak, Pavlo
Kruchinenko, Vitaliy
Lamy, Philippe
Levasseur-Regourd, A.-Ch.
Lodders, Katharina
Lugaro, Maria
Lyon, Ian
Makalkin, Andrei
Mann, Ingrid
Maris, Michele
Martinez-Frias, Jesus
Marvin, Ursula
Mason, John
Mawet, Dimitri
Meisel, David
Misconi, Nebil
Moor, Attila
Murray, Carl
NAGAHARA, Hiroko
NAKAMURA, Takuji
NAKAZAWA, Kiyoshi
Napier, William
Nuth, Joseph
Pecina, Petr
Pellinen-Wannberg, Asta
Politi, Romolo
Poole, Graham
Rendtel, Juergen
Rickman, Hans
Ripken, Hartmut
Santos-Sanz, Pablo
Sasaki, Sho
Sekanina, Zdenek
SHANG, Hsien
Shrbeny, Lukáš
Soberman, Robert
Spurny, Pavel
Steel, Duncan
Stewart-Mukhopadhyay, S.
Svestka, Jiri
Svoren, Jan
Tatum, Jeremy
Tedesco, Edward
Toshihiro, Kasuga
Tosi, Federico
Toth, Juraj
Trigo-Rodriguez, Josep
Valsecchi, Giovanni
Vaubaillon, Jérémie

Voloschuk, Yuri
WATANABE, Jun-ichi
Webster, Alan
Weinberg, Jerry
Williams, Iwan
Woolfson, Michael
Yair, Yoav
YAMAMOTO, Masayuki
Yeomans, Donald
Zadnik, Marjan
Zender, Joe
ZHANG, Xiaoxiang
ZHAO, Haibin

Division IX Commission 25 Astronomical Photometry & Polarimetry

President: Eugene F. Milone
Vice-President: Alistair Walker

Organizing Committee Members:

Anthony-Twarog, Barbara
Bastien, Pierre
Knude, Jens
Kurtz, Donald
Menzies, John
Mironov, Aleksey
QIAN, Shengbang

Members:

Ables, Harold
Ahumada, Javier
Aigrain, Suzanne
AKITAYA, Hiroshi
Albrecht, Rudolf
Alecian, Evelyne
Anandaram, Mandayam
Andreuzzi, Gloria
Angel, J.
Angione, Ronald
Ashok, N.
Aspin, Colin
Aungwerojwit, Amornrat
Baliyan, Kiran
Balona, Luis
Baran, Andrzej
Barnes III, Thomas
Barrett, Paul
Barrientos, Luis
Baume, Gustavo
Bellazzini, Michele
Berdyugin, Andrei
Bessell, Michael
Birkmann, Stephan
Bjorkman, Jon
Borisova, Ana
Borra, Ermanno
Braithwaite, Jonathan
Breger, Michel
Brown, Douglas
Brown, Thomas
Buser, Roland
CAI, Mingsheng
Cantiello, Michele
Canto Martins, Bruno
Carciofi, Alex
Carney, Bruce
Carter, Brian
Castelaz, Micheal
Cesetti, Mary
Chadid, Merieme
CHEN, An-Le
Cioni, Maria-Rosa
Clark, David
Clem, James
Clocchiatti, Alejandro
Connolly, Leo
Copin, Yannick
Coyne, S.J, George
Crawford, David
Cuypers, Jan
Dahn, Conard
DAI, Zhibin
DAISAKU, Nogami
Danford, Stephen
Demory, Brice-Olivier
Deshpande, M.
Dolan, Joseph
DOU, Jiangpei
Dubout, Renee
Ducati, Jorge
Ducourant, Christine
Edwards, Paul
Elkin, Vladimir
Elmhamdi, Abouazza
Fabiani, Sergio
Fabregat, Juan
Fabrika, Sergei
Feinstein, Alejandro
Fernandez Lajus, Eduardo
Fernie, J.
Fluri, Dominique
Forte, Juan
Galadi-Enriquez, David
Garrison, Robert
Gehrz, Robert
Genet, Russell
Gerbaldi, Michèle
Ghosh, Swarna
Gilliland, Ronald
Gillon, Michaël
Giorgi, Edgard
Glass, Ian
Golay, Marcel
Graham, John
Grauer, Albert
Grenon, Michel
Grewing, Michael
Grundahl, Frank
Guetter, Harry
Guglielmino, Salvatore
Gulbis, Amanda
Hackman, Thomas
Hauck, Bernard
Hensberge, Herman
Hilditch, Ronald
Hubrig, Swetlana
Huovelin, Juhani
Hyland, Harry
Ioannou, Zacharias
Irwin, Alan
Ivezic, Zeljko
Jeffers, Sandra
Jerzykiewicz, Mikolaj
JIANG, Zhibo

Joshi, Umesh
Karoff, Christoffer
Kasiviswanathan, Sankara-subramanian
Kazlauskas, Algirdas
Kebede, Legesse
Keller, Stefan
KENTARO, Matsuda
Kepler, S.
Kilkenny, David
KIM, Seung-Lee
King, Ivan
Kornilov, Victor
Kospal, Ágnes
Kulkarni, Prabhakar
Kurucz, Robert
Landstreet, John
Langlois, Maud
Laskarides, Paul
Laugalys, Vygandas
Lazauskaite, Romualda
Lemke, Michael
Lenzen, Rainer
Leroy, Jean-Louis
LI, Qingkang
LI, Min
Linde, Peter
Lockwood, G.
Lub, Jan
Luna, Homero
Magnier, Eugene
Maitzen, Hans
Manfroid, Jean
Manset, Nadine
Markkanen, Tapio
Marraco, Hugo
Marsden, Stephen
Martinez, Peter
Martinez Roger, Carlos
Maslennikov, Kirill
Mason, Paul
Mathys, Gautier
Mayer, Pavel
McDavid, David
McLean, Ian
Mendoza, V.
Metcalfe, Travis
Miller, Joseph
Milone, Eugene
Mironov, Aleksey
Moffett, Thomas
Moitinho, André
Monaco, Lorenzo
Mourard, Denis
Munari, Ulisse
Narbutis, Donatas
Naylor, Tim
Neiner, Coralie
Nicolet, Bernard
Nikoghosyan, Elena
NOGUCHI, Kunio
Notni, Peter
Oblak, Edouard
Oestreicher, Roland
Orsatti, Ana
PAK, Soojong
Parimucha, Stefan
Pavani, Daniela
Pedreros, Mario
Pel, Jan
Penny, Alan
Perrin, Marshall
Petit, Pascal
Pfeiffer, Raymond
Philip, A.G.
Piirola, Vilppu
Platais, Imants
Pokrzywka, Bartlomiej
Pulone, Luigi
QIAN, Shengbang
Rank-Lueftinger, Theresa
Rao, Pasagada
Raveendran, A.
Rawlings, Mark
Reglero Velasco, Victor
Reshetnyk, Volodymyr
Rhee, Jaehyon
Robb, Russell
Robinson, Edward
Rodrigues, Claudia
Romanyuk, Yaroslav
Rostopchina, Alla
Sabin, Laurence
Santos Agostinho, Rui
Santos-Sanz, Pablo
Schiller, Stephen
Schuster, William
SEKIGUCHI, Kazuhiro
Sen, Asoke
Shankland, Paul
Shawl, Stephen
Shoyoqubov, Shoayub
Simons, Douglas
Snik, Frans
Snowden, Michael
Stagg, Christopher
Steinlin, Uli
Stockman Jr, Hervey
Stone, Remington
Stonkut?, Rima
Straizys, Vytautas
Stritzinger, Maximilian
Subramaniam, Annapurni
Sudzius, Jokubas
Sullivan, Denis
Szkody, Paula
Szymanski, Michal
Tandon, S.
Taranova, Olga
Tas, Günay
Tedds, Jonathan
Thompson, Rodger
Thurston, Mark
Tokunaga, Alan
Tolbert, Charles
Townsend, Richard
Turcu, Vlad
Ueta, Toshiya
UMEDA, Hideyuki
Ureche, Vasile
Uslenghi, Michela
Vaughan, Arthur
Verma, R.
Vidotto, Aline
Voloshina, Irina
Vrba, Frederick
Walker, William
Warren Jr, Wayne
Weiss, Werner
Weistrop, Donna
Wesselius, Paul
Wheatley, Peter
White, Nathaniel
Wielebinski, Richard
Willstrop, Roderick
Winiarski, Maciej
Wramdemark, Stig
YAMASHITA, Yasumasa
YAO, Yongqiang
Young, Andrew

Yudin, Ruslan
YUJI, Ikeda
Zdanavičius, Justas
ZHU, Liying
Zoccali, Manuela
Zwintz, Konstanze

Division IV Commission 26 Double & Multiple Stars

President: Jose-Angel Docobo
Vice-President: Brian D. Mason

Organizing Committee Members:

Arenou, Frédéric
Balega, Yurij
Oswalt, Terry
Pourbaix, Dimitri
Scardia, Marco
Scarfe, Colin
Tamazian, Vakhtang

Members:

Abt, Helmut
Ahumada, Javier
AK, Tansel
Allen, Christine
Andrade, Manuel
Anosova, Joanna
Antokhina, Eleonora
Arenou, Frédéric
Argyle, Robert
Armstrong, John
Aungwerojwit, Amornrat
Bagnuolo Jr, William
Bailyn, Charles
Batten, Alan
Beavers, Willet
Beklen, Elif
Boden, Andrew
Bonneau, Daniel
Boyajian, Tabetha
Brandner, Wolfgang
Brosche, Peter
Brown, David
Budaj, Jan
Carciofi, Alex
CHEN, Wen Ping
Culver, Roger
Cvetkovic, Zorica
DAISAKU, Nogami
De Becker, Michaël
De Cat, Peter
de Mink, Selma
de Val-Borro, Miguel
Dominis Prester, Dijana
Dukes Jr., Robert
Dunham, David
Eldridge, John
Elkin, Vladimir
Falceta-Goncalves, Diego
Fekel, Francis
Fernandes, João
Ferrer, Osvaldo
Fors, Octavi
Fox-Machado, Lester
Franz, Otto
Fredrick, Laurence
Freitas Mourao, Ronaldo
Gandolfi, Davide
Gatewood, George
Gaudenzi, Silvia
Gavras, Panagiotis
Geller, Aaron
Genet, Russell
Geyer, Edward
Ghez, Andrea
Goncalves, Denise
Grundstrom, Erika
Gun, Gulnur
Hakkila, Jon
Halbwachs, Jean-Louis
Hartigan, Patrick
Hartkopf, William
HE, Jinhua
Heacox, William
Hershey, John
Hidayat, Bambang
Hill, Graham
Hillwig, Todd
Hindsley, Robert
Horch, Elliott
Hummel, Christian
Hummel, Wolfgang
Ianna, Philip
Ireland, Michael
Izmailov, Igor
Izzard, Robert
Jahreiss, Hartmut
Jassur, Davoud
JEON, Young Beom
Johnston, Helen
Jurdana-Sepic, Rajka
Kafka, Styliani (Stella)
Kazantseva , Liliya
Kisseleva-Eggleton, L.
Kitsionas, Spyridon
Kley, Wilhelm
Koehler , Rainer
Kouwenhoven, M.B.N.
Kroupa, Pavel
Kubat, Jiri
Latham, David
Lattanzi, Mario
Lee, William
LEE, Jae Woo
LEE, Chung-Uk
Leinert, Christoph
Lepine, Sebastien
Levato, Orlando
LIM, Jeremy
Ling, Josefina
Lippincott Zimmerman, S.
LIU, Michael
Lyubchik, Yuri
Maddison, Sarah
Maiz Apellaniz, Jesús
Malkov, Oleg
Malogolovets, Evgeny
Marsakova, Vladislava
Martayan, Christophe

Martin, Eduardo
Mathieu, Robert
Mawet, Dimitri
McAlister, Harold
McBride, Vanessa
McDavid, David
Mennickent, Ronald
Middleton, Christopher
Mikkola, Seppo
Mikolajewski, Maciej
Millour, Florentin
Mohan, Chander
Morbidelli, Roberto
Morel, Pierre-Jacques
Morrell, Nidia
Negueruela, Ignacio
Neuhaeuser, Ralph
Nitschelm, Christian
Nuernberger, Dieter
Orlov, Victor
Oswalt, Terry
Parimucha, Stefan
Pauls, Thomas
Pereira, Claudio
Perets, Hagai
Peterson, Deane
Petr-Gotzens, Monika
Pietrukowicz, Pawel
Pollacco, Don
Popovic, Georgije
Pourbaix, Dimitri
Poveda, Arcadio
Prieto, Cristina
Prieur, Jean-Louis
Prsa, Andrej
Rastegaev, Denis
Roberts Jr, Lewis
Rodrigues de Oliveira F., I.
Russell, Jane
Ruzdjak, Domagoj
Sagar, Ram
Scardia, Marco
Scarfe, Colin
Schmidtke, Paul
Schoeller, Markus
Shakht, Natalia
Shatsky, Nicolai
Simon, Michal
Sinachopoulos, Dimitris
Skokos, Charalambos
Smak, Jozef
Smith, J.
Soderhjelm, Staffan
Sowell, James
Stein, John
Sterzik, Michael
Sudar, Davor
Szabados, Laszlo
Tamazian, Vakhtang
Tango, William
Tarasov, Anatolii
Teixeira, Paula
Terquem, Caroline
Titov, Vladimir
Tomasella, Lina
Torres, Guillermo
Trimble, Virginia
Tsygankov, Sergey
Turner, Nils
Udry, Stephane
Upgren, Arthur
Valtonen, Mauri
van Altena, William
van der Bliek, Nicole
van der Hucht, Karel
van Dessel, Edwin
Vanko, Martin
Vaz, Luiz Paulo
Vennes, Stephane
WANG, Jiaji
Weis, Edward
WEN, Linqing
Zasche, Petr
Zavala, Robert
Zheleznyak, Alexander
ZHU, Liying
Zinnecker, Hans

Division V Commission 27 Variable Stars

President: Gerald Handler
Vice-President: Karen Pollard

Organizing Committee Members:

Bedding, Timothy
Catelan, Márcio
Cunha, Margarida
Eyer, Laurent
Jeffery, Christopher
Kepler, S.
Kolenberg, Katrien
Martinez, Peter
Mkrtichian, David
Olah, Katalin
Somasundaram, Seetha

Members:

Aerts, Conny
Aigrain, Suzanne
Airapetian, Vladimir
Aizenman, Morris
AK, Tansel
Albinson, James
Albrow, Michael
Alencar, Silvia
Alfaro, Emilio
Alpar, Mehmet
Amado Gonzalez, Pedro
ANDO, Hiroyasu
Andrievsky, Sergei
Andronov, Ivan
Antipin, Sergei
Antonello, Elio
Antonyuk, Kirill
Antov, Alexandar
Arellano Ferro, Armando
Arentoft, Torben
Arias, Maria
Arkhipova, Vera
Asteriadis, Georgios
Aungwerojwit, Amornrat
Avgoloupis, Stavros
Baade, Dietrich
Baglin, Annie
Balman, Solen
Balona, Luis
Baran, Andrzej
Baransky, Olexander
Barban, Caroline
Barnes III, Thomas
Bartolini, Corrado
Barway, Sudhashu
Barwig, Heinz
Baskill, Darren
Bastien, Pierre
Bauer, Wendy
Bazot, Michael
Beaulieu, Jean-Philippe
Bedding, Timothy
Bedogni, Roberto
Belkacem, Kevin
Belmonte Aviles, J. A.
Belserene, Emilia
Belvedere, Gaetano
Benko, Jozsef
Berdnikov, Leonid
Bersier, David
Berthomieu, Gabrielle
Bessell, Michael
Bianchini, Antonio
Bjorkman, Karen
Bolton, Charles
Bond, Howard
Bopp, Bernard
Borczyk, Wojciech
Borisova, Ana
Bortoletto, Alexandre
Boyd, David
Bradley, Paul
Breger, Michel
Briquet, Maryline
Brown, Douglas
Bruch, Albert
Bruntt, Hans
Buccino, Andrea
Burki, Gilbert
Burwitz, Vadim
Busa, Innocenza
Busko, Ivo
Butkovskaya, Varvara
Butler, Christopher
Buzasi, Derek
Cacciari, Carla
Caldwell, John
Cameron, Andrew
CAO, Huilai
Carciofi, Alex
Carrier, Fabien
Casares, Jorge
Catchpole, Robin
Catelan, Márcio
Cenko, Stephen
Chadid, Merieme
Chaplin, William
CHEN, An-Le
CHEN, Alfred
Cherchneff, Isabelle
Cherepashchuk, Anatolij
CHOU, Yi
Christensen-Dalsgaard, J.
Christie, Grant
Cioni, Maria-Rosa
Clement, Christine
Clementini, Gisella
Clocchiatti, Alejandro
Cohen, Martin
Connolly, Leo
Contadakis, Michael
Cook, Kem
Corral, Luis
Corsico, Alejandro
Costa, Vitor

Cottrell, Peter
Coulson, Iain
Crause, Lisa
Creech-Eakman, Michelle
Cunha, Margarida
Cutispoto, Giuseppe
Cuypers, Jan
D'Amico, Nicolo'
DAI, Zhibin
DAISAKU, Nogami
Dall'Ora, Massimo
Danford, Stephen
Daszynska-Daszkiewicz, J.
De Becker, Michaël
De Cat, Peter
de Ridder, Joris
Delgado, Antonio
Demers, Serge
DENG, LiCai
Depagne, Éric
Derekas, Aliz
Deupree, Robert
Di Mauro, Maria Pia
Donahue, Robert
Dorokhova, Tetyana
Downes, Ronald
Dukes Jr., Robert
Dunlop, Storm
Dziembowski, Wojciech
Ederoclite, Alessandro
Edwards, Paul
Edwards, Suzan
Efremov, Yurij
Eggenberger, Patrick
Elkin, Vladimir
Engelbrecht, Chris
Esenoglu, Hasan
Evans, Nancy
Evans, Aneurin
Evans, Dafydd
Evren, Serdar
Fadeyev, Yurij
Feast, Michael
Ferland, Gary
Fernandez Lajus, Eduardo
Fernie, J.
Figer, Donald
Fokin, Andrei
Formiggini, Lilliana
Fox-Machado, Lester
Frew, David
Fromang, Sebastien
FU, Hsieh-Hai
FU, Jian-Ning
FUJIWARA, Tomoko
Gahm, Goesta
Gal-Yam, Avishay
Galis, Rudolf
Gameiro, Jorge
Gamen, Roberto
Garrido, Rafael
Gavras, Panagiotis
Gay, Pamela
Genet, Russell
Gershberg, R.
Geyer, Edward
Gieren, Wolfgang
Gies, Douglas
Gillet, Denis
Glagolevskij, Yurij
Gondoin, Philippe
Gosset, Eric
Gough, Douglas
Goupil, Marie-Jose
Graham, John
Grasberg, Ernest
Green, Daniel
Grinin, Vladimir
Groenewegen, Martin
Groh, Jose
Grundstrom, Erika
Grygar, Jiri
Guerrero, Gianantonio
Guinan, Edward
Gun, Gulnur
Gunthardt, Guillermo
GUO, Jianheng
Guzik, Joyce
Haas, Martin
Hackman, Thomas
Hackwell, John
Haefner, Reinhold
Haisch, Bernard
Halbwachs, Jean-Louis
Hallinan, Gregg
Hamdy, M.
Handler, Gerald
HAO, Jinxin
Harmanec, Petr
Hawley, Suzanne
Heiser, Arnold
Hempelmann, Alexander
Henden, Arne
Hesser, James
Hill, Henry
Hintz, Eric
Hojaev, Alisher
Horner, Scott
Houdek, Gunter
Houk, Nancy
Howell, Steve
Huenemoerder, David
Humphreys, Elizabeth
Hutchings, John
Iben Jr, Icko
Iijima, Takashi
Ikonnikova, Natalia
Ireland, Michael
Ishida, Toshihito
Ismailov, Nariman
ITA, Yoshifusa
Ivezic, Zeljko
Jablonski, Francisco
Janik, Jan
Jankov, Slobodan
Jeffers, Sandra
JEON, Young Beom
Jerzykiewicz, Mikolaj
Jetsu, Lauri
Jewell, Philip
Jha, Saurabh
JIANG, Biwei
JIN, Zhenyu
Joner, Michael
Jurcsik, Johanna
Kaeufl, Hans Ulrich
Kafka, Styliani (Stella)
Kalomeni, Belinda
KAMBE, Eij i
KANAMITS , Osamu
Kanbur, Shashi
Karitskaya, Evgeniya
Karovska, Margarita
Karp, Alan
Katsova, Maria
Kaufer, Andreas
Kawaler, Steven
Kaye, Anthony
Kazarovets, Elena
Keller, Stefan

Kervella, Pierre
Kilkenny, David
KIM, Tu Whan
KIM, Chulhee
KIM, Seung-Lee
KIM, Young-Soo
Kiplinger, Alan
Kippenhahn, Rudolf
Kiss, Laszlo
Kjeldsen, Hans
Kjurkchieva, Diana
Kochukhov, Oleg
Koen, Marthinus
Kollath, Zoltan
Komzik, Richard
Konstantinova-Antova, R.
Konacki, Grzegorz
Korhonen, Heidi
Kospal, Ágnes
Kovari, Zsolt
Krautter, Joachim
Kreiner, Jerzy
Krisciunas, Kevin
Krzesinski, Jerzy
Kubiak, Marcin
Kudryavtseva, Nadezhda
Kuhi, Leonard
Kunjaya, Chatief
KURAYAMA, Tomoharu
Kurtz, Donald
Lago, Maria
Lampens, Patricia
Landolt, Arlo
Laney, Clifton
Lanza, Antonino
Larionov, Valeri
Laskarides, Paul
Lawlor, Timothy
Lawson, Warrick
Lazaro, Carlos
Le Bertre, Thibaut
Lebzelter, Thomas
LEE, Jae Woo
LEE, Myung Gyoon
LEE, Hyun-chul
LEE, Jae Woo
Letarte, Bruno
Leung, Kam
LI, Yan
LI, Zhiping

Liermann, Adriane
Little-Marenin, Irene
LIU, Jifeng
Lloyd, Christopher
Lockwood, G.
Longmore, Andrew
Lopez, De
Lorenz-Martins, Silvia
Lub, Jan
Machado Folha, Daniel
Macri, Lucas
Madore, Barry
Maeder, Andre
Malov, Igor
Mantegazza, Luciano
Marchev, Dragomir
Marconi, Marcella
Margrave Jr, Thomas
Markoff, Sera
Marsakova, Vladislava
Martayan, Christophe
Martic, Milena
Martinez, Peter
Mason, Paul
Mateu , Cecilia
Mathias, Philippe
MATSUMOTO, Katsura
MATSUNAGA, Noriyuki
Matthews, Jaymie
Mauche, Christopher
Mazumdar, Anwesh
McGraw, John
Melikian, Norair
Mennickent, Ronald
Messina, Sergio
Michel, Eric
Miglio, Andrea
Mikolajewski, Maciej
Millour, Florentin
Milone, Eugene
Milone, Luis
Minikulov, Nasridin
Mkrtichian, David
Moffett, Thomas
Mohan, Chander
Molenda-Zakowicz, Joanna
Monard, Libert
Montalban, Josefina
Monteiro, Mario Joao
Montgomery, Michele

Morales Rueda, Luisa
Morel, Thierry
Morrison, Nancy
Moskalik, Pawe?
Mosoni, Laszlo
Mukai, Koji
Murdin, Paul
Nardetto, Nicolas
Nather, R.
Naylor, Tim
Naze, Yael
Neiner, Coralie
Neustroev, Vitaly
NGEOW, Chow Choong
Niarchos, Panagiotis
Niemczura, Ewa
NIINUMA, Kotaro
Nikoghosyan, Elena
Nitschelm, Christian
NITTA, Atsuko
Nota, Antonella
Nugis, Tiit
O'Donoghue, Darragh
O'Toole, Simon
Ogloza, Waldemar
Olah, Katalin
Oliveira, Alexandre
Olivier, Enrico
Oluseyi, Hakeem
Ostensen, Roy
Oswalt, Terry
OTSUKI, Kaori
Pal, András
Panei, Jorge
Papaloizou, John
Paparo, Margit
Parimucha, Stefan
PARK, Byeong-Gon
Parsamyan, Elma
Parthasarathy, Mudumba
Pat-El, Igal
Patat, Ferdinando
Paterno, Lucio
Pavlovski, Kresimir
Pazhouhesh, Reza
Pearson, Kevin
Penny, Matthew
Percy, John
Perez Hernandez, Fernando
Petersen, J.

Petit, Pascal
Petrov, Peter
Pettersen, Bjørn
Pietrukowicz, Pawel
Piirola, Vilppu
Pijpers, Frank
Plachinda, Sergei
Plavchan, Jr., Peter
Pollacco, Don
Pont, Frédéric
Pop, Alexandru
Pop, Vasile
Price, Charles
Pricopi, Dumitru
Pringle, James
Pritzl, Barton
Provost, Janine
Pugach, Alexander
Pustynski, Vladislav-V.
QIAN, Shengbang
Rank-Lueftinger, Theresa
Ransom, Scott
Rao, N.
Ratcliff, Stephen
Reale, Fabio
Reiners, Ansgar
Reinsch, Klaus
Renson, P.
Rey, Soo-Chang
Rivinius, Thomas
Robinson, Edward
Rodriguez, Eloy
Romanov, Yuri
Rosenbush, Alexander
Rountree, Janet
Russev, Ruscho
Ruzdjak, Domagoj
Sachkov, Mikhail
Safari, Hossein
Saha, Abhijit
SAKAMOTO, Tsuyoshi
SAKON, Itsuki
Samus, Nikolay
Sandmann, William
Sareyan, Jean-Pierre
Sasselov, Dimitar
Schaefer, Bradley
Schlegel, Eric
Schmidt, Edward
Schmidtobreick, Linda
Schuh, Sonja
Schwarzenberg-Czerny, A.
Schwope, Axel
Scuflaire, Richard
Seeds, Michael
Selam, Selim
Shahbaz, Tariq
Shakhovskaya, Nadejda
Shenavrin, Victor
Sherwood, William
Silvotti, Roberto
Singh, Harinder
Sivakoff, Gregory
Skinner, Stephen
Smak, Jozef
Smeyers, Paul
Smirnova, Olesja
Smith, Myron
Smith, Horace
Smolec, Radoslaw
Sodor, Ádám
Somasundaram, Seetha
Soszynski, Igor
Southworth, John
Stachowski, Grzegorz
Starrfield, Sumner
Stellingwerf, Robert
Stepien, Kazimierz
Sterken, Christiaan
Stoyanov, Kiril
Stringfellow, Guy
Strom, Stephen
Sudar, Davor
Szabados, Laszlo
Szabo, Robert
Szatmary, Karoly
Szekely, Péter
Szkody, Paula
TAKATA, Masao
Tammann, Gustav
TAMURA, Shin'ichi
Tarasova, Taya
Tas, Günay
Teixeira, Paula
Templeton, Matthew
Tessema, Solomon
Thurston, Mark
TOMINAGA, Nozomu
Tomov, Toma
Townsend, Richard
Traulsen, Iris
Tremko, Jozef
Tsvetkov, Milcho
Tsvetkova, Katja
Tsygankov, Sergey
Turcu, Vlad
Turner, David
Tutukov, Aleksandr
Tylenda, Romuald
Udovichenko, Sergei
UEMURA, Makoto
UKITA, Nobuharu
Ulla Miguel, Ana
Usher, Peter
Uslenghi, Michela
Utrobin, Victor
Vaccaro, Todd
Valeev, Azamat
Valtier, Jean-Claude
Van Doorsselaere, Tom
van Genderen, Arnoud
Van Hoolst, Tim
Vaz, Luiz Paulo
Ventura, Rita
Viotti, Roberto
Vivas, Anna
Vogt, Nikolaus
Voloshina, Irina
von Braun, Kaspar
Votruba, Viktor
Waelkens, Christoffel
Walker, Merle
Walker, William
Walker, Edward
Wallerstein, George
WANG, Xunhao
Warner, Brian
Watson, Robert
Webbink, Ronald
Wehlau, Amelia
Weis, Kerstin
Weiss, Werner
Welch, Douglas
Wesson, Roger
Wheatley, Peter
Whitelock, Patricia
Williamon, Richard
Willson, Lee Anne
Wing, Robert
Wittkowski, Markus

Wood, Peter
Worters, Hannah
Woudt, Patrick
XIONG, Da Run
Yakut, Kadri
Yuce, Kutluay
YUJI , Ikeda
Zamanov, Radoslav
Zejda, Miloslav
ZHANG, Xiaobin
ZHANG, Chengmin
ZHU, Liying
Zijlstra, Albert
Zola, Stanislaw
Zsoldos, Endre
Zuckerman, Benjamin
Zwintz, Konstanze

Division VIII Commission 28 Galaxies

President: Roger L. Davies
Vice-President: John S. Gallagher III

Organizing Committee Members:

Courteau, Stéphane
Dekel, Avishai
Franx, Marijn
Jog, Chanda
Jogee, Shardha
Karachentseva, Valentina
Knapp, Gillian
Kraan-Korteweg, Renée
Leibundgut, Bruno
NAKAI, Naomasa
Narlikar, Jayant
Rubio, Monica
Tacconi, Linda
Terlevich, Elena

Members:

Aalto, Susanne
Ables, Harold
Adler, David
Afanas'ev, Viktor
Aghaee, Alireza
Aguero, Estela
Aguilar, Luis
Aharonian, Felix
Ahmad, Farooq
Ajhar, Edward
Akashi, Muhammad
AKIYAMA, Masayuki
Aldaya, Victor
Alexander, Tal
Allen, Ronald
Allington-Smith, Jeremy
Alloin, Danielle
Almaini, Omar
Aloisi, Alessandra
Alonso, Maria
Alonso, Maria
Alonso-Herrero, Almudena
Ammons, Stephen
Amram, Philippe
Andernach, Heinz
Anderson, Joseph
ANN, Hong-Bae
Anosova, Joanna
Anton, Sonia
Antonelli, Lucio Angelo
Antoniou, Vallia
Aoki, Kentaro
Aparicio, Antonio
Aragon-Salamanca, Alfonso
Ardeberg, Arne
Aretxaga, Itziar
Argo, Megan
Arkhipova, Vera
Arnaboldi, Magda
Artamonov, Boris
Athanassoula, Evangelie
Aussel, Hervé
Avila-Reese, Vladimir
AYANI, Kazuya
Azzopardi, Marc
BABA, Junichi
Bachev, Rumen
Baddiley, Christopher
Baes, Maarten
BAI, Jinming
Bailey, Mark
Bajaja, Esteban
Baker, Andrew
Baldwin, Jack
Balkowski-Mauger, Chantal
Ballabh, Goswami
Balogh, Michael
Bamford, Steven
Banhatti, Dilip
Bannikova, Elena
Barbon, Roberto
Barcons, Xavier
Barkhouse, Wayne
Barnes, David
Barrientos, Luis
Barth, Aaron
Barthel, Peter
Barton, Elizabeth
Barway, Sudhashu
Bassino, Lilia
Battaner, Eduardo
Battinelli, Paolo
Bauer, Amanda
Baum, Stefi
Bautista, Manuel
Bayet, Estelle
Beaulieu, Sylvie
Beck, Rainer
Beck, Sara
Beckmann, Volker
Begeman, Kor
Bender, Ralf
Benedict, George
Benetti, Stefano
Benitez, Erika
Bensby, Thomas
Bentz, Misty
Berczik, Peter
Bergeron, Jacqueline
Berkhuijsen, Elly
Berta, Stefano
Bertola, Francesco
Bettoni, Daniela
BIAN, Yulin
Bianchi, Simone
Biermann, Peter
Bignall, Hayley
Bijaoui, Albert
Binette, Luc
Binggeli, Bruno
Binney, James
Biretta, John

Birkinshaw, Mark
Bjornsson, Claes-Ingvar
Blakeslee, John
Bland-Hawthorn, Jonathan
Blitz, Leo
Block, David
Blumenthal, George
Bodaghee, Arash
Boeker, Torsten
Boissier, Samuel
Boisson, Catherine
Boksenberg, Alec
Boles, Thomas
Bolzonella, Micol
Bomans, Dominik
Bon, Natasa
Bongiovanni, Angel
Boquien, Médéric
Borne, Kirk
Bosma, Albert
Bot, Caroline
Bouwens, Rychard
Bowen, David
Bower, Gary
Braine, Jonathan
Braithwaite, Jonathan
Brammer, Gabriel
Braun, Robert
Bravo-Alfaro, Hector
Brecher, Kenneth
Bressan, Alessandro
Bridges, Terry
Briggs, Franklin
Brinchmann, Jarle
Brinks, Elias
Brodie, Jean
Bromberg, Omer
Brosch, Noah
Brough, Sarah
Brouillet, Nathalie
Brown, Thomas
Brown, Michael
Brunner, Robert
Bruzual, Gustavo
Bryant, Julia
Buat, Véronique
Buote, David
Burbidge, Eleanor
Bureau, Martin
Burgarella, Denis
Burkert, Andreas
Burns, Jack
Busarello, Giovanni
Buta, Ronald
Butcher, Harvey
Byrd, Gene
BYUN, Yong Ik
Cabanac, Remi
CAI, Michael
Calderon, Jesús
Calura, Francesco
Calzetti, Daniela
Campusano, Luis
Cannon, Russell
Cannon, John
Cantiello, Michele
Canzian, Blaise
CAO, Xinwu
CAO, Li
Caon, Nicola
Capaccioli, Massimo
Cappellari, Michele
Caproni, Anderson
Caputi, Karina
Caretta, Cesar
Carigi, Leticia
Carollo, Marcella
Carrillo, Rene
Carswell, Robert
Carter, David
Casasola, Viviana
Casoli, Fabienne
Cattaneo, Andrea
Cayatte, Véronique
Cellone, Sergio
Cepa, Jordi
Cesetti, Mary
CHA, Seung-Hoon
CHAE, Kyu Hyun
Chakrabarti, Sandip
Chamaraux, Pierre
CHANG, Ruixiag
Charmandaris, Vassilis
Chatterjee, Tapan
Chatzichristou, Eleni
Chavushyan, Vahram
Chelliah Subramonian, S.
Chelouche, Doron
Chemin, Laurent
CHEN, Jiansheng
CHEN, Yang
CHEN, Lin-wen
CHEN, Dongni
Chiappini, Cristina
CHIBA, Masashi
Chincarini, Guido
CHOU, Chih-Kang
CHOU, Mei-Yin
Choudhury, Tirthankar
CHU, Yaoquan
Chugai, Nikolaj
CHUN, Sun
Cid Fernandes, Roberto
Cioni, Maria-Rosa
Ciotti, Luca
Ciroi, Stefano
Clark, David
Clavel, Jean
Clementini, Gisella
Cohen, Ross
Colbert, Edward
Colina, Luis
Comeron, Sébastien
Comte, Georges
Conselice, Christopher
Conti, Alberto
Contopoulos, George
Cook, Kem
Corbin, Michael
Corsini, Enrico
Corwin Jr, Harold
Cote, Stéphanie
Cote, Patrick
Couch, Warrick
Courbin, Frederic
Courtois, Helene
Courvoisier, Thierry
Couto da Silva, Telma
Cowsik, Ramanath
Coziol, Roger
Cracco, Valentina
Crane, Philippe
Crawford, Carolin
Cress, Catherine
Cristobal, David
Croston, Judith
Croton, Darren
Csabai, Istvan
CUI, Wenyuan
Cunniffe, John

Cypriano, Eduardo
d'Odorico, Sandro
D'Onofrio, Mauro
da Costa, Luiz
Da Rocha, Cristiano
DAISUKE, Iono
Dalla Bonta, Elena
Dallacasa, Daniele
Danks, Anthony
Dannerbauer, Helmut
Dantas, Christine
Dasyra, Kalliopi
Davidge, Timothy
Davies, Rodney
Davies, Benjamin
Davies, Roger
Davis, Marc
Davis, Timothy
De Blok, Erwin
de Bruyn, A.
de Carvalho, Reinaldo
de Diego Onsurbe, Jose
de Grijs, Richard
de Jong, Roelof
De Lucia, Gabriella
de Mello, Duilia
de Propris, Roberto
de Rijcke, Sven
de Silva, Gayandhi
de Swardt, Bonita
de Zeeuw, Pieter
Dejonghe, Herwig
Demarco, Ricardo
Demers, Serge
DENG, Zugan
Dennefeld, Michel
Dettmar, Ralf-Juergen
Devost, Daniel
Diaferio, Antonaldo
Diaz, Angeles
Diaz, Ruben
Diaz-Santos, Tanio
Dickey, John
Dietrich, Matthias
DOI, Mamoru
Dokuchaev, Vyacheslav
Dole, Hervé
Dominguez, Mariano
Dominis Prester, Dijana
Donas, Jose
Donea, Alina
DONG, Xiao-Bo
Donner, Karl
Donzelli, Carlos
Dopita, Michael
Dottori, Horacio
Dovciak, Michal
Doyon, René
Dressel, Linda
Dressler, Alan
Drinkwater, Michael
Driver, Simon
Duc, Pierre-Alain
Dufour, Reginald
Dultzin-Hacyan, Deborah
Dumont, Anne-Marie
Dunne, Loretta
Durret, Florence
Duval, Marie-France
Eales, Stephen
Edelson, Rick
Edmunds, Michael
Edwards, Louise
Efstathiou, George
Ehle, Matthias
Einasto, Jaan
Ekers, Ronald
Ellis, Simon
Ellison, Sara
Elmegreen, Debra
Elvis, Martin
Elyiv, Andrii
Emsellem, Eric
English, Jayanne
ENOKI, Motohiro
Espey, Brian
Evans, Robert
Fabbiano, Giuseppina
Faber, Sandra
Fabricant, Daniel
Falceta-Goncalves, Diego
Falco, Emilio
Falcon Barroso, Jesus
Fall, S.
Famaey, Benoit
FAN, Junhui
Farrell, Sean
Fasano, Giovanni
Fathi, Kambiz
Faure, Cécile
Feain, Ilana
Feast, Michael
Feinstein, Carlos
Feitzinger, Johannes
Ferguson, Annette
Ferland, Gary
Ferrarese, Laura
Ferrari, Fabricio
Ferreras, Ignacio
Ferrini, Federico
Field, George
Filippenko, Alexei
Fletcher, Andrew
Flin, Piotr
Florido, Estrella
Floyd, David
Foltz, Craig
Forbes, Duncan
Ford, Holland
Ford Jr, W.
Foschini, Luigi
Foucaud, Sébastien
Fouqué, Pascal
Fraix-Burnet, Didier
Francis, Paul
Freedman, Wendy
Freeman, Kenneth
Fricke, Klaus
Fried, Josef
Frogel, Jay
FUJITA, Yutaka
FUKUGITA, Masataka
FUNATO, Yoko
Funes, José
Furlanetto, Steven
Gadotti, Dimitri
Gaensler, Bryan
Gallagher, Sarah
Gallart, Carme
Gallazzi, Anna
Gallego, Jesús
Galletta, Giuseppe
Gallimore, Jack
Ganguly, Rajib
GAO, Yu
GAO, Jian
Garcia-Lorenzo, Maria
Gardner, Jonathan
Garilli, Bianca
Gavignaud, Isabelle

Gay, Pamela
Gelderman, Richard
Geller, Margaret
Gentile, Gianfranco
Georgiev, Tsvetan
Gerhard, Ortwin
Ghigo, Francis
Ghosh, P.
Giacani, Elsa
Giani, Elisabetta
Gibson, Brad
Gigoyan, Kamo
Gilbank, David
Giovanardi, Carlo
Giovanelli, Riccardo
Giroletti, Marcello
Gitti, Myriam
Glass, Ian
Glazebrook, Karl
Godlowski, Wlodzimierz
Goicoechea, Luis
Gomez, Haley
Gonzalez Delgado, Rosa
Gonzalez-Serrano, J. I.
Goodrich, Robert
Gordon, Karl
Gorgas, Garcia
Goss, W. Miller
GOTO, Tomotsugu
Gottesman, Stephen
Graham, John
Graham, Alister
Granato, Gian Luigi
Gray, Meghan
Grebel, Eva
Gregg, Michael
Greve, Thomas
Griffiths, Richard
Grillmair, Carl
Griv, Evgeny
Gronwall, Caryl
Grove, Lisbeth
Grupe, Dirk
GU, Qiusheng
GU, Minfeng
Gunn, James
Gunthardt, Guillermo
Guseva, Natalia
Gutierrez, Carlos
Gyulbudaghian, Armen
Haas, Martin
Hagen-Thorn, Vladimir
Haghi, Hosein
Hakopian, Susanna
HAMABE, Masaru
Hambaryan, Valeri
Hammer, François
HAN, Cheongho
HANAMI, Hitoshi
HAO, Lei
HARA, Tetsuya
Hardy, Eduardo
Harms, Richard
Harnett, Julienne
Hasan, Hashima
Hashimoto, Yasuhiro
HATTORI, Makoto
HATTORI, Takashi
Hatziminaoglou, Evanthia
Hau, George
Haugboelle, Troels
Hayes, Matthew
HE, XiangTao
Heald, George
Heckman, Timothy
Heidt, Jochen
Heinz, Sebastian
Held, Enrico
Helou, George
Henning, Patricia
Henry, Richard
Hensler, Gerhard
Heraudeau, Philippe
Hernandez, Xavier
Hess, Kelley
Hicks, Amalia
Hickson, Paul
Hintzen, Paul
HIRASHITA, Hiroyuki
Hjalmarson, Ake
Hjorth, Jens
Ho, Luis
Hodge, Paul
Hoekstra, Hendrik
Hoenig, Sebastian
Holz, Daniel
Hopkins, Andrew
Hopp, Ulrich
Horellou, Cathy
Hornschemeier, Ann
Hornstrup, Allan
HOU, Jinliang
Houdashelt, Mark
Hough, James
HU, Fuxing
HUANG, Keliang
Huang, Jiasheng
Huchtmeier, Walter
Huertas-Company, Marc
Huettemeister, Susanne
Hughes, David
Humphreys, Roberta
Humphreys, Elizabeth
Hunstead, Richard
Hunt, Leslie
Hunter, James
Huynh, Minh
HWANG, Chorng-Yuan
Ibata, Rodrigo
ICHIKAWA, Takashi
ICHIKAWA, Shin-ichi
Idiart, Thais
Ilic, Dragana
Iliev, Ilian
Illingworth, Garth
IM, Myungshin
IMANISHI, Masatoshi
Impey, Christopher
Infante, Leopoldo
INOUE, Akio
Irwin, Judith
ISHIMARU, Yuhri
Israel, Frank
Ivezic, Zeljko
Ivison, Robert
IWAMURO, Fumihide
IWATA, Ikuru
IYE, Masanori
Izotov, Yuri
Izotova, Iryna
Jablonka, Pascale
Jachym, Pavel
Jaffe, Walter
Jahnke, Knud
JANG, Minwhan
Jarrett, Thomas
Jerjen, Helmut
JIANG, Ing-Guey
Jimenez-Vicente, Jorge
Johansson, Peter

Johnston, Helen
Johnston-Hollitt, Melanie
Jones, Paul
Jones, Thomas
Jones, Christine
Jordan, Andrés
Jorgensen, Inger
Joshi, Umesh
Jovanovic, Predrag
Joy, Marshall
Jungwiert, Bruno
Junkes, Norbert
Junkkarinen, Vesa
Junor, William
Kaisin, Serafim
KAJISAWA, Masaru
Kalloglian, Arsen
KAMENO, Seiji
Kandalyan, Rafik
Kanekar, Nissim
KANEKO, Noboru
KANG, Xi
Karachentsev, Igor
KAROJI, Hiroshi
Karouzos, Marios
KASHIKAWA, Nobunari
Kaspi, Shai
Kassim, Namir
Katgert, Peter
Katsiyannis, Athanassios
Kauffmann, Guinevere
Kaufman, Michele
Kaviraj, Sugata
Kawada, Mitsunobu
KAWAKATU, Nozomu
Keel, William
Keenan, Ryan
Keeney, Brian
Kellermann, Kenneth
Kelly, Brandon
Kemp, Simon
Kennicutt, Robert
Kent, Brian
Keshet, Uri
Khachikian, Edward
Khanna, Ramon
Khare, Pushpa
Khosroshahi, Habib
Kilborn, Virginia
KIM, Dong Woo
KIM, Sungsoo
KIM, Minsun
KIM, Sang Chul
KIM, Woong-Tae
KIM, Ji Hoon
Kimball, Amy
King, Ivan
Kinman, Thomas
Kirshner, Robert
Kissler-Patig, Markus
Klein, Ulrich
Knapen, Johan
Knezek, Patricia
Kniazev, Alexei
Knudsen, Kirsten
KO, Chung-Ming
Kobayashi, Chiaki
KOBAYASHI, Masakazu
Koch, Andreas
Kochhar, Rajesh
KODAIRA, Keiichi
KODAMA, Tadayuki
Kogoshvili, Natela
Kollatschny, Wolfram
KOMIYAMA, Yutaka
KONG, Xu
Kontizas, Evangelos
Kontorovich, Victor
Koo, David
Koopmans, Leon
Koratkar, Anuradha
Koribalski, Bärbel
Kormendy, John
Kotilainen, Jari
Kotulla, Ralf
Kovacevic, Jelena
Kozlowski, Szymon
Krabbe, Angela
Kraft, Ralph
Krajnovic, Davor
Krause, Marita
Krishna, Gopal
Kriwattanawong, Wichean
Kron, Richard
Krumholz, Mark
Kudryavtseva, Nadezhda
KUMAI, Yasuki
Kunchev, Peter
Kunert-Bajraszewska, M.
KUNO, Nario
Kunth, Daniel
Kuntschner, Harald
Kunz, Martin
Kuzio de Naray, Rachel
La Barbera, Francesco
La Franca, Fabio
La Mura, Giovanni
Labbe, Ivo
Lagache , Guilaine
Lake, George
Lal, Dharam
Lancon, Ariane
Lanfranchi, Gustavo
Lapi, Andrea
Larsen, Søren
Larson, Richard
Laurikainen, Eija
Layzer, David
Le Fèvre, Olivier
Leao, João Rodrigo
Lebron, Mayra
LEE, Myung Gyoon
LEE, Sang-Sung
LEE, Hyun-chul
LEE, Joon Hyeop
Leeuw, Lerothodi
Lehnert, Matthew
Lehto, Harry
Lenc, Emil
Lequeux, James
Levenson, Nancy
Levine, Robyn
LI, Ji
LIANG, Yanchun
Lilly, Simon
LIM, Jeremy
Lima Neto, Gastao
LIN, Weipeng
LIN, Yen-Ting
LIN, Lihwai
Lindblad, Per
Linden-Vørnle, Michael
Lintott, Chris
Lo, Fred K. Y.
Lobo, Catarina
Lokas, Ewa
Londrillo, Pasquale
Longo, Giuseppe
Lopes, Paulo
Lopez, Ericson

Lopez Aguerri, Jose Alfonso
Lopez Cruz, Omar
Lopez Hermoso, Maria
Lopez-Sanchez, Angel
Lord, Steven
Loubser, Ilani
Loup, Cecile
Lowenthal, James
Lu, Limin
LU, Youjun
Lucero, Danielle
Lugger, Phyllis
Luminet, Jean-Pierre
LUO, Ali
Lutz, Dieter
Lynden-Bell, Donald
Lynds, Beverly
Lynds, Roger
MA, Jun
Macalpine, Gordon
Maccagni, Dario
Maccarone, Thomas
Macchetto, Ferdinando
Maciejewski, Witold
Mackey, Alasdair
Mackie, Glen
Macquart, Jean-Pierre
Madden, Suzanne
Madore, Barry
Magorrian, Stephen
Magrini, Laura
Magris, Gladis
Mahtessian, Abraham
Maier, Christian
Mainieri, Vincenzo
Maiolino, Roberto
Makarov, Dmitry
Makarova, Lidia
Malagnini, Maria
Malhotra, Sageeta
Mann, Robert
Mannucci, Filippo
Marcelin, Michel
Marco, Olivier
Marconi, Alessandro
Marin-Franch, Antonio
Marino, Antonietta
Markoff, Sera
Marleau, Francine
Marquez, Isabel
Marr, Jonathon
Marston, Anthony
Martin, Rene
Martin, Maria
Martin, Crystal
Martin, Sergio
Martinet, Louis
Martinez , Vicent
Martini, Paul
Marziani, Paola
Masegosa, Josefa
Masters, Karen
MATSUMURA, Tomotake
MATSUSHITA, Kyoko
Matthews, Lynn
Mauersberger, Rainer
Maurice, Eric
Mayya, Divakara
Mazzarella, Joseph
McBreen, Brian
McCracken, Henry
McGaugh, Stacy
McKean, John
McMillan, Paul
McNeil, Stephen
Mediavilla, Evencio
Mehlert, Dörte
Meier, David
Meikle, William
Meisenheimer, Klaus
Melbourne, Jason
Melnik, Anna
Melnyk, Olga
Mendes de Oliveira, C.
Menon, T.
Mercurio, Amata
Merluzzi, Paola
Merrifield, Michael
Metevier, Anne
Meusinger, Helmut
Meyer, Martin
Meyer, Angela
Meza, Andres
Mihov, Boyko
Miley, George
Miller, Joseph
Miller, Hugh
Miller, Richard
Miller, Neal
Miller, Eric
Milvang-Jensen, Bo
MINOWA, Yosuke
Mirabel, Igor
Miroshnichenko, Alla
Misawa, Toru
MIZUNO, Takao
Moiseev, Alexei
Moles Villamate, Mariano
Molinari, Emilio
Molla, Mercedes
Monaco, Pierluigi
Monard, Libert
Moodley, Kavilan
Moody, Joseph
Morelli, Lorenzo
MORI, Masao
MOROKUMA, Tomoki
MOTOHARA, Kentaro
Mould, Jeremy
Mourao, Ana Maria
Mueller, Volker
Mueller , Andreas
Mujica, Raul
Mukhopadhyay, Banibrata
Mulchaey, John
Muller, Erik
Muller, Sebastien
Munoz Tunon, Casiana
MURAOKA, Kazuyuki
MURAYAMA, Takashi
Murphy, Michael
Murray, Stephen
Mushotzky, Richard
Muzzio, Juan
Nagao, Tohru
NAGASHIMA, Masahiro
Nair, Sunita
NAKANISHI, Kouichiro
NAKANISHI, Hiroyuki
NAKATA, Fumiaki
Nakos, Theodoros
Namboodiri, P.
Napolitano, Nicola
Narbutis, Donatas
Navarro, Julio
Nedialkov, Petko
Nesvadba, Nicole
NGUYEN, Lan
Nikolajuk, Marek
Ninkovic, Slobodan

Nipoti, Carlo
Nishikawa, Ken-Ichi
Nityananda, Rajaram
NOGUCHI, Masafumi
Noll, Stefan
Norman, Colin
Noterdaeme, Pasquier
Nucita, Achille
Nulsen, Paul
Nuza, Sebastian
O'Connell, Robert
O'Dea, Christopher
Ocvirk, Pierre
Oemler Jr, Augustus
Oey, Sally
Ogando, Ricardo
OHTA, Kouji
OKA, Tomoharu
OKAMOTO, Takashi
OKAMURA, Sadanori
Olling, Robert
Olofsson, Kjell
Omizzolo, Alessandro
Oosterloo, Thomas
Orienti, Monica
Origlia, Livia
Ostlin, Göran
Ostorero, Luisa
Ostriker, Eve
OTA, Naomi
OTSUKI, Kaori
Ott, Juergen
OUCHI, Masami
Ovcharov, Evgeni
Owers, Matthew
OYA, Shin
OYABU, Shinki
Page, Mathew
PAK, Soojong
Palmer, Philip
Palumbo, Giorgio
Panessa, Francesca
Pannuti, Thomas
Pantoja, Carmen
Papayannopoulos, Th.
PARK, Jang Hyun
Parker, Quentin
Pastoriza, Miriani
Patton, David
Paturel, Georges
Pearce, Frazer
Pedrosa, Susana
Peel, Michael
Peimbert, Manuel
Peletier, Reynier
Pellegrini, Silvia
Pello, Roser
PENG, Qingyu
PENG, Eric
Pentericci, Laura
Perea-Duarte, Jaime
Perez, Fournon
Perez-Gonzalez, Pablo
Perez-Torres, Miguel
Peroux, Céline
Peters, William
Peterson, Charles
Peterson, Bradley
Petit, Jean-Marc
Petrosian, Artaches
Petrov, Georgi
Petuchowski, Samuel
Pfenniger, Daniel
Philipp, Sabine
Phillipps, Steven
Phillips, Mark
Pihlström, Ylva
Pikichian, Hovhannes
Piotrovich, Mikhail
Pipino, Antonio
Pirzkal, Norbert
Pisano, Daniel
Pizzella, Alessandro
Plana, Henri
Pogge, Richard
Poggianti, Bianca
Polletta, Maria del Carmen
Polyachenko, Evgeny
Pompei, Emanuela
Pooley, David
Popescu, Cristina
Popović, Luka
Portinari, Laura
Poveda, Arcadio
Povic, Mirjana
Prabhu, Tushar
Pracy, Michael
Prandoni, Isabella
Press, William
Prieto, Almudena
Prires Martins, Lucimara
Pritchet, Christopher
Proctor, Robert
Pronik, Iraida
Pronik, Vladimir
Proust, Dominique
Prugniel, Philippe
Puech, Mathieu
Puerari, Ivânio
Pulatova, Nadiia
Pustilnik, Simon
Puxley, Phil
Puzia, Thomas
QIN, Yi-Ping
Quinn, Peter
Rafanelli, Piero
Raiteri, Claudia
Rampazzo, Roberto
Ranalli, Piero
Rand, Richard
Rasmussen, Jesper
Rauch, Michael
Ravindranath, Swara
Raychaudhury, Somak
Read, Andrew
Read, Justin
Reboul, Henri
Recchi, Simone
Rector, Travis
Reddy, Bacham
Rejkuba, Marina
Rekola, Rami
Rephaeli, Yoel
Reshetnikov, Vladimir
Reunanen, Juha
Revaz, Yves
Revnivtsev, Mikhail
Rey, Soo-Chang
Reyes, Reinabelle
Reynolds, Cormac
Ribeiro, André Luis
Richard, Johan
Richer, Harvey
Richstone, Douglas
Richter, Philipp
Ridgway, Susan
Risaliti, Guido
Rix, Hans-Walter
Rizzi, Luca
Robert, Carmelle

Roberts, Morton
Roberts, Timothy
Roberts Jr, William
Rodrigues de Oliveira F., I.
Roediger, Elke
Romano, Patrizia
Romeo, Alessandro
Romero-Colmenero, E.
Rosa, Michael
Rosa Gonzalez, Daniel
Rosado, Margarita
Rose, James
Rothberg, Barry
Rots, Arnold
Rozas, Maite
Rubin, Vera
Rudnicki, Konrad
Ruzicka, Adam
Ryder, Stuart
Sackett, Penny
Sadler, Elaine
Sadun, Alberto
SAITOH, Takayuki
Saiz, Alejandro
Sakai, Shoko
SAKON, Itsuki
Sala, Ferran
Salvador-Sole, Eduardo
Samurović, Srdjan
Sanahuja Parera, Blai
Sanchez-Blazquez, Patricia
Sancisi, Renzo
Sanders, Robert
Sanders, David
Sanroma, Manuel
Sansom, Anne
Santiago, Basilio
Santos-Lleo, Maria
Sapre, Ashok
Saracco, Paolo
Sarazin, Craig
Sasaki, Toshiyuki
SASAKI, Minoru
Saslaw, William
Saucedo Morales, Julio
Saviane, Ivo
SAWA, Takeyasu
Sawicki, Marcin
Scaramella, Roberto
Schaerer, Daniel
Schaye, Joop
Schechter, Paul
Schmidt, Maarten
Schmitt, Henrique
Schmitz, Marion
Schroder, Anja
Schucking, Engelbert
Schwarz, Ulrich
Schweizer, François
Scodeggio, Marco
Scorza , Cecilia
Scoville, Nicholas
Seigar, Marc
Sellwood, Jerry
Semelin, Benoit
Sempere, Maria
SEON, Kwang il
Sergeev, Sergey
Serjeant, Stephen
Serote Roos, Margarida
Setti, Giancarlo
Severgnini, Paola
SHAN, Hongguang
Shapovalova, Alla
Sharma, Prateek
Sharp, Nigel
Sharples, Ray
Shastri, Prajval
Shaver, Peter
Shaya, Edward
SHEN, Zhiqiang
SHEN, Juntai
Sherwood, William
Shields, Gregory
Shields, Joseph
SHIMASAKU, Kazuhiro
SHIRAHATA, Mai
SHIRASAKI, Yuji
Shostak, G.
Shukurov, Anvar
Siebenmorgen, Ralf
Siebert, Arnaud
Sigurdsson, Steinn
Sil'chenko, Olga
Sillanpaa, Aimo
Silva, David
Silva, Laura
Simic, Sasa
Simkin, Susan
Singh, Kulinder Pal
Siopis, Christos
Sivakoff, Gregory
Skillman, Evan
Slavcheva-Mihova, Lyuba
Slezak, Eric
Smail, Ian
Smecker-Hane, Tammy
Smirnova, Aleksandrina
Smith, Malcolm
Smith, Haywood
Smith, Eric
Soares, Domingos Savio
Sobouti, Yousef
SOHN, Young Jong
Soltan, Andrzej
SONG, Liming
SORAI , Kazuo
Soria, Roberto
Sparks, William
Spinoglio, Luigi
Spinrad, Hyron
Sridhar, Seshadri
Srinivasan, Ganesan
Stacy, Athena
Stadel, Joachim
Statler, Thomas
Staveley-Smith, Lister
Steenbrugge, Katrien
Steiman-Cameron, Thomas
Steinbring, Eric
Stiavelli, Massimo
Stirpe, Giovanna
Stoehr, Felix
Stone, Remington
Stonkutė, Rima
Storchi-Bergmann, Thaisa
Stott, John
Strauss, Michael
Strom, Richard
Strom, Robert
Strubbe, Linda
Stuik, Remko
SU, Cheng-yue
Subramaniam, Annapurni
SUGAI, Hajime
Sulentic, Jack
Sullivan, Mark
Sullivan, III, Woodruff
Sundin, Maria
SUSA, Hajime

Sutherland, Ralph
Tacconi-Garman, Lowell
Tagger, Michel
TAKADA, Masahiro
TAKAGI, Toshinobu
TAKAHASHI, Rohta
TAKASHI, Hasegawa
TAKATA, Tadafumi
TAKATO, Naruhisa
TAKEUCHI, Tsutomu
TAKIZAWA, Motokazu
Tamm, Antti
Tammann, Gustav
TANAKA, Masayuki
TANAKA, Ichi Makoto
TANIGUCHI, Yoshiaki
Tantalo, Rosaria
TASHIRO, Makoto
Taylor, James
Telles, Eduardo
Tempel, Elmo
Temporin, Sonia
Teng, Stacy
Tenjes, Peeter
Tepper Garcia, Thorsten
Terlevich, Roberto
Terzian, Yervant
Teuben, Peter
Teyssier, Romain
Theis, Christian
Thoene, Christina
Thomasson, Magnus
Thonnard, Norbert
Thornley, Michele
Thuan, Trinh
Tifft, William
Tikhonov, Nikolai
Tilanus, Remo
Tissera, Patricia
Tisserand , Patrick
Toft, Sune
Tolstoy, Eline
Tomita, Akihiko
Toomre, Alar
Torres-Papaqui, Juan
Tovmassian, Hrant
TOYAMA, Kiyotaka
Traat, Peeter
Trager, Scott
Tremaine, Scott
Tremonti, Christy
Trenti, Michele
Tresse, Laurence
Treu, Tommaso
Trimble, Virginia
Trinchieri, Ginevra
TRIPPE, Sascha
Trujillo Cabrera, Ignacio
TSAI, An-Li
TSUCHIYA, Toshio
Tsvetkov, Dmitry
Tuffs, Richard
Tugay, Anatoliy
Tully, Richard
Turner, Edwin
Tyson, John
Tyul'bashev, Sergei
Tzanavaris, Panayiotis
Ulrich, Marie-Helene
Urbanik, Marek
Uslenghi, Michela
Utrobin, Victor
Vaccari, Mattia
Valcheva, Antoniya
Valdes Parra, Jose
Valentijn, Edwin
Vallenari, Antonella
Valluri, Monica
Valotto, Carlos
Valtchanov, Ivan
Valtonen, Mauri
van Albada, Tjeerd
van den Bergh, Sidney
van der Hulst, Jan
van der Kruit, Pieter
van der Laan, Harry
van der Marel, Roeland
van Driel, Wim
van Eymeren, Janine
van Gorkom, Jacqueline
van Kampen, Eelco
van Moorsel, Gustaaf
van Woerden, Hugo
Van Zee, Liese
Vansevicius, Vladas
Varela Lopez, Jesús
Varma, Ram
Vasta, Magda
Vaughan, Simon
Vauglin, Isabelle
Vavilova, Iryna
Vazdekis, Alexandre
Vega, Olga
Veilleux, Sylvain
Venters, Tonia
Vercellone, Stefano
Verdes-Montenegro, L.
Verdoes Kleijn, Gijsbert
Vergani, Daniela
Verma, Aprajita
Vermeulen, Rene
Viel, Matteo
Vigroux, Laurent
Viironen, Kerttu
Villata, Massimo
Vivas, Anna
Vlahakis, Catherine
Vlasyuk, Valerij
Voit, Gerard
Vollmer, Bernd
Volonteri, Marta
Vrtilek, Jan
WADA, Keiichi
Wadadekar, Yogesh
Wagner, Stefan
Wagner, Alexander
WAKAMATSU, Ken-Ichi
Walker, Mark
Walter, Fabian
Walterbos, Rene
WANAJO, Shinya
WANG, Yiping
WANG, Tinggui
WANG, Huiyuan
WANG, Hong-Guang
WANG, Junzhi
WANG, Yu
Ward, Martin
Weedman, Daniel
WEI, Jianyan
Weilbacher, Peter
Weiler, Kurt
Welch, Gary
Westera, Pieter
Westmeier, Tobias
White, Simon
Whiting, Matthew
Whitmore, Bradley
Wielebinski, Richard
Wielen, Roland

Wiita, Paul
Wilcots, Eric
Wild, Wolfgang
Williams, Robert
Williams, Theodore
Williams, Barbara
Willis, Jon
Wills, Beverley
Wills, Derek
Wilson, Gillian
Windhorst, Rogier
Winkler, Hartmut
Winter, Lisa
Wise, Michael
Wisotzki, Lutz
Wofford, Aida
Wold, Margrethe
Wong, Tony
WOO, Jong-Hak
Woosley, Stanford
Worrall, Diana
Woudt, Patrick
Wozniak, Hervé
Wrobel, Joan
WU, Xue-bing
WU, Hong
WU, Wentao
WU, Jianghua
Wu, Yanling
Wulandari, Hesti
Wunsch, Richard
Wynn-Williams, Gareth
Xanthopoulos, Emily
XIA, Xiao-Yang
Xilouris, Emmanouel
XU, Dawei
XUE, Suijian
YAGI, Masafumi
Yakovleva, Valerija
YAMADA, Yoshiyuki
YAMADA, Toru
YAMADA, Shimako
YAMAGATA, Tomohiko
YAMAUCHI, Aya
YI, Sukyoung
YONEHARA, Atsunori
YOSHIDA, Michitoshi
YOSHIKAWA, Kohji
Yun, Min
Zaggia, Simone
Zamorano, Jaime
Zaritsky, Dennis
Zaroubi, Saleem
Zasov, Anatoly
Zeilinger, Werner
Zepf, Stephen
Zezas, Andreas
ZHANG, Xiaolei
ZHANG, Yang
ZHANG, JiangShui
ZHANG, Jingyi
ZHANG, Bo
ZHANG, Fenghui
Zhang, Yuying
ZHENG, XianZhong
ZHOU, Youyuan
ZHOU, Xu
ZHOU, Hongyan
ZHOU, Jianjun
ZHU, Ming
Zibetti, Stefano
Ziegler, Bodo
Ziegler, Harald
Zinn, Robert
Zirm, Andrew
ZOU, Zhenlong
Zwaan, Martin

Division IV Commission 29 Stellar Spectra

President: Nikolai E. Piskunov
Vice-President: Katia Cunha

Organizing Committee Members:

AOKI, Wako
Asplund, Martin
Bohlender, David
Carpenter, Kenneth
Melendez, Jorge
Rossi, Silvia
Smith, Verne
Soderblom, David
Wahlgren, Glenn

Members:

Abia, Carlos
Abt, Helmut
Adelman, Saul
Afram, Nadine
Aikman, G.
Airapetian, Vladimir
Ake III, Thomas
Alcala, Juan Manuel
Alecian, Georges
Alencar, Silvia
Allende Prieto, Carlos
Andretta, Vincenzo
Andreuzzi, Gloria
Annuk, Kalju
Antoniou, Vallia
AOKI, Wako
Appenzeller, Immo
Ardila, David
Aret, Anna
Arias, Maria
Arkharov, Arkadij
Atac, Tamer
Audard, Marc
Aufdenberg, Jason
Baade, Dietrich
Bagnulo, Stefano
Bakker, Eric
Baliunas, Sallie
Ballereau, Dominique
Balman, Solen
Banerjee, Dipankar
Barber, Robert
Barbuy, Beatriz
Barklem, Paul
Baron, Edward
Basri, Gibor
Batalha, Celso
Bauer, Wendy
Beckman, John
Beers, Timothy
Beiersdorfer, Peter
Bellas-Velidis, Ioannis
Bensby, Thomas
Bertone, Emanuele
Bessell, Michael
Biazzo, Katia
Bikmaev, Ilfan
Boehm, Torsten
Boesgaard, Ann
Boggess, Albert
Bohlender, David
Bon, Natasa
Bond, Howard
Bonifacio, Piercarlo
Bopp, Bernard
Borczyk, Wojciech
Bouvier, Jerôme
Bragaglia, Angela
Brandi, Elisande
Breysacher, Jacques
Brickhouse, Nancy
Briot, Danielle
Brown, Douglas
Bruhweiler, Frederick
Bruning, David
Bruntt, Hans
Bues, Irmela
Burkhart, Claude
Busa, Innocenza
Butkovskaya, Varvara
Butler, Keith
Canto Martins, Bruno
Carlin, Jeffrey
Carney, Bruce
Carretta, Eugenio
Carter, Bradley
Casassus, Simon
Catala, Claude
Catanzaro, Giovanni
Catchpole, Robin
Cayrel, Roger
Cesetti, Mary
Chadid, Merieme
Chavez-Dagostino, Miguel
CHEN, Alfred
CHEN, Yuqin
CHOU, Mei-Yin
Cidale, Lydia
Claudi, Riccardo
Cohen, David
Collet, Remo
Conti, Peter
Corbally, Christopher
Cornide, Manuel
Cottrell, Peter
Cowley, Anne
Cowley, Charles
CUI, Wenyuan
Cure, Michel
da Silva, Licio
Dacic, Miodrag
Daflon, Simone
DAISAKU, Nogami
Dall, Thomas
Damineli Neto, Augusto

Davies, Benjamin
de Castro, Elisa
de Laverny, Patrick
del Peloso, Eduardo
Deleuil, Magali
Depagne, Éric
Derekas, Aliz
Doppmann, Gregory
Dougados, Catherine
Dragunova, Alina
Drake, Natalia
Dufour, Patrick
Duncan, Douglas
Dworetsky, Michael
Edwards, Suzan
Elkin, Vladimir
Elmhamdi, Abouazza
Faraggiana, Rosanna
Feast, Michael
Felenbok, Paul
Fernandez-Figueroa, M.
Figer, Donald
Fitzpatrick, Edward
Foing, Bernard
Foy, Renaud
Franchini, Mariagrazia
Francois, Patrick
Frandsen, Soeren
Freire Ferrero, Rubens
Friel, Eileen
Fullerton, Alexander
Gamen, Roberto
Garcia, Lopez
Garcia, Miriam
Garcia-Hernandez, D.
Garmany, Katy
Garrison, Robert
Gautier, Daniel
Gehren, Thomas
Gerbaldi, Michèle
Gershberg, R.
Gesicki, Krzysztof
Giampapa, Mark
Giovannelli, Franco
Glagolevskij, Yurij
Glazunova, Ljudmila
Goebel, John
Gomboc, Andreja
Gonzalez, Guillermo
Gopka, Vera
Gorbaneva, Tatyana
Goswami, Aruna
Grady, Carol
Gratton, Raffaele
Gray, David
Griffin, R. Elizabeth
Griffin, Roger
Grundstrom, Erika
GU, Sheng-hong
Gustafsson, Bengt
Hackman, Thomas
HAN, Inwoo
Hanson, Margaret
Hanuschik, Reinhard
Harmer, Charles
Harmer, Dianne
Hartman, Henrik
Hartmann, Lee
HASHIMOTO, Osamu
Hearnshaw, John
Heber, Ulrich
Heiter, Ulrike
Henrichs, Hubertus
Hessman, Frederic
Hill, Grant
Hillier, John
Hinkle, Kenneth
HIRAI, Masanori
HIRATA, Ryuko
Hoeflich, Peter
Honda, Satoshi
HORAGUCHI, Toshihiro
Houk, Nancy
Houziaux, Leo
Hron, Josef
HU, Zhong wen
Hubert-Delplace, A.-M.
Hubrig, Swetlana
Huenemoerder, David
Hyland, Harry
Israelian, Garik
Ivans, Inese
IZUMIURA, Hideyuki
Jankov, Slobodan
Jehin, Emmanuel
Johnson, Hollis
Johnson, Jennifer
Johnson, Christian
Jordan, Carole
Josselin, Eric
Kaeufl, Hans Ulrich
Kawka, Adela
Kipper, Tonu
Klochkova, Valentina
Koch, Andreas
Kochukhov, Oleg
KODAIRA, Keiichi
KOGURE, Tomokazu
Kolka, Indrek
Kordi, Ayman
Korn, Andreas
Korotin, Sergey
Kotnik-Karuza, Dubravka
Koubsky, Pavel
Kovachev, Bogomil
Kovacs, József
Kovtyukh, Valery
Krempec-Krygier, Janina
Kucinskas, Arunas
Kwok, Sun
Lago, Maria
Lagrange, Anne-Marie
Laird, John
Lambert, David
Lamers, Henny
Lamontagne, Robert
Landstreet, John
Lanz, Thierry
Le Contel, Jean-Michel
Leao, Joao Rodrigo
Lebre, Agnes
Leckrone, David
LEE, Jae Woo
LEE, Hyun-chul
LEE, Chung-Uk
Leedjarv, Laurits
Lester, John
Letarte, Bruno
Leushin, Valerij
Levato, Orlando
LI, Ji
LIANG, Yanchun
LIANG, Guiyun
Liebert, James
Little-Marenin, Irene
LIU, Michael
LIU, Jifeng
LIU, Guoqing
LIU, Yujuan
Lobel, Alex

Lodders, Katharina
Lubowich, Donald
Lucatello, Sara
Luck, R.
Lugaro, Maria
Lundstrom, Ingemar
LUO, Ali
Lyubimkov, Leonid
Magain, Pierre
Magazzu, Antonio
Magrini, Laura
Maillard, Jean-Pierre
Mainzer, Amy
Maitzen, Hans
Malaroda, Stella
Manteiga Outeiro, Minia
Marilli, Ettore
Marsden, Stephen
Martinez Fiorenzano, Aldo
Massey, Philip
Mathys, Gautier
Matsuura, Mikako
Mazzali, Paolo
McDavid, David
McGregor, Peter
McSwain, Mary
Megessier, Claude
Melo, Claudio
Merlo, David
Mickaelian, Areg
Mikuláek, Zden?k
Moffat, Anthony
Molaro, Paolo
Monaco, Lorenzo
Monin, Dmitry
Montes, David
Moos, Henry
Morel, Thierry
Morossi, Carlo
Morrison, Nancy
Napiwotzki, Ralf
Nazarenko, Victor
Naze, Yael
Neiner, Coralie
Niedzielski, Andrzej
Nielsen, Krister
Niemczura, Ewa
Nieva, Maria
Nilsson, Hampus
NISHIMURA, Shiro
Norris, John
North, Pierre
Nugis, Tiit
O'Neal, Douglas
O'Toole, Simon
OKAZAKI, Atsuo
OTSUKA, Masaaki
OTSUKI, Kaori
Oudmaijer, Rene
Owocki, Stanley
Pakhomov, Yury
Parsons, Sidney
Parthasarathy, Mudumba
Pavani, Daniela
Pavlenko, Yakov
Peters, Geraldine
Peterson, Ruth
Petit, Pascal
Pilachowski, Catherine
Pintado, Olga
Plez, Bertrand
Polcaro, V.
Polidan, Ronald
Polosukhina-Chuvaeva, N.
Porto de Mello, Gustavo
Primas, Francesca
Prinja, Raman
Prires Martins, Lucimara
Querci, Monique
Raassen, Ion
Rank-Lueftinger, Theresa
Rao, N.
Rashkovskij, Sergey
Rastogi, Shantanu
Rauw, Gregor
Rawlings, Mark
Rebolo, Rafael
Reddy, Bacham
Rego, Fernandez
Reid, Warren
Reimers, Dieter
Reiners, Ansgar
Rettig, Terrence
Rhee, Jaehyon
Ringuelet, Adela
Rivinius, Thomas
Romanyuk, Iosif
Rose, James
Rossi, Corinne
Rossi, Silvia
Rutten, Robert
Ryan, Sean
Ryde, Nils
Sachkov, Mikhail
SADAKANE, Kozo
Saffe, Carlos
SAKON, Itsuki
Sanchez Almeida, Jorge
Sanwal, Basant
Sareyan, Jean-Pierre
Sarre, Peter
Sasso, Clementina
Sbordone, Luca
Schild, Rudolph
Schroeder, Klaus
Schuh, Sonja
Schuler, Simon
Seggewiss, Wilhelm
Selam, Selim
Shetrone, Matthew
SHI, Huoming
SHI, Jianrong
Shimansky, Vladislav
Sholukhova, Olga
Shore, Steven
Simic, Zoran
Simon, Theodore
Simon-Diaz, Sergio
Singh, Mahendra
Singh, Harinder
Slechta, Miroslav
Smalley, Barry
Smith, Myron
Smith, Graeme
Smith, Verne
Snow, Theodore
Sonneborn, George
Spite, François
Spite, Monique
St-Louis, Nicole
Stateva, Ivanka
Stawikowski, Antoni
Stecher, Theodore
Steffen, Matthias
Stencel, Robert
Sundqvist, Jon
Suntzeff, Nicholas
Swings, Jean-Pierre
Szeifert, Thomas
TAKADA-HIDAI, M.

TAKAHASHI, Hidenori
TAKASHI, Hasegawa
Talavera, Antonio
Tantalo, Rosaria
Tautvaisiene, Graina
Thévenin, Frédéric
Todt, Helge
Tomasella, Lina
Tomov, Toma
Torrejon, José Miguel
Torres-Papaqui, Juan
Tripathi, Durgesh
Ulla Miguel, Ana
Ulyanov, Oleg
Usenko, Igor
Utrobin, Victor
UTSUMI, Kazuhiko
Valdivielso, Luisa
Valeev, Azamat
Valenti, Jeff
Valtier, Jean-Claude
Valyavin, Gennady
van der Hucht, Karel
Van Doorsselaere, Tom
van Eck, Sophie
Van Winckel, Hans
van't Veer-Menneret, C.
Vasta, Magda
Vasu-Mallik, Sushma
Vennes, Stéphane
Verdugo, Eva
Verheijen, Marc
Vilhu, Osmi
Viotti, Roberto
Vladilo, Giovanni
Vogt, Nikolaus
Vogt, Steven
Vreux, Jean
Wade, Gregg
Wahlgren, Glenn
Wallerstein, George
WANG, Feilu
Wegner, Gary
Whelan, Emma
Williams, Peredur
Wing, Robert
Wolff, Sidney
Wood, Brian
Worters, Hannah
Wright, Nicholas
Wyckoff, Susan
YAMASHITA, Yasumasa
YOSHIOKA, Kazuo
Yuce, Kutluay
Yushkin, Maxim
Zaggia, Simone
Zapatero-Osorio, Maria R.
ZHANG, Huawei
ZHANG, Bo
ZHANG, Haotong
ZHANG, Yanxia
ZHU, Zh[illegible]
Zoccali, Manuela
Zorec, Juan
Zverko, Juraj

Division IX Commission 30 Radial Velocities

President: Guillermo Torres
Vice-President: Dimitri Pourbaix

Organizing Committee Members:

Marcy, Geoffrey
Mathieu, Robert
Mazeh, Tsevi
Minniti, Dante
Moutou, Claire
Pepe, Francesco
Turon, Catherine
Zwitter, Tomaž

Members:

Abt, Helmut
Andersen, Johannes
Arnold, Richard
Balona, Luis
Batten, Alan
Beavers, Willet
Beers, Timothy
Beuzit, Jean-Luc
Borczyk, Wojciech
Boyajian, Tabetha
Breger, Michel
Buchhave, Lars
Burki, Gilbert
Butkovskaya, Varvara
Butler, Paul
Cardoso Santos, Nuno
Carney, Bruce
Chadid, Merieme
Chemin, Laurent
CHEN, Yuqin
Cochran, William
Couto da Silva, Telma
Crampton, David
Crifo, Françoise
da Costa, Luiz
Davis, Robert
Davis, Marc
de Medeiros, Jose
De Souza Pellegrini, Paulo
Derekas, Aliz
Dravins, Dainis
Dubath, Pierre
Elkin, Vladimir
FAN, Yufeng
Fekel, Francis
Foltz, Craig
Forveille, Thierry
Gandolfi, Davide
Garcia, Beatriz
Geller, Aaron
Gilmore, Gerard
Giovanelli, Riccardo
Gnedin, Yurij
Gonzalez, Jorge
Gray, David
Griffin, Roger
Hakopian, Susanna
Halbwachs, Jean-Louis
HAN, Inwoo
Hearnshaw, John
Hewett, Paul
Hilditch, Ronald
Hill, Graham
Howard, Andrew
Hrivnak, Bruce
HU, Zhong wen
Hube, Douglas
Hubrig, Swetlana
Ibata, Rodrigo
Imbert, Maurice
Irwin, Alan
Karachentsev, Igor
Konacki, Maciej
Kovacs, József
Krabbe, Angela
Latham, David
Levato, Orlando
Lewis, Brian
Lindgren, Harri
LIU, Yujuan
Lo Curto, Gaspare
Lokas, Ewa
Lovis, Christophe
Marschall, Laurence
Martinez Fiorenzano, Aldo
Maurice, Eric
Mayor, Michel
McMillan, Robert
Meibom, Soren
Melnick, Gary
Meylan, Georges
Mink, Jessica
Mkrtichian, David
Monaco, Lorenzo
Morrell, Nidia
Naef, Dominique
Napolitano, Nicola
NARITA, Norio
Ogando, Ricardo
Pepe, Francesco
Perrier-Bellet, Christian
Peterson, Ruth
Philip, A.G.
Preston, George
Rastorguev, Alexey
Ratnatunga, Kavan
Reid, Warren
Romanov, Yuri
Royer, Frédéric
Rubenstein, Eric
Rubin, Vera
Sachkov, Mikhail
SAKAMOTO, Tsuyoshi
Samus, Nikolay
Scarfe, Colin
Schroder, Anja
Siebert, Arnaud
Sivan, Jean-Pierre

Smith, Myron
Solivella, Gladys
Stefanik, Robert
Stickland, David
Strauss, Michael
Suntzeff, Nicholas
Szabados, Laszlo
Tokovinin, Andrei
Tomasella, Lina
Tonry, John
van Dessel, Edwin
Verschueren, Werner
Vinko, Jozsef
Walker, Gordon
Wegner, Gary
Willstrop, Roderick
Wittenmyer, Robert
XIAO, Dong
YANG, Stephenson
Zaggia, Simone
ZHANG, Haotong

Division I Commission 31 Time

President: Richard N. Manchester
Vice-President: Mizuhiko HOSOKAWA

Organizing Committee Members:

Arias, Elisa
Petit, Gérard
Tuckey, Philip
ZHANG, Shougang
Zharov, Vladimir

Members:

Abele, Maris
AHN, Youngsook
Alley, Carrol
ARAKIDA, Hideyoshi
Archinal, Brent
Becker, Werner
Boehm, Johannes
Breakiron, Lee
Brentjens, Michiel
Brumberg, Victor
Bruyninx, Carine
CAI, Yong
Carter, William
CHOU, Yi
Defraigne, Pascale
Dehant, Véronique
Dick, Wolfgang
Dickey, Jean
DONG, Shaowu
Douglas, R.
DU, Lan
Fallon, Frederick
Fliegel, Henry
Foschini, Luigi
FUJIMOTO, Masa Katsu
Gambis, Daniel
GAO, Yuping
Gonzalez, Gabriela
Guinot, Bernard
GUO, Ji
HAN, Tianqi
HANADO, Yuko
Hobbs, George
Hobiger, Thomas
HU, Yonghui
HUA, Yu
Ivanov, Dmitrii
Jaeggi, Adrian
JIN, WenJing
KAKUTA, Chuichi
Klepczynski, William
Kolaczek, Barbara
Koshelyaevsky, Nikolay
Kovalevsky, Jean
Kwok, Sun
Lammers, Uwe
Le Poncin-Lafitte, Ch.
LI, xiaohui
Lieske, Jay
LIU, Tao
LU, BenKui
LU, Xiaochun
Luck, John
MA, Lihua
Maciesiak, Krzysztof
Manchester, Richard
McCarthy, Dennis
Melbourne, William
Mendes, Virgilio
Minazzoli, Olivier
Morgan, Peter
Mueller, Ivan
Newhall, X.
Paquet, Paul
Petrov, Sergey
Pilkington, John
Pineau des Forêts, G.
Potapov, Vladimir
Pugliano, Antonio
Ray, James
Ray, Paul
Robertson, Douglas
Rodin, Alexander
Rushton, Anthony
Sheikh, Suneel
Smylie, Douglas
Stappers, Benjamin
SUN, Fuping
Thomas, Claudine
van Leeuwen, Joeri
Vernotte, François
Vicente, Raimundo
Vilinga, Jaime
WANG, Yulin
Wilkins, George
Wooden, William
WU, Guichen
WU, Haitao
XIE, Yi
YANG, Xuhai
Yatskiv, Yaroslav
YE, Shuhua
ZHANG, Shougang
ZHANG, Weiqun
ZHANG, Haotong
ZHENG, Yong

Division VII Commission 33 Structure & Dynamics of the Galactic System

President: Rosemary F. Wyse
Vice-President: Birgitta Nordström

Organizing Committee Members:

Bland-Hawthorn, Jonathan
Efremov, Yurij
Feltzing, Sofia
Flynn, Chris
Grindlay, Jonathan
Minniti, Dante

Members:

Aarseth, Sverre
Acosta Pulido, Jose
Afanas'ev, Viktor
Aguilar, Luis
AK, Serap
Alcobe, Santiago
Allende Prieto, Carlos
Altenhoff, Wilhelm
Ambastha, Ashok
Andersen, Johannes
Ardeberg, Arne
Ardi, Eliani
Arifyanto, Mochamad
Arnold, Richard
Asanok, Kitiyanee
Asteriadis, Georgios
BABA, Junichi
Babusiaux, Carine
BAEK, Chang Hyun
Baier, Frank
Balazs, Lajos
Balbus, Steven
Balcells, Marc
Banhatti, Dilip
Barberis, Bruno
Bartasiute, Stanislava
Bash, Frank
Baud, Boudewijn
Bauer, Amanda
Bellazzini, Michele
Benjamin, Robert
Bensby, Thomas
Berkhuijsen, Elly
Bienayme, Olivier
Binney, James
Blitz, Leo
Bloemen, Hans
Blommaert, Joris
Bobylev, Vadim
Bodaghee, Arash
Bon, Edi
Borka Jovanovic, Vesna
Brand, Jan
Bronfman, Leonardo
Brown, Warren
Burke, Bernard
Burton, W.
Butler, Ray
Caldwell, John
Cane, Hilary
CAO, Zhen
Caretta, Cesar
Carlin, Jeffrey
Carollo, Daniela
Carpintero, Daniel
Carrasco, Luis
Cesarsky, Diego
Cesarsky, Catherine
Cesetti, Mary
CHA, Seung-Hoon
Chapman, Jessica
Chemin, Laurent
CHEN, Li
CHEN, Yuqin
CHIBA, Masashi
CHOU, Mei-Yin
Christodoulou, Dimitris
Churchwell, Edward
Cincotta, Pablo
Cioni, Maria-Rosa
Clemens, Dan
Comeron, Sébastien
Comins, Neil
Contopoulos, George
Corradi, Romano
Costa, Edgardo
Crampton, David
Crawford, David
Creze, Michel
Cropper, Mark
Croton, Darren
Cubarsi, Rafael
Cudworth, Kyle
Cuperman, Sami
Dalla Bonta, Elena
Dambis, Andrei
Dauphole, Bertrand
Davies, Rodney
Dawson, Peter
de Jong, Teije
de Silva, Gayandhi
Dejonghe, Herwig
Dekel, Avishai
Diaferio, Antonaldo
Diaz, Ruben
Dickel, Helene
Dickel, John
Dickman, Robert
Djorgovski, Stanislav
do Nascimento, José
Downes, Dennis
Drilling, John
Drimmel, Ronald
Ducati, Jorge
Ducourant, Christine

Egret, Daniel
Einasto, Jaan
Elmegreen, Debra
ESAMDIN, Ali
Esguerra, Jose Perico
ESIMBEK, Jarken
Evangelidis, E.
Faber, Sandra
Fathi, Kambiz
Feast, Michael
Feitzinger, Johannes
Ferguson, Annette
Ferrari, Fabricio
Figueras, Francesca
Foster, Tyler
Freeman, Kenneth
FUJIMOTO, Masa Katsu
FUJIWARA, Takao
Ganguly, Rajib
Garzon, Francisco
Genzel, Reinhard
Gerhard, Ortwin
Gilmore, Gerard
Goldreich, Peter
Gomez, Ana
Gordon, Mark
Gottesman, Stephen
Grayzeck, Edwin
Green, Anne
Green, James
Grenon, Michel
Grillmair, Carl
GU, Minfeng
Gupta, Sunil
HABE, Asao
Habing, Harm
Haghi, Hosein
Hakkila, Jon
HANAMI, Hitoshi
Hanson, Margaret
Hartkopf, William
Hawkins, Michael
Hayli, Abraham
Haywood, Misha
Heiles, Carl
Helmi, Amina
Herbst, William
Hernandez-Pajares, Manuel
Hetem Jr., Annibal
Heyl, Jeremy
HONMA, Mareki
HOZUMI, Shunsuke
Hron, Josef
Humphreys, Roberta
Humphreys, Elizabeth
Ibata, Rodrigo
IGUCHI, Osamu
IKEDA, Norio
IKEUCHI, Satoru
INAGAKI, Shogo
ISHIHARA, Daisuke
Israel, Frank
Ivezic, Zeljko
IYE, Masanori
Jablonka, Pascale
Jachym, Pavel
Jackson, Peter
Jahreiss, Hartmut
Jalali, Mir Abbas
Jasniewicz, Gerard
JIANG, Dongrong
JIANG, Ing-Guey
Johansson, Peter
Jones, Derek
Kaisin, Serafim
Kalnajs, Agris
KANG, Yong-Hee
KATO, Shoji
Khovritchev, Maxim
KIM, Sungsoo
KIM, Sang Chul
KIM, Woong-Tae
King, Ivan
Kinman, Thomas
Knapp, Gillian
Korchagin, Vladimir
Kormendy, John
Krajnovic, Davor
Kucinskas, Arunas
Kulsrud, Russell
Kutuzov, Sergej
Laloum, Maurice
Larson, Richard
Latham, David
LEE, Sang-Gak
LEE, Hyung-Mok
LEE, Kang Hwan
LEE, Myung Gyoon
Lépine, Sébastien
LI, Jinzeng
Liebert, James
LIN, Qing
Lindblad, Per
Lokas, Ewa
LU, Youjun
LUO, Ali
MacConnell, Darrell
Mackey, Alasdair
Maier, Christian
Majumdar, Subhabrata
Manchester, Richard
Mandel, Ilya
Marin-Franch, Antonio
Marochnik, Leonid
Martin, Christopher
Martinet, Louis
Martinez Delgado, David
Martos, Marco
Mateu, Cecilia
MATSUNAGA, Noriyuki
Matteucci, Francesca
Mayor, Michel
McBride, Vanessa
McClure-Griffiths, Naomi
McGregor, Peter
McMillan, Paul
Melnik, Anna
Mendez Bussard, Rene
Merrifield, Michael
Mezger, Peter
Migaszewski, Cezary
Mikkola, Seppo
Miller, Richard
Mirabel, Igor
Mishurov, Yury
Miszalski, Brent
Moffat, Anthony
Mohammed, Ali
Moitinho, André
Monet, David
Morales Rueda, Luisa
Moreno Lupianez, Manuel
Morris, Mark
Morris, Rhys
Muench, Guido
NAKASATO, Naohito
Namboodiri, P.
Napolitano, Nicola
Narbutis, Donatas
Nelemans, Gijs

Nelson, Alistair
Newberg, Heidi
Nikiforov, Igor
Ninkovic, Slobodan
Norman, Colin
Nota, Antonella
Oblak, Edouard
Ocvirk, Pierre
Oey, Sally
OH, Kap Soo
Oja, Tarmo
Ojha, Devendra
OKA, Tomoharu
OKUDA, Haruyuki
Olano, Carlos
Oluseyi, Hakeem
Orlov, Victor
Ortiz, Roberto
Ostorero, Luisa
Ostriker, Jeremiah
Ostriker, Eve
Palmer, Patrick
Palous, Jan
Pandey, A.
Pandey, Birendra
Papayannopoulos, Theodoros
PARK, Byeong-Gon
Parmentier, Geneviève
Patsis, Panos
Pauls, Thomas
Peimbert, Manuel
PENG, Eric
Perek, Luboš
Perets, Hagai
Perryman, Michael
Pesch, Peter
Philip, A.G.
Pier, Jeffrey
Pietrukowicz, Pawel
Pirzkal, Norbert
Polyachenko, Evgeny
Portinari, Laura
Price, R.
Rabolli, Monica
Raharto, Moedji
Rastegaev, Denis
Ratnatunga, Kavan
Read, Justin
Recio-Blanco, Alejandra
Reid, Iain
Reif, Klaus
Reylé, Céline
Rhee, Jaehyon
Rho, Jeonghee
Rich, Robert
Richter, Philipp
Riegel, Kurt
Roberts, Morton
Roberts Jr, William
Rocha-Pinto, Hélio
Rodrigues de Oliveira F., I.
Rubin, Vera
Ruelas-Mayorga, R.
Ruiz, Maria Teresa
Ruzicka, Adam
Rybicki, George
Saar, Enn
SAKAMOTO, Tsuyoshi
Sakano, Masaaki
Sala, Ferran
Sanchez Doreste, Néstor
Sanchez-Saavedra, M.
Sandqvist, Aage
Santiago, Basilio
Santillan, Alfredo
Sanz, Jaume
Sargent, Annelia
Schechter, Paul
Schmidt, Maarten
Schoedel, Rainer
Seggewiss, Wilhelm
Seimenis, John
Sellwood, Jerry
Serabyn, Eugene
SHAN, Hongguang
Shane, William
SHEN, Juntai
SHI, Huoming
SHU, Frank
Siebert, Arnaud
Sigalotti, Leonardo
Simonson, S.
Sobouti, Yousef
SONG, Liming
SONG, Qian
Sotnikova, Natalia
Soubiran, Caroline
Sparke, Linda
Spergel, David
Spiegel, Edward
Stecker, Floyd
Steiman-Cameron, Thomas
Steinlin, Uli
Stoehr, Felix
Stone, Jennifer
Strobel, Andrzej
Strubbe, Linda
SU, Cheng-yue
Subramaniam, Annapurni
SUMI, Takahiro
Surdin, Vladimir
Sygnet, Jean-Francois
TAKASHI, Hasegawa
Tammann, Gustav
TANAKA, Ichi Makoto
Tempel, Elmo
Teyssier, Romain
The, Pik-Sin
Thomas, Claudine
TIAN, Wenwu
Tinney, Christopher
Tobin, William
TOMISAKA , Kohji
Toomre, Alar
Toomre, Juri
Torra, Jordi
Torres-Papaqui, Juan
TOSA, Makoto
TSENG, Yao-Huan
TSUJIMOTO, Takuji
Turon, Catherine
Upgren, Arthur
Urquhart, James
Valluri, Monica
Valtonen, Mauri
van der Kruit, Pieter
van Woerden, Hugo
Vandervoort, Peter
Varela Perez, Antonia
Vega, E.
Venugopal, V.
Vergne, María
Verschuur, Gerrit
Villas da Rocha, Jaime
Vivas, Anna
Volkov, Evgeni
Volonteri, Marta
Voroshilov, Volodymyr
Wachlin, Felipe

Wagner, Alexander
Weaver, Harold
Weistrop, Donna
Whiteoak, John
Whittet, Douglas
Wielebinski, Richard
Wielen, Roland
Woltjer, Lodewijk
Wong, Tony
Woodward, Paul
Wouterloot, Jan
Wramdemark, Stig
Wright, Nicholas
Wulandari, Hesti
Wunsch, Richard
Wyse, Rosemary
XU, Ye
YAMADA, Shimako
YAMAGATA, Tomohiko
YIM, Hong Suh
YOSHII, Yuzuru
YU, Qingjuan
Zachilas, Loukas
Zaggia, Simone
ZHANG, Haotong
ZHOU, Jianjun
ZHU, Qingfeng
Zoccali, Manuela

Division VI Commission 34 Interstellar Matter

President: You-Hua Chu
Vice-President: Sun Kwok

Organizing Committee Members:

Breitschwerdt, Dieter
Burton, Michael
Cabrit, Sylvie
Caselli, Paola
de Gouveia Dal Pino, E.
Evans, Neal
Ferland, Gary
Henning, Thomas
Juvela, Mika
KOO, Bon-Chul
Lizano, Susana
Rozyczka, Michal
TSUBOI, Masato
YANG, Ji

Members:

Aannestad, Per
Abgrall, Hervé
Acker, Agnes
Adams, Fred
AIKAWA, Yuri
Akashi, Muhammad
Al-Mostafa, Zaki
Alcolea, Javier
Altenhoff, Wilhelm
Alves, João
Ambrocio-Cruz, Silvia
Andersen, Anja
Andersen, Morten
Andersson, B.-G.
Andronov, Ivan
Anglada, Guillem
Arbutina, Bojan
Ardila, David
Arkhipova, Vera
Arny, Thomas
Arthur, Jane
ASANO, Katsuaki
Asanok, Kitiyanee
Audard, Marc
Azcarate, Diana
Baars, Jacob
BABA, Junichi
Babkovskaia, Natalia
Bachiller, Rafael
BAEK, Chang Hyun
Baker, Andrew
Ballesteros-Paredes, Javier
Balser, Dana
Baluteau, Jean-Paul
Bania, Thomas
Barlow, Michael
Barnes, Aaron
Baryshev, Andrey
Bash, Frank
Basu, Shantanu
Baudry, Alain
Bautista, Manuel
Bayet, Estelle
Becklin, Eric
Beckman, John
Beckwith, Steven
Bedogni, Roberto
Benisty, Myriam
Bergeron, Jacqueline
Bergin, Edwin
Bergman, Per
Bergstrom, Lars
Berkhuijsen, Elly
Bernat, Andrew
Bertout, Claude
Bhat, Ramesh
Bhatt, H.
Bianchi, Luciana
Bieging, John
Bignall, Hayley
Bignell, R.
Binette, Luc
Black, John
Blair, Guy
Blair, William
Bless, Robert
Blitz, Leo
Bloemen, Hans
Bobrowsky, Matthew
Bocchino, Fabrizio
Bochkarev, Nikolai
Bode, Michael
Bodenheimer, Peter
Boggess, Albert
Bohlin, Ralph
Boisse, Patrick
Boland, Wilfried
Bontemps, Sylvain
Boquien, Médéric
Bordbar, Gholam
Borka Jovanovic, Vesna
Borkowski, Kazimierz
Bot, Caroline
Boulanger, François
Boumis, Panayotis
Bourke, Tyler
Bouvier, Jerôme
Bowen, David
Brand, Jan
Brand, Peter
Briceno, Cesar
Bromage, Gordon
Brooks, Kate
Brouillet, Nathalie
Bruhweiler, Frederick
Bujarrabal, Valentin
Burke, Bernard
Burton, W.
Bychkov, Konstantin
Bykov, Andrei
Bzowski, Maciej
CAI, Kai

Cambresy, Laurent
Cami, Jan
Caplan, James
Cappa de Nicolau, Cristina
Capriotti, Eugene
Capuzzo Dolcetta, Roberto
Carretti, Ettore
Carruthers, George
Casasola, Viviana
Casassus, Simon
Caselli, Paola
Castaneda, Héctor
Castelletti, Gabriela
Cattaneo , Andrea
Cazaux, Stéphanie
Ceccarelli, Cecilia
Cecchi-Pestellini, Cesare
Centurion Martin, Miriam
Cernicharo, José
Cerruti Sola, Monica
Cersosimo, Juan
Cesarsky, Diego
Cesarsky, Catherine
CHA, Seung-Hoon
Chandra, Suresh
Chelouche, Doron
CHEN, Yang
CHEN, Yafeng
CHEN, Huei-Ru
CHEN, Xuefei
CHENG, Kwang
Cherchneff, Isabelle
Chevalier, Roger
CHIHARA, Hiroki
Chini, Rolf
Christopoulou, P.-E.
Chu, You-Hua
Churchwell, Edward
Ciardullo, Robin
Cichowolski, Silvina
Ciroi, Stefano
Clark, Frank
Clarke, David
Clegg, Robin
Codella, Claudio
Coffey, Deirdre
Cohen, Marshall
Colangeli, Luigi
Collin, Suzy
Combes, Françoise
Corbelli, Edvige
Corradi, Wagner
Corradi, Romano
Costantini, Elisa
Costero, Rafael
Cowie, Lennox
Cox, Donald
Cox, Pierre
Coyne, S.J, George
Cracco, Valentina
Crane, Philippe
Crawford, Ian
Crovisier, Jacques
Cuesta Crespo, Luis
Cunningham, Maria
d'Hendecourt, Louis
d'Odorico, Sandro
Dahn, Conard
Dale, James
Danks, Anthony
Danly, Laura
Dannerbauer, Helmut
Davies, Rodney
Davis, Christopher
Davis, Timothy
de Almeida, Amaury
De Avillez, Miguel
De Bernardis, Paolo
De Buizer, James
de Gregorio-Monsalvo, I.
de Jong, Teije
de La Noë, Jerome
De Marco, Orsola
Decourchelle, Anne
DEGUCHI, Shuji
Deharveng, Lise
Deiss, Bruno
Dennefeld, Michel
Dewdney, Peter
Dias da Costa, Roberto
Diaz, Ruben
Diaz-Santos, Tanio
Dib, Sami
Dickel, Helene
Dickel, John
Dickey, John
Dieleman, Pieter
Dinerstein, Harriet
Dinh, Trung
Dionatos, Odysseas
Disney, Michael
Djamaluddin, Thomas
Docenko, Dmitrijs
Dokuchaev, Vyacheslav
Dokuchaeva, Olga
Dominik, Carsten
Dopita, Michael
Dorschner, Johann
Dottori, Horacio
Dougados, Catherine
Downes, Dennis
Draine, Bruce
Dreher, John
Dubner, Gloria
Dubout, Renee
Dudorov, Aleksandr
Dufour, Reginald
Duley, Walter
Dunne, Loretta
Dupree, Andrea
Dutrey, Anne
Duvert, Gilles
Dwarkadas, Vikram
Dwek, Eli
Edwards, Suzan
Egan, Michael
Ehlerova, Soňa
Eisloeffel, Jochen
Elia, Davide
Elitzur, Moshe
Elliott, Kenneth
Ellison, Sara
Elmegreen, Bruce
Elmegreen, Debra
Emerson, James
Encrenaz, Pierre
ESAMDIN, Ali
Escalante, Vladimir
ESIMBEK, Jarken
Esipov, Valentin
Esteban, César
Evans, Aneurin
Falceta-Goncalves, Diego
Falgarone, Edith
Falize, Emeric
Falle, Samuel
Federman, Steven
Feitzinger, Johannes
Felli, Marcello
Fendt, Christian

Ferlet, Roger
Fernandes, Amadeu
Ferrière, Katia
Ferrini, Federico
Fesen, Robert
Fiebig, Dirk
Field, George
Field, David
Fierro, Julieta
Figer, Donald
Fischer, Jacqueline
Flannery, Brian
Fleck, Robert
Fletcher, Andrew
Florido, Estrella
Flower, David
Folini, Doris
Ford, Holland
Forster, James
Franco, Gabriel Armando
Franco, José
Fraschetti, Federico
Fraser, Helen
Freimanis, Juris
Frew, David
Fridlund, Malcolm
Frisch, Priscilla
Fromang, Sebastien
Fuente, Asuncion
Fukuda, Naoya
FUKUI, Yasuo
Fuller, Gary
Furniss, Ian
Furuya, Ray
Gaensler, Bryan
Galli, Daniele
GAO, Yu
GAO, Jian
Garay, Guido
Garcia, Paulo
Garcia-Hernandez, D.
Garcia-Lario, Pedro
Garcia-Segura, Guillermo
Garnett, Donald
Gathier, Roel
Gaume, Ralph
Gaustad, John
Gay, Jean
Geballe, Thomas
Genzel, Reinhard
Gerard, Eric
Gerin, Maryvonne
Gezari, Daniel
Ghanbari, Jamshid
Giacani, Elsa
Gibson, Steven
Giovanelli, Riccardo
Glover, Simon
Goddi, Ciriaco
Godfrey, Peter
Goebel, John
Goldes, Guillermo
Goldreich, Peter
Goldsmith, Donald
Golovatyj, Volodymyr
Gomez, Gonzalez
Goncalves, Denise
Gonzales-Alfonso, Eduardo
Goodman, Alyssa
Gordon, Mark
Gordon, Karl
Gosachinskij, Igor
Goss, W. Miller
Graham, David
Granato, Gian Luigi
Gredel, Roland
Green, James
Gregorio-Hetem, Jane
Greisen, Eric
Grewing, Michael
Guelin, Michel
Guertler, Joachin
Guesten, Rolf
Guilloteau, Stéphane
Gull, Theodore
Gunthardt, Guillermo
GUO, Jianheng
Habing, Harm
Hackwell, John
Haisch Jr, Karl
HANAMI, Hitoshi
HAO, Lei
Harrington, J.
Harris, Alan
Harris-Law, Stella
Hartl, Herbert
Hartquist, Thomas
Harvey, Paul
Hatchell, Jennifer
Haverkorn, Marijke
Hayashi, Saeko
Haynes, Raymond
HE, Jinhua
Hébrard, Guillaume
Hecht, James
Heikkila, Arto
Heiles, Carl
Heinz, Sebastian
Helfer, H.
Helmich, Frank
Helou, George
Heng, Kevin
Henkel, Christian
Henney, William
Hernandez, Jesús
Herpin, Fabrice
Hersant, Franck
Heydari-Malayeri, M.
Heyer, Mark
Hidayat, Bambang
Higgs, Lloyd
Hildebrand, Roger
Hillenbrand, Lynne
Hippelein, Hans
HIRANO, Naomi
Hiriart, David
HIROMOTO, Norihisa
HIROSE, Shigenobu
Hjalmarson, Ake
Hobbs, Lewis
Hollenbach, David
Hollis, Jan
HONDA, Mitsuhiko
HONG, Seung-Soo
Hora, Joseph
Horacek, Jiri
Horns, Dieter
Houde, Martin
Houziaux, Leo
Hudson, Reggie
Huggins, Patrick
Hutchings, John
Hutsemekers, Damien
Hyung, Siek
IKEDA, Norio
Ikonnikova, Natalia
Ilin, Vladimir
INOUE, Akio
INUTSUKA, Shu-ichiro
Irvine, William

ISHIHARA, Daisuke
ISHII, Miki
Israel, Frank
ITOH, Yoichi
Jackson, James
Jacoby, George
Jacq, Thierry
Jaffe, Daniel
Jahnke, Knud
Jenkins, Edward
JIANG, Zhibo
Jimenez-Vicente, Jorge
JIN, Zhenyu
Joblin, Christine
Johnson, Fred
Johnston, Kenneth
Johnstone, Douglas
Jones, Christine
Jones, David
Jorgensen, Uffe
Jorgensen, Jes
Jourdain de Muizon, Marie
Jura, Michael
Just, Andreas
Justtanont-Liseau, Kay
Kafatos, Menas
Kaftan, May
KAIFU, Norio
Kalenskii, Sergei
Kaler, James
KAMAYA, Hideyuki
KAMAZAKI, Takeshi
KAMEGAI, Kazuhisa
Kamp, Inga
Kanekar, Nissim
Kantharia, Nimisha
Karakas, Amanda
Kassim, Namir
Kaviraj, Sugata
Kawada, Mitsunobu
Keene, Jocelyn
Kegel, Wilhelm
Keheyan, Yeghis
Kemper, Francesca
Kennicutt, Robert
KIM, Jongsoo
KIM, Woong-Tae
KIM, Ji Hoon
KIMURA, Toshiya
King, David
Kirkpatrick, Ronald
Kirshner, Robert
Kiss, Csaba
Klessen, Ralf
Knacke, Roger
Knapp, Gillian
Knezek, Patricia
Knude, Jens
KO, Chung-Ming
KOBAYASHI, Naoto
Kohoutek, Lubos
KOIKE, Chiyoe
KONDO, Yoji
KONG, Xu
Koornneef, Jan
Korpela, Eric
Kospal, Ágnes
KOZASA, Takashi
Krabbe, Angela
Krajnovic, Davor
Kramer, Busaba
Krautter, Joachim
Kravchuk, Sergei
Kreysa, Ernst
Krishna, Swamy
Krumholz, Mark
KUAN, Yi-Jehng
KUDOH, Takahiro
Kuiper, Thomas
Kuiper, Rolf
Kulhanek, Petr
Kumar, C.
Kunth, Daniel
Kutner, Marc
Kwitter, Karen
Kylafis, Nikolaos
Lada, Charles
LAI, Shih-Ping
Laloum, Maurice
Langer, William
Latter, William
Laureijs, Rene
Laurent, Claudine
Lauroesch, James
Lazarian, Alexandre
Lazio, Joseph
Leao, Joao Rodrigo
Lebron, Mayra
LEE, Hee Won
LEE, Dae Hee
LEE, Myung Gyoon
LEE, Jung-Won
LEE, Jeong-Eun
Lefloch, Bertrand
Léger, Alain
Lehtinen, Kimmo
Leisawitz, David
Lépine, Jacques
Lequeux, James
Leto, Giuseppe
LI, Jinzeng
LIANG, Yanchun
Ligori, Sebastiano
Likkel, Lauren
Liller, William
Limongi, Marco
LIN, Weipeng
Linnartz, Harold
Lis, Dariusz
Liseau, René
Liszt, Harvey
LIU, Sheng-Yuan
LIU, Xiaowei
Lloyd, Myfanwy
Lo, Fred K. Y.
Lo, Wing-Chi Nadia
Lockman, Felix
Lodders, Katharina
Loinard, Laurent
Lopez Garcia, Jose
Louise, Raymond
Lovas, Francis
Lozinskaya, Tatjana
Lucas, Robert
Lucero, Danielle
Lynds, Beverly
Lyon, Ian
MA, Jun
Mac Low, Mordecai-Mark
Maciel, Walter
MacLeod, John
Madsen, Gregory
Magrini, Laura
Maier, Christian
MAIHARA, Toshinori
MAKIUTI, Sin'itirou
Malbet, Fabien
Mallik, D.
Mampaso, Antonio
Manchado, Arturo

Manchester, Richard
Manfroid, Jean
Mantere, Maarit
Mardones, Diego
Maret, Sébastien
Marleau, Francine
Marston, Anthony
Martin, Peter
Martin, Robert
Martin, Christopher
Martin, Sergio
Martin-Pintado, Jesus
Masson, Colin
Mather, John
Mathews, William
Mathis, John
MATSUHARA, Hideo
MATSUMOTO, Tomoaki
MATSUMURA, Masafumi
MATSUMURA, Tomotake
Mattila, Kalevi
Mauersberger, Rainer
McCall, Marshall
McCall, Benjamin
McClure-Griffiths, Naomi
McCombie, June
McCray, Richard
McGehee, Peregrine
McGregor, Peter
McKee, Christopher
McNally, Derek
Meaburn, John
Mebold, Ulrich
Meier, Robert
Meixner, Margaret
Mellema, Garrelt
Melnick, Gary
Mendez, Roberto
Mennella, Vito
Menon, T.
Menten, Karl
Menzies, John
Meszaros, Peter
Meyer, Martin
Mezger, Peter
Millar, Thomas
Miller, Joseph
Miller, Eric
Milne, Douglas
Minier, Vincent
MINN, Young-Ki
Minter, Anthony
Miszalski, Brent
Mitchell, George
MIYAMA, Syoken
Mo, Jinger
Monin, Jean-Louis
Montmerle, Thierry
Mookerjea, Bhaswati
Moor, Attila
Moore, Marla
Moreno-Corral, Marco
Moriarty-Schieven, Gerald
Morris, Mark
Morris, Patrick
Morton, Donald
Mosoni, Laszlo
Mouschovias, Telemachos
Muench, Guido
Mufson, Stuart
Mulas, Giacomo
Muller, Erik
Muller, Sebastien
Murthy, Jayant
Myers, Philip
NAGAHARA, Hiroko
NAGATA, Tetsuya
NAKADA, Yoshikazu
NAKAGAWA, Takao
NAKAMOTO, Taishi
NAKAMURA, Fumitaka
NAKANO, Makoto
NAKANO, Takenori
Nammahachak, Suwit
Natta, Antonella
Nazé, Yaël
Nesvadba, Nicole
Nguyen-Quang, Rieu
Nikoghosyan, Elena
Nikolic, Silvana
NISHI, Ryoichi
NOMURA, Hideko
Nordh, Lennart
Norman, Colin
Noterdaeme, Pasquier
Nuernberger, Dieter
Nulsen, Paul
Nussbaumer, Harry
Nuth, Joseph
O'Dell, Charles
O'Dell, Stephen
Oberst, Thomas
Oey, Sally
OHTANI, Hiroshi
OKA, Tomoharu
OKUDA, Haruyuki
OKUMURA, Shin-ichiro
Olofsson, Hans
Omont, Alain
OMUKAI, Kazuyuki
ONAKA, Takashi
Onello, Joseph
Orlando, Salvatore
Osborne, John
Oskinova, Lidia
Ostriker, Eve
OTSUKA, Masaaki
Ott, Juergen
Oudmaijer, Rene
Pagani, Laurent
Pagano, Isabella
PAK, Soojong
Palla, Francesco
Palmer, Patrick
Palumbo, Maria Elisabetta
Panagia, Nino
Pandey, Birendra
Pankonin, Vernon
PARK, Yong Sun
Parker, Quentin
Parker, Eugene
Paron, Sergio
Parthasarathy, Mudumba
Pauls, Thomas
Pavlyuchenkov, Yaroslav
Pecker, Jean-Claude
Peeters, Els
Peimbert, Manuel
Pellegrini, Silvia
Pena, Miriam
Pendleton, Yvonne
PENG, Qingyu
Penzias, Arno
Péquignot, Daniel
Perault, Michel
Persi, Paolo
Persson, Carina
Peters, William
Petrosian, Vahe
Petuchowski, Samuel

Philipp, Sabine
Phillips, Thomas
Pihlström, Ylva
Pilbratt, Göran
Pilling, Sergio
Pineau des Forêts, G.
Pinte, Christophe
Plume, René
Poeppel, Wolfgang
Pongracic, Helen
Pontoppidan, Klaus
Porceddu, Ignazio
Pottasch, Stuart
Pound, Marc
Pouquet, Annick
Prasad, Sheo
Preite Martinez, Andrea
Price, R.
Price, Daniel
Prochaska, Jason
Pronik, Iraida
Prusti, Timo
Puget, Jean-Loup
Ramirez, José
Ramos-Larios, Gerardo
Ranalli, Piero
Rastogi, Shantanu
Ratag, Mezak
Rathborne, Jill
Rawlings, Jonathan
Rawlings, Mark
Raymond, John
Recchi, Simone
Redman, Matthew
Reid, Michael
Reipurth, Bo
Rengarajan, Thinniam
Rengel, Miriam
Reshetnyk, Volodymyr
Reyes, Rafael
Reynolds, Ronald
Reynolds, Cormac
Reynoso, Estela
Rho, Jeonghee
Richter, Philipp
Rickard, Lee
Roberge, Wayne
Roberts, Douglas
Roberts Jr, William
Robinson, Garry
Roche, Patrick
Rodrigues, Claudia
Rodriguez, Luis
Rodriguez, Monica
Roediger, Elke
Roelfsema, Peter
Roeser, Hans-peter
Roger, Robert
Rogers, Alan
Roman-Zuniga, Carlos
Rosa, Michael
Rosado, Margarita
Rosolowsky, Erik
Rouan, Daniel
Rowell, Gavin
Roxburgh, Ian
Ryabov, Michael
Sabbadin, Franco
Sabin, Laurence
Sahu, Kailash
SAIGO, Kazuya
Sakano, Masaaki
SAKON, Itsuki
Salama, Farid
Salinari, Piero
Salomé, Philippe
Salter, Christopher
Samodurov, Vladimir
Sanchez Doreste, Néstor
Sanchez-Saavedra, M.
Sancisi, Renzo
Sandell, Göran
Sandqvist, Aage
Sankrit, Ravi
Sarazin, Craig
Sargent, Annelia
Sarma, N.
Sarre, Peter
SATO, Fumio
SATO, Shuji
Savage, Blair
Savedoff, Malcolm
Scalo, John
Scherb, Frank
Schilke, Peter
Schlemmer, Stephan
Schmid-Burgk, J.
Schroder, Anja
Schure, Klara
Schwarz, Ulrich
Scoville, Nicholas
SEKI, Munezo
Sellgren, Kristen
Sembach, Kenneth
Semenov, Dmitry
Sen, Asoke
SEON, Kwang il
Shadmehri, Mohsen
Shalchi, Andreas
Shane, William
Shapiro, Stuart
Sharma, Prateek
Shaver, Peter
Shawl, Stephen
Shchekinov, Yuri
Shematovich, Valerij
Sherwood, William
Shields, Gregory
SHIMOIKURA, Tomomi
Shipman, Russell
Shmeld, Ivar
SHU, Frank
Shull, John
Shull, Peter
Shustov, Boris
Siebenmorgen, Ralf
Sigalotti, Leonardo
Silich, Sergey
Silk, Joseph
Silva, Laura
Silvestro, Giovanni
Simon-Diaz, Sergio
Sitko, Michael
Sivan, Jean-Pierre
Skilling, John
Skulskyj, Mychajlo
Slane, Patrick
Sloan, Gregory
Smirnova, Tatiana
Smith, Peter
Smith, Craig
Smith, Michael
Smith, Tracy
Smith, Randall
Smith, Robert
Snell, Ronald
Snow, Theodore
Sobolev, Andrey
Sofia, Ulysses
Sofia, Sabatino

SOFUE, Yoshiaki
Solc, Martin
Somerville, William
SONG, In Ok
Spaans, Marco
Stahler, Steven
Stanga, Ruggero
Stanghellini, Letizia
Stanimirovic, Snezana
Stapelfeldt, Karl
Stark, Ronald
Stasinska, Grazyna
Stecher, Theodore
Stecklum, Bringfried
Stenholm, Björn
Stone, James
Stone, Jennifer
Strom, Richard
SUH, Kyung Won
Suleymanova, Svetlana
SUNG, Hyun-Il
SUSA, Hajime
Sutherland, Ralph
SUZUKI, Tomoharu
Swade, Daryl
Sylvester, Roger
Szczerba, Ryszard
TACHIHARA, Kengo
Tafalla, Mario
TAKAHASHI, Junko
TAKAHASHI, Hidenori
TAKANO, Toshiaki
TAMURA, Motohide
TAMURA, Shin'ichi
TANAKA, Masuo
Tantalo, Rosaria
Teixeira, Paula
Tenorio-Tagle, Guillermo
Tepper Garcia, Thorsten
TERADA, Yukikatsu
Terzian, Yervant
Testi, Leonardo
Teyssier, Romain
Thaddeus, Patrick
The, Pik-Sin
Thoene, Christina
Thompson, A.
Thonnard, Norbert
Thronson Jr, Harley
Tilanus, Remo
Tokarev, Yurij
Torrelles, Jose-Maria
Torres-Peimbert, Silvia
Tosi, Monica
Tothill, Nicholas
Trammell, Susan
Treffers, Richard
Tremonti, Christy
Trinidad, Miguel
TSAI, An-Li
Turner, Kenneth
Tyul'bashev, Sergei
Ueta, Toshiya
Ulrich, Marie-Helene
Urosevic, Dejan
Urquhart, James
USUDA, Tomonori
van de Steene, Griet
van den Ancker, Mario
van der Hulst, Jan
van der Laan, Harry
van der Tak, Floris
van Dishoeck, Ewine
van Gorkom, Jacqueline
van Loon, Jacco
van Woerden, Hugo
VandenBout, Paul
Varshalovich, Dmitrij
Vasta, Magda
Vavrek, Roland
Vazquez, Roberto
Vega, Olga
Velazquez, Pablo
Verdoes Kleijn, Gijsbert
Verheijen, Marc
Verma, Aprajita
Verner, Ekaterina
Verschuur, Gerrit
Viala, Yves
Viallefond, Francois
Vidal-Madjar, Alfred
Viegas, Sueli
Viironen, Kerttu
Vijh, Uma
Vilchez, Jose
Villaver, Eva
Vink, Jacco
Viti, Serena
Vlahakis, Catherine
Vlemmings, Wouter
Voit, Gerard
Volk, Kevin
Vorobyov, Eduard
Voronkov, Maxim
Voshchinnikov, Nikolai
Vrba, Frederick
Wakelam, Valentine
Wakker, Bastiaan
Walker, Gordon
Walmsley, C.
Walsh, Wilfred
Walsh, Andrew
Walton, Nicholas
WANG, Hongchi
Wang, Q. Daniel
WANG, Jun-Jie
WANG, Hong-Guang
WANG, Junzhi
WANG, Min
WANG, Chen
Wannier, Peter
Ward-Thompson, Derek
Wardle, Mark
Watt, Graeme
Weaver, Harold
Weiler, Kurt
Weinberger, Ronald
Wendt, Martin
Wesselius, Paul
Wesson, Roger
Weymann, Ray
Whelan, Emma
White, Glenn
White, Richard
Whitelock, Patricia
Whiteoak, John
Whittet, Douglas
Whitworth, Anthony
Wickramasinghe, N.
Wiebe, Dmitri
Wild, Wolfgang
Wilkin, Francis
Williams, David
Williams, Robert
Williams, Robin
Willis, Allan
Willner, Steven
Wilson, Robert
Wilson, Thomas
Wilson, Christine

Winnberg, Anders
Witt, Adolf
Wolff, Michael
Wolfire, Mark
Wolstencroft, Ramon
Wolszczan, Alexander
Woltjer, Lodewijk
Wong, Tony
Wood, Brian
Woodward, Paul
Woolf, Neville
Wootten, Henry
Wouterloot, Jan
Wright, Edward
Wunsch, Richard
Wynn-Williams, Gareth
YABUSHITA, Shin
YAMADA, Masako
YAMAMOTO, Satoshi
YAMAMOTO, Hiroaki
YAMAMURA , Issei
YAMASHITA , Takuya
YAN, Jun
York, Donald
Yorke, Harold
YOSHIDA , Shigeomi
YUI, Yukari
Yun, Joao
Zavagno, Annie
Zealey, William
Zeilik, Michael
ZENG, Qin
ZHANG, JiangShui
ZHANG, Jingyi
ZHOU, Jianjun
ZHU, Wenbai
ZHU, Qingfeng
Zibetti, Stefano
Zijlstra, Albert
Zinchenko, Igor
Zuckerman, Benjamin

Division IV Commission 35 Stellar Constitution

President: Corinne Charbonnel
Vice-President: Marco Limongi

Organizing Committee Members:

Fontaine, Gilles
Isern, Jordi
Lattanzio, John
Leitherer, Claus
van Loon, Jacco
Weiss, Achim
Yungelson, Lev

Members:

Adams, Mark
Aizenman, Morris
Angelov, Trajko
Antia, H.
Appenzeller, Immo
ARAI, Kenzo
Arentoft, Torben
Argast, Dominik
ARIMOTO, Nobuo
Arnett, W.
Arnould, Marcel
Audouze, Jean
Baglin, Annie
Barnes, Sydney
Basu, Sarbani
Baym, Gordon
Bazot, Michael
Becker, Stephen
Belkacem, Kevin
Belmonte Aviles, J. A.
Benz, Willy
Bergeron, Pierre
Bertelli, Gianpaolo
Berthomieu, Gabrielle
Bisnovatyi-Kogan, G.
Blaga, Cristina
Bludman, Sidney
Bodenheimer, Peter
Bombaci, Ignazio
Bono, Giuseppe
Boss, Alan
Braithwaite, Jonathan
Brassard, Pierre
Bravo, Eduardo
Bressan, Alessandro
Brown, David
Browning, Matthew
Brownlee, Robert
Bruenn, Stephen
Brun, Allan
Busso, Maurizio
Callebaut, Dirk
Caloi, Vittoria
Campbell, Simon
Canal, Ramon
Caputo, Filippina
Carson, T.
Castor, John
Chaboyer, Brian
Chabrier, Gilles
Chamel, Nicolas
CHAN, Kwing
Chan, Roberto
Charbonnel, Corinne
Charpinet, Stéphane
Chechetkin, Valerij
Chiosi, Cesare
Chitre, Shashikumar
Christensen-Dalsgaard, J.
Cohen, Judith
Connolly, Leo
Corsico, Alejandro
Cowan, John
Cristallo, Sergio
Das, Mrinal
Daszynska-Daszkiewicz, J.
Davies, Benjamin
de Greve, Jean-Pierre
de Jager, Cornelis
de Loore, Camiel
de Medeiros, Jose
de Mink, Selma
de Silva, Gayandhi
Dearborn, David
Deinzer, W.
Deliyannis, Constantine
Demarque, Pierre
Denisenkov, Pavel
Deupree, Robert
Di Mauro, Maria Pia
Dluzhnevskaya, Olga
Domiciano de Souza, A.
Dupuis, Jean
Durisen, Richard
Dziembowski, Wojciech
Edwards, Suzan
Eggenberger, Patrick
Eggleton, Peter
Ekstrom Garcia N., Sylvia
Eldridge, John
Elmhamdi, Abouazza
Engelbrecht, Chris
ERIGUCHI, Yoshiharu
Fadeyev, Yurij
Faulkner, John
Flannery, Brian
Fontaine, Gilles
Forbes, J.
Fossat, Eric
Foukal, Peter
Fox-Machado, Lester
FUJIMOTO, Masayuki
Gabriel, Maurice
Gallino, Roberto
Garcia, Domingo
Garcia, Miriam
Gautschy, Alfred
Georgy, Cyril

Geroyannis, Vassilis
Giannone, Pietro
Gimenez, Alvaro
Girardi, Leo
Giridhar, Sunetra
Glatzmaier, Gary
Goedhart, Sharmila
Gomez, Haley
Goncalves, Denise
Goriely, Stephane
Goswami, Aruna
Gough, Douglas
Goupil, Marie-José
Graham, Eric
Greggio, Laura
Groh, Jose
Guenther, David
Guzik, Joyce
HACHISU, Izumi
Hammond, Gordon
HAN, Zhanwen
HASHIMOTO, Masa-aki
Heger, Alexander
Henry, Richard
Hernanz, Margarita
Hillier, John
Hirschi, Raphael
Hollowell, David
Homeier, Derek
Huggins, Patrick
Humphreys, Roberta
Iben Jr, Icko
Iliev, Ilian
Imbroane, Alexandru
Imshennik, Vladimir
Isern, Jordi
ISHIHARA, Daisuke
ISHIZUKA, Toshihisa
ITOH, Naoki
Ivanova, Natalia
IWAMOTO, Nobuyuki
Izzard, Robert
James, Richard
Jones, David
Jorgensen, Jes
Jose, Jordi
Kaehler, Helmuth
Kalirai, Jason
Kaminker, Alexander
Kapyla, Petri
KATO, Mariko
Kervella, Pierre
KIGUCHI, Masayoshi
Kippenhahn, Rudolf
Kiziloglu, Nilgun
Knoelker, Michael
Kochhar, Rajesh
Koester, Detlev
Konar, Sushan
Kosovichev, Alexander
Kovetz, Attay
Kucinskas, Arunas
Kuiper, Rolf
Kumar, Shiv
Kwok, Sun
Labay, Javier
Lamb, Susan
Lamb Jr, Donald
Lamzin, Sergei
Langer, Norbert
Laskarides, Paul
Lasota-Hirszowicz, J.-P.
Lebovitz, Norman
Lebreton, Yveline
LEE, Thyphoon
Lee, William
Leitherer, Claus
Lepine, Jacques
LI, Qingkang
Liebendoerfer, Matthias
Lignieres, François
Limongi, Marco
Linnell, Albert
Littleton, John
LIU, Guoqing
Livio, Mario
Lucatello, Sara
Lugaro, Maria
MAEDA, Keiichi
Maeder, Andre
Maheswaran, M.
Mallik, D.
Marin-Franch, Antonio
Martayan, Christophe
Mathis, Stephane
Matteucci, Francesca
Mazumdar, Anwesh
Mazurek, Thaddeus
Mazzitelli, Italo
McDavid, David
Melbourne, Jason
Mendes, Luiz
Mestel, Leon
Meyer-Hofmeister, Eva
Meynet, meynet
Michaud, Georges
Miszalski, Brent
Mitalas, Romas
MIYAJI, Shigeki
Moellenhoff, Claus
Mohan, Chander
Moiseenko, Sergey
Monaghan, Joseph
Monier, Richard
Montalban, Josefina
Monteiro, Mario Joao
Moore, Daniel
Morgan, John
Moskalik, Pawe?
Moss, David
Mowlavi, Nami
Nadyozhin, Dmitrij
NAGATAKI, Shigehiro
NAKAMURA, Takashi
NAKANO, Takenori
NAKAZAWA, Kiyoshi
Narasimha, Delampady
NARITA, Shinji
Nelemans, Gijs
Newman, Michael
Noels, Arlette
NOMOTO, Ken'ichi
O'Toole, Simon
Odell, Andrew
OKAMOTO, Isao
Oliveira, Joana
OSAKI, Yoji
Ostriker, Jeremiah
Oswalt, Terry
Oudmaijer, Rene
Palacios, Ana
Pamyatnykh, Alexey
Panei, Jorge
Papaloizou, John
Pearce, Gillian
Phillips, Mark
Pinotsis, Antonis
Pongracic, Helen
Pontoppidan, Klaus
Poveda, Arcadio

Prentice, Andrew
Prialnik, Dina
Proffitt, Charles
Provost, Janine
Pulone, Luigi
Raedler, K.
Ramadurai, Souriraja
Rauscher, Thomas
Ray, Alak
Rayet, Marc
Reeves, Hubert
Reisenegger, Andreas
Renzini, Alvio
Reyniers, Maarten
Richard, Olivier
Ritter, Hans
Rizzi, Luca
Roca Cortes, Teodoro
Roxburgh, Ian
Ruiz-Lapuente, María
Sackmann, Inge
SAIO, Hideyuki
Salaris, Maurizio
Santos, Filipe
Sarna, Marek
SATO, Katsuhiko
Savedoff, Malcolm
Savonije, Gerrit
Scalo, John
Schatten, Kenneth
Schoenberner, Detlef
Schuler, Simon
Schutz, Bernard
Scuflaire, Richard
Seidov, Zakir
Sengbusch, Kurt
SHIBAHASHI, Hiromoto
SHIBATA, Yukio
Shustov, Boris
Siess, Lionel
Sigalotti, Leonardo
Signore, Monique
Sills, Alison
Silvestro, Giovanni
Simon-Diaz, Sergio
Sion, Edward
Smeyers, Paul
Smith, Robert
Smolec, Radoslaw
Sobouti, Yousef
Sofia, Sabatino
Sparks, Warren
Spiegel, Edward
Sreenivasan, S.
Stancliffe, Richard
Starrfield, Sumner
Stellingwerf, Robert
Stergioulas, Nikolaos
Stringfellow, Guy
Strittmatter, Peter
Suda, Takuma
SUGIMOTO, Daiichiro
Sundqvist, Jon
Sweigart, Allen
Taam, Ronald
TAKAHARA, Mariko
Thielemann, Friedrich-Karl
Todt, Helge
Tohline, Joel
TOMINAGA, Nozomu
Toomre, Juri
Tornambe, Amedeo
Townsend, Richard
Trimble, Virginia
TruranJr, James
Tscharnuter, Werner
Turck-Chièze, Sylvaine
Tutukov, Aleksandr
Ubeda, Leonardo
UCHIDA, Juichi
ud-Doula, Asif
Ulrich, Roger
UNNO, Wasaburo
Utrobin, Victor
van den Heuvel, Edward
van Horn, Hugh
van Loon, Jacco
van Riper, Kenneth
VandenBerg, Don
Vauclair, Gérard
Ventura, Paolo
Vilhu, Osmi
Vilkoviskij, Emmanuil
Vink, Jorick
Ward, Richard
Weaver, Thomas
Webbink, Ronald
Weiss, Nigel
Weiss, Achim
Wheeler, J.
Willson, Lee Anne
Wilson, Robert
Winkler, Karl-Heinz
Wood, Peter
Wood, Matthew
Woosley, Stanford
XIONG, Da Run
YAMADA, Shimako
YAMAOKA, Hitoshi
YI, Sukyoung
Yorke, Harold
YOSHIDA, Shin'ichirou
YOSHIDA, Takashi
Yungelson, Lev
Yushkin, Maxim
Zahn, Jean-Paul
Ziolkowski, Janusz

Division IV Commission 36 Theory of Stellar Atmospheres

President: Martin Asplund
Vice-President: Joachim Puls

Organizing Committee Members:

Allende Prieto, Carlos
Ayres, Thomas
Berdyugina, Svetlana
Gustafsson, Bengt
Hubeny, Ivan
Ludwig, Hans
Mashonkina, Lyudmila
Randich, Sofia

Members:

Afram, Nadine
Altrock, Richard
Andretta, Vincenzo
Ardila, David
Aret, Anna
Arpigny, Claude
Atanackovic, Olga
Aufdenberg, Jason
Avrett, Eugene
Baade, Dietrich
Baird, Scott
Baliunas, Sallie
Balona, Luis
Barber, Robert
Barbuy, Beatriz
Baschek, Bodo
Basri, Gibor
Bennett, Philip
Berdyugina, Svetlana
Bernat, Andrew
Bertone, Emanuele
Bertout, Claude
Biazzo, Katia
Bingham, Robert
Blanco, Carlo
Bless, Robert
Blomme, Ronny
Bodo, Gianluigi
Boehm-Vitense, Erika
Boesgaard, Ann
Bonifacio, Piercarlo
Bopp, Bernard
Brown, Alexander
Brown, Douglas
Brown, David
Browning, Matthew
Buchlin, Eric
Bues, Irmela
Busa, Innocenza
Butkovskaya, Varvara
Cameron, Andrew
Carbon, Duane
Carson, T.
Cassinelli, Joseph
Castelli, Fiorella
Castor, John
Catala, Claude
Catalano, Franco
Cayrel, Roger
CHAN, Kwing
CHEN, Peisheng
Christlieb , Norbert
Chugai, Nikolaj
Cidale, Lydia
Cohen, David
Collet, Remo
Conti, Peter
Cowley, Charles
Cram, Lawrence
Cruzado, Alicia
Cugier, Henryk
CUI, Wenyuan
Cuntz, Manfred
Cure, Michel
Dall , Thomas
Daszynska-Daszkiewicz, J.
Day Jones, Avril
de Koter, Alex
Decin, Leen
Deliyannis, Constantine
Depagne, Éric
Dimitrijevic, Milan
Donati, Jean-Francois
Doyle MRIA, John
Drake, Stephen
Dravins, Dainis
Dreizler, Stefan
Duari, Debiprosad
Dufton, Philip
Dupree, Andrea
Edvardsson, Bengt
Elmhamdi, Abouazza
Eriksson, Kjell
Evangelidis, E.
Faraggiana, Rosanna
Faurobert, Marianne
Feigelson, Eric
Ferreira, Joao
Fitzpatrick, Edward
Fluri, Dominique
Fontaine, Gilles
Fontenla, Juan
Forveille, Thierry
Foy, Renaud
Freire Ferrero, Rubens
Fremat, Yves
Freytag, Bernd
Frisch, Hélène
Frisch, Uriel
Froeschle, Christiane
Gail, Hans-Peter
Gallino, Roberto
Garcia, Lopez
Garcia, Miriam
Gebbie, Katharine

Gesicki, Krzysztof
Giampapa, Mark
Gonzalez, Jean-Francois
Goswami, Aruna
Gough, Douglas
Grant, Ian
Gratton, Raffaele
Gray, David
Grevesse, Nicolas
Grinin, Vladimir
Groh, Jose
Guedel, Manuel
Gussmann, Ernst-August
Haberreiter, Margit
Haisch, Bernard
Hamann, Wolf-Rainer
Harper, Graham
Hartmann, Lee
Harutyunian, Haik
Heasley, James
Heber, Ulrich
Heiter, Ulrike
Hempel, Marc
Heyl, Jeremy
Hill, Vanessa
Hillier, John
Ho, Wynn
Hoare, Melvin
Hoeflich, Peter
Hoefner, Susanne
Holzer, Thomas
Homeier, Derek
Hui bon Hoa, Alain
Hutchings, John
Ignace, Richard
Ignjatovic, Ljubinko
Ivanov, Vsevolod
Jahn, Krzysztof
Jankov, Slobodan
Jatenco-Pereira, Vera
Jevremovic, Darko
Johnson, Hollis
Jones, Carol
Jordan, Stefan
Judge, Philip
Kalkofen, Wolfgang
Kamp, Lucas
Kamp, Inga
Kandel, Robert
Kapyla, Petri
Karp, Alan
Kasparova, Jana
Katsova, Maria
Kervella, Pierre
Kiselman, Dan
Klein, Richard
KOBAYASHI, Masakazu
Kochukhov, Oleg
KODAIRA, Keiichi
Koester, Detlev
Kolesov, Aleksandr
KONDO, Yoji
Kontizas, Evangelos
Korcakova, Daniela
Korn, Andreas
Kraus, Michaela
Krikorian, Ralph
Krishna, Swamy
Krticka, Jiri
Kubat, Jiri
Kucinskas, Arunas
Kuhi, Leonard
Kumar, Shiv
Kupka, Friedrich
Kurucz, Robert
Lambert, David
Lamers, Henny
Lanz, Thierry
LEE, Jae Woo
LEE, Jun
Leenaarts, Jorrit
Leibacher, John
Leitherer, Claus
LI, Ji
Liebert, James
Linnell, Albert
Linsky, Jeffrey
LIU, Yujuan
Lobel, Alex
Loskutov, Viktor
Lucatello, Sara
Luck, R.
Ludwig, Hans
Luo, Qinghuan
Luttermoser, Donald
Lyubimkov, Leonid
Machado, Maria
Madej, Jerzy
Magazzu, Antonio
Magnan, Christian
Maguire, Kate
Marley, Mark
Martins, Fabrice
Mashonkina, Lyudmila
Massaglia, Silvano
Mathys, Gautier
Mauas, Pablo
Medupe, Rodney
Merlo, David
Michaud, Georges
Mihajlov, Anatolij
Molaro, Paolo
Montalban, Josefina
Morel, Thierry
Morris, Patrick
Muench, Guido
Musielak, Zdzislaw
Mutschlecner, Joseph
Nagirner, Dmitrij
Najarro de la Parra, F.
Narasimha, Delampady
Nardetto, Nicolas
NARIAI, Kyoji
Niemczura, Ewa
Nieva, Maria
Nikoghossian, Arthur
NISHIMURA, Masayoshi
Nordlund, Aake
Nowotny-Schipper, Walter
Oskinova, Lidia
Owocki, Stanley
Pacharin-Tanakun, P.
Pakhomov, Yury
Pandey, Birendra
Pavlenko, Yakov
Pecker, Jean-Claude
Peraiah, Annamaneni
Peters, Geraldine
Pinsonneault, Marc
Pintado, Olga
Pinto, Philip
Piskunov, Nikolai
Plez, Bertrand
Pogodin, Mikhail
Pottasch, Stuart
Przybilla, Norbert
Pustynski, Vladislav-V.
Puzia, Thomas
Querci, Monique
Rachkovsky, D.

Ramsey, Lawrence
Randich, Sofia
Rangarajan, K.
Rauch, Thomas
Reale, Fabio
Reid, Warren
Reimers, Dieter
Reiners, Ansgar
Rostas, François
Rovira, Marta
Rozanska, Agata
Rucinski, Slavek
Rutten, Robert
Rybicki, George
Sachkov, Mikhail
Saffe, Carlos
SAIO, Hideyuki
Sakhibullin, Nail
Sapar, Arved
Sapar, Lili
Sarre, Peter
Sasselov, Dimitar
Sasso, Clementina
Sauty, Christophe
Savanov, Igor
Sbordone, Luca
Schaerer, Daniel
Scharmer, Goeran
Schmid-Burgk, J.
Schmutz, Werner
Schoenberner, Detlef
Scholz, M.
Schrijver, Karel
Sedlmayer, Erwin
Sengupta, Sujan
SHANG, Hsien
SHI, Jianrong
Shimansky, Vladislav
Shine, Richard
Shipman, Harry
Short, Christopher
Sigut, T. A. Aaron
Simon, Theodore
Simon-Diaz, Sergio
Simonneau, Eduardo
Skumanich, Andrew
Sneden, Chris
Socas-Navarro, Hector
Soderblom, David
Spiegel, Edward
Spite, François
Spite, Monique
Spruit, Hendrik
Stauffer, John
Stee, Philippe
Steffen, Matthias
Stein, Robert
Stepien, Kazimierz
Stern, Robert
Strom, Stephen
Stuik, Remko
Sundqvist, Jon
TAKEDA, Yoichi
Thejll, Peter
Thompson, Rodger
Todt, Helge
Toomre, Juri
Townsend, Richard
TSUJI, Takashi
Ulmschneider, Peter
UNNO, Wasaburo
Utrobin, Victor
Vakili, Farrokh
Valeev, Azamat
van't Veer-Menneret, C.
Vardavas, Ilias
Vasu-Mallik, Sushma
Vaughan, Arthur
Velusamy, T.
Vennes, Stephane
Vidotto, Aline
Vieytes, Mariela
Viik, Tõnu
Vilhu, Osmi
Vink, Jorick
Walter, Frederick
WATANABE, Tetsuya
Waters, Laurens
Weber, Stephen
Werner, Klaus
White, Richard
Wickramasinghe, N.
Willson, Lee Anne
Woehl, Hubertus
Wolff, Sidney
Yanovitskij, Edgard
Yorke, Harold
Zacs, Laimons
Zahn, Jean-Paul
ZHANG, Bo

Division VII Commission 37 Star Clusters & Associations

President: Bruce G. Elmegreen
Vice-President: Giovanni Carraro

Organizing Committee Members:

Cannon, Russell
Cudworth, Kyle
Da Costa, Gary
de Grijs, Richard
DENG, LiCai
Lada, Charles
Lee, Young-Wook
Minniti, Dante
Sarajedini, Ata
Tosi, Monica

Members:

Aarseth, Sverre
Ahumada, Javier
Ahumada, Andrea
Aigrain, Suzanne
Ajhar, Edward
AK, Serap
AK, Tansel
Akeson, Rachel
Alfaro, Emilio
Alksnis, Andrejs
Allen, Christine
Allen, Lori
Alves, Virgínia
Anderson, Joseph
Andreuzzi, Gloria
Antoniou, Vallia
Aparicio, Antonio
Arifyanto, Mochamad
Armandroff, Taft
Aurière, Michel
Bailyn, Charles
Balazs, Bela
Barmby, Pauline
Barrado y Navascues, David
Bartasiute, Stanislava
Bastian, Nathan
Baume, Gustavo
Baumgardt, Holger
Beck, Sara
Bekki, Kenji
Bellazzini, Michele
Benacquista, Matthew
Biazzo, Katia
Bijaoui, Albert
Blum, Robert
Boily, Christian
Bonatto, Charles
Bosch, Guillermo
Bragaglia, Angela
Brandner, Wolfgang
Brown, Anthony
Buonanno, Roberto
Burderi, Luciano
Burkhead, Martin
Butler, Ray
Buzzoni, Alberto
Byrd, Gene
Calamida, Annalisa
Callebaut, Dirk
Caloi, Vittoria
Campbell, Simon
Cantiello, Michele
Canto Martins, Bruno
Caputo, Filippina
Capuzzo Dolcetta, Roberto
Carney, Bruce
Carretta, Eugenio
Chaboyer, Brian
Charbonnel, Corinne
Chavarria-K, Carlos
CHEN, Huei-Ru
CHEN, Li
CHENG, Kwang
Chiosi, Cesare
Christian, Carol
CHUN, Mun-suk
Claria, Juan
Clark, David
Clementini, Gisella
Colin, Jacques
Corral, Luis
Covino, Elvira
Crause, Lisa
Cropper, Mark
D'Amico, Nicolo'
D'Antona, Francesca
Da Costa, Gary
Dale, James
Danford, Stephen
Dapergolas, Anastasios
Daube-Kurzemniece, Ilga
Davies, Melvyn
Davies, Benjamin
De Marchi, Guido
de Mink, Selma
de Silva, Gayandhi
Demarque, Pierre
Demers, Serge
DENG, LiCai
Diaz-Santos, Tanio
Dionatos, Odysseas
Djupvik, Anlaug Amanda
Dluzhnevskaya, Olga
Dougados, Catherine
Downes Wallace, Juan
Drissen, Laurent
Durrell, Patrick
Eastwood, Kathleen
Eldridge, John
Fall, S.
Feinstein, Alejandro
Figer, Donald

Forbes, Douglas
Forte, Juan
Fox-Machado, Lester
Friel, Eileen
FUKUSHIGE, Toshiyuki
Fusi-Pecci, Flavio
Gandolfi, Davide
Garcia, Beatriz
Geffert, Michael
Geller, Aaron
Giersz, Miroslav
Giorgi, Edgard
Glushkova, Elena
Golay, Marcel
Gouliermis, Dimitrios
Gratton, Raffaele
Green, Elizabeth
Grillmair, Carl
Grundahl, Frank
Guetter, Harry
Haghi, Hosein
Haisch Jr, Karl
Hanes, David
Hanson, Margaret
Harris, Gretchen
Harris, Hugh
Hayes, Matthew
Heggie, Douglas
Herbst, William
Hesser, James
Heudier, Jean-Louis
Hilker, Michael
Hillenbrand, Lynne
Hills, Jack
Hodapp, Klaus
Huensch, Matthias
Hut, Piet
Iben Jr, Icko
IKEDA, Norio
Illingworth, Garth
ITOH, Yoichi
Ivanova, Natalia
Janes, Kenneth
JEON, Young Beom
JIANG, Zhibo
Johnson, Christian
Joshi, Umesh
Kaisin, Serafim
Kalirai, Jason
Kamp, Lucas
Karakas, Amanda
Kaviraj, Sugata
KIM, Sungsoo
KIM, Seung-Lee
KIM, Sang Chul
King, Ivan
Kitsionas, Spyridon
KO, Chung-Ming
Koch, Andreas
Kontizas, Evangelos
Kontizas, Mary
Kotulla, Ralf
Kouwenhoven, M.B.N.
Kroupa, Pavel
Krumholz, Mark
Kun, Maria
Kundu, Arunav
Kurtev, Radostin
Lada, Charles
Landolt, Arlo
Lapasset, Emilio
Larsson-Leander, Gunnar
Laugalys, Vygandas
Laval, Annie
LEE, Jae Woo
LEE, Kang Hwan
LEE, Hyun-chul
LEE, Young-Wook
Leisawitz, David
Leonard, Peter
LI, Jinzeng
LIU, Michael
Lodieu, Nicolas
Loktin, Alexhander
Lu, Phillip
Lucatello, Sara
Lynden-Bell, Donald
Maccarone, Thomas
Mackey, Alasdair
Maeder, Andre
Magrini, Laura
Maiz Apellaniz, Jesús
Makalkin, Andrei
MAKINO, Junichiro
Mamajek, Eric
Mandel, Ilya
Marco, Amparo
Mardling, Rosemary
Marin-Franch, Antonio
Marino, Antonietta
Markkanen, Tapio
Markov, Haralambi
Marraco, Hugo
Marsden, Stephen
Martinez Roger, Carlos
Martins, Donald
McGehee, Peregrine
Meibom, Soren
Melnik, Anna
Menon, T.
Menzies, John
Meyer, Michael
Meylan, Georges
Milone, Eugene
Moehler, Sabine
Mohan, Vijay
Moitinho, André
Monaco, Lorenzo
Montalban, Josefina
Moor, Attila
Mould, Jeremy
Muminov, Muydinjon
Muzzio, Juan
Narbutis, Donatas
Navone, Hugo
Naylor, Tim
Nemec, James
Nesci, Roberto
Neuhaeuser, Ralph
Nikoghosyan, Elena
Ninkov, Zoran
Nota, Antonella
Nuernberger, Dieter
Oey, Sally
OGURA, Katsuo
Oliveira, Joana
Origlia, Livia
Ortolani, Sergio
Oskinova, Lidia
OTSUKI, Kaori
Oudmaijer, Rene
Palacios, Ana
Pandey, A.
PARK, Byeong-Gon
Parmentier, Geneviève
Parsamyan, Elma
Patten, Brian
Paunzen, Ernst
Pavani, Daniela
Pedreros, Mario

PENG, Eric
Penny, Alan
Perez, Mario
Peterson, Charles
Petrovskaya, Margarita
Peykov, Zvezdelin
Phelps, Randy
Piatti, Andrés
Pietrukowicz, Pawel
Pilachowski, Catherine
Piskunov, Anatolij
Platais, Imants
Pooley, David
Porras Juárez, Bertha
Portegies Zwart, Simon
Poveda, Arcadio
Price, Daniel
Pritchet, Christopher
Pulone, Luigi
Puzia, Thomas
Quanz, Sascha
Raimondo, Gabriella
Rathborne, Jill
Ravindranath, Swara
Rebull, Luisa
Renzini, Alvio
Rey, Soo-Chang
Rhee, Jaehyon
Richard, Olivier
Richer, Harvey
Richtler, Tom
Robberto, Massimo
Rodrigues de Oliveira F., I.
Roman-Zuniga, Carlos
Rothberg, Barry
Rountree, Janet
Rowell, Gavin
Royer, Pierre
Russeva, Tatjana
Sagar, Ram
Samus, Nikolay
Sanchez Bejar, Victor
Sanchez Doreste, Néstor
Sanders, Walter
Santiago, Basilio
Santos Jr., Joao
Sarajedini, Ata
Schoedel, Rainer
Scholz, Alexander
Schuler, Simon
Schweizer, François
Seitzer, Patrick
Semkov, Evgeni
Shawl, Stephen
Sher, David
SHI, Huoming
SHU, Chenggang
Skinner, Stephen
Smith, Graeme
Smith, J.
Song, Inseok
Southworth, John
Spurzem, Rainer
Stauffer, John
Stringfellow, Guy
Stuik, Remko
Subramaniam, Annapurni
SUGIMOTO, Daiichiro
SUNG, Hwankyung
Suntzeff, Nicholas
Szekely, Péter
Tadross, Ashraf
TAKAHASHI, Koji
TAKAHASHI, Hidenori
TAKASHI, Hasegawa
Tas, Günay
Terranegra, Luciano
Thoul, Anne
Thurston, Mark
Tikhonov, Nikolai
Tornambe, Amedeo
Tosi, Monica
Tremonti, Christy
Trenti, Michele
Tripicco, Michael
Trullols, I.
Tsvetkov, Milcho
Tsvetkova, Katja
Turner, David
Twarog, Bruce
Ubeda, Leonardo
Upgren, Arthur
van Altena, William
van den Berg, Maureen
van den Bergh, Sidney
van der Bliek, Nicole
VandenBerg, Don
Vazquez, Ruben
Veltchev, Todor
Ventura, Paolo
Verschueren, Werner
Vesperini, Enrico
von Hippel, Theodore
Walker, Merle
Walker, Gordon
Warren Jr, Wayne
Weaver, Harold
Wehlau, Amelia
Wielen, Roland
Wofford, Aida
Worters, Hannah
Wramdemark, Stig
Wright, Nicholas
WU, Hsin-Heng
WU, Zhen-Yu
Wulandari, Hesti
Yakut, Kadri
YI, Sukyoung
YIM, Hong Suh
YUMIKO, Oasa
Zaggia, Simone
Zakharova, Polina
Zapatero-Osorio, M. Rosa
Zdanavičius, Justas
ZHANG, Fenghui
ZHAO, Jun Liang
ZHU, Qingfeng
Zinn, Robert
Zoccali, Manuela
Zwintz, Konstanze

Division X Commission 40 Radio Astronomy

President: A. Russell Taylor
Vice-President: Jessica Mary Chapman

Organizing Committee Members:

Carilli, Christopher
Dubner, Gloria
Garrett, Michael
Goss, W. Miller
Hills, Richard
HIRABAYASHI, Hisashi
Rodriguez, Luis
Shastri, Prajval
Torrelles, Jose-Maria

Members:

Abraham, Péter
Ade, Peter
Akujor, Chidi
Alberdi, Antonio
Alexander, Joseph
Alexander, Paul
Allen, Ronald
Aller, Margo
Aller, Hugh
Altenhoff, Wilhelm
Altunin, Valery
Ambrosini, Roberto
AN, TAO
Andernach, Heinz
Anglada, Guillem
Antonova, Antoaneta
Argo, Megan
Arnal, Edmundo
Asanok, Kitiyanee
Asareh, Habibolah
Aschwanden, Markus
Aubier, Monique
Augusto, Pedro
Baan, Willem
Baars, Jacob
Baath, Lars
Babkovskaia, Natalia
Bachiller, Rafael
Bailes, Matthew
Bajaja, Esteban
Bajkova, Anisa
Baker, Joanne
Baker, Andrew
Balasubramanian, V.
Balasubramanyam, Ramesh
Ball, Lewis
Bally, John
Balonek, Thomas
Banhatti, Dilip
Barrow, Colin
Bartel, Norbert
Barthel, Peter
Bartkiewicz, Anna
Barvainis, Richard
Baryshev, Andrey
Bash, Frank
Basu, Kaustuv
Baudry, Alain
Bauer, Amanda
Baum, Stefi
Bayet, Estelle
Beasley, Anthony
Beck, Rainer
Beck, Sara
Bekki, Kenji
Beklen, Elif
Benaglia, Paula
Benn, Chris
Bennett, Charles
Benz, Arnold
Bergman, Per
Berkhuijsen, Elly
Bhandari, Rajendra
Bhat, Ramesh
Bieging, John
Biermann, Peter
Bietenholz, Michael
Bigdeli, Mohsen
Biggs, James
Bignall, Hayley
Bignell, R.
Biraud, François
Biretta, John
Birkinshaw, Mark
Blair, David
Blandford, Roger
Bloemhof, Eric
Blundell, Katherine
Boboltz, David
Bock, Douglas
Bockelée-Morvan, D.
Bolatto, Alberto
Bondi, Marco
Boonstra, Albert
Booth, Roy
Borisov, Galin
Borka Jovanovic, Vesna
Bosch-Ramon, Valenti
Bot, Caroline
Bouton, Ellen
Bower, Geoffrey
Branchesi, Marica
Bregman, Jacob
Brentjens, Michiel
Breton, Rene
Bridle, Alan
Brinks, Elias
Britzen, Silke
Broderick, John
Bronfman, Leonardo
Brooks, Kate
Brouw, Willem
Brown, Jo-Anne
Brown, Joanna
Browne, Ian

Brunetti, Gianfranco
Brunthaler, Andreas
Bryant, Julia
Bujarrabal, Valentin
Burderi, Luciano
Burke, Bernard
Campbell, Robert
Campbell-Wilson, Duncan
Caproni, Anderson
Carlqvist, Per
Caroubalos, Constantinos
Carretti, Ettore
Carvalho, Joel
Casasola, Viviana
casassus, simon
Casoli, Fabienne
Cassano, Rossella
Castelletti, Gabriela
Castets, Alain
Cecconi, Baptiste
Celotti, Anna Lisa
Cernicharo, Jose
CH, Ishwara Chandra
CHAN, Kwing
Chandler, Claire
Charlot, Patrick
CHEN, Yongjun
CHEN, Huei-Ru
CHEN, Xuefei
CHEN, Zhiyuan
CHEN, Zhijun
Chengalur, Jayaram
CHIKADA, Yoshihiro
CHIN, Yi-nan
Chini, Rolf
CHO, Se Hyung
Choudhury, Tirthankar
CHUNG, Hyun-Soo
Chyzy, Krzysztof
Cichowolski, Silvina
Ciliegi, Paolo
Clark, Barry
Clark, David
Clark, Frank
Clegg, Andrew
Clemens, Dan
Cohen, Marshall
Coleman, Paul
Colomer, Francisco
Combes, Françoise
Combi, Jorge
Condon, James
Conklin, Edward
Contreras, Maria
Conway, John
Corbel, Stéphane
Cordes, James
Cotton Jr, William
Courtois, Helene
Crane, Patrick
Crawford, Fronefield
Croft, Steve
Croston, Judith
Crovisier, Jacques
Crutcher, Richard
Cunningham, Maria
D'Amico, Nicolo'
Dagkesamansky, Rustam
DAISHIDO, Tsuneaki
DAISUKE, Iono
Dallacasa, Daniele
Dannerbauer, Helmut
Davies, Rodney
Davis, Michael
Davis, Robert
Davis, Richard
Davis, Timothy
de Bergh, Catherine
De Bernardis, Paolo
de Gregorio-Monsalvo, I.
de Jager, Cornelis
de La Noë, Jerome
de Lange, Gert
de Petris, Marco
de Ruiter, Hans
de Vicente, Pablo
Deller, Adam
Deshpande, Avinash
Despois, Didier
Dewdney, Peter
Dhawan, Vivek
Diamond, Philip
Dickel, Helene
Dickel, John
Dickey, John
Dickman, Robert
Dionatos, Odysseas
DOBASHI, Kazuhito
Dodson, Richard
Doubinskij, Boris
Dougherty, Sean
Downes, Dennis
Downs, George
Drake, Frank
Drake, Stephen
Dreher, John
Dubner, Gloria
Duffett-Smith, Peter
Dutrey, Anne
Dwarakanath, K.
Dyson, Freeman
Eales, Stephen
Edelson, Rick
Edwards, Philip
Ehle, Matthias
Ekers, Ronald
Elia, Davide
Ellingsen, Simon
Emerson, Darrel
Emonts, Bjorn
Epstein, Eugene
Erickson, William
ESAMDIN, Ali
ESIMBEK, Jarken
Ewing, Martin
EZAWA, Hajime
Facondi, Silvia
Falcke, Heino
Fanaroff, Bernard
Fanti, Roberto
Farrell, Sean
Faulkner, Andrew
Feain, Ilana
Fedotov, Leonid
Feigelson, Eric
Feldman, Paul
Felli, Marcello
Feretti, Luigina
Fernandes, Francisco
Ferrari, Attilio
Ferrusca, Daniel
Fey, Alan
Field, George
Figueiredo, Newton
Filipovic, Miroslav
Fletcher, Andrew
Florkowski, David
Foley, Anthony
Fomalont, Edward
Fort, David

Forveille, Thierry
Fouque, Pascal
Frail, Dale
Frater, Robert
Frey, Sandor
Friberg, Per
Fuerst, Ernst
FUKUI, Yasuo
Gabanyi, Krisztina
Gabuzda, Denise
Gaensler, Bryan
Gallego, Juan Daniel
Gallimore, Jack
Gangadhara, R.T.
GAO, Yu
Garay, Guido
Garrett, Michael
Garrington, Simon
Gasiprong, Nipon
Gaume, Ralph
Gawronski, Marcin
GENG, Lihong
Gentile, Gianfranco
Genzel, Reinhard
Gérard, Eric
Gergely, Tomas
Gervasi, Massimo
Ghigo, Francis
Ghosh, Tapasi
Gil, Janusz
Gimenez, Alvaro
Gioia, Isabella
Giroletti, Marcello
Gitti, Myriam
Goddi, Ciriaco
Goedhart, Sharmila
Gomez, Gonzalez
Gomez Fernandez, Jose
GONG, Biping
Gopalswamy, Natchimuthuk
Gordon, Mark
Gordon, Chris
Gorschkov, Aleksandr
Gosachinskij, Igor
Goss, W. Miller
Gottesman, Stephen
Gower, Ann
Graham, David
Green, David
Green, Anne
Green, James
Gregorini, Loretta
Gregorio-Hetem, Jane
Gregory, Philip
Grewing, Michael
GU, Xuedong
GU, Minfeng
Gubchenko, Vladimir
Guelin, Michel
Guesten, Rolf
Guilloteau, Stéphane
Gulkis, Samuel
Gull, Stephen
GUO, Jianheng
Gupta, Yashwant
Gurvits, Leonid
Gwinn, Carl
Hall, Peter
Hallinan, Gregg
HAN, JinLin
Hanasz, Jan
HANDA, Toshihiro
Hanisch, Robert
Hankins, Timothy
Hardee, Philip
Harnett, Julienne
Harris, Daniel
HASEGAWA, Tetsuo
Hatziminaoglou, Evanthia
Haverkorn, Marijke
Hayashi, Masahiko
Haynes, Raymond
Haynes, Martha
HAZUMI, Masashi
HE, Jinhua
Heald, George
Heeralall-Issur, Nalini
Heiles, Carl
Heinz, Sebastian
Helou, George
Henkel, Christian
Herpin, Fabrice
Hess, Kelley
Hessels, Jason
Hewish, Antony
Hibbard, John
Higgs, Lloyd
HIRAMATSU, Masaaki
HIROTA, Tomoya
Hjalmarson, Ake
Ho, Paul
Ho, Wynn
Hoang, Binh
Hobbs, George
Hofner, Peter
Hogbom, Jan
Hogg, David
Hojaev, Alisher
Hollis, Jan
HONG, Xiaoyu
Hopkins, Andrew
Horiuchi, Shinji
Hotan, Aidan
Howard III, William
HUANG, Hui-Chun
Huchtmeier, Walter
Hughes, Philip
Hughes, David
Humphreys, Elizabeth
Hunstead, Richard
Huynh, Minh
HWANG, Chorng-Yuan
Ibrahim, Zainol Abidin
IGUCHI, Satoru
IKEDA, Norio
Ikhsanov, Robert
Iliev, Ilian
Ilin, Gennadii
IMAI, Hiroshi
INATANI, Junji
INOUE, Makoto
Ipatov, Aleksandr
Irvine, William
ISHIGURO, Masato
Israel, Frank
Ivanov, Dmitrii
IWATA, Takahiro
Jachym, Pavel
Jackson, Carole
Jackson, Neal
Jacq, Thierry
Jaffe, Walter
Jamrozy, Marek
Janssen, Michael
Jauncey, David
Jelic, Vibor
Jewell, Philip
JIANG, Zhibo
JIN, Zhenyu
Johnston, Kenneth

Johnston, Helen
Johnston-Hollitt, Melanie
Jones, Paul
Jones, Dayton
Josselin, Eric
JUNG, Jae-Hoon
Kaftan, May
Kaidanovski, Mikhail
KAIFU, Norio
KAKINUMA, Takakiyo
Kalberla, Peter
Kaltman, Tatyana
KAMAZAKI, Takeshi
KAMEGAI, Kazuhisa
KAMENO, Seiji
KAMEYA, Osamu
Kandalyan, Rafik
Kanekar, Nissim
Kardashev, Nicolay
Karouzos, Marios
Kassim, Namir
KASUGA, Takashi
Kaufmann, Pierre
KAWABE , Ryohei
KAWAGUCHI, Kentarou
KAWAMURA, Akiko
Kedziora-Chudczer, Lucyna
Kellermann, Kenneth
Kent, Brian
Keshet, Uri
Kesteven, Michael
Khaikin, Vladimir
Khodachenko, Maxim
Kijak, Jaroslaw
Kilborn, Virginia
Killeen, Neil
KIM, Hyun-Goo
KIM, Tu Whan
KIM, Kwang tae
KIM, Sang Joon
KIM, Ji Hoon
Kimball, Amy
Kislyakov, Albert
Kitaeff, Vyacheslav
Klein, Ulrich
Klein, Karl
Knudsen, Kirsten
KOBAYASHI, Hideyuki
Kocharovsky, Vitaly
KODA, Jin
KOHNO, Kotaro
KOJIMA, Masayoshi
Kolomiyets, Svitlana
Kondratiev, Vladislav
Konovalenko, Alexander
Kopylova, Yulia
Korpela, Eric
Korzhavin, Anatoly
Kovalev, Yuri
Kovalev, Yuri
KOYAMA, Yasuhiro
Kraft, Ralph
Kramer, Michael
Kramer, Busaba
Kreysa, Ernst
Krichbaum, Thomas
Krishna, Gopal
Krishnan, Thiruvenkata
Kronberg, Philipp
Krugel, Endrik
KUAN, Yi-Jehng
Kudryavtseva, Nadezhda
Kuijpers, H.
Kuiper, Thomas
Kulkarni, Prabhakar
Kulkarni, Vasant
Kulkarni, Shrinivas
Kumkova, Irina
Kundt, Wolfgang
Kunert-Bajraszewska, M.
KURAYAMA, Tomoharu
Kus, Andrzej
Kutner, Marc
Kwok, Sun
La Franca, Fabio
Lada, Charles
LAI, Shih-Ping
Laing, Robert
Lal, Dharam
Landecker, Thomas
Landt, Hermine
Lang, Kenneth
Langer, William
Langston, Glen
LaRosa, Theodore
Lasenby, Anthony
Lawrence, Charles
Leahy, J.
Lebron, Mayra
LEE, Youngung
LEE, Chang Won
LEE, Yong Bok
LEE, Sang-Sung
LEE, Jung-Won
LEE, Jeong-Eun
Lefloch, Bertrand
Lehnert, Matthew
Lenc, Emil
Lépine, Jacques
Lequeux, James
Lesch, Harald
Lestrade, Jean-François
Li, Hong-Wei
LI, Zhi
LIANG , Shiguang
Likkel, Lauren
Lilley, Edward
LIM, Jeremy
Lindqvist, Michael
Lis, Dariusz
Liseau, René
Lister, Matthew
LIU, Xiang
LIU, Sheng-Yuan
Lo, Fred K. Y.
Lo, Wing-Chi Nadia
Lockman, Felix
Loiseau, Nora
Longair, Malcolm
Loubser, Ilani
Loukitcheva, Maria
Lovell, James
Lozinskaya, Tatjana
Lubowich, Donald
Lucero, Danielle
Luks, Thomas
Luque-Escamilla, Pedro
Lyne, Andrew
Lytvynenko, Leonid
MA, Guanyi
Macchetto, Ferdinando
MacDonald, Geoffrey
MacDonald, James
Machalski, Jerzy
Maciesiak, Krzysztof
Mack, Karl-Heinz
MacLeod, John
MAEHARA, Hideo
Malofeev, Valery
Manchester, Richard

Mandolesi, Nazzareno
Mantovani, Franco
MAO, Rui-Qing
Maran, Stephen
Marcaide, Juan-Maria
Mardones, Diego
Mardyshkin, Vyacheslav
Marecki, Andrzej
Markoff, Sera
Marscher, Alan
Marshalov, Dmitriy
Marti, Josep
Martin, Robert
Martin, Christopher
Martin, Sergio
Martin-Pintado, Jesus
Marvel, Kevin
Masheder, Michael
Maslowski, Jozef
Mason, Paul
Massaro, Francesco
Masson, Colin
Masters, Karen
Matsakis, Demetrios
MATSUO, Hiroshi
MATSUSHITA, Satoki
Matthews, Brenda
Mattila, Kalevi
Matveenko, Leonid
Mauersberger, Rainer
McConnell, David
McCulloch, Peter
McKean, John
McKenna Lawlor, Susan
McMullin, Joseph
Mebold, Ulrich
Meier, David
Melikidze, Giorgi
Menon, T.
Menten, Karl
Meyer, Martin
Mezger, Peter
Michalec, Adam
Migenes, Victor
Mikhailov, Andrey
Miley, George
Miller, Neal
Milne, Douglas
Mirabel, Igor
Miroshnichenko , Alla
Mitchell, Kenneth
Miyawaki, Ryosuke
MIYAZAKI, Atsushi
MIYOSHI, Makoto
MIZUNO, Akira
MIZUNO, Norikazu
Moellenbrock III, George
Moffett, David
Momjian, Emmanuel
MOMOSE, Munetake
Montmerle, Thierry
Morabito, David
Moran, James
Morison, Ian
MORIYAMA, Fumio
Morras, Ricardo
Morris, David
Morris, Mark
Moscadelli, Luca
Mosoni, Laszlo
Muller, Erik
Muller, Sebastien
Mundy, Lee
MURAOKA, Kazuyuki
MURATA, Yasuhiro
Murphy, Tara
Mutel, Robert
Muxlow, Thomas
Myers, Philip
Nadeau, Daniel
Nagnibeda, Valerij
NAKANO, Takenori
NAKASHIMA, Jun-ichi
Nammahachak, Suwit
Neeser, Mark
Nesvadba, Nicole
Nguyen-Quang, Rieu
Nicastro, Luciano
Nice, David
Nicolson, George
NIINUMA, Kotaro
Nikolic, Silvana
NISHIO, Masanori
Norris, Raymond
Nuernberger, Dieter
Nuza, Sebastian
O'Dea, Christopher
O'Sullivan, John
Odonoghue, Aileen
OGAWA, Hideo
Ohashi, Nagayoshi
OHISHI, Masatoshi
Ojha, Roopesh
OKA, Tomoharu
OKUMURA, Sachiko
Olberg, Michael
Onel, Hakan
ONISHI , Toshikazu
Oozeer, Nadeem
Orchiston, Wayne
Orienti, Monica
Otmianowska-Mazur, K.
Ott, Juergen
Owen, Frazer
Özel, Mehmet
Özeren, Ferhat
Padman, Rachael
Palmer, Patrick
Panessa, Francesca
Pankonin, Vernon
Paragi, Zsolt
Paredes Poy, Josep
Parijskij, Yurij
PARK, Yong Sun
Parma, Paola
Paron, Sergio
Parrish, Allan
Pasachoff, Jay
Pashchenko, Mikhail
Patel, Nimesh
Pauls, Thomas
Pavlyuchenkov, Yaroslav
Pearson, Timothy
Peck, Alison
Pedersen, Holger
Pedlar, Alan
Peel, Michael
PENG, Bo
PENG, Qingyu
Penzias, Arno
Perez, Fournon
Perez-Torres, Miguel
Perley, Richard
Perozzi, Ettore
Persson, Carina
Peters, William
Petrova, Svetlana
Pettengill, Gordon
PHAM, Diep
Philipp, Sabine

Phillips, Thomas
Phillips, Christopher
Pick, Monique
PING, Jinsong
Pisano, Daniel
Pitkin, Matthew
Planesas, Pere
Pogrebenko, Sergei
Polatidis, Antonios
Pompei, Emanuela
Ponsonby, John
Pooley, Guy
Porcas, Richard
Porras Juárez, Bertha
Potapov, Vladimir
Prandoni, Isabella
Preston, Robert
Preuss, Eugen
Price, R.
Pshirkov, Maxim
Puschell, Jeffery
Pushkarev, Alexander
Puxley, Phil
Radford, Simon
Rahimov, Ismail
Ransom, Scott
Rao, A.
Rathborne, Jill
Ray, Tom
Ray, Paul
Readhead, Anthony
Redman, Matthew
Reich, Wolfgang
Reid, Mark
Reif, Klaus
Reyes, Francisco
Reynolds, John
Reynolds, Cormac
RHEE, Myung Hyun
Ribo, Marc
Richer, John
Rickard, Lee
Ridgway, Susan
Rioja, Maria
Rizzo, Jose
Roberts, Morton
Roberts, David
Robertson, James
Robertson, Douglas
Rodriguez, Luis
Roeder, Robert
Roelfsema, Peter
Roennaeng, Bernt
Roeser, Hans-peter
Roger, Robert
Rogers, Alan
Rogstad, David
Romanov, Andrey
Romero, Gustavo
Romney, Jonathan
Rosa Gonzalez, Daniel
Rosolowsky, Erik
Rubio-Herrera, Eduardo
Rudnick, Lawrence
Rudnitskij, Georgij
Rushton, Anthony
Russell, Jane
Rydbeck, Gustaf
Rys, Stanislaw
Sadler, Elaine
Saikia, Dhruba
SAKAMOTO, Seiichi
Salomé, Philippe
Salter, Christopher
Samodurov, Vladimir
Sandell, Göran
Sanders, David
Sargent, Annelia
Saripalli, Lakshmi
Sarma, N.
Sarma, Anuj
Sastry, Ch.
SATO, Fumio
Savolainen, Tuomas
Sawada, Tsuyoshi
SAWADA-SATOH, Satoko
Sawant, Hanumant
Scalise Jr, Eugenio
Schilizzi, Richard
Schilke, Peter
Schlickeiser, Reinhard
Schmidt, Maarten
Schroder, Anja
Schuch, Nelson
Schure, Klara
Schwarz, Ulrich
Scott, Paul
Seaquist, Ernest
Seielstad, George
SEKIDO, Mamoru
SEKIMOTO, Yutaro
Semenov, Dmitry
Sese, Rogel Mari
SETA, Masumichi
Setti, Giancarlo
Seymour, Nicholas
Shaffer, David
SHANG, Hsien
Shaposhnikov, Vladimir
Shaver, Peter
SHEN, Zhiqiang
Shepherd, Debra
Shevgaonkar, R.
SHIBATA, Katsunori
SHIMOIKURA, Tomomi
Shinnaga, Hiroko
Shmeld, Ivar
Shone, David
SHU, Fengchun
Shulga, Valerii
Sieber, Wolfgang
Singal, Ashok
Sinha, Rameshwar
Siringo, Giorgio
Skillman, Evan
Slade, Martin
Slee, O.
Smirnova, Tatiana
Smith, Francis
Smith, Dean
Smith, Niall
Smolentsev, Sergej
Smol'kov, Gennadij
Snellen, Ignas
Sobolev, Yakov
Sodin, Leonid
SOFUE , Yoshiaki
SOMANAH, R.
SONG, Qian
Soria, Roberto
Sorochenko, Roman
Spencer, Ralph
Spencer, John
Sramek, Richard
Sridharan, Tirupati
Stairs, Ingrid
Stanghellini, Carlo
Stappers, Benjamin
Steffen, Matthias
Stewart, Paul

Stil, Jeroen
Storey, Michelle
Strom, Richard
Strukov, Igor
Subrahmanya, C.
Subrahmanyan, Ravi
SUGITANI, Koji
Suleymanova, Svetlana
Sullivan, III, Woodruff
SUNADA, Kazuyoshi
Surkis, Igor
Swarup, Govind
Swenson Jr, George
Szomoru, Arpad
Szymczak, Marian
TAKABA, Hiroshi
TAKANO, Toshiaki
TAKANO, Shuro
Tammi, Joni
TAN, Baolin
Tapping, Kenneth
Tarter, Jill
TATEMATSU, Ken'ichi
Taylor, A.
te Lintel Hekkert, Peter
Teng, Stacy
Terzian, Yervant
Theureau, Gilles
Thomasson, Peter
Thompson, A.
Thum, Clemens
TIAN, Wenwu
Tingay, Steven
Tiplady, Adrian
Tofani, Gianni
Tolbert, Charles
Tornikoski, Merja
Torrelles, Jose-Maria
TOSAKI, Tomoka
Tovmassian, Hrant
Trigilio, Corrado
Trinidad, Miguel
TRIPPE, Sascha
Tritton, Keith
Troland, Thomas
Trushkin, Sergey
TSAI, An-Li
TSUBOI, Masato
Tsutsumi, Takahiro
Tuccari, Gino
Turner, Kenneth
Turner, Jean
Tyul'bashev, Sergei
Tzioumis, Anastasios
Udaya, Shankar
Ulrich, Marie-Helene
Ulvestad, James
Ulyanov, Oleg
Umana, Grazia
UMEMOTO, Tomofumi
Unwin, Stephen
Urama, Johnson
Urosevic, Dejan
Urquhart, James
Uson, Juan
Vaccari, Mattia
Vakoch, Douglas
Val'tts, Irina
Vallee, Jacques
Valtaoja, Esko
Valtonen, Mauri
van der Hulst, Jan
van der Kruit, Pieter
van der Laan, Harry
van der Tak, Floris
van Driel, Wim
van Gorkom, Jacqueline
van Kampen, Eelco
van Langevelde, Huib
van Leeuwen, Joeri
van Woerden, Hugo
VandenBout, Paul
Vats, Hari
Vaughan, Alan
Velusamy, T.
Venturi, Tiziana
Venugopal, V.
Verheijen, Marc
Verkhodanov, Oleg
Vermeulen, Rene
Verschuur, Gerrit
Verter, Frances
Vestergaard, Marianne
Vilas, Faith
Vilas-Boas, José
Vivekanand, M.
Vlahakis, Catherine
Vlemmings, Wouter
Vogel, Stuart
Volvach, Alexander
Voronkov, Maxim
Wadadekar, Yogesh
WAJIMA, Kiyoaki
Walker, Robert
Wall, Jasper
Wall, William
Walmsley, C.
Walsh, Wilfred
Walsh, Andrew
WANG, Shouguan
WANG, Na
WANG, Hong-Guang
WANG, Shujuan
WANG, Jingyu
WANG, Junzhi
WANG, Guangli
WANG, Min
WANG, Chen
Wannier, Peter
Ward-Thompson, Derek
Wardle, John
Warmels, Rein
Warner, Peter
Watson, Robert
Wayth, Randall
Wehrle, Ann
Wei, Mingzhi
WEI, Erhu
Weigelt, Gerd
Weiler, Kurt
Weiler, Edward
Welch, William
WEN, Linqing
Wendt, Harry
WENLEI, Shan
Westmeier, Tobias
White, Glenn
Whiteoak, John
Whiting, Matthew
Wicenec, Andreas
Wickramasinghe, N.
Wielebinski, Richard
Wiik, Kaj
Wiklind, Tommy
Wild, Wolfgang
Wilkinson, Peter
Willis, Anthony
Wills, Beverley
Wills, Derek
Willson, Robert

Wilner, David
Wilson, Robert
Wilson, Thomas
Wilson, William
Windhorst, Rogier
Winnberg, Anders
Wise, Michael
Witzel, Arno
Wolleben, Maik
Wolszczan, Alexander
Woltjer, Lodewijk
Wong, Tony
Woodsworth, Andrew
Wootten, Henry
Wright, Alan
Wrobel, Joan
WU, Yuefang
WU, Xinji
Wucknitz, Olaf
YAMAMOTO, Hiroaki
YANG, Ji
YANG, Zhigen
YAO, Qijun
YE, Shuhua
Yin, Qi-Feng
YONEKURA, Yoshinori
Yusef-Zadeh, Farhad
Zainal Abidin, Zamri
Zaitsev, Valerij
Zanichelli, Alessandra
Zannoni, Mario
Zarka, Philippe
Zavala, Robert
Zensus, J-Anton
ZHANG, Jian
ZHANG, Xizhen
ZHANG, Hongbo
ZHANG, Qizhou
ZHANG, JiangShui
ZHANG, Jingyi
ZHANG, Haiyan
ZHANG, Yong
ZHANG, Chengmin
ZHANG, Zhibin
ZHAO, Jun-Hui
Zheleznyak, Alexander
Zheleznyakov, Vladimir
ZHENG, Xinwu
ZHENG, Xiaonian
ZHOU, Jianfeng
ZHOU, Jianjun
ZHOU, Xia
ZHU, LiChun
ZHU, Wenbai
ZHU, Ming
ZHU, Qingfeng
Zieba, Stanislaw
Zinchenko, Igor
Zlobec, Paolo
Zlotnik, Elena
Zuckerman, Benjamin
Zwaan, Martin
Zylka, Robert

Division XII Commission 41 History of Astronomy

President: Clive L.N. Ruggles
Vice-President: Rajesh Kochhar

Organizing Committee Members:

Corbin, Brenda
de Jong, Teije
DeVorkin, David
NAKAMURA, Tsuko
Norris, Raymond
Pigatto, Luisa
SOMA, Mitsuru
Sterken, Christiaan
Sun, Xiaochun
Zsoldos, Endre

Members:

Abalakin, Viktor
Abt, Helmut
Acharya, Bannanje
AHN, Youngsook
Alves, Virgínia
Ansari, S.M.
Antonello, Elio
Arifyanto, Mochamad
Ashok, N.
Babu, G.S.D.
Babul, Arif
Badolati, Ennio
Bailey, Mark
Ball, Lewis
Ballabh, Goswami
Balyshev, Marat
Baneke, David
Barlai, Katalin
Batten, Alan
Bennett, Jim
Berendzen, Richard
Bertola, Francesco
Bessell, Michael
Bhatia, Vishnu
Bhatt, H.
Bhattacharjee, Pijush
Bien, Reinhold
Bishop, Roy
Boerngen, Freimut
Bon, Edi
Bonoli, Fabrizio
Botez, Elvira
Bougeret, Jean-Louis
Bouton, Ellen
Bowen, David
Brecher, Kenneth
Bretones, Paulo
Brooks, Randall
Brosche, Peter
Brouw, Willem
Burman, Ronald
CAI, Kai
Campana, Riccardo
Cannon, Russell
Caplan, James
CHANG, Heon-Young
Chapman, Jessica
CHIN, Yi-nan
Chinnici, Ileana
Choudhary, Debi Prasad
Chung, Eduardo
Clifton, Gloria
CUI, Zhenhua
CUI, Shizhu
Dadic, Zarko
Danezis, Emmanuel
Das, P.
Davies, Rodney
Davis, A. E. L.
Davoust, Emmanuel
de Jong, Teije
Debarbat, Suzanne
DeVorkin, David
Dick, Steven
Dick, Wolfgang
Dluzhnevskaya, Olga
Dorokhova, Tetyana
Dorschner, Johann
Duffard, René
Dutil, Yvan
Dworetsky, Michael
Edwards, Philip
Ehgamberdiev, Shuhrat
Engels, Dieter
Espenak, Fred
Esteban, César
Evans, Robert
Ferlet, Roger
Fernie, J.
Feulner, Georg
Field, J. V.
Fierro, Julieta
Firneis, Maria
Flin, Piotr
Florides, Petros
Fluke, Christopher
Freeman, Kenneth
Freitas Mourao, Ronaldo
Frew, David
Funes, José
Gabor, Pavel
Gangui, Alejandro
Gavrilov, Mikhail
Geffert, Michael
Geyer, Edward
Gingerich, Owen
Glass, Ian
Goss, W. Miller
Graham, John
Green, Daniel
Green, Anne
Green, David
Griffin, R. Elizabeth
Gurshtein, Alexander
Gussmann, Ernst-August

Hadrava, Petr
HAN, Wonyong
Hasan, S. Sirajul
HASEGAWA, Ichiro
Haubold, Hans
Haupt, Hermann
Hayli, Abraham
Haynes, Raymond
Haynes, Roslynn
Hearnshaw, John
Heddle, Douglas
Helou, George
Hemenway, Mary
Herrmann, Dieter
Hidayat, Bambang
HIRAI, Masanori
Hockey, Thomas
Hoeg, Erik
Hollow, Robert
Holmberg, Gustav
Hopkins, Andrew
Huan, Nguyen
Hunstead, Richard
HWANG, Chorng-Yuan
Hysom, Edmund
Hyung, Siek
Ibrahim, Alaa
Jafelice, Luiz
Jahreiss, Hartmut
Jauncey, David
JEONG, Jang-Hae
JIANG, Xiaoyuan
Jimenez-Vicente, Jorge
Jones, Paul
Kapoor, Ramesh
Keay, Colin
Keller, Hans-Ulrich
Kellermann, Kenneth
Kepler, S.
Kerschbaum, Franz
KIM, Chun-Hwey
KIM, Yonggi
KIM, Yong Cheol
KIM, Young-Soo
KIM, Sang Hyuk
KIM, Sang Chul
Kippenhahn, Rudolf
Knight, Matthew
Kollerstrom, Nicholas
Kolomiyets, Svitlana
Komonjinda, Siramas
Koribalski, Bärbel
Kosovichev, Alexander
Kovacs, József
Krajnovic, Davor
Kreiner, Jerzy
Krisciunas, Kevin
Krishnan, Thiruvenkata
Krupp, Edwin
Lanfranchi, Gustavo
Lang, Kenneth
Las Vergnas, Olivier
Launay, Françoise
Le Guet Tully, Françoise
LEE, Eun Hee
LEE, Woo baik
LEE, Yong Sam
LEE, Yong Bok
LEE, Ki-Won
Lerner, Michel-Pierre
Leung, Kam
Levy, Eugene
LI, Yong
LI, Min
Liller, William
Liritzis, Ioannis
LIU, Ciyuan
Locher, Kurt
Lomb, Nicholas
Longo, Giuseppe
Lopes, Rosaly
Lopez, Alejandro
Luminet, Jean-Pierre
Maiz Apellaniz, Jesús
Malin, David
Mallamaci, Claudio
Malville, J.
Manchester, Richard
Marco, Olivier
Mason, Brian
McConnell, David
McKenna Lawlor, Susan
Meadows, A.
Meech, Karen
Mendillo, Michael
Menon, T.
Mickaelian, Areg
Mickelson, Michael
Milne, Douglas
Molnar, Michael
Nadal, Robert
NAKAMURA, Tsuko
Narlikar, Jayant
Nazé, Yaël
Nefedyev, Yury
NHA, Il Seong
Nicolaidis, Efthymios
NING, Xiaoyu
Nitschelm, Christian
Nussbaumer, Harry
OH, Kyu-Dong
Ohashi, Nagayoshi
Olivier, Enrico
Olowin, Ronald
Oproiu, Tiberiu
Orchiston, Wayne
Osorio, José
Pang, Kevin
Papathanasoglou, Dimitrios
Pasachoff, Jay
Pati, Ashok
Pecker, Jean-Claude
Peterson, Charles
Pettersen, Bjørn
Pfleiderer, Jorg
PHAN, Dong
Pilbratt, Göran
Pineda de Carias, Maria
Pinigin, Gennadiy
Polcaro, V.
Polyakhova, Elena
Pozhalova, Zhanna
Preston, Robert
Proverbio, Edoardo
Rafferty, Theodore
Ray, Tom
Reboul, Henri
Robertson, James
Rubio-Herrera, Eduardo
Ruggles, Clive
Ryder, Stuart
Saucedo Morales, Julio
Schaefer, Bradley
Schechner, Sara
Schmadel, Lutz
Schmidt, Maarten
Schnell, Anneliese
Seck, Friedrich
Seggewiss, Wilhelm
Shank, Michael

Shankland, Paul
Shaver, Peter
Shelton, Ian
Shingareva, Kira
Shore, Steven
Shukre, C.
Sigismondi, Costantino
Signore, Monique
Sima, Zdislav
Simonia, Irakli
Sinachopoulos, Dimitris
Singh, Jagdev
Slechta, Miroslav
Slee, O.
Smith, Malcolm
Sobouti, Yousef
Solc, Martin
SOMA, Mitsuru
Soonthornthum, B.
Souchay, Jean
Stathopoulou, Maria
Steel, Duncan
Steele, John
Steinle, Helmut
Stephenson, F.
Sterken, Christiaan
Stoev, Alexey
Storey, Michelle
Subramanian, K.
Sullivan, III, Woodruff
Swarup, Govind
Swerdlow, Noel
Szabados, Laszlo
Tammann, Gustav
TAMURA, Shin'ichi
TANIKAWA, Kiyotaka
Taub, Liba
Terzian, Yervant
Theodossiou, Efstratios
Tignalli, Horacio
Tobin, William
Trimble, Virginia
Tripathy, Sushanta
Tsvetkov, Milcho
Urama, Johnson
Usher, Peter
Vahia, Mayank
Vakoch, Douglas
Valdes Parra, Jose
van Gent, Robert
van Woerden, Hugo
Vass, Gheorghe
Vats, Hari
Vaughan, Alan
Vavilova, Iryna
Venkatakrishnan, P.
Verdun, Andreas
Viollier, Raoul
Voigt, Hans
Volyanska, Margaryta
Wainscoat, Richard
WANG, Guangchao
Watson, Frederick
Weiss, Werner
Weiss, John
Wendt, Harry
White, Graeme
Whiteoak, John
Wielen, Roland
Wilkins, George
Williams, Thomas
Wilson, Curtis
Wolfschmidt, Gudrun
Woudt, Patrick
Wright, Alan
XIONG, Jianning
YAMAOKA, Hitoshi
YANG, Hong-Jin
Yau, Kevin
Yeomans, Donald
Zanini, Valeria
Zeilik, Michael
ZHOU, Yonghong
Zsoldos, Endre

Division V Commission 42 Close Binary Stars

President: Ignasi Ribas
Vice-President: Mercedes T. Richards

Organizing Committee Members:

Bradstreet, David
Harmanec, Petr
Hilditch, Ronald
Kaluzny, Janusz
Mikolajewska, Joanna
Munari, Ulisse
Niarchos, Panagiotis
Nordstrom, Birgitta
Olah, Katalin
Pribulla, Theodor
Scarfe, Colin
Sion, Edward
Torres, Guillermo

Members:

AK, Tansel
Akashi, Muhammad
Al-Naimiy, Hamid
Andersen, Johannes
Andronov, Ivan
Antokhin, Igor
Antonopoulou, Evgenia
Anupama, G.
Aquilano, Roberto
Arbutina, Bojan
Arefiev, Vadim
Aungwerojwit, Amornrat
Awadalla, Nabil
BABA, Hajime
Babkovskaia, Natalia
Bailyn, Charles
Balman, Solen
Baptista, Raymundo
Baran, Andrzej
Barkin, Yuri
Barone, Fabrizio
Bartolini, Corrado
Batten, Alan
Bear, Ealeal
Bell, Steven
Benacquista, Matthew
Bianchi, Luciana
Blair, William
Blundell, Katherine
Boffin, Henri
Bolton, Charles
Bonazzola, Silvano
Bopp, Bernard
Borisov, Nikolay
Bortoletto, Alexandre
Boyd, David
Boyle, Stephen
Bozic, Hrvoje
Bradstreet, David
Brandi, Elisande
Broglia, Pietro
Brown, David
Brownlee, Robert
Bruch, Albert
Bruhweiler, Frederick
Budaj, Jan
Budding, Edwin
Bunner, Alan
Burderi, Luciano
Burikham, Piyabut
Busa, Innocenza
Busso, Maurizio
Callanan, Paul
Canalle, Joao
Chambliss, Carlson
Chapman, Robert
Chaty, Sylvain
Chaubey, Uma
CHEN, An-Le
CHEN, Xuefei
Cherepashchuk, Anatolij
Chochol, Drahomir
CHOI, Kyu Hong
CHOI, Chul-Sung
CHOU, Yi
Ciardi, David
Claria, Juan
Cornelisse, Remon
Corradi, Romano
Cowley, Anne
Crause, Lisa
Cropper, Mark
CUI, Wenyuan
Cutispoto, Giuseppe
D'Amico, Nicolo'
D'Antona, Francesca
DAI, Zhibin
Dall, Thomas
Day Jones, Avril
de Greve, Jean-Pierre
de Loore, Camiel
de Mink, Selma
Del Santo, Melania
Delgado, Antonio
Demircan, Osman
Derekas, Aliz
Diaz, Marcos
Dobrotka, Andrej
Dobrzycka, Danuta
Dorfi, Ernst
Dougherty, Sean
Drechsel, Horst
Dubus, Guillaume
Dupree, Andrea
Durisen, Richard
Duschl, Wolfgang
Eaton, Joel
Ederoclite, Alessandro
Eggleton, Peter
Eldridge, John
Elias II, Nicholas
Engelbrecht, Chris

Etzel, Paul
Eyres, Stewart
Fabiani, Sergio
Fabrika, Sergei
Falize, Emeric
Farrell, Sean
Faulkner, John
Fekel, Francis
Ferluga, Steno
Fernandez Lajus, Eduardo
Ferrario, Lilia
Ferrer, Osvaldo
Flannery, Brian
Fors, Octavi
Frank, Juhan
Fredrick, Laurence
Gaensicke, Boris
Gallagher III, John
Gamen, Roberto
Garcia, Lia
García de María, Juan
Garcia-Lorenzo, Maria
Garmany, Katy
Gasiprong, Nipon
Geldzahler, Barry
Geyer, Edward
Giannone, Pietro
Gies, Douglas
Giovannelli, Franco
Goldman, Itzhak
Gomboc, Andreja
GONG, Biping
Gonzalez, Gabriela
Gonzalez Martinez Pais, I.
Gosset, Eric
Groot, Paul
Grygar, Jiri
GU, Wei-Min
Guinan, Edward
Gulliver, Austin
Gun, Gulnur
Gunn, Alastair
GUO, Jianheng
Haas, Martin
Hadrava, Petr
Hakala, Pasi
Hallinan, Gregg
HANAWA, Tomoyuki
Hantzios, Panayiotis
Harmanec, Petr
Hassall, Barbara
Haswell, Carole
HAYASAKI, Kimitake
HE, Jinhua
Hegedues, Tibor
Hellier, Coel
Helt, Bodil
Hempelmann, Alexander
Hensler, Gerhard
Hill, Graham
Hills, Jack
Hillwig, Todd
HIROSE, Shigenobu
Hoard, Donald
Holmgren, David
Holt, Stephen
Honeycutt, R.
HORIUCHI, Ritoku
Hric, Ladislav
Hrivnak, Bruce
Hube, Douglas
Hutchings, John
Ibanoglu, Cafer
Ikhsanov, Nazar
Imamura, James
Imbert, Maurice
Ioannou, Zacharias
Ivanova, Natalia
Izzard, Robert
Jasniewicz, Gerard
Jeffers, Sandra
JEONG, Jang-Hae
JIN, Zhenyu
Jin, Liping
Jones, David
Jonker, Peter
Joss, Paul
Kafka, Styliani (Stella)
Kaitchuck, Ronald
Kalomeni, Belinda
Kaluzny, Janusz
KANG, Young Woon
Karami, Kayoomars
Karetnikov, Valentin
KATO, Taichi
KAWABATA, Shusaku
KENJI, Nakamura
Kenny, Harold
Kenyon, Scott
KIM, Chun-Hwey
KIM, Ho-il
KIM, Young-Soo
KIM, Woong-Tae
King, Andrew
Kjurkchieva, Diana
Kley, Wilhelm
Kolb, Ulrich
Kolesnikov, Sergey
Komonjinda, Siramas
Konacki, Maciej
KONDO, Yoji
Koubsky, Pavel
Kraicheva, Zdravka
Krautter, Joachim
Kreiner, Jerzy
Kreykenbohm , Ingo
Kriwattanawong, Wichean
Kruchinenko, Vitaliy
Kruszewski, Andrzej
Kudashkina, Larisa
Kwee, K.
Lacy, Claud
Lamb Jr, Donald
Landolt, Arlo
Lapasset, Emilio
Larsson, Stefan
Larsson-Leander, Gunnar
LEE, Woo baik
LEE, Yong Sam
Lee, William
LEE, Jae Woo
LEE, Chung-Uk
Leedjarv, Laurits
Leung, Kam
LI, Ji
LI, Zhi
LI, Lifang
LIM, Jeremy
LIN, Yi-qing
Linnell, Albert
Linsky, Jeffrey
LIU, Qingzhong
LIU, Jifeng
Livio, Mario
Lucy, Leon
Luque-Escamilla, Pedro
MacDonald, James
Maceroni, Carla
Malasan, Hakim
Malkov, Oleg

Mandel, Ilya
Manimanis, Vassilios
Mardirossian, Fabio
Marilli, Ettore
Markoff, Sera
Markworth, Norman
Marsh, Thomas
Martayan, Christophe
Mason, Paul
Mathieu, Robert
Mayer, Pavel
Mazeh, Tsevi
McCluskey Jr, George
Meibom, Soren
Meintjes, Petrus
Melia, Fulvio
Meliani, Mara
Mennickent, Ronald
Mereghetti, Sandro
Meyer-Hofmeister, Eva
Mezzetti, Marino
Migaszewski, Cezary
Mikolajewska, Joanna
Mikulášek, Zdeněk
Milano, Leopoldo
Millour, Florentin
Milone, Eugene
MINESHIGE, Shin
Miszalski, Brent
MIYAJI, Shigeki
Mochnacki, Stefan
Monard, Libert
Montgomery, Michele
Morales Rueda, Luisa
Morgan, Thomas
Morrell, Nidia
Mouchet, Martine
Munari, Ulisse
Mutel, Robert
NAKAMURA, Yasuhisa
NAKAO, Yasushi
NARIAI, Kyoji
Nather, R.
Naylor, Tim
Neff, James
Nelemans, Gijs
Nelson, Burt
Neustroev, Vitaly
Newsom, Gerald
NHA, Il Seong
Niarchos, Panagiotis
Nitschelm, Christian
Norton, Andrew
Ogloza, Waldemar
OH, Kyu-Dong
OKAZAKI, Akira
Olah, Katalin
Oliveira, Alexandre
Olson, Edward
OSAKI, Yoji
Özeren, Ferhat
Ozkan, Mustafa
Pandey, Uma
Panei, Jorge
Parimucha, Stefan
PARK, Hong-Seo
Parthasarathy, Mudumba
Patkos, Laszlo
Pavlenko, Elena
Pavlovski, Kresimir
Pearson, Kevin
Peters, Geraldine
Piirola, Vilppu
Pojmański, Grzegorz
Polidan, Ronald
Pollacco, Don
Pooley, David
Popov, Sergey
Postnov, Konstantin
Potter, Stephen
Pretorius, Magaretha
Pribulla, Theodor
Pringle, James
Prokhorov, Mikhail
Prsa, Andrej
Pustynski, Vladislav-V.
QIAO, Guojun
Rafert, James
Rahunen, Timo
Ramsey, Lawrence
Ransom, Scott
Rao, Pasagada
Rasio, Frederic
Reglero Velasco, Victor
Rey, Soo-Chang
Ricker, Paul
Riles, Keith
Ringwald, Frederick
Ritter, Hans
Robb, Russell
Robertson, John
Robinson, Edward
Rodrigues, Claudia
Rovithis-Livaniou, Helen
Roxburgh, Ian
Ruffert, Maximilian
Russo, Guido
SAIJO, Keiichi
Samec, Ronald
Sarty, Gordon
Savonije, Gerrit
Schartel, Norbert
Schiller, Stephen
Schmid, Hans
Schmidtke, Paul
Schmidtobreick, Linda
Schober, Hans
Seggewiss, Wilhelm
Selam, Selim
Semeniuk, Irena
Shafter, Allen
Shahbaz, Tariq
Shakura, Nikolaj
Shaviv, Giora
Shimansky, Vladislav
SHU, Frank
Sima, Zdislav
Simmons, John
Sistero, Roberto
Skopal, Augustin
Smak, Jozef
Smith, Robert
Soderhjelm, Staffan
Solheim, Jan
SONG, Liming
Southworth, John
Sowell, James
Sparks, Warren
Stachowski, Grzegorz
Stagg, Christopher
Stanishev, Vallery
Starrfield, Sumner
Steiman-Cameron, Thomas
Steiner, Joao
Stencel, Robert
Sterken, Christiaan
Stoyanov, Kiril
Stringfellow, Guy
Sudar, Davor
SUGIMOTO, Daiichiro

Szkody, Paula
Taam, Ronald
TAKAHASHI, Rohta
TAN, Huisong
Tas, Günay
Tauris, Thomas
Teays, Terry
TERADA, Yukikatsu
Terrell, Dirk
Tessema, Solomon
Torres, Guillermo
Tout, Christopher
Tremko, Jozef
Trimble, Virginia
Turolla, Roberto
Tutukov, Aleksandr
Ulla Miguel, Ana
Unda-Sanzana, Eduardo
Ureche, Vasile
Vaccaro, Todd
van den Berg, Maureen
van den Heuvel, Edward
van Hamme, Walter
Vaz, Luiz Paulo
Vennes, Stephane
Vierdayanti, Kiki
Vilhu, Osmi
Voloshina, Irina
Wachter, Stefanie
Wade, Richard
Walder, Rolf
Walker, William
WANG, Xunhao
WANG, Bo
Ward, Martin
Warner, Brian
Webbink, Ronald
Weiler, Edward
Wesson, Roger
Wheatley, Peter
Wheeler, J.
White II, James
Williamon, Richard
Williams, Robert
Williams, Glen
Wilson, Robert
Wittenmyer, Robert
Worters, Hannah
XUE, Li
YAMAOKA, Hitoshi
YAMASAKI, Atsuma
YOON, Tae-Seog
YU, Cong
Yuce, Kutluay
Zakirov, Mamnum
Zamanov, Radoslav
Zavala, Robert
Zeilik, Michael
Zejda, Miloslav
ZHANG, Er-Ho
ZHANG, Bo
ZHANG, Shu
Zharikov, Sergey
Zhilkin, Andrey
ZHU, Liying
Ziolkowski, Janusz
Zola, Stanislaw
Zwitter, Tomaž

Division XI Commission 44 Space & High Energy Astrophysics

President: Christine Jones

Organizing Committee Members:

Braga, João
Brosch, Noah
Gurvits, Leonid
Helou, George
Howarth, Ian

Members:

Abramowicz, Marek
Acharya, Bannanje
Acton, Loren
Aghaee, Alireza
Agrawal, P.
Aguiar, Odylio
Aharonian, Felix
Ahluwalia, Harjit
Ahmad, Imad
Alexander, Joseph
Allington-Smith, Jeremy
Amati, Lorenzo
Andersen, Bo Nyborg
Antonelli, Lucio Angelo
Antoniou, Vallia
Apparao, K.
ARAFUNE, Jiro
Arefiev, Vadim
Arnaud, Monique
Arnould, Marcel
Arons, Jonathan
ASANO, Katsuaki
Aschenbach, Bernd
Asvarov, Abdul
Audard, Marc
Audley, Michael
Audouze, Jean
Augereau, Jean-Charles
AWAKI, Hisamitsu
Axelsson, Magnus
AYA, Bamba
Ayres, Thomas
Baan, Willem
Bailyn, Charles
Balikhin, Michael
Baliunas, Sallie
Balman, Solen
Baring, Matthew
Barkhouse, Wayne
Barrantes, Marco
Barret, Didier
Barstow, Martin
Baskill, Darren
Baym, Gordon
Bazzano, Angela
Becker, Robert
Becker, Werner
Beckmann, Volker
Begelman, Mitchell
Behar, Ehud
Beiersdorfer, Peter
Beklen, Elif
Belloni, Tomaso
Bender, Peter
Benedict, George
Benford, Gregory
Bennett, Charles
Bennett, Kevin
Benvenuto, Omar
Bergeron, Jacqueline
Bernardini, Federico
Berta, Stefano
Beskin, Gregory
Beskin, Vasily
Bhattacharjee, Pijush
Bhattacharya, Dipankar
Bhattacharyya, Sudip
Bianchi, Luciana
Bianchi, Stefano
Bicknell, Geoffrey
Biermann, Peter
Bigdeli, Mohsen
Bignami, Giovanni
Bingham, Robert
Birkmann, Stephan
Blandford, Roger
Bleeker, Johan
Bless, Robert
Blinnikov, Sergey
Bloemen, Hans
Blondin, John
Bludman, Sidney
Bocchino, Fabrizio
Bodaghee, Arash
Boer, Michel
Boggess, Albert
Boggess, Nancy
Bohlin, Ralph
Boksenberg, Alec
Bonazzola, Silvano
Bonnet, Roger
Bonnet-Bidaud, Jean-Marc
Bonometto, Silvio
Borka Jovanovic, Vesna
Borozdin, Konstantin
Bosch-Ramon, Valenti
Bougeret, Jean-Louis
Bowyer, C.
Bradley, Arthur
Braithwaite, Jonathan
Branchesi, Marica
Brandt, Soeren
Brandt, John
Brandt, William
Brecher, Kenneth
Brenneman, Laura
Breslin, Ann
Breton, Rene
Brinkman, Bert
Bromberg, Omer
Brown, Alexander

Bruhweiler, Frederick
Bruner, Marilyn
Brunetti, Gianfranco
Bumba, Vaclav
Bunner, Alan
Buote, David
Burderi, Luciano
Burenin, Rodion
Burikham, Piyabut
Burke, Bernard
Burrows, David
Burrows, Adam
Bursa, Michal
Butler, Christopher
Caccianiga, Alessandro
CAI, Michael
CAI, Mingsheng
Camenzind, Max
Campana, Riccardo
Campbell, Murray
Cannon, Kipp
CAO, Li
Cappi, Massimo
Caputi, Karina
Caraveo, Patrizia
Cardenas, Rolando
Cardini, Daniela
Carlson, Per
Carpenter, Kenneth
Casandjian, Jean-Marc
Cash Jr, Webster
Cassano, Rossella
Cassé, Michel
Castro-Tirado, Alberto
Cavaliere, Alfonso
Celotti, Anna Lisa
Cenko, Stephen
Cesarsky, Catherine
Chakrabarti, Sandip
Chakraborty, Deo
CHANG, Hsiang-Kuang
CHANG, Heon-Young
Channok, Chanruangrit
Chapman, Robert
Chapman, Sandra
Charles, Philip
Chartas, George
Chechetkin, Valerij
Chelliah Subramonian, S.
Chelouche, Doron
CHEN, Lin-wen
Chenevez, Jérôme
CHENG, Kwongsang
Chernyakova, Maria
Cheung, Cynthia
Chian, Abraham
Chiappetti, Lucio
CHIKAWA, Michiyuki
Chitre, Shashikumar
Chochol, Drahomir
CHOE, Gwangson
CHOI, Chul-Sung
CHOU, Yi
Churazov, Eugene
Ciotti, Luca
Clark, George
Clark, Thomas
Clark, David
Clay, Roger
Cohen, David
Collin, Suzy
Comastri, Andrea
Condon, James
Contopoulos, Ioannis
Corbel, Stéphane
Corbet, Robin
Corbett, Ian
Corcoran, Michael
Cordova, France
Cornelisse, Remon
Costantini, Elisa
Courvoisier, Thierry
Cowie, Lennox
Cowsik, Ramanath
Crannell, Carol
Crocker, Roland
Cropper, Mark
Croston, Judith
Croton, Darren
Cruise, Adrian
Cuadra, Jorge
Cui, Wei
Culhane, John
Cunniffe, John
Curir, Anna
Cusumano, Giancarlo
D'Ammando, Filippo
da Costa, Antonio
da Silveira, Enio
Dadhich, Naresh
DAI, Zigao
Dalla Bonta, Elena
DAmico, Flavio
Darriulat, Pierre
Davidson, William
Davis, Michael
Davis, Robert
Dawson, Bruce
De Becker, Michaël
de Felice, Fernando
de Jager, Cornelis
de Martino, Domitilla
de Ugarte Postigo, Antonio
Del Santo, Melania
Del Zanna, Luca
Della Ceca, Roberto
Dempsey, Robert
den Herder, Jan-Willem
DeNisco, Kenneth
Dennerl, Konrad
Dennis, Brian
Dermer, Charles
Di Cocco, Guido
Diaz Trigo, Maria
Digel, Seth
Disney, Michael
Dokuchaev, Vyacheslav
Dolan, Joseph
Domingo, Vicente
Dominis Prester, Dijana
Donea, Alina
DONG, Xiao-Bo
DOTANI, Tadayasu
DOU, Jiangpei
Dovciak, Michal
Downes, Turlough
Drake, Frank
Drury, Luke
Dubus, Guillaume
Duorah, Hira
Dupree, Andrea
Durouchoux, Philippe
Easther, Richard
Edelson, Rick
Edwards, Paul
Edwards, Philip
Ehle, Matthias
Eichler, David
Eilek, Jean
Elvis, Martin

Elyiv, Andrii
Emanuele, Alessandro
Ensslin, Torsten
ESAMDIN, Ali
ESIMBEK, Jarken
Ettori, Stefano
Eungwanichayapant, Anant
Evans, Daniel
Fabian, Andrew
Fabiani, Sergio
Fabricant, Daniel
Falize, Emeric
Faraggiana, Rosanna
Farrell, Sean
Fatkhullin, Timur
Faure, Cécile
Fazio, Giovanni
Feldman, Paul
Fender, Robert
Fendt, Christian
Ferrari, Attilio
Field, George
Fisher, Philip
Fishman, Gerald
Florido, Estrella
Foing, Bernard
Fomin, Valery
Fonseca Gonzalez, Maria
Forman, William
Foschini, Luigi
Franceschini, Alberto
Frandsen, Soeren
Frank, Juhan
Fransson, Claes
Fraschetti, Federico
Fredga, Kerstin
Fruscione, Antonella
FUJIMOTO, Shin-ichiro
FUJITA, Mitsutaka
Furniss, Ian
Fyfe, Duncan
Gabriel, Alan
Gaensler, Bryan
Gaisser, Thomas
Gal-Yam, Avishay
Galeotti, Piero
Galloway, Duncan
Gammie, Charles
Gangadhara, R.T.
GAO, Yu
Garmire, Gordon
Gaskell, C.
Gastaldello, Fabio
Gathier, Roel
Gehrels, Neil
Gendre, Bruce
Georgantopoulos, Ioannis
Gezari, Daniel
Ghia, Piera Luisa
Ghirlanda, Giancarlo
Ghisellini, Gabriele
Giacconi, Riccardo
Gioia, Isabella
Giroletti, Marcello
Gitti, Myriam
Goicoechea, Luis
Goldsmith, Donald
Goldwurm, Andrea
Gomboc, Andreja
Gomez, Haley
Gomez de Castro, Ana
GONG, Biping
Gonzalez, Gabriela
Gordon, Chris
Gotthelf, Eric
Gotz, Diego
Grebenev, Sergei
Greenhill, John
Gregorio, Anna
Grenier, Isabelle
Grewing, Michael
Greyber, Howard
Griffiths, Richard
Grindlay, Jonathan
Grosso, Nicolas
GU, Minfeng
Guessoum, Nidhal
Gull, Theodore
Gumjudpai, Burin
Gun, Gulnur
Gunn, James
Gutierrez, Carlos
Guziy, Sergiy
Hakkila, Jon
Halevin, Alexandros
Hameury, Jean-Marie
Hanna, David
Hannah, Iain
Hannikainen, Diana
Hardcastle, Martin
Harms, Richard
Harris, Daniel
Harvey, Paul
Harwit, Martin
Hasan, Hashima
HATSUKADE, Isamu
Haubold, Hans
Haugboelle, Troels
Hauser, Michael
Hawkes, Robert
Hawking, Stephen
HAYAMA, Kazuhiro
Haymes, Robert
Heger, Alexander
Heinz, Sebastian
Heise, John
Helfand, David
Hempel, Marc
Heng, Kevin
Henoux, Jean-Claude
Henriksen, Richard
Henry, Richard
Hensberge, Herman
Heyl, Jeremy
Hicks, Amalia
Hill, Adam
Hjalmarsdotter, Linnea
Ho, Wynn
Hoffman, Jeffrey
Holberg, Jay
Holloway, Nigel
Holt, Stephen
Holz, Daniel
Hora, Joseph
Horandel, Jörg
Horns, Dieter
Hornschemeier, Ann
Hornstrup, Allan
Houziaux, Leo
Hoyng, Peter
HSU, Rue-Ron
HUANG, YongFeng
Huang, Jiasheng
Huber, Martin
Hulth, Per
Hurley, Kevin
Hutchings, John
HWANG, Chorng-Yuan
Ibrahim, Alaa
ICHIMARU, Setsuo

Ikhsanov, Nazar
Illarionov, Andrei
Imamura, James
Imhoff, Catherine
in't Zand, Johannes
INOUE, Makoto
INOUE, Hajime
IOKA, Kunihito
Ipser, James
ISHIDA, Manabu
Israel, Werner
ITOH, Masayuki
Jackson, John
Jaffe, Walter
Jakobsson, Pall
Jamar, Claude
Janka, Hans
Jaranowski, Piotr
Jenkins, Edward
JI, Li
Johns, Bethany
Jokipii, Jack
Jones, Thomas
Jonker, Peter
Jordan, Carole
Jordan, Stuart
Joss, Paul
Kafatos, Menas
Kalemci, Emrah
KAMENO, Seiji
Kaminker, Alexander
KANEDA, Hidehiro
Kaper, Lex
Kapoor, Ramesh
Karakas, Amanda
Karami , Kayoomars
Karpov, Sergey
Kaspi, Victoria
Kasturirangan, K.
Katarzynski, Krzysztof
KATO, Tsunehiko
KATO, Yoshiaki
Katsova, Maria
Katz, Jonathan
KAWAI, Nobuyuki
Kellermann, Kenneth
Kellogg, Edwin
Kelly, Brandon
Kembhavi, Ajit
KENJI, Nakamura
Kessler, Martin
Khumlumlert, Thiranee
Killeen, Neil
KIM, Yonggi
KIM, Minsun
Kimble, Randy
KINUGASA, Kenzo
Kirk, John
KIYOSHI, Hayashida
Klinkhamer, Frans
Klose, Sylvio
Knapp, j.knapp
KO, Chung-Ming
Kobayashi, Shiho
Koch-Miramond, Lydie
Kohmura, Takayoshi
KOIDE , Shinji
KOJIMA, Yasufumi
Kokubun, Motohide
Kolb, Edward
KONDO, Yoji
KONDO, Masaaki
Kong, Albert
KOSHIBA, Masatoshi
KOSUGI, George
Koupelis, Theo
Kovar, Jiří
KOYAMA, Katsuji
Kozma, Cecilia
Kraft, Ralph
Kretschmar, Peter
Kreykenbohm , Ingo
Krittinatham, Watcharawuth
Kryvdyk, Volodymyr
Kudryavtseva, Nadezhda
Kuiper, Lucien
Kulsrud, Russell
KUMAGAI, Shiomi
Kuncic, Zdenka
Kundt, Wolfgang
Kunz, Martin
Kurt, Vladimir
KUSUNOSE, Masaaki
La Franca, Fabio
La Mura, Giovanni
Lagache, Guilaine
Lal, Dharam
Lamb, Frederick
Lamb, Susan
Lamb Jr, Donald
Lamers, Henny
Lampton, Michael
Lapi, Andrea
Lapington, Jonathan
Lattimer, James
Lea, Susan
Leckrone, David
LEE, Wo-Lung
Lee, William
LEE, Sang-Sung
Leighly, Karen
Lemaire, Philippe
Levenson, Nancy
Levin, Yuri
Levine, Robyn
Lewin, Walter
LI, Tipei
LI, Xiangdong
LI, Min
LI, Li-Xin
Liang, Edison
LIN, Xuan-bin
Linsky, Jeffrey
Liu, Bifang
LIU, Jifeng
LIU, Guoqing
LIU, Siming
Loaring, Nicola
Lochner, James
Long, Knox
Longair, Malcolm
Lopes de Oliveira, R.
Loubser, Ilani
Lovelace, Richard
LU, Tan
LU, Jufu
LU, Fangjun
LU, Ye
LU, Youjun
Luest, Reimar
Luminet, Jean-Pierre
Luo, Qinghuan
Luque-Escamilla, Pedro
Lutovinov, Alexander
Lynden-Bell, Donald
Lyubarsky, Yury
MA, YuQian
Maccacaro, Tommaso
Maccarone, Thomas
Macchetto, Ferdinando

MACHIDA, Mami
Maciesiak, Krzysztof
Maggio, Antonio
Maguire, Kate
Mainieri, Vincenzo
Majumdar, Subhabrata
Makarov, Valeri
Malesani, Daniele
Malitson, Harriet
Malkan, Matthew
Manara, Alessandro
Mandolesi, Nazzareno
Mangano, Vanessa
Maran, Stephen
Marar, T.
Maricic, Darije
Marino, Antonietta
Markoff, Sera
Marov, Mikhail
Marranghello, Guilherme
Martinez-Bravo, Oscar
Martinis, Mladen
MASAI, Kuniaki
Masnou, Jean-Louis
Mason, Glenn
Massaro, Francesco
Mather, John
MATSUMOTO, Ryoji
MATSUMOTO, Hironori
MATSUOKA, Masaru
MATSUSHITA, Kyoko
Matt, Giorgio
Matz, Steven
Mazurek, Thaddeus
McBreen, Brian
McBride, Vanessa
McCluskey Jr, George
McCray, Richard
McWhirter, R.
Medina, Jose
Meier, David
Meiksin, Avery
Melatos, Andrew
Melia, Fulvio
Melikidze, Giorgi
Melnick, Gary
Melnyk, Olga
Melrose, Donald
Mendez, Mariano
Mereghetti, Sandro
Merlo, David
Mestel, Leon
Meszaros, Peter
Meyer, Friedrich
Meyer, Jean-Paul
Micela, Giuseppina
Miller, Michael
Miller, Guy
Miller, John
Miller, Eric
Mineo, Teresa
Miroshnichenko, Alla
MIYAJI, Shigeki
Miyaji, Takamitsu
Miyata, Emi
MIZUMOTO, Yoshihiko
MIZUNO, Yosuke
MIZUTANI, Kohei
Moderski, Rafal
Molla, Mercedes
Monet, David
Moodley, Kavilan
MOON, Shin-Haeng
Moos, Henry
Morgan, Thomas
MORI, Koji
MORI, Masaki
MOROKUMA, Tomoki
Morsony, Brian
Morton, Donald
Mota, David
Motch, Christian
MOTIZUKI, Yuko
Mourao, Ana Maria
Mukhopadhyay, Banibrata
Mulchaey, John
MURAKAMI, Hiroshi
MURAKAMI, Toshio
MURAYAMA, Hitoshi
Murdock, Thomas
Murtagh, Fionn
Murthy, Jayant
NAGATAKI, Shigehiro
Nakar, Ehud
NAKAYAMA, Kunji
Neff, Susan
Ness, Norman
Neuhaeuser, Ralph
Neupert, Werner
Neustroev, Vitaly
Nichols, Joy
Nicollier, Claude
Nielsen, Krister
Nikolajuk, Marek
NISHIMURA, Osamu
NITTA, Shin-ya
Nityananda, Rajaram
NOMOTO, Ken'ichi
Norci, Laura
Nordh, Lennart
Norman, Colin
Noyes, Robert
Nulsen, Paul
Nymark, Tanja
O'Brien, Paul
O'Connell, Robert
O'Sullivan, Denis
OGAWARA, Yoshiaki
Okeke, Pius
OKUDA , Toru
Olthof, Henk
Onken, Christopher
OOHARA, Ken-ichi
Oozeer, Nadeem
Orellana, Mariana
Orford, Keith
Orio, Marina
Orlandini, Mauro
Orlando, Salvatore
Osborne, Julian
Oskinova, Lidia
Osten, Rachel
Ostriker, Jeremiah
Ostrowski, Michal
Ott, Juergen
Owen, Tobias
Owers, Matthew
OZAKI, Masanobu
Özel, Mehmet
Paciesas, William
Page, Clive
Page, Mathew
PAK, Soojong
Paltani, Stéphane
Palumbo, Giorgio
Pandey, Uma
Panessa, Francesca
Papadakis, Iossif
Papitto, Alessandro
Paragi, Zsolt

Pareschi, Giovanni
PARK, Myeong Gu
Parker, Eugene
Patten, Brian
Paul, Biswajit
Pavlov, George
Peacock, Anthony
Pearce, Mark
Pearson, Kevin
Pellegrini, Silvia
Pellizza, Leonardo
PENG, Qiuhe
PENG, Qingyu
Perez, Mario
Perez-Gonzalez, Pablo
Perola, Giuseppe
Perry, Peter
Peters, Geraldine
Peterson, Bruce
Peterson, Laurence
Peterson, Bradley
Pethick, Christopher
Petkaki, Panagiota
Petro, Larry
Petrosian, Vahe
Phillips, Kenneth
Pian, Elena
Pinkau, K.
Pinto, Philip
Pipher, Judith
Piran, Tsvi
Piro, Luigi
Pitkin, Matthew
Polidan, Ronald
Polletta, Maria del Carmen
Pooley, David
Popov, Sergey
Porquet, Delphine
Pottschmidt, Katja
Pounds, Kenneth
Poutanen, Juri
Pozanenko, Alexei
Prasanna, A.
Preuss, Eugen
Produit, Nicolas
Protheroe, Raymond
Prouza, Michael
Prusti, Timo
Pshirkov, Maxim
Pustilnik, Lev
QU, Jinlu
Raiteri, Claudia
Ramadurai, Souriraja
Ramirez, Jose
Ranalli, Piero
Rao, Arikkala
Rasmussen, Ib
Rasmussen, Jesper
Ray, Paul
Raychaudhury, Somak
Rea, Nanda
Reale, Fabio
Rees, Martin
Reeves, Hubert
Reiprich, Thomas
Reisenegger, Andreas
Reitze, David
Rengarajan, Thinniam
Revnivtsev, Mikhail
Rhoads, James
Ricker, Paul
Riles, Keith
Risaliti, Guido
Robba, Natale
Roberts, Timothy
Roman, Nancy
Romano, Patrizia
Roming, Peter
Rosendhal, Jeffrey
Rosner, Robert
Rossi, Elena
Rovero, Adrián
Rowell, Gavin
Rozanska , Agata
Rubino-Martin, J. A.
Rubio-Herrera, Eduardo
Ruder, Hanns
Ruffini, Remo
Ruffolo, David
Russell, Alexander
Ruszkowski, Mateusz
Rutledge, Robert
Sabau-Graziati, Lola
Safi-Harb, Samar
Sagdeev, Roald
Sahlen, Martin
Saiz, Alejandro
Sakano, Masaaki
Sakelliou, Irini
Sanchez, Norma
Sanders, Gary
Sanders III, Wilton
Santos-Lleo, Maria
Sari, Re'em
Saslaw, William
SATO, Katsuhiko
Savage, Blair
Savedoff, Malcolm
Sazonov, Sergey
Sbarufatti, Boris
Scargle, Jeffrey
Schaefer, Gerhard
Schartel, Norbert
Schatten, Kenneth
Schilizzi, Richard
Schmitt, Juergen
Schnopper, Herbert
Schreier, Ethan
Schulz, Norbert
Schure, Klara
Schwartz, Daniel
Schwartz, Steven
Sciortino, Salvatore
Seielstad, George
Selvelli, Pierluigi
Semerak, Oldrich
SEON, Kwang il
Sequeiros, Juan
Setti, Giancarlo
Severgnini, Paola
Seward, Frederick
Shahbaz, Tariq
Shakhov, Boris
Shakura, Nikolaj
Shalchi, Andreas
Sharma, Prateek
Shaver, Peter
Shaviv, Giora
SHEN, Zhiqiang
SHIBAI, Hiroshi
Shibanov, Yuri
SHIBAZAKI, Noriaki
Shields, Gregory
SHIGEYAMA, Toshikazu
SHIMURA, Toshiya
SHIN, Watanabe
SHIRASAKI, Yuji
Shoemaker, David
Shukre, C.
Shustov, Boris

Signore, Monique
Sikora, Marek
Silvestro, Giovanni
Simic, Sasa
Simon, Vojtech
Simon, Paul
Sims, Mark
Simunac, Kristin
Skilling, John
Skinner, Stephen
Skjaeraasen, Olaf
Slany, Petr
Smale, Alan
Smida, Radomír
Smith, Bradford
Smith, Peter
Smith, Linda
Smith, Nigel
Snow, Theodore
Sofia, Sabatino
Sokolov, Vladimir
Somasundaram, Seetha
SONG, Qian
Sonneborn, George
Sood, Ravi
Soria, Roberto
Spallicci di Filottrano, A.
Sreekumar, Parameswaran
Srinivasan, Ganesan
Srivastava, Dhruwa
Staubert, Rüdiger
Stecher, Theodore
Stecker, Floyd
Steigman, Gary
Steiner, Joao
Stencel, Robert
Stephens, S.
Stergioulas, Nikolaos
Stern, Robert
Stevens, Ian
Stier, Mark
Still, Martin
Stockman Jr, Hervey
Stoehr, Felix
Straumann, Norbert
Stringfellow, Guy
Strohmayer, Tod
Strong, Ian
Struminsky, Alexei
Stuchlik, Zdenek
Sturrock, Peter
SU, Cheng-yue
Subr, Ladislav
Suleimanov, Valery
SUMIYOSHI, Kosuke
SUN, Wei-Hsin
Sunyaev, Rashid
SUZUKI, Hideyuki
Swank, Jean
Tagliaferri, Gianpiero
TAKAHARA, Fumio
TAKAHASHI, Masaaki
TAKAHASHI, Tadayuki
TAKAHASHI, Rohta
TAKEI, Yoh
Tammi, Joni
Tanaka, Yasuo
TASHIRO, Makoto
TATEHIRO, Mihara
Tavecchio, Fabrizio
Tempel, Elmo
Templeton, Matthew
Teng, Stacy
TERADA, Yukikatsu
TERASHIMA, Yuichi
Terrell, James
Tessema, Solomon
Thoene, Christina
Thorne, Kip
Thronson Jr, Harley
TIAN, Wenwu
TOMIMATSU, Akira
TOMINAGA, Nozomu
Torok, Gabriel
Torres, Diego
Torres, Carlos Alberto
Tovmassian, Hrant
Traub, Wesley
Tresse, Laurence
Trimble, Virginia
Truemper, Joachim
TruranJr, James
Trussoni, Edoardo
TSAI, An-Li
TSUGUYA, Naito
Tsujimoto, Masahiro
TSUNEMI, Hiroshi
TSURU, Takeshi
Tsuruta, Sachiko
Tsygan, Anatolij
Tsygankov, Sergey
Tuerler, Marc
Tylka, Allan
UEDA, Yoshihiro
Ulyanov, Oleg
URATA, Yuji
Uslenghi, Michela
Usov, Vladimir
Uttley, Philip
Vahia, Mayank
Valiviita, Jussi-Pekka
Valtonen, Mauri
van den Berg, Maureen
van den Heuvel, Edward
van der Hucht, Karel
van der Walt, Diederick
van Duinen, R.
van Putten, Maurice
van Riper, Kenneth
Vaughan, Simon
Venter, Christo
Venters, Tonia
Vercellone, Stefano
Vestergaard, Marianne
Vial, Jean-Claude
Vidal-Madjar, Alfred
Vierdayanti, Kiki
Vignali, Cristian
Vikhlinin, Alexey
Vilhu, Osmi
Villata, Massimo
Vink, Jacco
Viollier, Raoul
Viotti, Roberto
Voelk, Heinrich
Volonteri, Marta
Vrtilek, Saeqa
Wagner, Alexander
Wagner, Robert
Walker, Helen
Wanas, Mamdouh
Wandel, Amri
WANG, Shouguan
WANG, Zhenru
Wang, Yi-ming
WANG, jiancheng
WANG, Ding-Xiong
WANG, Hong-Guang
WANG, Feilu
WANG, Shujuan

WANG, Shiang-Yu
WANG, Chen
WANG, Sen
Watanabe, Ken
WATARAI, Kenya
Watts, Anna
Waxman, Eli
Weaver, Kimberly
Weaver, Thomas
Webster, Adrian
Wehrle, Ann
WEI, Daming
Weiler, Kurt
Weiler, Edward
Weinberg, Jerry
Weisskopf, Martin
Wells, Donald
WEN, Linqing
Wesselius, Paul
Wheatley, Peter
Wheeler, J.
Whitcomb, Stanley
White, Nicholas
Wijers, Ralph
Wijnands, Rudy
Will, Clifford
Willis, Allan
Willner, Steven
Wilms, Jörn
Wilson, Gillian
Winkler, Christoph
Winter, Lisa
Wise, Michael
Wolfendale FRS, Sir Arnold
Wolstencroft, Ramon
Wolter, Anna
Woltjer, Lodewijk
WOO, Jong-Hak
Worrall, Diana
WU, Shaoping
WU, Jiun-Huei
WU, Xue-Feng
Wunner, Guenter
XU, Renxin
XU, Dawei
Yadav, Jagdish
Yakut, Kadri
Yamada, Shoichi
Yamasaki, Tatsuya
YAMASAKI, Noriko
YAMASHITA, Kojun
YAMAUCHI, Makoto
YAMAUCHI, Shigeo
Yock, Philip
YOICHIRO, Suzuki
YOKOYAMA, Takaaki
YONETOKU, Daisuke
YOSHIDA, Atsumasa
YU, Wang
YU, Wenfei
YU, Cong
YU, Qingjuan
YUAN, Ye-fei
Yuan, Weimin
YUAN, Feng
Zacchei, Andrea
Zamorani, Giovanni
Zane, Silvia
Zannoni, Mario
Zarnecki, John
Zdziarski, Andrzej
Zezas, Andreas
ZHANG, William
ZHANG, Jialu
ZHANG, Shuang Nan
ZHANG, Li
ZHANG, JiangShui
ZHANG, Jingyi
ZHANG, You-Hong
ZHANG, Yanxia
ZHANG, Jie
ZHANG, Yuying
ZHANG, Chengmin
ZHANG, Shu
ZHANG, Zhibin
ZHENG, Wei
ZHENG, Xiaoping
ZHOU, Jianfeng
ZHOU, Xia
Zombeck, Martin
Zwintz, Konstanze

Division IV Commission 45 Stellar Classification

President: Richard O. Gray
Vice-President: Birgitta Nordström

Organizing Committee Members:

Burgasser, Adam
Eyer, Laurent
Gupta, Ranjan
Hanson, Margaret
Irwin, Michael
Soubiran, Caroline

Members:

AK, Serap
Allende Prieto, Carlos
Ardeberg, Arne
Arellano Ferro, Armando
Babu, G.S.D.
Baglin, Annie
Bartkevicius, Antanas
Bear, Ealeal
Buchhave, Lars
Buser, Roland
Cherepashchuk, Anatolij
Christy, James
Claria, Juan
Corral, Luis
Cowley, Anne
Crawford, David
Creech-Eakman, Michelle
Dal Ri Barbosa, Cassio
Drilling, John
Eglitis, Ilgmars
Egret, Daniel
Eyer, Laurent
Faraggiana, Rosanna
Feast, Michael
Feltzing, Sofia
Fitzpatrick, Edward
FUKUDA, Ichiro
Gamen, Roberto
Garcia, Miriam
Garmany, Katy
Garrison, Robert
Gerbaldi, Michele
Geyer, Edward
Giorgi, Edgard
Giridhar, Sunetra
Gizis, John
Glagolevskij, Yurij
Golay, Marcel
Goswami, Aruna
Grenon, Michel
Grosso, Monica
Guetter, Harry
Gupta, Ranjan
Hanson, Margaret
Hauck, Bernard
Houk, Nancy
Humphreys, Roberta
Irwin, Michael
KATO, Ken-ichi
Kurtanidze, Omar
Kurtz, Michael
Kurtz, Donald
Lasala Jr., Gerald
Lattanzio, John
Laugalys, Vygandas
LEE, Sang-Gak
Lepine, Sebastien
Levato, Orlando
LI, Jinzeng
Lobel, Alex
Lu, Phillip
LUO, Ali
Luri, Xavier
Lutz, Julie
MacConnell, Darrell
MAEHARA, Hideo
Maiz Apellaniz, Jesús
Malagnini, Maria
Malaroda, Stella
Mendoza, V.
Montes, David
Morossi, Carlo
Morrell, Nidia
Nicolet, Bernard
North, Pierre
Notni, Peter
Oja, Tarmo
Olsen, Erik
Osborn, Wayne
Oswalt, Terry
Pakhomov, Yury
Paletou, Frédéric
Parsons, Sidney
Philip, A.G.
Pizzichini, Graziella
Preston, George
Prsa, Andrej
Pulone, Luigi
Reid, Warren
Rizzi, Luca
Roman, Nancy
Rountree, Janet
Schild, Rudolph
Shore, Steven
Shvelidze, Teimuraz
Sion, Edward
Smith, J.
Steinlin, Uli
Straizys, Vytautas
Strobel, Andrzej
TAKAHASHI, Hidenori
Ubeda, Leonardo
Ueta, Toshiya
Upgren, Arthur
von Hippel, Theodore
Walborn, Nolan
Walker, Gordon
Warren Jr, Wayne

Weaver, William
Weiss, Werner
Williams, John
Wing, Robert
Wright, Nicholas
WU, Hsin-Heng
Wyckoff, Susan
YAMASHITA, Yasumasa
YUMIKO, Oasa
Zdanavičius, Kazimeras
Zdanavičius, Justas
ZHANG, Yanxia

Division XII Commission 46 Astronomy Education & Development

President: Rosa M. Ros
Vice-President: John B. Hearnshaw

Organizing Committee Members:

Balkowski-Mauger, Chantal
de Greve, Jean-Pierre
Deustua, Susana
Garcia, Beatriz
Gerbaldi, Michele
Guinan, Edward
Haubold, Hans
Kochhar, Rajesh
Malasan, Hakim
Marschall, Laurence
Metaxa, Margarita
Miley, George
Morrell, Nidia
Pasachoff, Jay
Percy, John
Tolbert, Charles
Torres-Peimbert, Silvia
White II, James
Urban, Sean

Members:

Acker, Agnes
Aghaee, Alireza
Aguilar, Maria
Airapetian, Vladimir
Al-Naimiy, Hamid
Albanese, Lara
Alexandrov, Yuri
Alonso-Herrero, Almudena
Alsabti, Abdul Athem
Alvarez, Rodrigo
Alvarez-Pomares, Oscar
Alves, Virgínia
Anandaram, Mandayam
Andersen, Johannes
Andrews, Frank
Ansari, S.M.
Arbutina, Bojan
Arcidiacono, Carmelo
Arellano Ferro, Armando
Arion, Douglas
Asanok, Kitiyanee
Aslan, Zeki
Aubier, Monique
Babu, G.S.D.
Badescu, Octavian
BAEK, Chang Hyun
Bailey, Katherine
Barclay, Charles
BARET, Bruny
Barlow, Nadine
Barrantes, Marco
Barthel, Peter
Baskill, Darren
Batten, Alan
Berger, Jean-Philippe
Bernabeu, Guillermo
Birlan, Mirel
Bobrowsky, Matthew
Bojurova, Eva
Booth, Roy
Botez, Elvira
Braes, Lucien
Brammer, Gabriel
Bretones, Paulo
Brieva, Eduardo
Brosch, Noah
Budding, Edwin
Cabanac, Rémi
CAI, Michael
Calvet, Nuria
Cannon, Wayne
Capaccioli, Massimo
Caretta, Cesar
Carter, Brian
Cassan, Arnaud
Celebre, Cynthia
Chakrabarti, Supriya
Chamcham, Khalil
CHEN, An-Le
CHEN, Alfred
CHEN, Lin-wen
CHEN, Xinyang
CHEN, Dongni
Chernyakova, Maria
Christensen, Lars
Christlieb, Norbert
Chung, Eduardo
Ciroi, Stefano
Clarke, David
Coffey, Deirdre
Cohen, David
Colafrancesco, Sergio
Conti, Alberto
Cora, Alberto
Corbally, Christopher
Cottrell, Peter
Couper, Heather
Courtois, Hélène
Couto da Silva, Telma
Covone, Giovanni
Cracco, Valentina
Crawford, David
CUI, Zhenhua
Cui, Wei
Cunningham, Maria
Dall'Ora, Massimo
Daniel, Jean-Yves
Danner, Rolf
Darhmaoui, Hassane
Darriulat, Pierre
de Swardt, Bonita
Del Santo, Melania
Delsanti, Audrey

Demircan, Osman
DeNisco, Kenneth
Devaney, Martin
Diego, Francisco
Dole, Herve
Donahue, Megan
Doran, Rosa
Ducati, Jorge
Dukes Jr., Robert
Duval, Marie-France
Dworetsky, Michael
Eastwood, Kathleen
El Eid, Mounib
Esguerra, Jose Perico
Eze, Romanus
Falceta-Goncalves, Diego
Feitzinger, Johannes
Fernandez, Julio
Fernandez-Figueroa, M.
Ferrusca, Daniel
Fienberg, Richard
Fierro, Julieta
Figueiredo, Newton
Fleck, Robert
Floyd, David
Forbes, Douglas
Forero Villao, Vicente
Frew, David
FU, Hsieh-Hai
Gabriel, Carlos
Gallino, Roberto
Gangui, Alejandro
Ganguly, Rajib
Gasiprong, Nipon
Gavrilov, Mikhail
Gay, Pamela
Gill, Peter
Gimenez, Alvaro
Gingerich, Owen
Girard, Julien
Gomez, Edward
Govender, Kevindran
Gray, Richard
Gregorio, Anna
Gregorio-Hetem, Jane
Grundstrom, Erika
Guglielmino, Salvatore
Haque, Shirin
Haubold, Hans
Havlen, Robert
Hemenway, Mary
Heudier, Jean-Louis
Heydari-Malayeri, M.
Hicks, Amalia
Hidayat, Bambang
Hobbs, George
Hockey, Thomas
Hoenig, Sebastian
Hollow, Robert
Horn, Martin
Hotan, Aidan
Houziaux, Leo
HSU, Rue-Ron
Huan, Nguyen
Huertas-Company, Marc
Huettemeister, Susanne
Hughes, Stephen
Ibrahim, Alaa
Impey, Christopher
Inglis, Michael
ISHIZAKA, Chiharu
Izzard, Robert
Jafelice, Luiz
Johnston-Hollitt, Melanie
Jubier, Xavier
Kablak, Nataliya
Kalemci, Emrah
Karetnikov, Valentin
Karttunen, Hannu
Kay, Laura
Keeney, Brian
Keller, Hans-Ulrich
Khan, J
Khodachenko, Maxim
Kiasatpour, Ahmad
Kikwaya Eluo, J.-B.
Kim, Yoo Jea
Klinglesmith III, Daniel
Knight, Matthew
Koechlin, Laurent
Kolenberg, Katrien
Kolka, Indrek
Kolomiyets, Svitlana
Komonjinda, Siramas
KONG, Xu
Kotulla, Ralf
Kouwenhoven, M.B.N.
KOZAI, Yoshihide
Kramer, Busaba
Kreiner, Jerzy
Krishna, Gopal
Krupp, Edwin
KUAN, Yi-Jehng
Kuiper, Rolf
Lago, Maria
Lai, Sebastiana
Lanciano, Nicoletta
Lanfranchi, Gustavo
Las Vergnas, Olivier
LEE, Kang Hwan
LEE, Yong Bok
Letarte, Bruno
Leung, Kam
Levato, Orlando
LI, Min
LIN, Weipeng
LIN, Chuang-Jia
Linden-Vørnle, Michael
Little-Marenin, Irene
LIU, Xiaoqun
Lomb, Nicholas
Lopes de Oliveira, R.
Loubser, Ilani
Lowenthal, James
Luck, John
Maciel, Walter
Maddison, Ronald
Madjarska, Maria
Madsen, Claus
Mahoney, Terence
Malasan, Hakim
Mallamaci, Claudio
Mancini, Dario
Marco, Olivier
Marshalov, Dmitriy
Martinet, Louis
Martinez, Peter
Martinez-Bravo, Oscar
Massey, Robert
Maza, José
Mazumdar, Anwesh
McKinnon, David
McNally, Derek
Meidav, Meir
Melbourne, Jason
Merlo, David
Meyer, Michael
MIZUNO, Takao
Montgomery, Michele
Moreels, Guy

Murphy, John
Najid, Nour-Eddine
Nammahachak, Suwit
Nandi, Dibyendu
Narlikar, Jayant
Navone, Hugo
Nayar, S.R.Prabhakaran
NGUYEN, Phuong
NGUYEN, Khanh
NGUYEN, Lan
Nguyen-Quang, Rieu
NHA, Il Seong
Nicolson, Iain
Ninkovic, Slobodan
Noels, Arlette
Norton, Andrew
Nymark, Tanja
Oberst, Thomas
Odman, Carolina
Odonoghue, Aileen
Oja, Heikki
Okeke, Pius
Olive, Don
Olsen, Hans
Oluseyi, Hakeem
Oozeer, Nadeem
Orchiston, Wayne
Ortiz Gil, Amelia
Osborn, Wayne
Osorio, José
Oswalt, Terry
Özeren, Ferhat
Pandey, Uma
Pantoja, Carmen
Parenti, Susanna
Pat-El, Igal
Penston, Margaret
Percy, John
Perez-Torres, Miguel
Perozzi, Ettore
Perrin, Marshall
PHAM, Diep
PHAN, Dong
Picazzio, Enos
Pompea, Stephen
Popov, Sergey
Porras Juárez, Bertha
Povic, Mirjana
Price, Charles
Proverbio, Edoardo
Pustilnik, Lev
Radeva, Veselka
Ramadurai, Souriraja
Rassat, Anais
Ravindranath, Swara
Reboul, Henri
Reid, Michael
Rekola, Rami
Reyes, Reinabelle
Rijsdijk, Case
Roberts, Morton
Roca Cortes, Teodoro
Rojas, Gustavo
Rosa Gonzalez, Daniel
Rosenzweig-Levy, Patrica
Routly, Paul
Rozanska , Agata
Rubio-Herrera, Eduardo
Russo, Pedro
Sabra, Bassem
Saenz, Eduardo
Saffari, Reza
Safko, John
Samodurov, Vladimir
Sampson, Russell
Sanchez-Blazquez, Patricia
Sandqvist, Aage
Sandrelli, Stefano
Saraiva, Maria de Fatima
Sattarov, Isroil
Saucedo Morales, Julio
Sawicki, Marcin
Schleicher, David
Schroeder, Daniel
Seaton, Daniel
Seeds, Michael
Sese, Rogel Mari
Shelton, Ian
Shipman, Harry
Sigismondi, Costantino
SIHER, El Arbi
Simons, Douglas
Slater, Timothy
Smail, Ian
Smith, Francis
Sobreira, Paulo
Solheim, Jan
Soriano, Bernardo
Stachowski, Grzegorz
Stefl, Vladimir
Stenholm, Björn
Stoev, Alexey
Stonkut?, Rima
Straizys, Vytautas
Strubbe, Linda
Svestka, Jiri
Swarup, Govind
Tessema, Solomon
Tignalli, Horacio
Torres, Jesus Rodrigo
Touma, Jihad
TRAN, Ha
Trinidad, Miguel
Tugay, Anatoliy
Ubeda, Leonardo
UEDA, Haruhiko
Ueta, Toshiya
Ugolnikov, Oleg
Ulla Miguel, Ana
Unda-Sanzana, Eduardo
Urama, Johnson
Valdes Parra, Jose
van den Heuvel, Edward
van Santvoort, Jacques
Vauclair, Sylvie
Verma, Aprajita
Vierdayanti, Kiki
Vilinga, Jaime
Vilks, Ilgonis
Villar Martin, Montserrat
Vinuales Gavin, Ederlinda
Voelzke, Marcos
Vujnovic, Vladis
Wadadekar, Yogesh
Walker, Constance
Walsh, Wilfred
WANG, Shouguan
Ward, Richard
West, Richard
Whelan, Emma
Whitelock, Patricia
Williamon, Richard
Willmore, A.
Winter, Lisa
Wittenmyer, Robert
Wolfschmidt, Gudrun
XIE, Xianchun
XIONG, Jianning
Yair, Yoav
YE, Shuhua

YIM, Hong Suh
YUMIKO , Oasa
Zadnik, Marjan
Zakirov, Mamnum
Zealey, William
Zeilik, Michael
ZHANG, You-Hong
ZHANG, Yong
ZHANG, Yang
ZHAO, Jun Liang
ZHENG, Xiaonian

Division VIII Commission 47 Cosmology

President: Thanu Padmanabhan
Vice-President: Brian P. Schmidt

Organizing Committee Members:

Bunker, Andrew
Campusano, Luis
Charlot, Stephane
Ciardi, Benedetta
da Costa, Luiz
JING, Yipeng
Koekemoer, Anton
Koo, David
Lahav, Ofer
Le Fèvre, Olivier
Scott, Douglas
SUTO, Yasushi

Members:

Abbas, Ummi
Abu Kassim, Hasan
Adami, Christophe
Adams, Jenni
Aghaee, Alireza
Aharonian, Felix
Ajhar, Edward
Akashi, Muhammad
Alard, Christophe
Alcaniz, Jailson
Alimi, Jean-Michel
Allan, Peter
Allington-Smith, Jeremy
Almaini, Omar
Amendola, Luca
Ammons, Stephen
Andersen, Michael
Andreani, Paola
Aretxaga, Itziar
Argueso, Francisco
Atrio Barandela, Fernando
Audouze, Jean
Aussel, Hervé
Avelino, Pedro
AZUMA, Takahiro
Babul, Arif
Baddiley, Christopher
Bagla, Jasjeet
Bahcall, Neta
Bajtlik, Stanislaw
Baker, Andrew
Balbi, Amedeo
Balland, Christophe
Bamford, Steven
Banday, Anthony
Banhatti, Dilip
Bannikova, Elena
Barberis, Bruno
Barbuy, Beatriz
Bardeen, James
Bardelli, Sandro
Barger, Amy
Barkana, Rennan
Barkhouse, Wayne
Barrow, John
Bartelmann, Matthias
Barthel, Peter
Barton, Elizabeth
Baryshev, Andrey
Basa, Stéphane
Bassett, Bruce
Basu, Kaustuv
Battye, Richard
Bechtold, Jill
Beckman, John
Beesham, Aroonkumar
Behar, Ehud
Belinski, Vladimir
Bennett, Charles
Bennett, David
Bergeron, Jacqueline
Bergvall, Nils
Berta, Stefano
Bertola, Francesco
Bertschinger, Edmund
Betancor Rijo, Juan
Bharadwaj, Somnath
Bhavsar, Suketu
Bianchi, Simone
Bicknell, Geoffrey
Bignami, Giovanni
Binetruy, Pierre
Birkinshaw, Mark
Biviano, Andrea
Bjaelde, Ole
Blakeslee, John
Blanchard, Alain
Bleyer, Ulrich
Bludman, Sidney
Blundell, Katherine
Boehringer, Hans
Boksenberg, Alec
Bolzonella, Micol
Bond, John
Bongiovanni, Angel
Boquien, Médéric
Borgani, Stefano
Boschin, Walter
Bouchet, François
Bouwens, Rychard
Bowen, David
Boyle, Brian
Branchesi, Marica
Brecher, Kenneth
Bridle, Sarah
Brinchmann, Jarle
Brough, Sarah
Brown, Michael
Brunner, Robert
Bryant, Julia
Buote, David
Burikham, Piyabut

Burns, Jack
Cabanac, Rémi
CAI, Mingsheng
Calvani, Massimo
CAO, Li
Cappi, Alberto
Caputi, Karina
Cardenas, Rolando
Carr, Bernard
Carretti, Ettore
Cassano, Rossella
Castagnino, Mario
Cattaneo , Andrea
Cavaliere, Alfonso
Cesarsky, Diego
CHAE, Kyu Hyun
CHANG, Kyongae
CHANG, Heon-Young
Chelliah Subramonian, S.
CHEN, DaMing
CHEN, Jiansheng
CHEN, Hsiao-Wen
CHEN, Lin-wen
CHEN, Pisin
CHEN, Xuelei
CHEN, Dongni
CHENG, Fuzhen
Chiang, Hsin
CHIBA, Takeshi
Chincarini, Guido
Chodorowski, Michal
Choudhury, Tirthankar
Christensen, Lise
CHU, Yaoquan
Ciliegi, Paolo
Claeskens, Jean-François
Claria, Juan
Clarke, Tracy
Clarkson, Chris
Clocchiatti, Alejandro
Clowe, Douglas
Clowes, Roger
Cocke, William
Cohen, Ross
Colafrancesco, Sergio
Cole, Shaun
Coles, Peter
Colless, Matthew
Colombi, Stephane
Condon, James
Conti, Alberto
Cooray, Asantha
Copin, Yannick
Cora, Sofia
Corsini, Enrico
Courbin, Frederic
Courteau, Stephane
Courtois, Hélène
Covone, Giovanni
Crane, Patrick
Crane, Philippe
Crawford, Steven
Cristiani, Stefano
Croom, Scott
Crosta, Mariateresa
Croton, Darren
Csabai, Istvan
Cui, Wei
Curran, Stephen
Cypriano, Eduardo
D'Odorico, Valentina
Da Costa, Gary
Da Rocha, Cristiano
Dadhich, Naresh
Dahle, Haakon
Daigne, Frederic
DAISUKE, Iono
Dalla Bonta, Elena
Danese, Luigi
Dannerbauer, Helmut
Das, P.
Davidson, William
Davies, Paul
Davies, Roger
Davis, Michael
Davis, Marc
Davis, Tamara
De Bernardis, Paolo
de Lapparent, Valérie
de Lima, José
De Lucia, Gabriella
de Petris, Marco
de Ruiter, Hans
de Silva, Lindamulage
de Ugarte Postigo, Antonio
de Zotti, Gianfranco
Dekel, Avishai
Del Popolo, Antonino
Dell'Antonio, Ian
Demarco, Ricardo
Demianski, Marek
Desert, François-Xavier
Deustua, Susana
Dhurandhar, Sanjeev
Diaferio, Antonaldo
Diaz, Ruben
Diaz-Santos, Tanio
Dietrich , Jörg
Djorgovski, Stanislav
Dobbs, Matt
Dobrzycki, Adam
Dole, Hervé
Dominguez, Mariano
DONG, Xiao-Bo
Dressler, Alan
Drinkwater, Michael
Dultzin-Hacyan, Deborah
Dunlop, James
Dunsby, Peter
Dyer, Charles
Eales, Stephen
Easther, Richard
Edsjo, Joakim
Efstathiou, George
Einasto, Jaan
Elgaroy, Oystein
Elizalde, Emilio
Ellis, George
Ellis, Richard
Ellison, Sara
Elvis, Martin
Elyiv, Andrii
Enginol, Turan
Ettori, Stefano
Eungwanichayapant, Anant
Faber, Sandra
Fall, S.
FAN, Zuhui
Fassnacht, Christopher
Fatkhullin, Timur
Faure, Cécile
Fedeli, Cosimo
Fedorova, Elena
FENG, Long Long
Ferreira, Pedro
Field, George
Figueiredo, Newton
Filippenko, Alexei
Florides, Petros
Focardi, Paola

Fong, Richard
Ford, Holland
Forman, William
Foucaud, Sébastien
Fouqué, Pascal
Fox, Andrew
Franceschini, Alberto
Frenk, Carlos
Friaca, Amancio
Frutos-Alfaro, Francisco
FUKUGITA, Masataka
FUKUI, Takao
Furlanetto, Steven
Fuzfa, Andre
Fynbo, Johan
Gallazzi, Anna
Gangui, Alejandro
Ganguly, Rajib
Garilli, Bianca
Garrison, Robert
Gastaldello, Fabio
Gavignaud, Isabelle
Geller, Margaret
GENG, Lihong
Ghirlanda, Giancarlo
Giallongo, Emanuele
Gilbank, David
Gioia, Isabella
Gitti, Myriam
Glazebrook, Karl
Glover , Simon
Goicoechea, Luis
Goldsmith, Donald
GONG, Biping
Gonzalez Sanchez, A.
Goobar, Ariel
Gordon, Chris
Goret, Philippe
Gosset, Eric
Gottloeber, Stefan
GOUDA, Naoteru
Govinder, Keshlan
Granato, Gian Luigi
Gray, Richard
Gray, Meghan
Green, Anne
Gregorio, Anna
Greve, Thomas
Greyber, Howard
Griest, Kim
Gudmundsson, Einar
Gumjudpai, Burin
Gunn, James
Gutierrez, Carlos
Guzzo, Luigi
Haas, Martin
Haehnelt, Martin
Hagen, Hans-Juergen
Hall, Patrick
Hamilton, Andrew
Hanna, David
Hannestad, Steen
Hansen, Frode
Hardy, Eduardo
Harms, Richard
Hau, George
Haugboelle, Troels
Hawking, Stephen
Hayes, Matthew
HAZUMI, Masashi
HE, XiangTao
Heavens, Alan
Heinamaki, Pekka
Hellaby, Charles
Heller, Michael
Hendry, Martin
Henriksen, Mark
Hernandez, Xavier
Hernandez-Monteagudo, C.
Hervik, Sigbjorn
Hewett, Paul
Heyrovsky, David
Hicks, Amalia
HIDEKI, Asada
Hirv, Anti
Hnatyk, Bohdan
Hoekstra, Hendrik
Holz, Daniel
Hu, Esther
HU, Hongbo
Huang, Jiasheng
Hudson, Michael
Huertas-Company, Marc
Hughes, David
Hutsi, Gert
Huynh, Minh
HWANG, Jai-chan
HWANG, Chorng-Yuan
Ibata, Rodrigo
Icke, Vincent
IKEUCHI, Satoru
Iliev, Ilian
IM, Myungshin
Impey, Christopher
INADA, Naohisa
Iovino, Angela
ISHIHARA, Hideki
Ivanov, Pavel
IWATA, Ikuru
Iyer, Balasubramanian
Jahnke, Knud
Jakobsson, Pall
Jannuzi, Buell
Jaroszynski, Michal
Jauncey, David
Jaunsen, Andreas
Jedamzik, Karsten
Jelic, Vibor
Jensen, Brian
Jetzer, Philippe
Jha, Saurabh
Johnston-Hollitt, Melanie
Jones, Bernard
Jones, Heath
Jones, Christine
Jordan, Andrés
Jovanovic, Predrag
Junkkarinen, Vesa
Kaisin, Serafim
KAJINO, Toshitaka
Kaminker, Alexander
Kanekar, Nissim
KANG, Hyesung
KANG, Xi
Kapoor, Ramesh
Karachentsev, Igor
Karami, Kayoomars
Karouzos, Marios
KATO, Shoji
Kauffmann, Guinevere
Kaul, Chaman
Kausch, Wolfgang
Kaviraj, Sugata
KAWABATA, Kiyoshi
KAWASAKI, Masahiro
KAYO, Issha
Kellermann, Kenneth
Kembhavi, Ajit
Keshet, Uri
Khare, Pushpa

Khmil, Sergiy
KIM, Jik
King, Lindsay
Kirilova, Daniela
KIYOTOMO, Ichiki
Kneib, Jean-Paul
KOBAYASHI, Masakazu
KODAMA, Hideo
Koivisto, Tomi
Kokkotas, Konstantinos
Kolb, Edward
Kompaneets, Dmitrij
Koopmans, Leon
Kormendy, John
Kovalev, Yuri
Kovetz, Attay
KOZAI, Yoshihide
Kozlovsky, Ben
Kozlowski, Szymon
Krasinski, Andrzej
Kriss, Gerard
Kristiansen, Jostein
Kudrya, Yury
Kunth, Daniel
Kunz, Martin
KUSAKABE, Motohiko
La Barbera, Francesco
La Franca, Fabio
Lacey, Cédric
Lachièze-Rey, Marc
Lagache, Guilaine
Lake, Kayll
Lake, George
Lanfranchi, Gustavo
Lapi, Andrea
Larionov, Mikhail
Lasota-Hirszowicz, J.-P.
Layzer, David
Leao, Joao Rodrigo
LEE, Wo-Lung
Lehnert, Matthew
Lequeux, James
Leubner, Manfred
Levin, Yuri
Levine, Robyn
Lewis, Geraint
LI, Li-Xin
LIAN, Luo
LIANG, Yanchun
Liddle, Andrew
Liebscher, Dierck-E
Lilje, Per
Lilly, Simon
LIN, Weipeng
LIN, Yen-Ting
LIOU, Guo Chin
LIU, Yongzhen
Lokas, Ewa
Lombardi, Marco
Longair, Malcolm
Longo, Giuseppe
Lonsdale, Carol
Lopes, Paulo
Lopez, Sebastian
Lopez-Corredoira, Martin
Loveday, Jon
Lowenthal, James
LU, Tan
LU, Youjun
Lubin, Lori
Lukash, Vladimir
Luminet, Jean-Pierre
Lynden-Bell, Donald
MA, Jun
Maartens, Roy
Maccagni, Dario
MacCallum, Malcolm
Mackey, Alasdair
Maddox, Stephen
MAEDA, Kei-ichi
Maguire, Kate
Maharaj, Sunil
Maia, Marcio
Maier, Christian
Mainieri, Vincenzo
Majumdar, Subhabrata
Malesani, Daniele
Mamon, Gary
Mandolesi, Nazzareno
Mangalam, Arun
Mann, Robert
Manrique, Alberto
Mansouri, Reza
Mao, Shude
Maoz, Dan
Marano, Bruno
Mardirossian, Fabio
Marek, John
Maris, Michele
Marleau, Francine
Marranghello, Guilherme
Martinez-Gonzalez, E.
Martinis, Mladen
Martins, Carlos
Masters, Karen
Mather, John
MATSUMOTO, Toshio
MATSUMURA, Tomotake
Matzner, Richard
McCracken, Henry
McKean, John
Mellier, Yannick
Melnyk, Olga
Melott, Adrian
Meneghetti, Massimo
Menendez-Delmestre, Karin
Merighi, Roberto
Meszaros, Peter
Meszaros, Attila
Meyer, David
Meyer, Martin
Meylan, Georges
Meza, Andres
Mezzetti, Marino
Miralda-Escude, Jordi
Miralles, Joan-Marc
Miranda, Oswaldo
Misner, Charles
Miyazaki, Satoshi
Miyoshi, Shigeru
Mo, Houjun
Mohr, Joseph
Molla, Mercedes
Monaco, Pierluigi
Moodley, Kavilan
Moore, Ben
Moreau, Olivier
MOROKUMA, Tomoki
Mortsell, Edvard
Moscardini, Lauro
Mota, David
Motta, Veronica
Mourao, Ana Maria
Muanwong, Orrarujee
Muecket, Jan
Mueller, Andreas
Mukhopadhyay, Banibrata
Muller, Richard
MURAKAMI, Izumi
MURAYAMA, Hitoshi

Murphy, Michael
Murphy, John
NAKAMICHI, Akika
NAMBU, Yasusada
Narasimha, Delampady
Nardetto, Nicolas
Narlikar, Jayant
Naselsky, Pavel
Nasr-Esfahani, Bahram
Neves de Araujo, Jose
NGUYEN, Lan
Nicoll, Jeffrey
Noerdlinger, Peter
NOH, Hyerim
Norman, Colin
Norman, Dara
Noterdaeme, Pasquier
Nottale, Laurent
Novikov, Igor
Novosyadlyj, Bohdan
Novotny, Jan
Nozari, Kourosh
Nuza, Sebastian
O'Connell, Robert
Ocvirk, Pierre
Oemler Jr, Augustus
Ogando, Ricardo
Oliver, Sebastian
Olowin, Ronald
Oscoz, Alejandro
Ostorero, Luisa
OTA, Naomi
OYA, Shin
Ozsvath, Istvan
Page, Don
Paragi, Zsolt
Parnovsky, Sergei
Partridge, Robert
Pecker, Jean-Claude
Pedersen, Kristian
Pedrosa, Susana
Peebles, P.
Peel, Michael
Pello, Roser
Pen, Ue-Li
Pentericci, Laura
Penzias, Arno
Perez-Garrido, Antonio
Peroux, Céline
Perryman, Michael
Persides, Sotirios
Persson, Carina
Peterson, Bruce
Peterson, Bradley
Petitjean, Patrick
Petrosian, Vahe
PHAM, Diep
Pimbblet, Kevin
Pitkin, Matthew
Plionis, Manolis
Podolsky, Jiri
Polletta, Maria del Carmen
Pompei, Emanuela
Popescu, Nedelia
Portinari, Laura
Power, Chris
Poznanski, Dovi
Prandoni, Isabella
Premadi, Premana
Press, William
Puetzfeld, Dirk
Puget, Jean-Loup
Puy, Denis
QIN, Bo
Rahvar, Sohrab
Ramella, Massimo
Ranalli, Piero
Rassat, Anais
Rauch, Michael
Ravindranath, Swara
Read, Justin
Rebolo, Rafael
Reboul, Henri
Rees, Martin
Reeves, Hubert
Reiprich, Thomas
Reisenegger, Andreas
Reitze, David
Revnivtsev, Mikhail
Rey, Soo-Chang
Reyes, Reinabelle
Rhodes, Jason
Riazi, Nematollah
Riazuelo, Alain
Ribeiro, Marcelo
Ribeiro, André Luis
Richard, Johan
Richter, Philipp
Ricker, Paul
Ricotti, Massimo
Ridgway, Susan
Rindler, Wolfgang
Roberts, David
Rocca-Volmerange, Brigitte
Roeder, Robert
Romano-Diaz, Emilio
Romeo, Alessandro
Romer, Anita
Rosa Gonzalez, Daniel
Rosquist, Kjell
Rothberg, Barry
Rottgering, Huub
Rowan-Robinson, Michael
Roxburgh, Ian
Rubin, Vera
Rubino-Martin, J. A.
Rudnick, Lawrence
Rudnicki, Konrad
Ruffini, Remo
Ruszkowski, Mateusz
Ruzicka, Adam
Saar, Enn
Saffari, Reza
Sahlen, Martin
Sahni, Varun
Saiz, Alejandro
Salvador-Sole, Eduardo
Salzer, John
Santos-Lleo, Maria
Sapar, Arved
Saracco, Paolo
SASAKI, Misao
SASAKI, Shin
SATO, Shinji
SATO, Katsuhiko
SATO, Humitaka
SATO, Jun'ichi
Saviane, Ivo
Sawicki, Marcin
Sazhin, Mikhail
Scaramella, Roberto
Schartel, Norbert
Schaye, Joop
Schechter, Paul
Schindler, Sabine
Schmidt, Maarten
Schneider, Donald
Schneider, Jean
Schneider, Peter
Schneider, Raffaella

Schramm, Thomas
Schuch, Nelson
Schucking, Engelbert
Scodeggio, Marco
Seielstad, George
Semerak, Oldrich
Sergijenko, Olga
Serjeant, Stephen
SETO, Naoki
Setti, Giancarlo
Severgnini, Paola
Seymour, Nicholas
Shandarin, Sergei
Shanks, Thomas
SHAO, Zhengyi
Sharp, Nigel
Shaver, Peter
Shaviv, Giora
Shaya, Edward
SHIBATA, Masaru
Shimon, Meir
SHIRAHATA, Mai
SHIRASAKI, Yuji
Siebenmorgen, Ralf
Signore, Monique
Silk, Joseph
Silva, Laura
Siringo, Giorgio
Sistero, Roberto
Smail, Ian
Smette, Alain
Smith, Rodney
Smith, Nigel
Smoot III, George
Sokolowski, Lech
Sollerman, Jesper
SONG, Doo-Jong
SONG, Yong-Seon
Souradeep, Tarun
Spinoglio, Luigi
Spyrou, Nicolaos
Squires, Gordon
Srianand, Raghunathan
Stacy, Athena
Stadel, Joachim
Stecker, Floyd
Steigman, Gary
Stoehr, Felix
Stolyarov, Vladislav
Storrie-Lombardi, Lisa
Stott, John
Straumann, Norbert
Strauss, Michael
Stritzinger, Maximilian
Struble, Mitchell
Strukov, Igor
Stuchlik, Zdenek
Stuik, Remko
Subrahmanya, C.
SUGINOHARA, Tatsushi
SUGIYAMA, Naoshi
Suhhonenko, Ivan
Sullivan, Mark
Sunyaev, Rashid
Surdej, Jean
SUSA, Hajime
Sutherland, William
Szalay, Alex
Szydlowski, Marek
TAGOSHI, Hideyuki
TAKADA, Masahiro
TAKAHARA, Fumio
Tammann, Gustav
TANABE, Kenji
TANAKA, Masayuki
Tarter, Jill
TARUYA, Atsushi
TATEKAWA, Takayuki
Taylor, Angela
Tempel, Elmo
Temporin, Sonia
Tepper Garcia, Thorsten
Teyssier, Romain
Thanjavur, Karunananth
Thoene, Christina
Thuan, Trinh
Tifft, William
Tipler, Frank
Toffolatti, Luigi
Toft, Sune
TOMIMATSU, Akira
TOMITA, Kenji
Tonry, John
Tormen, Giuseppe
TOTANI, Tomonori
Tozzi, Paolo
Tremaine, Scott
Trenti, Michele
Tresse, Laurence
Treu, Tommaso
Trevese, Dario
Trimble, Virginia
Trotta, Roberto
Trujillo Cabrera, Ignacio
Tsamparlis, Michael
Tugay, Anatoliy
Tully, Richard
Turner, Edwin
Turner, Michael
Turnshek, David
Tyson, John
Tytler, David
Tyul'bashev, Sergei
UEDA, Haruhiko
Ugolnikov, Oleg
UMEMURA, Masayuki
Uson, Juan
Vaccari, Mattia
Vagnetti, Fausto
Vaisanen, Petri
Valdivielso, Luisa
Valiviita, Jussi-Pekka
Valls-Gabaud, David
Valluri, Monica
Valtchanov, Ivan
van der Laan, Harry
van Eymeren, Janine
van Haarlem, Michiel
van Kampen, Eelco
Vedel, Henrik
Vega, Olga
Venters, Tonia
Verdoes Kleijn, Gijsbert
Vergani, Daniela
Verma, Aprajita
Vestergaard, Marianne
Vettolani, Giampaolo
Viana, Pedro
Viel, Matteo
Vishniac, Ethan
Vishveshwara, C.
Voit, Gerard
Volonteri, Marta
von Borzeszkowski, H.
Waddington, Ian
Wagner, Robert
Wagoner, Robert
Wainwright, John
Wambsganss, Joachim
Wanas, Mamdouh

WANG, Huiyuan
WANG, Yu
Watson, Darach
Webb, Tracy
Webster, Adrian
Weilbacher, Peter
Weinberg, Steven
Wendt, Martin
Wesson, Paul
West, Michael
White, Simon
Whiting, Alan
Widrow, Lawrence
Will, Clifford
Willis, Jon
Wilson, Gillian
Windhorst, Rogier
Wold, Margrethe
Woltjer, Lodewijk
WOO, Jong-Hak
Woszczyna, Andrzej
Wright, Edward
WU, Xiangping
WU, Jiun-Huei
WU, Wentao
WU, Jianghua
WU, Xue-Feng
Wucknitz, Olaf
Wulandari, Hesti
Wyithe, Stuart
XIANG, Shouping
XU, Dawei
YAMAMOTO, Hiroaki
YASUDA, Naoki
Yesilyurt, Ibrahim
YI, Sukyoung
YOICHIRO, Suzuki
YOKOYAMA, Jun'ichi
YONETOKU, Daisuke
YOSHIDA, Hiroshi
YOSHII, Yuzuru
YOSHIKAWA, Kohji
YOSHIOKA, Satoshi
YU, Qingjuan
Yushchenko, Alexander
Zacchei, Andrea
Zamorani, Giovanni
Zanichelli, Alessandra
Zannoni, Mario
Zaroubi, Saleem
ZHANG, Tong-Jie
ZHANG, Jialu
Zhang, Yuying
ZHAO, Donghai
ZHOU, Youyuan
ZHOU, Hongyan
ZHU, Xingfeng
Zhuk, Alexander
Zibetti, Stefano
Zieba, Stanislaw
Zirm, Andrew
ZOU, Zhenlong
Zucca, Elena

Division II Commission 49 Interplanetary Plasma & Heliosphere

President: Natchimuthuk Gopalswamy
Vice-President: Ingrid Mann, P. K. Manoharan

Organizing Committee Members:

Briand, Carine
Lallement, Rosine
Lario, David
SHIBATA, Kazunari
Webb, David

Members:

Ahluwalia, Harjit
Andretta, Vincenzo
Balikhin, Michael
Banerjee, Dipankar
Barnes, Aaron
Barrantes, Marco
Barrow, Colin
Barta, Miroslav
Benz, Arnold
Bertaux, Jean-Loup
Blandford, Roger
Bochsler, Peter
Bonnet, Roger
Brandt, John
Briand, Carine
Browning, Philippa
Bruno, Roberto
Burlaga, Leonard
Buti, Bimla
Cairns, Iver
Cecconi, Baptiste
Chakrabarti, Supriya
Channok, Chanruangrit
Chapman, Sandra
Chassefiere, Eric
Chitre, Shashikumar
CHOE, Gwangson
CHOU, Chih-Kang
Couturier, Pierre
Cramer, Neil
Cuperman, Sami
Daglis, Ioannis
Dalla, Silvia
Damé, Luc
Dasso, Sergio
de Jager, Cornelis
De Keyser, Johan
de Toma, Giuliana
Del Zanna, Luca
Dorotovic, Ivan
Dryer, Murray
Duldig, Marcus
Durney, Bernard
Eviatar, Aharon
Fahr, Hans
Fernandes, Francisco
Feynman, Joan
Fichtner, Horst
Field, George
Foullon, Claire
Fraenz, Markus
Fraschetti, Federico
Frutos-Alfaro, Francisco
Galvin, Antoinette
Gangadhara, R.T.
Gedalin, Michael
Gleisner, Hans
Goldman, Martin
Gosling, John
Grzedzielski, Stanislaw
Guglielmino, Salvatore
Habbal, Shadia
HAYASHI, Keiji
Heynderickx, Daniel
Hollweg, Joseph
Holzer, Thomas
Huber, Martin
Humble, John
HWANG, Junga
INAGAKI, Shogo
Ivanov, Evgenij
Jacobs, Carla
Jokipii, Jack
KAKINUMA, Takakiyo
Kasiviswanathan, S.
Keller, Horst
Khan, J.
KO, Chung-Ming
KOJIMA, Masayoshi
Kozak, Lyudmyla
Kretzschmar, Matthieu
Krittinatham, Watcharawuth
Lai, Sebastiana
Lallement, Rosine
Landi, Simone
Lapenta, Giovanni
Lario, David
Levy, Eugene
Li, Bo
LIU, Siming
Livadiotis, George
Lotova, Natalja
Luest, Reimar
Lundstedt, Henrik
MA, Guanyi
MacQueen, Robert
Malara, Francesco
Mangeney, André
Manoharan, P.
Marsch, Eckart
Mason, Glenn
Matsuura, Oscar
Mavromichalaki, Helen
Meister, Claudia
Melrose, Donald
Mestel, Leon
Moncuquet, Michel
Morabito, David

Moussas, Xenophon
MUNETOSHI, Tokumaru
Nandi, Dibyendu
Nickeler, Dieter
Nozawa, Satoshi
OH, Suyeon
Pandey, Birendra
Parenti, Susanna
Paresce, Francesco
Parhi, Shyamsundar
Parker, Eugene
Perkins, Francis
Pflug, Klaus
Podesta, John
Pustilnik, Lev
Quemerais, Eric
Raadu, Michael
Readhead, Anthony
Reinard, Alysha
Reshetnyk, Volodymyr
Riley, Pete
Ripken, Hartmut
Rosa, Reinaldo
Rosner, Robert
Roth, Ilan
Roxburgh, Ian
Ruffolo, David
Russell, Christopher
Russell, Alexander
Sagdeev, Roald
Saiz, Alejandro
Sarris, Emmanuel
Sastri, Hanumath
Sawyer, Constance
Scherb, Frank
Schindler, Karl
Schreiber, Roman
Schwartz, Steven
Seaton, Daniel
Setti, Giancarlo
Shalchi, Andreas
Shea, Margaret
SHIBATA, Kazunari
Simunac, Kristin
Smith, Dean
Srivastava, Nandita
Struminsky, Alexei
Sturrock, Peter
Suess, Steven
Tritakis, Basil
Tyul'bashev, Sergei
Usoskin, Ilya
Vainshtein, Leonid
Vandas, Marek
Verheest, Frank
Viall, Nicholeen
Vidotto, Aline
Vinod, S.
Voitsekhovska, Anna
Vucetich, Héctor
Wang, Yi-ming
WATANABE, Takashi
WATARI, Shinichi
Weller, Charles
Wood, Brian
Wu, Shi
Wu, Chin-Chun
YANG, Jing
YANG, Lei
Yesilyurt, Ibrahim
YI, Yu
Zhang, Jie
Zhang, Tielong
Zharkova, Valentina

Division XII Commission 50 Protection of Existing & Potential Observatory Sites

President: Wim van Driel
Vice-President: Richard F. Green

Organizing Committee Members:

Alvarez del Castillo, E.
Blanco, Carlo
Crawford, David
Metaxa, Margarita
OHISHI, Masatoshi
Sullivan, III, Woodruff
Tzioumis, Anastasios

Members:

Alvarez del Castillo, E.
Ardeberg, Arne
Baan, Willem
Baddiley, Christopher
Baskill, Darren
Bazzano, Angela
Benkhaldoun, Zouhair
Bensammar, Slimane
Blanco, Carlo
Boonstra, Albert
Brown, Robert
Cabanac, Rémi
Carraminana, Alberto
Carrasco, Bertha
Cayrel, Roger
Clegg, Andrew
Colas, François
Costero, Rafael
Coyne, S.J, George
Crawford, David
Davis, Donald
Dukes Jr., Robert
Edwards, Paul
FAN, Yufeng
Garcia-Lorenzo, Maria
Gergely, Tomas
Green, Richard
Haenel, Andreus
Helmer, Leif
Hempel, Marc
Hidayat, Bambang
Ilin, Gennadii
Ilyasov, Sabit
Ioannou, Zacharias
Keeney, Brian
Kolomanski, Sylwester
Kontizas, Evangelos
Kontizas, Mary
Kovalevsky, Jean
KOZAI, Yoshihide
Kramer, Busaba
Leibowitz, Elia
Lewis, Brian
Lomb, Nicholas
Malin, David
Mancini, Dario
Markkanen, Tapio
Masciadri, Elena
Mattig, W.
McNally, Derek
Mendoza-Torress, J.-E.
Menzies, John
Mitton, Jacqueline
Murdin, Paul
Nelson, Burt
Osorio, José
Owen, Frazer
Özel, Mehmet
Ozisik, Tuncay
Pankonin, Vernon
Percy, John
Pound, Marc
Sanchez, Francisco
Sanders, Gary
Schilizzi, Richard
Sese, Rogel Mari
Shetrone, Matthew
Siebenmorgen, Ralf
Smith, Francis
Smith, Malcolm
Smith, Robert
Stencel, Robert
Storey, Michelle
Sullivan, III, Woodruff
Suntzeff, Nicholas
Tremko, Jozef
Tzioumis, Anastasios
Upgren, Arthur
van den Bergh, Sidney
van Driel, Wim
Vernin, Jean
Walker, Merle
WANG, Xunhao
Whiteoak, John
Woolf, Neville
YANO, Hajime
ZHANG, Haiyan
ZHENG, Xiaonian

Division III Commission 51 Bio-Astronomy

President: William M. Irvine
Vice-President: Pascale Ehrenfreund

Organizing Committee Members:

Cosmovici, Cristiano
Kwok, Sun
Levasseur-Regourd, A.-C.
Morrison, David
Udry, Stéphane

Members:

Al-Naimiy, Hamid
Alibert, Yann
Allard, France
Almar, Ivan
Alonso Sobrino, Roi
Alsabti, Abdul Athem
ANDO, Hiroyasu
Apai, Daniel
Balazs, Bela
Balbi, Amedeo
Ball, John
Bania, Thomas
Barbieri, Cesare
Basu, Kaustuv
Beckman, John
Beckwith, Steven
Beebe, Reta
Benest, Daniel
Bennett, David
Berendzen, Richard
Berger, Jean-Philippe
Biraud, François
Bless, Robert
Bond, Ian
Bouchet, Patrice
Bowyer, C.
Boyce, Peter
Brandeker, Alexis
Bretones, Paulo
Broderick, John
Buccino, Andrea
Burke, Bernard
CAI, Kai
Calvin, William
Campusano, Luis
Cardenas, Rolando
Cardoso Santos, Nuno
Cassan, Arnaud
Chaisson, Eric
Chung, Eduardo
Cirkovic, Milan
Connes, Pierre
Corbet, Robin
Cosmovici, Cristiano
Cottin, Hervé
Coude du Foresto, Vincent
Couper, Heather
Coustenis, Athena
Cuesta Crespo, Luis
Cunningham, Maria
Cuntz, Manfred
da Silveira, Enio
Daigne, Gerard
Davis, Michael
De Becker, Michaël
de Jager, Cornelis
de Loore, Camiel
Deeg, Hans
Despois, Didier
Dick, Steven
Domiciano de Souza, A.
Dorschner, Johann
Doubinskij, Boris
Downs, George
Drake, Frank
Dutil, Yvan
Dyson, Freeman
Ehrenreich, David
Ellis, George
Epstein, Eugene
Evans, Neal
Fazio, Giovanni
Feldman, Paul
Field, George
Firneis, Maria
Fisher, Philip
Fraser, Helen
Fredrick, Laurence
Freire Ferrero, Rubens
FUJIMOTO, Masa Katsu
Gatewood, George
Ghigo, Francis
Gillon, Michaël
Giovannelli, Franco
Golden, Aaron
Goldsmith, Donald
Gott, J.
Goudis, Christos
Gregory, Philip
Gulkis, Samuel
Gunn, James
Haisch, Bernard
Hale, Alan
Hart, Michael
Hershey, John
Heudier, Jean-Louis
HIRABAYASHI, Hisashi
Hoang, Binh
Hogbom, Jan
Hollis, Jan
Horowitz, Paul
Howard, Andrew
Hsieh, Henry
Hunter, James
Hysom, Edmund
Irvine, William
Israel, Frank
Kafatos, Menas
KAIFU, Norio
Kane, Stephen
Kardashev, Nicolay

Kaufmann, Pierre
Kawada, Mitsunobu
Keay, Colin
Keheyan, Yeghis
Keller, Hans-Ulrich
Kellermann, Kenneth
Kilston, Steven
Klahr , Hubert
Kocer, Dursun
Koeberl, Christian
Korpela, Eric
Ksanfomality, Leonid
KUAN, Yi-Jehng
Kuiper, Thomas
Kwok, Sun
Lammer, Helmut
Lamontagne, Robert
Laufer, Diana
Lazio, Joseph
LEE, Sang-Gak
Leger, Alain
Lilley, Edward
Lippincott Zimmerman, S.
LIU, Sheng-Yuan
Lodieu, Nicolas
Lyon, Ian
Margrave Jr, Thomas
Marov, Mikhail
Martin, Maria
Martinez-Frias, Jesus
Matsakis, Demetrios
MATSUDA, Takuya
Mayor, Michel
McAlister, Harold
McDonough, Thomas
Melott, Adrian
Mendoza, V.
Merin Martin, Bruno
MINH, Young-Chol
MINN, Young-Ki
Minniti, Dante
Mirabel, Felix
Mokhele, Khotso
Moore, Marla
Morris, Mark
Morrison, David
Mousis, Olivier
Muller, Richard
Naef, Dominique
NARITA, Norio
Nelson, Robert
Neuhaeuser, Ralph
Niarchos, Panagiotis
Norris, Raymond
Nuth, Joseph
Ostriker, Jeremiah
Owen, Tobias
Palle Bago, Enric
Parijskij, Yurij
Pascucci, Ilaria
Perek, Luboš
Pilling, Sergio
Pollacco, Don
Ponsonby, John
Quanz, Sascha
Quintana, Jose
Quirrenbach, Andreas
Rawlings, Mark
Rees, Martin
Reyes-Ruiz, Mauricio
Rodriguez, Luis
Rowan-Robinson, Michael
Russell, Jane
Sanchez Bejar, Victor
Sancisi, Renzo
Sarre, Peter
Scargle, Jeffrey
Schild, Rudolph
Schneider, Jean
Schober, Hans
Schuch, Nelson
Seielstad, George
Semenov, Dmitry
Shostak, G.
Sims, Mark
Singh, Harinder
Sivaram, C.
Snyder, Lewis
SOFUE, Yoshiaki
SONG, In Ok
Sozzetti, Alessandro
Stein, John
Sterzik, Michael
Stone, Remington
Straizys, Vytautas
Sturrock, Peter
Sullivan, III, Woodruff
TAKABA, Hiroshi
TAKADA-HIDAI, M.
Tarter, Jill
Tavakol, Reza
Tedesco, Edward
Tejfel, Victor
Terzian, Yervant
Thaddeus, Patrick
TIAN, Feng
Tolbert, Charles
Tovmassian, Hrant
Trimble, Virginia
Turner, Kenneth
Turner, Edwin
Udry, Stéphane
Vakoch, Douglas
Vallee, Jacques
Valtaoja, Esko
Varshalovich, Dmitrij
Vauclair, Gérard
Vazquez, Manuel
Vazquez, Roberto
Venugopal, V.
Verschuur, Gerrit
Vogt, Nikolaus
von Braun, Kaspar
von Hippel, Theodore
Vukotic, Branislav
Wallis, Max
Walsh, Wilfred
Walsh, Andrew
Wandel, Amri
Watson, Frederick
Welch, William
Wesson, Paul
Wielebinski, Richard
Williams, Iwan
Willson, Robert
Wilson, Thomas
Wolstencroft, Ramon
Womack, Maria
Wright, Ian
Wright, Alan
XU, Weibiao
YE, Shuhua
Zadnik, Marjan
Zapatero-Osorio, Maria R.
Zuckerman, Benjamin

Division I Commission 52 Relativity in Fundamental Astronomy

President: Gérard Petit
Vice-President: Michael H. Soffel

Organizing Committee Members:

Brumberg, Victor
Capitaine, Nicole
Fienga, Agnès
FUKUSHIMA, Toshio
Guinot, Bernard
HUANG, Cheng
Klioner, Sergei
Mignard, François
Seidelmann, P.
Wallace, Patrick

Members:

Abbas, Ummi
Aharonian, Felix
Antonelli, Lucio Angelo
ARAKIDA, Hideyoshi
Bazzano, Angela
Boucher, Claude
Bromberg, Omer
Brumberg, Victor
Burikham, Piyabut
Bursa, Michal
Calabretta, Mark
Cannon, Kipp
Capitaine, Nicole
CHAE, Kyu Hyun
Clarkson, Chris
Crosta, Mariateresa
de Felice, Fernando
Dominguez, Mariano
Efroimsky, Michael
Evans, Daniel
Fienga, Agnès
Foschini, Luigi
Frutos-Alfaro, Francisco
FUKUSHIMA, Toshio
Gangadhara, R.T.
Giorgini, Jon
Gonzalez, Gabriela
Gray, Norman
Guinot, Bernard
Gumjudpai, Burin
Hackman, Christine
Hernandez-Monteagudo, C.
Hilton, James
Ho, Wynn
Hobbs, David
Hobbs, George
Hohenkerk, Catherine
Holz, Daniel
Horn, Martin
HSU, Rue-Ron
HU, Hongbo
HUANG, Tianyi
HUANG, Cheng
Iorio, Lorenzo
Ivanov, Pavel
Kaplan, George
Kausch, Wolfgang
Khumlumlert, Thiranee
Klioner, Sergei
Koivisto, Tomi
Kovalevsky, Jean
Lammers, Uwe
Le Poncin-Lafitte, Ch.
LI, Li-Xin
Luzum, Brian
Maciesiak, Krzysztof
Manchester, Richard
Mandel, Ilya
Marranghello, Guilherme
McCarthy, Dennis
Melnyk, Olga
Mignard, François
Minazzoli, Olivier
MIZUNO, Yosuke
Morabito, David
Mota, David
Mueller, Andreas
Mukhopadhyay, Banibrata
NAGATAKI, Shigehiro
Orellana, Mariana
Osorio, José
Panessa, Francesca
Pireaux, Sophie
Pitjeva, Elena
Pitkin, Matthew
Podolsky, Jiri
Potapov, Vladimir
Ray, James
Reisenegger, Andreas
Reitze, David
Riles, Keith
Rosinska, Dorota
Rushton, Anthony
Saffari, Reza
Sanders, Gary
Schartel, Norbert
Seidelmann, P.
Shoemaker, David
Sigismondi, Costantino
Smirnova, Tatiana
Soria, Roberto
Standish, E.
Stergioulas, Nikolaos
TAKAHASHI, Rohta
van Leeuwen, Joeri
Vityazev, Veniamin
Wallace, Patrick
Watts, Anna
WEN, Linqing
WU, Jiun-Huei
Wucknitz, Olaf
XIE, Yi
Yakut, Kadri
ZHANG, Zhibin

Division III Commission 53 Extrasolar Planets (WGESP)

President: Alan Paul Boss
Vice-President: Alain Lecavelier des Etangs

Organizing Committee Members:

Bodenheimer, Peter
Cameron, Andrew
Jayawardhana, Ray
KOKUBO, Eiichiro
Mardling, Rosemary
Queloz, Didier
Rauer, Heike
ZHAO, Gang

Members:

Absil, Olivier
Adams, Fred
Aigrain, Suzanne
Alibert, Yann
Allan, Alasdair
Allard, France
Alonso Sobrino, Roi
Ammons, Stephen
ANDO, Hiroyasu
Apai, Daniel
Ardila, David
Artymowicz, Pawel
Augereau, Jean-Charles
Baines, Ellyn
Baran, Andrzej
Barge, Pierre
Baryshev, Andrey
Bear, Ealeal
Beaulieu, Jean-Philippe
Bennett, David
Berger, Jean-Philippe
Biazzo, Katia
Bodaghee, Arash
Bodenheimer, Peter
Bodewits, Dennis
Bond, Ian
Bonfils, Xavier
Borde, Pascal
Boss, Alan
Boyajian, Tabetha
Brandeker, Alexis
Brandner, Wolfgang
Briot, Danielle
Buccino, Andrea
Buchhave, Lars
Burrows, Adam
Cameron, Andrew
Canto Martins, Bruno
Cardoso Santos, Nuno
Casassus, Simon
Casoli, Fabienne
Cassan, Arnaud
Chabrier, Gilles
Chakrabarti, Supriya
Chakraborty, Deo
Chauvin, Gael
CHEN, Xinyang
Cieza, Lucas
Connolly Jr., Harold
Cosmovici, Cristiano
Cottini, Valeria
Coustenis, Athena
Creech-Eakman, Michelle
Cuesta Crespo, Luis
DAI, Zhibin
Dall, Thomas
Day Jones, Avril
de Val-Borro, Miguel
Deleuil, Magali
Delplancke, Francoise
Demory, Brice-Olivier
Dominis Prester, Dijana
DOU, Jiangpei
Downes Wallace, Juan
Ehrenreich, David
Emelyanenko, Vacheslav
Esenoglu, Hasan
FAN, Yufeng
Ferlet, Roger
Fernandez Lajus, Eduardo
Ferraz-Mello, Sylvio
Fletcher, Leigh
Ford, Eric
Fouque, Pascal
Fridlund, Malcolm
Fromang, Sebastien
FUKAGAWA, Misato
Gabor, Pavel
Gandolfi, Davide
Gawronski, Marcin
Geller, Aaron
Giani, Elisabetta
Gillon, Michaël
Girard, Julien
Golden, Aaron
Goldsmith, Donald
Gomez, Edward
Gregory, Philip
Grillmair, Carl
Haghighipour, Nader
Hallinan, Gregg
HAN, Inwoo
Helled, Ravit
Heng, Kevin
Hersant, Franck
HIROSE, Shigenobu
Homeier, Derek
HOU, Xiyun
Hovenier, J.
Howard, Andrew
HU, Zhong wen
Hubbard, William
Ianna, Philip
Ireland, Michael
Iro, Nicolas

Irvine, William
ITOH, Yoichi
Ivanov, Pavel
Janes, Kenneth
Jeffers, Sandra
JI, Jianghui
Jin, Liping
Johansen, Anders
Jura, Michael
Kabath, Petr
Kafka, Styliani (Stella)
Kaltenegger, Lisa
Kane, Stephen
Karoff, Christoffer
Kenworthy, Matthew
Kern, Pierre
Khodachenko, Maxim
KIM, Yoo Jea
KIM, Seung-Lee
KIM, Joo Hyeon
Kitiashvili, Irina
Klahr , Hubert
Kley, Wilhelm
Koeberl, Christian
KOKUBO, Eiichiro
Konacki, Maciej
Kostama, Veli-Petri
Kovacs, József
Kozak, Lyudmyla
Kozlowski, Szymon
Ksanfomality, Leonid
Kuerster, Martin
Lacour, Sylvestre
Lagage, Pierre-Olivier
Lammer, Helmut
Lamontagne, Robert
Langlois, Maud
Lanza, Antonino
Latham, David
Lawson, Peter
Lazio, Joseph
LEE, Myung Gyoon
LEE, Jae Woo
LEE, Chung-Uk
Léger, Alain
Li, Jian-Yang
Lin, Douglas
Lineweaver, Charles
Lintott, Chris
Lissauer, Jack
LIU, Michael
LIU, Yujuan
Livio, Mario
Lodieu, Nicolas
Longmore, Andrew
Lovis, Christophe
Malbet, Fabien
Mamajek, Eric
Mancini, Dario
Marchi, Simone
Mardling, Rosemary
Marley, Mark
Marois, Christian
Martinez Fiorenzano, Aldo
Marzari, Francesco
Masciadri, Elena
Mawet, Dimitri
Mayor, Michel
Mazeh, Tsevi
Meech, Karen
Meibom, Soren
Mendillo, Michael
Menzies, John
Merin Martin, Bruno
Meyer, Michael
Migaszewski, Cezary
Mignard, François
Millan Gabet, Rafael
Minniti, Dante
Moerchen, Margaret
Montes, David
Mookerjea, Bhaswati
Moor, Attila
Mordasini, Christoph
Mousis, Olivier
MURAKAMI, Naoshi
Naef, Dominique
Namouni, Fathi
NARITA, Norio
Neuhaeuser, Ralph
Niedzielski, Andrzej
Novakovic, Bojan
Oberst, Thomas
Owen, Tobias
Pal, András
Palle Bago, Enric
Papaloizou, John
PARK, Byeong-Gon
Patten, Brian
Penny, Alan
Penny, Matthew
Pepe, Francesco
Perets, Hagai
Perez, Mario
Perez-Garrido, Antonio
Perrin, Marshall
Persson, Carina
Pilat-Lohinger, Elke
Pilling, Sergio
Plavchan, Jr., Peter
Podolak, Morris
Pollacco, Don
Prsa, Andrej
Puxley, Phil
Puzia, Thomas
Quanz, Sascha
Queloz, Didier
Quirrenbach, Andreas
Rank-Lueftinger, Theresa
Rasio, Frederic
Reddy, Bacham
Reffert, Sabine
Reiners, Ansgar
Reyes-Ruiz, Mauricio
Richardson, Derek
Ridgway, Stephen
Roberts Jr, Lewis
Rojo, Patricio
Saffe, Carlos
Sahu, Kailash
Sanchez Bejar, Victor
Sari, Re'em
Sarre, Peter
SATO, Bunei
Schneider, Jean
Seiradakis, John
Selam, Selim
Sengupta, Sujan
Shankland, Paul
Silvotti, Roberto
Snik, Frans
Sozzetti, Alessandro
Stam, Daphne
Stapelfeldt, Karl
Sterzik, Michael
Stewart-Mukhopadhyay, S.
Stringfellow, Guy
Suli, Áron
Szabo, Gyula
TAKADA-HIDAI, M.

Tarter, Jill
TIAN, Feng
Traub, Wesley
Trimble, Virginia
Tsvetanov, Zlatan
Turner, Edwin
Udry, Stéphane
Unda-Sanzana, Eduardo
Valdivielso, Luisa
van Belle, Gerard
Vavrek, Roland
Vidal-Madjar, Alfred
Vidotto, Aline
Villaver, Eva
von Braun, Kaspar
von Hippel, Theodore
Wallace, James
WANG, Xiao-bin
WANG, Shiang-Yu
Whelan, Emma
Williams, Iwan
Winn, Joshua
Wittenmyer, Robert
Wolszczan, Alexander
Womack, Maria
WU, Zhen-Yu
XIAO, Dong
Yesilyurt, Ibrahim
YU, Cong
YU, Qingjuan
Yuce, Kutluay
YUMIKO, Oasa
YUTAKA, Shiratori
Zapatero-Osorio, M. Rosa
Zarka, Philippe
ZHANG, You-Hong
Zinnecker, Hans
Zucker, Shay
Zuckerman, Benjamin

Division IX Commission 54 Optical & Infrared Interferometry

President: Stephen T. Ridgway
Vice-President: Gerard T. van Belle

Organizing Committee Members:

Duvert, Gilles
Genzel, Reinhard
Haniff, Christopher
Hummel, Christian
Lawson, Peter
Monnier, John
Tuthill, Peter
Vakili, Farrokh

Members:

Absil, Olivier
Acke, Bram
Ambrocio-Cruz , Silvia
Arcidiacono, Carmelo
Aufdenberg, Jason
Augereau, Jean-Charles
Babkovskaia, Natalia
Baddiley, Christopher
Baines, Ellyn
Bakker, Eric
Bamford, Steven
Barrientos, Luis
Beklen, Elif
Bendjoya, Philippe
Benisty, Myriam
Benson, James
Berger, Jean-Philippe
Boonstra, Albert
Borde, Pascal
Boyajian, Tabetha
Brandner, Wolfgang
Bryant, Julia
Buscher, David
Carciofi, Alex
Cenko, Stephen
CHEN, Zhiyuan
CHEN, Xinyang
Ciliegi, Paolo
Coffey, Deirdre
Crause, Lisa
Crawford, Steven
Creech-Eakman, Michelle
Cuby, Jean-Gabriel
Cuillandre, Jean-Charles
Damé, Luc
Danchi, William
de Lange, Gert
de Petris, Marco
Delplancke, Francoise
Demory, Brice-Olivier
Dieleman, Pieter
Dionatos, Odysseas
Domiciano de Souza, A.
Dougados, Catherine
Dutrey, Anne
Duvert, Gilles
Eisner, Josh
Elias II, Nicholas
Faure, Cécile
Florido, Estrella
Gabanyi, Krisztina
Gabor, Pavel
Genzel, Reinhard
Giani, Elisabetta
Girard, Julien
GONG, Xuefei
Groh, Jose
GU, Xuedong
Haniff, Christopher
HAO, Lei
Hatziminaoglou, Evanthia
Hoenig, Sebastian
Hora, Joseph
HU, Zhong wen
Hummel, Christian
Hutter, Donald
IWATA, Ikuru
Jordan, Andrés
Jorgensen, Anders
Kalomeni, Belinda
Kawada, Mitsunobu
Kenworthy, Matthew
Kern, Pierre
Kervella, Pierre
KIM, Young-Soo
KIM, Ji Hoon
Koehler , Rainer
Konacki, Maciej
Kospal, Ágnes
Kraus, Stefan
Lacour, Sylvestre
Langlois, Maud
Lawson, Peter
Le Bouquin, Jean-Baptiste
Lebohec, Stephan
Lehnert, Matthew
Leisawitz, David
Mason, Brian
Mawet, Dimitri
Millan Gabet, Rafael
Millour, Florentin
Monnier, John
Mosoni, Laszlo
MURAKAMI, Naoshi
MURAYAMA, Hitoshi
NARDETTO, Nicolas
NGUYEN, Khanh
Oberst, Thomas
Ohnaka, Keiichi
Paumard, Thibaut
Percheron, Isabelle
Perez-Torres, Miguel
Pinte, Christophe
Pott, Jörg-Uwe
Rajagopal, Jayadev

Rastegaev, Denis
Ridgway, Stephen
Rivinius, Thomas
Rousset, Gérard
Schinckel, Antony
Schoedel, Rainer
Schuller, Peter
SHANG, Hsien
Tallon, Michel
Tallon-Bosc, Isabelle
Trippe, Sascha
Tristram, Konrad
Tuthill, Peter
Tycner, Christopher
Urquhart, James
Vakili, Farrokh
van Kampen, Eelco
Vanko, Martin
Vega, Olga
Verhoelst, Tijl
von Braun, Kaspar
Wallace, James
WANG, Guomin
WANG, Shen
WANG, Jingyu
WANG, Sen
Woillez, Julien
XIAO, Dong
YANG, Dehua
YUAN, Xiangyan
Zavala, Robert
ZHANG, You-Hong
ZHANG, Yong
ZHOU, Jianfeng

Division XII Commission 55 Communicating Astronomy with the Public

President: Dennis Crabtree
Vice-President: Lars Lindberg Christensen

Organizing Committee Members:

Alvarez-Pomares, Oscar
Arcand, Kimberly
Damineli Neto, Augusto
Green, Anne
Kembhavi, Ajit
Odman, Carolina
Reed, Sarah
Robson, Ian
Russo, Pedro
SEKIGUCHI, Kazuhiro
Whitelock, Patricia
ZHU, Jin
ZHU, Zi

Members:

Accomazzi, Alberto
Ajhar, Edward
Allan, Alasdair
Alvarez-Pomares, Oscar
Alves, Virgínia
Apai, Daniel
Argo, Megan
Arion, Douglas
Augusto, Pedro
Axelsson, Magnus
Babul, Arif
BAEK, Chang Hyun
Bailey, Katherine
Balbi, Amedeo
Bamford, Steven
Barlow, Nadine
Bartlett, Jennifer
Bauer, Amanda
Beck, Sara
Becker, Werner
Beckmann, Volker
Berrilli, Francesco
Botti, Thierry
Bretones, Paulo
Briand, Carine
Cassan, Arnaud
Cattaneo, Andrea
CHEN, Dongni
Chernyakova, Maria
Coffey, Deirdre
Conti, Alberto
Copin, Yannick
Cora, Alberto
Couto da Silva, Telma
Crabtree, Dennis
Croft, Steve
Crosta, Mariateresa
Cuesta Crespo, Luis
Cui, Wei
DAISUKE, Iono
Damineli Neto, Augusto
Darhmaoui, Hassane
de Grijs, Richard
de Lange, Gert
de Swardt, Bonita
Delsanti, Audrey
Demarco, Ricardo
DeNisco, Kenneth
Dominguez, Mariano
Donahue, Megan
Doran, Rosa
Easther, Richard
Ehle, Matthias
Ekstrom Garcia N., Sylvia
Engelbrecht, Chris
English, Jayanne
Esguerra, Jose Perico
Espenak, Fred
Feain, Ilana
Figueiredo, Newton
Forero Villao, Vicente
Gangui, Alejandro
GAO, Jian
Garcia-Lorenzo, Maria
Gavrilov, Mikhail
Gay, Pamela
George, Martin
Gilbank, David
Gills, Martins
Girard, Julien
Gomez, Edward
Govender, Kevindran
Green, Anne
Gumjudpai, Burin
Gunthardt, Guillermo
Hannah, Iain
Hau, George
Haverkorn, Marijke
Hempel, Marc
Heydari-Malayeri, M.
HIRAMATSU, Masaaki
Hjalmarsdotter, Linnea
Hollow, Robert
Horn, Martin
Ibrahim, Alaa
Iliev, Ilian
ITOH, Yoichi
Jafelice, Luiz
Jahnke, Knud
JEON, Young Beom
Jimenez-Vicente, Jorge
Johns, Bethany
Johnston, Helen
Jones, David
Kalemci, Emrah
Karouzos, Marios
Keeney, Brian
Kembhavi, Ajit
Kolomanski, Sylwester

Komonjinda, Siramas
Koschny, Detlef
Kostama, Veli-Petri
Kotulla, Ralf
Kristiansen, Jostein
Kuiper, Rolf
Las Vergnas, Olivier
Laufer, Diana
LEE, Jun
Letarte, Bruno
Levenson, Nancy
Lintott, Chris
Loaring, Nicola
Longo, Giuseppe
Lopes de Oliveira, R.
Lopez-Sanchez, Angel
Lowenthal, James
Madsen, Claus
Majumdar, Subhabrata
Marchi, Simone
Maricic, Darije
Martinez Delgado, David
Mason, Helen
Masters, Karen
Medina, Etelvina
Melbourne, Jason
Merin Martin, Bruno
Miller, Eric
Minier, Vincent
Morris, Patrick
Mourao, Ana Maria
Mueller, Andreas
Mujica, Raul
MURAYAMA, Hitoshi
NARITA, Norio
Nesvadba, Nicole
NGUYEN, Khanh
Niemczura, Ewa
Nota, Antonella
Nymark, Tanja
Ocvirk, Pierre
Odman, Carolina
Odonoghue, Aileen
Olive, Don
Olivier, Enrico
Onel, Hakan
Oozeer, Nadeem
Ortiz Gil, Amelia
Özeren, Ferhat
Ozisik, Tuncay
Palacios, Ana
Pantoja, Carmen
Pat-El, Igal
Pavani, Daniela
Peel, Michael
Perrin, Marshall
PHAN, Dong
Politi, Romolo
Price, Daniel
Price, Charles
Rassat, Anais
Reid, Michael
Rekola, Rami
Reyes, Reinabelle
Reynolds, Cormac
Roberts, Douglas
Rodriguez Hidalgo, Inés
Rojas, Gustavo
Roman-Zuniga, Carlos
Rozanska , Agata
Saffari, Reza
Sampson, Russell
Sandrelli, Stefano
Santos-Lleo, Maria
Sawicki, Marcin
Sese, Rogel Mari
Sharp, Nigel
Shelton, Ian
SHIMOIKURA, Tomomi
Simons, Douglas
Snik, Frans
Srivastava, Nandita
Stam, Daphne
Strubbe, Linda
SUMI, Takahiro
SUNG, Hyun-Il
TANAKA, Ichi Makoto
Templeton, Matthew
Teng, Stacy
Tignalli, Horacio
Tomasella, Lina
TRAN, Ha
Tugay, Anatoliy
Unda-Sanzana, Eduardo
Vakoch, Douglas
Valdivielso, Luisa
van Eymeren, Janine
Verdoes Kleijn, Gijsbert
Vergani, Daniela
Vierdayanti, Kiki
Villar Martin, Montserrat
Voit, Gerard
Wadadekar, Yogesh
Wagner, Robert
Walker, Constance
Walsh, Robert
WANG, Chen
Whitelock, Patricia
Wilson, Gillian
Winter, Lisa
WU, Jiun-Huei
WU, Zhen-Yu
YAN, Yihua
ZHU, Jin

Printed in the United States
by Baker & Taylor Publisher Services